This volume records the presentations and discussions at the Second Royal Society–Unilever Indo-UK Forum on "Dynamics of Complex Fluids" which was the culmination of the six-month programme on this topic organised at the Issac Newton Institute for Mathematical Sciences, Cambridge University.

The authors of this important volume present an up-to-date, wide-ranging view on developments in the analysis of complex fluid behaviour. Emphasis is placed upon the relation between small-scale structure and large-scale response: this brings together the approaches of molecular physics and continuum mechanics.

Experiments, constitutive models and computer simulations are combined to yield new insights into the flow behaviour of polymer melts and solutions, colloidal and neutral particle suspensions, and pastes and soils.

Proceedings of the Royal Society–Unilever Indo–UK Forum in Materials Science and Engineering

---

Published

Solid–Solid Interactions (1996)
ed. M. J. Adams, S. K. Biswas and B. J. Briscoe

Dynamics of Complex Fluids (1998)
ed. M. J. Adams, R. A. Mashelkar, J. R. A. Pearson and A. R. Rennie

Forthcoming

Responsive Inorganic Materials
ed. K. R. Seddon, C. J. Adams and R. Townsend

Structure and Dynamics of Materials in the Mesoscopic Domain
ed. M. Lal

DYNAMICS OF COMPLEX FLUIDS

Proceedings of the
Second Royal Society–Unilever Indo-UK Forum
in Materials Science and Engineering

Editors

M J Adams
Unilever Research, UK

R A Mashelkar
CSIR, India

J R A Pearson
Cambridge, UK

A R Rennie
Cavendish Laboratory, UK

Imperial College Press

The Royal Society

A co-publication of Imperial College Press and the The Royal Society

Published by

Imperial College Press
203 Electrical Engineering Building
Imperial College
London SW7 2BT

and

The Royal Society
6 Carlton House Terrace
London SW1Y 5AG

Distributed by

World Scientific Publishing Co. Pte. Ltd.

P O Box 128, Farrer Road, Singapore 912805

USA office: Suite 1B, 1060 Main Street, River Edge, NJ 07661

UK office: 57 Shelton Street, Covent Garden, London WC2H 9HE

Library of Congress Cataloging-in-Publication Data
Royal Society-Unilever Indo-UK Forum in Materials Science and
Engineering (2nd : 1998: Cavendish Laboratory, Cambridge, England)
Dynamics of complex fluids : proceedings of the Second Royal
Society-Unilever Indo-UK Forum in Materials Science and Engineering
/ editors, M. J. Adams . . . [et al.].
p. cm.
Includes index.
ISBN 1-86094-086-2 (alk. paper)
1. Complex fluids -- Fluid dynamics -- Congresses. I. Adams, M. J. II. Title.
QD549.2.C66R69 1998
620.1'064--dc21 98-23902
CIP

British Library Cataloguing-in-Publication Data
A catalogue record for this book is available from the British Library.

This book is printed on acid-free paper.

Printed in Singapore by Uto-Print

Foreword

This book represents a record of the work presented, and the subsequent discussions that took place, at the second Indo-UK Forum held at the Cavendish Laboratory, Cambridge. The first meeting on "Solid-Solid Interactions" was held at Imperial College, London. These meetings were intended to alternate between the United Kingdom and India, and to provide a basis for discussion between scientists in both countries who share a common approach in the fields of materials science and engineering. Two further meetings have been held on "Responsive Inorganic Materials" and "Structure and Dynamics of Materials in the Mesoscopic Domain". This series of Forums is made possible through the generosity of Unilever Plc and the support of the Royal Society. The present Forum was kindly supported by the Isaac Newton Institute for Mathematical Sciences at the University of Cambridge. The Institute had organised a six-month research programme on "Dynamics of Complex Fluids" and the aim of the Forum was to bring together the various themes and to discuss the current status and future directions of this area. I am pleased that Dr. Ramesh Mashelkar, Director-General of the Council of Scientific and Industrial Research in New Delhi, India, and an outstanding chemical engineer, was involved in this endeavour. I congratulate all the organisers on the success of this meeting, which brought together many of the leading scientists in the field of complex fluids both from India and the United Kingdom, and also from many other countries.

C. N. R. Rao
Bangalore, India

Preface

The second Royal Society-Unilever Indo-UK Forum in Materials Science and Engineering on the subject of complex fluids was planned for the same period that the Isaac Newton Institute (INI) for Mathematical Sciences had organised a six-month research programme on "Dynamics of Complex Fluids". At the suggestion of Sir Sam Edwards FRS, the Steering Committee of the Forums and the Organising Committee for the Institute programme agreed that a Forum be held at the end of the programme to act as a culmination of some of the activities that had taken place during the previous six months. This offered the unique opportunity of bringing together some of the best scientists from around the world, already resident at the Newton Institute, together with other distinguished scientists from India and the UK who were specialists in facets of complex fluids that had not been considered in the programme previously. The five-day format of the Forums, with extended presentation and discussion periods and also workshop and poster sessions, was considered ideal for the purpose of promoting an exchange of ideas. It is hoped that this proceedings volume will capture some of the spirit and lively discussions that took place during the time of the Forum which was held from 24–28 June 1996 at the Cavendish Laboratory in Cambridge.

As in the first proceedings volume (*Solid-Solid Interactions*, ed. M. J. Adams, S. K. Biswas and B. J. Briscoe, Imperial College Press-The Royal Society, 1996), the editors have taken the liberty of organising the structure of this volume in a way that is different from that adopted for the Forum. We hope that this will provide a more sensible basis for interlinking the various themes developed at the Forum. Consequently, we have not distinguished between plenary and workshop sessions. However, papers presented as posters have been included as shorter summaries. The chapters were written after the Forum had taken place in order to allow the contributors to reflect on the ideas that were generated during the formal and informal discussion periods. In many cases, they have recorded the formal discussion as a specific section while others considered it more appropriate to take into account the various exchanges that took place whilst writing their chapters. Their primary brief was to provide an overview of a particular area with extensive references that would be of value to those undertaking research in complex fluids. The final data of receipt of the chapters for this volume was June 1997 and inevitably many authors have taken account of developments since the Forum.

It is appropriate to record here our gratitude to the institutions and people who have made this venture possible. The support of Unilever Plc, the Royal Society, the Newton Institute and the Cavendish Laboratory were crucial; in particular, we would like to acknowledge the support of Dr Anne McLaren of the Royal Society and Drs Ashok Ganguly and Tony Lee of Unilever Research and Engineering Division. We would like to thank the Steering Committee for the Forums and the Organising Committee of the INI programme. The joint Chairmen of the former committee are Sir Sam Edwards (Cambridge) and Professor C. N. R. Rao (IIS Bangalore) with the following membership: Sir Hugh Ford (Imperial College), Dr A. McLaren (the Royal Society), Professor R. Townsend (Unilever Research) and Professor M. J. Adams (Unilever Research). Professors T. C. B. McLeish (Leeds), J. R. A. Pearson (Cambridge) and K. Walters (Aberystwyth) formed the Organising Committee for the INI programme. The Organising Committee for the Forum included Professor M. J. Adams, Professor S. K. Biswas (IIS Bangalore), Professor B. J. Briscoe (Imperial College), Dr M. R. Mackley (Cambridge), Professor T. C. B. McLeish, Dr R. A. Mashelkar (CSIR, New Delhi), Dr V. M. Naik (Unilever Research, India), Professor J. R. A. Pearson, Dr A. R. Rennie (Cambridge), Professor R. P. Townsend and Professor K. Walters.

Finally, there are many people we would like to thank for their contributions to both the Forum and this volume. This includes the Chairmen of the sessions, the speakers and the participants. In particular, we thank the administrator, Meg Staff at the Cavendish Laboratory, and the administrative staff of the INI for ensuring the smooth running of the meeting.

General Introduction

The complexity mentioned in the title of this book refers to the structural and rheological (deformational) nature of the fluids being considered. Neither simple liquids, like water or paraffin at room temperature, nor gases, like air and methane, will be considered here, nor will purely elastic solids, like spring steel or wood, because their behaviour is well known and easily described and explained. Fluids are to be understood as coherent materials that can be deformed continuously and permanently. Further definition is neither appropriate nor desirable, because the object of the text is to bring together ideas and techniques that have been applied to a wide class of materials, showing different properties in different contexts. Many materials can behave as solids, complex fluids or liquids depending on the conditions, typically temperature and stress, to which they are subjected; it is their complex fluid behaviour that will be discussed.

The list of materials to be considered is long. Some, such as soils, muds and pastes, are known and have been in use for centuries. Others, particularly polymers, liquid crystals and superplastic metals, have only been developed and studied within the last 50 years. Interest in studying such a wide range of complex materials has been spread over such a wide range of disciplines, each with its own way of working, that only partial understanding has been gained by any group. For instance, scientists have seldom been able to provide the simple models that enable engineers to design and control processes, while the latter have largely remained unaware of the insights gained by the former. The gathering of chemists, physicists, mathematicians and engineers to forge and discuss common interests was at the heart of the INI Programme and the RS/Unilever Forum. A secondary, but equally important, objective was to discuss the impact that modern computational methods have had, and will have, on our way of approaching problems.

There are many ways in which the papers presented could be classified. One way would be to start with experimental observations, continuing with model theories (leading to constitutive relations) and ending with computational simulation of complex flows. Another would be to distinguish between molecular and continuum approaches. A third way would be to move from purely fundamental approaches towards detailed applications. None of these has been chosen because the aim has throughout been to bring together all approaches and disciplines. In this volume, the pragmatic splitting into four parts was based on a natural distinction between polymer (viscoelastic) and particulate (viscoplastic) systems. This reflects our view that: (i) successful advances (as described in the papers published here) often require experimental, analytic and computational inputs to be used together on a single problem; (ii) both molecular and continuum considerations are needed to describe such behaviour; and (iii) problems in applications can have very fundamental explanations. On the other hand, the material in question usually defines the basic representation needed to examine its behaviour. The seeking of universality in material behaviour has not proved to be as profitable as concentrating on the special features of particular contexts, and bringing all techniques to bear on these particularities.

One generic feature of complex fluids is the wide range of length and time scales that determine the various aspects of their behaviour. With simple liquids, there is a large gap between the very small length and time scales characteristic of the small molecules that compose them, and the bulk length and time scales of the flows of these liquids that are of interest; the former control the simple constitutive parameters, density and viscosity, that are needed to predict the latter. With polymers, the range of time scales associated with molecular motions can overlap those characterising the bulk flows of interest. This leads to "history dependence" and viscoelasticity; it is also possible for macro-molecular structures to have length scales that overlap with the smallest scales of importance in bulk flows, such as surface roughness. With suspensions of identifiable particles, the overlap of length scales is obvious. A specific feature of complex flows is that the flow process itself can generate a range of length and time scales that is not clearly imposed by the imposed boundary conditions (driving forces). Turbulence in the flow of simple liquids or gases is the best known example, and is associated with dynamical instability and consequential inhomogeneities. In the case of complex fluids, such instability and inhomogeneity can arise for a variety of reasons, often the result of the interplay between mechanical and structural dynamics.

This interplay between large-scale and small-scale effects can to some extent be separated by introducing constitutive relations (rheological equations of state or REOS), which formally relate the stress and strain histories of fluid elements. Molecular and structural factors are assumed to define the REOS, which are sufficient to determine continuum behaviour. Implicit in this approach are characteristic length scales, below which continuum behaviour is not expected and above which molecular features are not discernible. This is in apparent contradiction to the notion of overlap. This paradoxical distinction can be resolved in terms of suitably defined averages, so that the continuum becomes a progressively more abstract notion as length scales are reduced, while the flows to which structural elements are subjected to become more unlikely as the structures grow in size. Alternatively, it can be recognised that all results are only asymptotically valid. This is well recognised by formal continuum theorists, who introduce physically unrealistic hypotheses such as rheological simplicity and affine deformation to obtain valuable approximations.

Fortunately, the polemical confrontations between intransigent proponents of the two approaches (a disappointing feature of the middle years of this century) have given way to a shared desire, evidenced in this collection of contributions, to combine the strengths and potential of each point of view. In part, this stems from the growing computational power of modern digital computers, which gives rise to the prospect of testing the merit of various theories and predicting the outcome of engineering processes by direct simulation. By their very nature, computations relate to discrete and not continuous variables. The discretisation of both space and time chosen for a particular calculation forces the user to relate these chosen values to those arising in either of the formal representations favoured by theorists. The range of computational techniques available to undertake simulation of mathematical models mirrors the range of models available for simulation. The computer has imposed a degree of intellectual discipline on those engineers and scientists who had previously, and successfully, restricted themselves in their work to simple "physical" arguments, and on those theorists who had restricted themselves to elaborating mathematical theories which were never able to provide precise numerical predictions.

The first part of this book deals wholly with bulk viscoelastic models for polymeric systems. McKinley's contribution shows how a purely continuum approach can combine experiment, constitutive modelling and simulation with particular reference to flow instability. Renardy's paper illustrates the power of mathematical analysis once the constitutive model is selected and Tanner's adds details about the computational techniques available for complex flow fields. The last four contributions are concerned with the potential for micro-macro simulation; the biggest advantage is that no continuum constitutive model needs to be specified. This is because a large number of individual "molecules" are subjected to convection and deformation (continuum variables) and Brownian (random molecular) forces, from which a continuum stress is obtained directly by averaging over the molecular conformations. Micro and macro computations are carried out sequentially at each time step so that no difficulty is associated with overlapping time scales. There remains an implied separation of length scales. The dynamics of the separate molecules are assumed to be known and only rather simple archetypal examples are employed.

The second part addresses many of the same issues, but concentrates more on the dynamics at the molecular level, where inter-atomic and long-range forces can be directly included in the analysis, and where the precise chemical and structural nature of the materials involved becomes dominant. Boundary effects at solid surfaces are specifically investigated by Harden and Cates who discuss adsorbed polymer layers, and by Tildesley who simulates boundary-layer lubrication (viewed as decreased friction) by amphiphiles. Larson's paper on surfactant solutions effectively introduces meso-scale structural elements where temperature and concentration, rather than deformation, is the independent (imposed) variable. Lele and Mashelkar consider the co-operative effects introduced in gels and concentrated solutions by transient cross-links, which clearly allows for structure at different length scales.

Part three examines the dynamics of suspensions on the structural scale: the importance of the individual particle is never lost, while the bulk flow remains homogeneous. Ramaswamy and Kumaran both consider sedimentation from the point of view of interactions between particles; statistical mechanics is the key technique employed and the problem tackled means that the fluid is explicitly two-phase. Goddard's paper on migrational instabilities is exceptional in that it is wholly continuum mechanical; it too considers a two-phase system. Kumar *et al.* consider chaotic behaviour of suspensions of rods in simple shear flow, though their results aim at defining single-phase continuum behaviour. Melrose *et al.* describe a computational approach to shear thickening in concentrated colloidal suspensions, showing the effect of inter-particle interactions and the role of ordering (clustering) on the shear viscosity. Clarke and Rennie give an experimental example of such micro-structural ordering in colloids.

The last part, on viscoplasticity, well surveyed by Piau, returns to analysis conducted at the continuum level, and largely empirical as far as constitutive relations are concerned. Boundary conditions play a dominant role in such flows; this makes rheometry, such as squeeze flow, a much more difficult matter in terms of interpretation than in the case of fluids showing no yield stress; this is discussed by Lawrence and Corfield and Adams *et al.* From a mathematical point of view, interfaces between yielded and unyielded material are unknowns in any flow field, and computational schemes have to bear this in mind. It could be argued that the conceptual simplicity of the ideal plastic solid has led to unnecessary mathematical difficulty in applying models based on it, and that realistic models for soft solids with much more structural complexity could be more suitable for calculations. However, experiments at small and full scale for geotechnical systems usually display localisation on length scales comparable with the experimental length scale, and this feature cannot be avoided; the case of finite element analysis is covered by Peric and Owen. Compressibility cannot be neglected in most of such systems as Smith and Choobbasti and Molenkamp show.

It is the absence of Brownian effects that makes rocks fundamentally different from polymer gels on our laboratory time scale, and is the underlying reason for what was offered as a pragmatic reason for subdividing our book.

CONTENTS

Foreword v

Preface vii

General Introduction ix

Part One: Viscoelasticity 1

Introduction 3

1. Extensional Rheology and Flow Instabilities in Elastic Polymer Solutions 6
   G. H. McKinley

2. Surface Instabilities During Extrusion of Linear Low Density Polyethylene 30
   R. P. G. Rutgers, M. R. Mackley and D. Gilbert

3. High Weissenberg Number Boundary Layers and Corner Singularities in Viscoelastic Flows 38
   M. Renardy

4. Computing with High Viscoelasticity 47
   R. I. Tanner

5. Brownian Dynamics Simulations in Polymer Physics 61
   H. C. Öttinger

6. **Two-Dimensional, Time-Dependent Viscoelastic Flow Calculations Using Molecular Models** **73**
M. Laso

7. **Macroscopic and Mesoscale Approaches to the Computer Simulation of Viscoelastic Flows** **88**
R. Keunings and P. Halin

8. **Simulation of the Flow of an Oldroyd-B Fluid Using Brownian Configuration Fields** **106**
M. A. Hulsen, A. P. G. van Heel and B. H. A. A. van den Brule

Part Two: Polymeric and Self-Assembled Systems **125**

Introduction **127**

9. **Role of Energetic Interactions in the Dynamics of Polymer Networks: Some New Suggestions** **131**
A. K. Lele and R. A. Mashelkar

10. **The Prediction of Universal Viscometric Functions for Dilute Polymer Solutions Under Theta Conditions** **155**
J. Ravi Prakash

11. **Polymer Liquids at High Shear Rates** **176**
G. Marrucci and G. Ianniruberto

12. **Deuterium NMR Investigations of Liquid Crystals During Shear Flow** **188**
C. Schmidt, S. Müller and H. Siebert

13. **Formation of Polymer Brushes** **193**
J. Wittmer and A. Johner

14. **Dynamics of Adsorbed Polymer Layers** **199**
J. L. Harden and M. E. Cates

15. **The Molecular Dynamics Simulation of Boundary-Layer Lubrication: Atomistic and Coarse-Grained Simulations** **213**
D. J. Tildesley

16. **Direct Simulation of Surfactant Solution Phase Behaviour** 230
R. G. Larson

Part Three: Particulate Dispersions 243

Introduction 245

17. **The Statistical Mechanics of Slow, Steady Sedimentation** 248
S. Ramaswamy

18. **Effect of Collisional Interactions on the Properties of Particle Suspensions** 263
V. Kumaran

19. **Migrational Instabilities in Particle Suspensions** 280
J. D. Goddard

20. **Review of Chaotic Behaviour of Suspensions of Slender Rods in Simple Shear Flow** 286
K. S. Kumar, S. Savithri and T. R. Ramamohan

21. **Shear Thickening in Model Concentrated Colloids: The Importance of Particle Surfaces and Order Change Through Thickening** 301
J. R. Melrose, J. H. van Vliet and R. C. Ball

22. **Micro-Structure of Colloidal Dispersions Under Flow** 315
S. M. Clarke and A. R. Rennie

23. **Measurement of the Flow Alignment of Clay Dispersions by Neutron Diffraction** 330
A. B. D. Brown, S. M. Clarke and A. R. Rennie

24. **The Rheology and Micro-Structure of Equine Blood** 338
R. M. de Roeck and M. R. Mackley

Part Four: Viscoplasticity 345

Introduction 347

25. Crucial Elements of Yield Stress Fluid Rheology **351**
J. M. Piau

26. Are Plug-Flow Regions Possible in Fluids Exhibiting a Yield Stress? **372**
M. M. Denn

27. Non-Viscometric Flow of Viscoplastic Materials: Squeeze Flow **379**
C. J. Lawrence and G. M. Corfield

28. The Wall Yield of Rate-Dependent Materials **394**
M. J. Adams, B. J. Briscoe, G. M. Corfield and C. J. Lawrence

29. Strain Localisation During the Axisymmetric Squeeze Flow of a Paste **399**
M. J. Adams, B. J. Briscoe, D. Kothari and C. J. Lawrence

30. Viscoplastic Approaches in Forming Processes: Phenomenological and Computational Aspects **405**
D. Perić and D. R. J. Owen

31. Computation of Large-Scale Viscoplastic Flows of Frictional Geotechnical Materials **425**
I. M. Smith

32. Modelling of Liquefaction and Flow of Water Saturated Soil **446**
F. Molenkamp, A. J. Choobbasti and A. A. R. Heshmati

33. Analysis of Behaviour of Sand at Very Large Deformation **469**
A. J. Choobbasti and F. Molenkamp

Author Index **475**

Subject Index **477**

DYNAMICS OF COMPLEX FLUIDS

Part One
Viscoelasticity

Introduction		3
Chapter 1.	Extensional Rheology and Flow Instabilities in Elastic Polymer Solutions *C. H. McKinley*	6
Chapter 2.	Surface Instabilities During Extrusion of Linear Low Density Polyethylene *R. P. G. Rutgers, M. R. Mackley and D. Gilbert*	30
Chapter 3.	High Weissenberg Number Boundary Layers and Corner Singularities in Viscoelastic Flows *M. Renardy*	38
Chapter 4.	Computing with High Viscoelasticity *R. I. Tanner*	47
Chapter 5.	Brownian Dynamics Simulations in Polymer Physics *H. C. Öttinger*	61
Chapter 6.	Two-Dimensional, Time-Dependent Viscoelastic Flow Calculations Using Molecular Models *M. Laso*	73
Chapter 7.	Macroscopic and Mesoscale Approaches to the Computer Simulation of Viscoelastic Flows *R. Keunings and P. Halin*	88

Chapter 8. Simulation of the Flow of an Oldroyd-B Fluid Using Brownian Configuration Fields 106
M. A. Hulsen, A. P. G. van Heel and B. H. A. A. van den Brule

Introduction

The papers in this section are concerned with relatively large-scale complex flows of viscoelastic fluids. The real fluids used in the experiments described or referred to are all polymeric, but this only reflects the fact that substantial viscoelasticity is usually associated in practice with the configurational entropy of polymer chains.

The first paper deals with instabilities that arise in shear and elongational flows with curved streamlines. The author shows how an understanding can be built up at different levels, using different approaches and concepts. First, there is a straightforward reliance on experimental observation, both to characterise the fluid in homogeneous controlled (viscometric) flows, and then to quantify the onset of symmetry-breaking instability in more complex flows. From the viscometric experiments, quantitative functional forms for the material functions arise: viscosity, first normal-stress-difference and transient Trouton ratio (elongational viscosity). Then comes the use of constitutive models, both continuum and molecular, parametrised by comparison with the viscometric functions, which allow more complex (inhomogeneous and unsteady) flows to be simulated analytically or computationally. The circle is closed by comparing the model predictions with experiment. The role of various dimensionless numbers, Görtler, Deborah, Weissenberg and a more recent one, M (that could justifiably now be christened the McKinley number), is explained, and their use in providing generic criteria for stability in curved flows is proposed.

The second paper (slightly expanded from a poster) provides experimental evidence for an exponential relation between the roughness of an LLDPE extrudate and the die wall shear stress.

Renardy's paper summarises two elegant asymptotic analyses that he carried out for the flow of viscoelastic fluids at high Weissenberg number Wi

(when elastic effects become dominant). He shows: (a) why stress boundary layers with a thickness $O(Wi^{-1})$ arise at a wall; and (b) what the dominant singularity is near a 270° re-entrant corner for both upper-convected Maxwell and Phan-Thien Tanner models.

The last five papers focus on computational techniques for the simulation of complex flows. Tanner treats the flow past a sharp-edged flat plate placed in a rectangular channel as a test problem, containing as it does a singularity in the flow at the sharp edge. The continuum MPTT model, which in general contains seven numerical parameters and includes *inter alia* the Newtonian, UCM, Oldroyd B and PTT models as special cases, is used in the computations. A comparison of his preferred SIMPLER-algorithm finite volume method with other continuum discretisation techniques has shown that the former works just as well at high Wi.

Öttinger provides a general introduction to a relatively new technique, (CONNFFESSIT), which he developed to simulate the flows of polymeric systems. Stochastic differential (Langevin) equations are used to follow the conformation of individual protopolymers (elastic dumbbells, bead/rod/spring chains) as they are convected and deformed affinely while subject to the random forces of Brownian dynamics. Suitable averages of these conformations yield a local continuum stress. His paper is largely concerned with the relative merits of the stochastic formulation and the earlier (equivalent) distribution-function approach which leads to Fokker–Planck equations.

Laso then shows just how the CONNFFESSIT approach is used to solve complex flow problems by interfacing the stochastic molecular description through the stress tensor to a standard discretised form of the continuum (mass and momentum) conservation equations. He shows how well it can match direct computation of the continuum equations (as in Tanner above) in cases where the molecular model (FENE-P) can be represented by a continuum constitutive equation, and how more realistic molecular models (for example, FENE) which have no simple continuum representation can just as easily be used without any computational penalty.

This micro-macro approach is examined in a very similar but more detailed fashion by Keunings and Halin, who use the same pair of molecular models, FENE-P and FENE, as Laso to illustrate and explain the significant differences obtained in an eccentric journal bearing flow. Both Laso, Keunings and Halin emphasise how apparently similar the FENE and FENE-P models are, and how different their predicted behaviour is, particularly in extensional flow. (This is important to the engineer, who is often interested in details of particular flows for particular real fluids and not just in generic qualitative behaviour, which is what physicists often seek to understand and explain.)

The last paper, by Hulsen, van Heel and van den Brule, introduces a variant on the CONNFFESSIT approach in order to reduce the computational demands of particle tracking. Instead of convecting discrete particles specified by their individual configuration vectors, each subject to its own history of (Brownian) Wiener forces, the authors use an ensemble of continuous configuration fields. Each field is then subjected to a uniform Brownian history as it is convected and deformed. The micro-macro interface, based here on averaging the ensemble of configurations to give the continuum stress, is the same in the two approaches. The test flow in this case is the start-up of channel flow past a circular cylinder. Very encouraging results were obtained for a dumbbell model which could be compared to corresponding results obtained by discretised continuum techniques for the exactly equivalent Oldroyd-B fluid.

Dynamics of Complex Fluids, pp. 6–29
ed. M. J. Adams, R. A. Mashelkar, J. R. A. Pearson & A. R. Rennie
Imperial College Press–The Royal Society, 1998

Chapter 1

Extensional Rheology and Flow Instabilities in Elastic Polymer Solutions

G. H. McKinley

Department of Engineering and Applied Sciences,
Harvard University, Cambridge MA 02138, USA
E-mail: gareth@mit.edu

Recent advances in the quantitative measurement and modelling of the rheology of elastic polymer solutions in both shear and extension are reviewed and the connection with the onset of purely elastic flow instabilities in torsional shear flows, pure extensional flows and more complex "mixed" two-dimensional flows is discussed. A dimensionless scaling criterion for the onset of elastic instabilities is introduced which can help understand the sensitivity of the numerically computed and experimentally measured stability boundaries to changes in both the flow geometry and the rheology of the test fluid. Finally, a simple heuristic argument is presented for interpreting the variation in the stability loci predicted by typical nonlinear differential viscoelastic constitutive models and for inter-relating the rheological material functions in steady shear and steady extensional flows at high Deborah numbers.

Introduction

The destabilising effects of fluid viscoelasticity on the steady flow of polymer melts and solutions are well-known [1, 2], and the past ten years have seen significant progress in both theoretical and experimental understanding of these instabilities [3]. The development of highly elastic "Boger fluids" of almost constant viscosity [4] has enabled experimentalists to focus on the dynamical effects of fluid elasticity in the absence of shear-thinning and inertial effects. Advances in quantitative techniques such as digital particle image velocimetry (DPIV), laser Doppler velocimetry and phase-modulated birefringence have permitted detailed characterisation of the spatio-temporal dynamics of both the steady base flow and the unstable flow that develops beyond a critical Deborah number. Complementary theoretical developments in the fields of numerical linear stability analyses and direct time-dependent computations mean that these observations can now be objectively compared with stability predictions obtained using constitutive models which provide a realistic description of the viscoelastic fluids employed in the experiments [5, 6]. The qualitative agreements between experiment and calculation are encouraging and suggest that the physical mechanisms governing the stability of viscoelastic flows are incorporated in simple constitutive equations.

Further progress towards an *a priori* predictive capability for the onset of flow instabilities now requires a quantitative and critical comparison of wave numbers, growth rates and neutral stability boundaries obtained from experimental observations and theoretical predictions. In addition to an understanding of the physical mechanisms leading to the onset of instability, this commonly leads to the need for marked improvements in the rheological characterisation of the test fluid and the constitutive model used to describe it. In addition to knowledge of the linear viscoelastic spectrum and the viscometric material functions measured in steady shear, the neutral stability boundaries of a flow are often found (by both experiment and computation) to be very sensitive to the extensional properties of the fluid, even in flows which, at least in the base state, are predominantly shearing in nature. The viscometric properties of the dilute polystyrene-based Boger fluid considered in the present study can be quantitatively described by the bead-spring model of Zimm [7]. However, such quasi-linear models are inadequate for describing the strain-hardening response of the fluid in uniaxial elongation, and finite extensibility of the chain must be incorporated. The present article surveys some of the recent experiments documenting "purely elastic" flow instabilities in several simple prototypical flow geometries, including simple shear flow in a cone-and-plate rheometer plus uniaxial elongation in a filament stretching device. We discuss a recently presented dimensionless criterion [8] that can be used to unify the physical effects and critical conditions required for purely elastic instabilities in each of the test geometries. This criterion incorporates both the presence of non-zero elastic normal stress differences in the fluid plus the magnitude of the streamline curvature in the flow, and can thus be thought of as the viscoelastic analogue of the Görtler number. Finally, we present some simple arguments for interpreting the sensitivity of the stability boundaries predicted by typical nonlinear viscoelastic constitutive models, and for inter-relating rheological properties in steady shear flows and extensional flows at high Deborah number.

Rheology of Ideal Elastic Fluids

Much has been written over the last 20 years about the rheology of "ideal elastic fluids" formulated from dilute (or semi-dilute) solutions of high molecular weight linear hydrocarbon polymers in viscous solvents. Molecular theories for dilute solutions suggest that the longest relaxation time of the chains is $\lambda_1 \sim [\eta]\,\eta_s\,M_w/RT$ and the first normal stress coefficient is $\Psi_1 \sim n\,k_B T\,\lambda_1{}^2$, where M_w is the molecular weight of the (monodisperse) chains, n is the number density, $[\eta]$ is the intrinsic viscosity of the molecules, and η_s is the viscosity of the suspending solvent. The viscoelastic properties of the solution may thus be varied over many orders of magnitude by simply using a viscous oligomeric solvent, ($\eta_s \sim 10$ Pa · s), in place of water or other low molecular weight organic solvents, ($\eta_s \sim 10^{-3}$ Pa · s). Early studies of similar systems were performed by Peterlin; however, these fluids are commonly termed "Boger fluids" after the pioneering work of Boger [4, 9]. It is not the purpose of the present paper to review this literature, and the evolution of the rheological modelling of these fluids from Maxwell to Oldroyd-B to nonlinear multi-mode model can be traced elsewhere [e.g. 10–14].

It is now commonly accepted that these highly elastic fluids are less ideal than desired and, as a result of polydispersity effects, finite chain extensibility and solvent-polymer interactions are only qualitatively described by single-mode dumbbell models (see [6, 15]). One of the study areas of the INI programme on complex fluids has been a re-examination of much of the published viscometric data for these elastic Boger fluids [16], and it appears that, at least for some of the more dilute ($c \ll c^*$) monodisperse fluids, the bead-rod spring model of Zimm can provide a good description of the linear viscoelastic properties. In Fig. 1, we show the elastic and viscous contributions to the dynamic material functions measured in small amplitude oscillatory shear flow for a 0.05 wt% solution of monodisperse polystyrene ($M_w = 2.25 \times 10^6$ g/mol) dissolved in oligomeric styrene [17]. From input

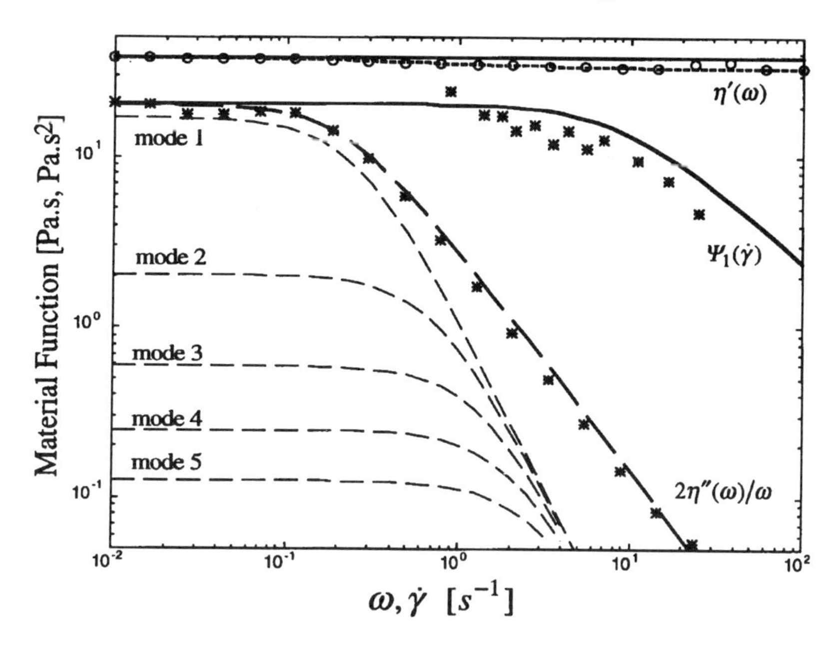

Fig. 1. Rheological properties of a monodisperse polystyrene Boger fluid in small amplitude oscillatory shear and steady shear flow. The hollow circles show the measured dynamic viscosity and the asterix indicate the elastic properties $2\eta''/\omega$ and Ψ_1. The dashed lines show the linear viscoelastic predictions of the Zimm bead-spring model plus the first five modal contributions. The solid lines show the predicted shear-thinning in $\eta(\dot{\gamma})$ and $\Psi_1(\dot{\gamma})$ as a result of finite molecular extensibility.

information about the solvent viscosity, the molecular weight and the concentration $c \equiv n\,M_w/N_A$ of the polystyrene chains (where $N_A = 6.02 \times 10^{23}$ is Avagadro's number), the Zimm model provides an accurate description of the plateau value of Ψ_{10} and the $\omega^{-4/3}$ scaling observed in the elastic rigidity $2\eta''/\omega$. The longest or "Zimm" relaxation time is predicted to be $\lambda_z \approx 3.8$ s, and is notably larger than the "average" relaxation time $\lambda_s \approx 1.9$ s estimated from the steady shear viscometric properties.

Simple Shear Rheology of "Ideal Elastic Fluids"

Much work remains to be performed in extending the *quantitative* rheological description of these supposedly "ideal" fluids to even simple shear flows. The finite extensibility of the chains, the spectrum of relaxation times, and the dominant hydrodynamic screening predicted by the Zimm model will all be important in describing the viscoelastic fluid properties in steady shear or large amplitude oscillatory shear flows. Single-mode FENE dumbbell models [18] can describe at least qualitatively the shear-thinning in $\Psi_1(\dot{\gamma})$ observed at large shear rates. However, the value of the dumbbell extensibility L is frequently anomalously small compared to that expected from molecular considerations (see [19, 20]), and incorrectly captures the transient response of the stresses [21, 22]. In contrast the Rouse/Zimm bead-spring chain models correctly capture the linear viscoelastic spectra of the fluids but the equilibrium pre-averaging of the configuration and the hydrodynamic interactions (if present) eliminates any shear-thinning in the first normal stress coefficient. More accurate analyses properly incorporating both effects are possible (see [7, 18]). However, they typically do not lead to closed form constitutive models suitable for regression to data or for use in two- and three-dimensional complex flow simulations.

The nonlinear variation in the viscometric data can be well described, however, by *ad hoc* multi-mode generalisations or simplifications of FENE-type models. As an example, in Fig. 1 we also show data for the first normal stress coefficient, $\Psi_1(\dot{\gamma})$, and the prediction of a recently proposed model [23] in which the polymer solution is treated as a non-interacting mixture of dumbbells of varying lengths, L_i (for modes $i = 1, 2, \ldots N_m$), with relaxation times, $\lambda_i = \lambda_z / i^{3/2}$, and finite extensibilities, $L_i = L/i^{1/2}$, that scale as expected from simple molecular arguments. Such a model is probably better suited to a polydisperse dilute solution, but the predictions in steady shear flow are essentially indistinguishable from other pre-averaged FENE chain models, such as the FENE-PM model of Wedgewood *et al.* [24]. Such models provide a framework for an increasingly realistic description of polymer solutions in steady and time-dependent flows by enabling the level of accuracy to be systematically varied without refitting the constitutive parameters, and by simply cutting off the resolution of the linear viscoelastic spectrum at a selected value of $i_{\max}$ (e.g. $i_{\max} = 1, 4, \ldots$).

Perhaps the most promising avenue for developing more realistic yet computationally tractable models is the recent explosive development of Brownian-dynamics-based techniques which can explore rapidly the approximations inherent in deriving closed-form constitutive equations (see Chap. 7 and [25, 26]). These techniques should provide ways of investigating more complex viscoelastic phenomena, such as conformation-dependent hydrodynamic interactions between the solvent and the polymer chains which may contribute to the largely unresolved experimental observations of the intermediate "plateau" or "knee" region observed in careful measurements of the first normal stress behaviour of many Boger fluids [12, 14].

Transient Extensional Rheology

Another recent development contributing to the quantitative experimental understanding of the viscoelasticity of polymer solutions is the development of *filament stretching rheometers.* Such devices, first developed by Sridhar and co-workers [27], permit quantitative measurements of the transient extensional viscosity in a uniaxially elongating filament of fluid from careful observations of the total tensile stress growth and radial evolution of the filament profile [28–30]. The discussion of the rheology and fluid dynamics of these devices formed the basis for many discussions at the INI workshop (for details, see [31]). The kinematics in such devices are spatially non-homogeneous as a result of the no-slip boundary conditions at either endplate (cf. Fig. 4 below).

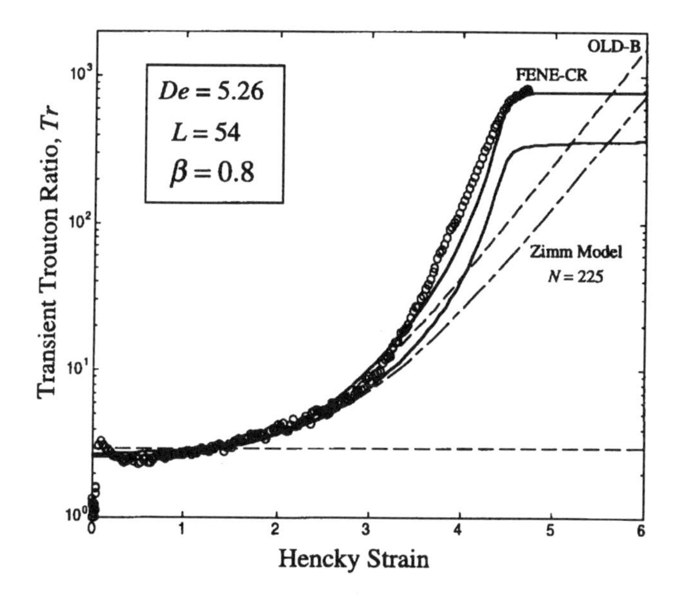

Fig. 2. The transient Trouton ratio showing the growth of the transient extensional stress difference in an elongating polystyrene filament at $De = 5.26$. The hollow symbols show the measured values for the 0.05 wt% PS solution, the solid line is the prediction of a FENE dumbbell model and the broken line is the prediction of a multi-mode FENE bead-spring chain model.

However, numerical calculations show that uniaxial extensional kinematics can be realised near the filament mid-plane, and such devices can therefore be used to extract the transient uniaxial elongational viscosity function $\bar{\eta}^+(\dot{\varepsilon}_0, t)$ for various polymer solutions. In Fig. 2 we show the transient Trouton ratio $Tr = \bar{\eta}^+(\dot{\varepsilon}_0, t)/\eta_0$ for the same dilute 0.05 wt% polystyrene Boger fluid shown in Fig. 1, plus a comparison with the single-mode FENE dumbbell model and the multi-mode FENE mixture model discussed above. It is clear that both models capture the material response with an initial plateau value of $Tr \approx 3\boldsymbol{\beta}$ where $\boldsymbol{\beta} = \eta_s \eta_0$ is the solvent contribution to the total viscosity, a pronounced strain hardening at Hencky strains $\varepsilon \geq 2$, and a slow approach to an asymptotic plateau value at $Tr \approx 1000$ for $\varepsilon \geq 5$. Surprisingly, the single-mode FENE dumbbell model provides a more quantitative fit of the data than the multi-mode model. This is because the total elastic modulus, $G\,(\sim nk_BT)$, is distributed in the multi-mode model over a spectrum of modes with decreasing relaxation times (and thus each shows a progressively decreasing strain hardening as the Deborah number, $De_i = \lambda_i \dot{\varepsilon}_0$, of each mode decreases). The simple FENE dumbbell-like response has also recently been shown to describe closely dynamical studies on the unravelling of single DNA chains [32].

Elastic and Viscous Contributions to Polymeric Stress

The deviation between the experimentally observed tensile stress and that predicted by the multi-mode FENE chain model in Fig. 2 suggests that there is an additional contribution to the total stress arising from hydrodynamic coupling between portions of the unravelling polymer chains. This transient effect does not appear to be captured correctly by the pre-averaged bead-spring chain models. There are numerous interesting consequences of this additional configuration-dependent stress contribution which are beginning to be explored by experiments monitoring the relaxation of the tensile stress difference following a cessation of the uniaxial elongation [17, 33, 34] and by Brownian dynamics calculations [35, 36]. This additional non-equilibrium stress arises because the conformation of the chains is not the "most probable" one predicted from a quasi-static random walk. The resulting stress in the chain is configuration-dependent and appears dissipative or "viscous" in nature rather than (entropically) elastic since it depends on the rate of stretching of the material. It should thus disappear rapidly upon cessation of the stretching. This dissipative contribution may be significant in resolving the discrepancies between the large non-Newtonian pressure drops (such as friction factors or Couette corrections) observed experimentally and computed numerically in complex flows involving extensional kinematics. It should also lead to pronounced hysteresis in measurements of the force-extension curves for the macro-molecules during uniaxial elongation and subsequent relaxation. This effect remains unexplored at present, but could be explored by independent measurements of the total polymeric stress difference and the average configuration $tr\{\boldsymbol{A}\}$ of the chains (where $\boldsymbol{A} = \langle \boldsymbol{RR} \rangle$ is the ensemble-averaged dyadic product of the end-to-end vector). For recent experimental and numerical work in this area, see Ref. [47].

Discussions throughout the workshop have shown that care must be taken in the decomposition of elastic and viscous contributions to the polymeric stress. To demonstrate this, it is worth noting that the asymptotic limit of the polymeric contribution to the extensional stress difference, $\Delta\tau \equiv (\tau_{zz} - \tau_{rr})$, of the FENE-P dumbbell model at high strains is of the form

$$\lim_{\substack{\varepsilon \to \infty \\ De \geq 1/2}} \Delta\tau^{+}(\dot{\varepsilon}_0, t) \to \Delta\tau_0 = 2nk_BT\, L^2\lambda_1\dot{\varepsilon}_0 \left[(1 - 3/L^2) - (2De)^{-1} + \ldots\right] , \quad (1)$$

where $nk_BT \equiv \eta_p/\lambda_1$ is the elastic modulus of the dilute solution of dumbbells and $De \equiv \lambda_1\dot{\varepsilon}_0$ is the Deborah number. This limiting stress itself appears "viscous" in nature (that is, it scales with $\dot{\varepsilon}_0$) even though the model incorporates only nonlinear entropic elastic effects! This result is a consequence of the limiting form of the nonlinear Warner spring near to full extension which predicts $f(tr\boldsymbol{A}) \to 2De$, that is, the limiting value of the dimensionless spring constant $f(tr\boldsymbol{A})$ increases linearly with the extension rate.

This nonlinearity in the spring connector force also leads to an extremely rapid relaxation of the elastic stress stored in the elongated dumbbells upon cessation of the homogeneous elongation. In fact, asymptotic calculations suggest that the tensile stress decays at short times, $\delta t^{-} = (t - t_f)$, following the cessation of stretching at a time t_f (corresponding to a final steady state strain of $\varepsilon_f = \dot{\varepsilon}_0 t_f$) as

$$\frac{\Delta\tau^{-}}{\Delta\tau_0} \approx \frac{(2De)^{-1}\exp\left((2De)^{-1}\right)\,\exp\left(-\left[2\,\delta t^{-}/\lambda_1 + (2De)^{-2}\right]^{1/2}\right)}{\left[2\,\delta t^{-}/\lambda_1 + (2De)^{-2}\right]^{1/2}} \quad (2)$$

for $\delta t^{-}/\lambda_1 \ll 1$, where $\Delta\tau_0$ is the limiting value of the steady state stress given by Eq. (1). For example, following the cessation of stretching at the Deborah number of $De = 5.26$ shown in Fig. 2, the FENE-P dumbbell model predicts that the polystyrene Boger fluid will lose 50% of its stress in an elapsed time of $\delta t^- \approx 48$ ms! This decay will appear "instantaneous" given the mechanical limitations of existing force transducer assemblies. Independent measurements of the configuration of the chains (using, for example, light scattering or birefringence) and the resulting stress appear to be necessary to discriminate systematically between the elastic and viscous contributions to the total polymeric stress.

Stress Relaxation Master Curves

A final item of rheological interest which can be noted from the form of the transient decay in the tensile stress difference in Eq. (2), and from similar asymptotic estimates of the stress relaxation at long times $\delta t^-/\lambda_1 \gg 1$, is that the stress relaxation profiles described by FENE-P type models at different De are in fact portions of a single "stress relaxation master curve". The construction of such a master curve is shown in Fig. 3 for the FENE-P model. The polymeric contributions to the stress growth and relaxation portions of individual transient elongational experiments at different De, but the same final Hencky strain, are shown in Fig. 3(a). In contrast to some recent experiments [34], the stress relaxation portions of these profiles are *not* self-similar when plotted as a function of the strain ε. This is in contrast to the corresponding stress growth functions, which roughly superpose — when scaled with stretch rate and plotted against the Hencky strain — to form a single extensional viscosity function, $\bar{\eta}^+(\dot{\varepsilon}_0 t)$ [17, 33]. However, the relaxation curves obtained following the cessation of elongation can be superimposed by using an *additive* shift factor, a_{De}, to rescale the actual elapsed time, $\delta t^{(i)}$, in an experiment at Deborah number, $De^{(i)}$, and form a reduced time according to

$$\frac{\delta t^{(red)}}{\lambda_1} \equiv \frac{\delta t^{(i)}}{\lambda_1} + a_{De} = \frac{\delta t^{(i)}}{\lambda_1} + \frac{1}{2}\left(\left(2De^{(i)}\right)^{-2} - \left(2De^{(ref)}\right)^{-2}\right), \tag{3}$$

where, as typical in the construction of master curves, a reference curve at a specific test condition (here denoted $De^{(ref)}$) must be chosen to shift the remaining curves onto.

Such relaxation master curves are a consequence of the form of the nonlinear FENE connector force, which is written as a unique function of only the *current configuration* of the dumbbells. In the absence of an external flow, the convection-diffusion equation for the evolution of the micro-structure

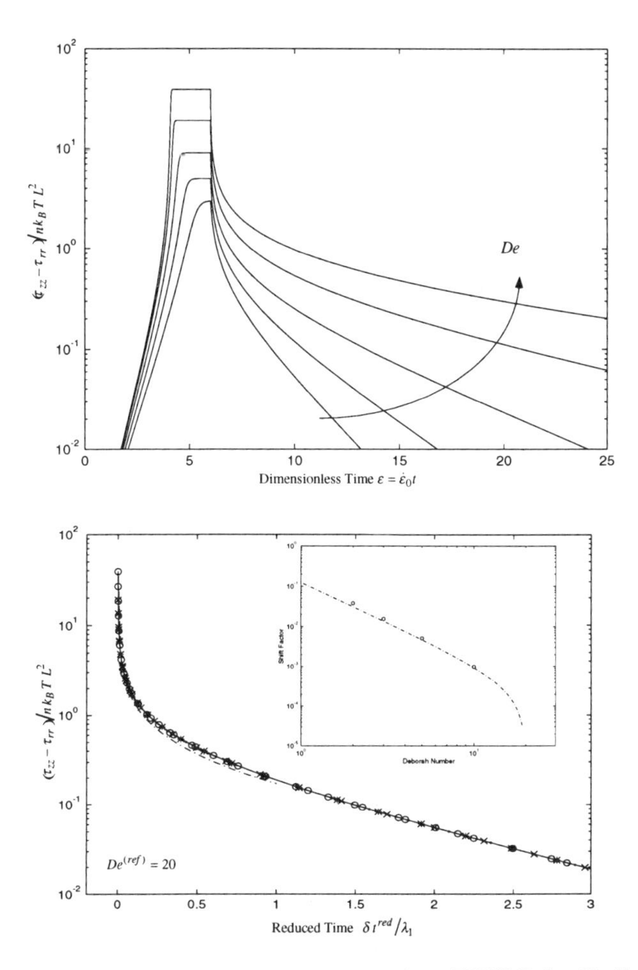

Fig. 3. Construction of a stress relaxation master curve for a FENE-P dumbbell model following the cessation of uniaxial elongation: (a) the relaxation profiles of the dimensionless scaled contribution to the polymeric stress computed for increasing values of $De = 2, 3, 5, 10 \,\&\, 20$ do not superimpose or have the same slope when scaled with $\dot{\varepsilon}_0 t$; (b) when plotted as a function of a reduced time, the profiles (indicated by different symbols) all form segments of a single master curve and can be shifted onto the master curve using a shift factor, a_{De}. The asymptotic estimate of Eq. (2) is shown by the dashed line. The shift factor is given in Eq. (3) and shown in the inset figure.

follows a single trajectory from a given initial configuration, and all of the initial states, $\Delta\tau_0(De^{(i)})$ given by Eq. (2), lie along this unique curve. The shift factor scales as $a_{De} \sim 1/{De^{(i)}}^2$ and is additive rather than multiplicative because the stress decay is exponential in nature. Similar master curves exist for the FENE-PM bead-spring chain model and numerical calculations with the bead-rod chain model have shown a similar universal relaxation response [36]. This response is expected to be quite general for all linear or nonlinear kinetic theory models in which the entropic elastic connector force is only a state function of the current configuration. Recent calculations have shown that stress relaxation profiles, following the cessation of stretching at lower final strains (that is, short of the asymptotic plateau region), also lie on this master curve through the use of an additional strain-dependent shift factor, $a_\varepsilon(\varepsilon_f, Dc)$. It will be very interesting to see if this nonlinear viscoelastic response is verified by stress relaxation experiments with dilute polymer solutions following filament stretching tests at different Deborah numbers.

Scaling of Elastic Instabilities

Although there are a number of outstanding issues in the rheometry of Boger fluids, some of which have been highlighted above, it is clear that the steady and transient rheology is at least reasonably well-described by simple dumbbell models. These models are computationally tractable, and much research has consequently focused on comparing experimental observations of the changes in the dynamics and stability of the motion of elastic liquids in simple or complex geometries with large-scale numerical simulations of the equations of motion. As Larson *et al.* [37] first showed, even the simple quasi-linear Oldroyd-B constitutive model appears to capture, at least qualitatively, the *purely elastic stability* that develops at vanishingly small Reynolds number in the circular Couette geometry. Purely elastic instabilities have also been observed experimentally in many complex two-dimensional flows, such as the flow through an axisymmetric contraction or the geometries shown schematically in Fig. 4. However, the lack of an analytic base solution for many two-dimensional flows indicates that, in the future, stability analyses in complex geometries will have to be performed through a numerical integration of the governing equations. In order to understand the expected geometric and rheological sensitivity of the resulting computations, and to separate numerical artifacts from true hydrodynamic instabilities, it would thus be helpful to have a framework in which to interpret the results.

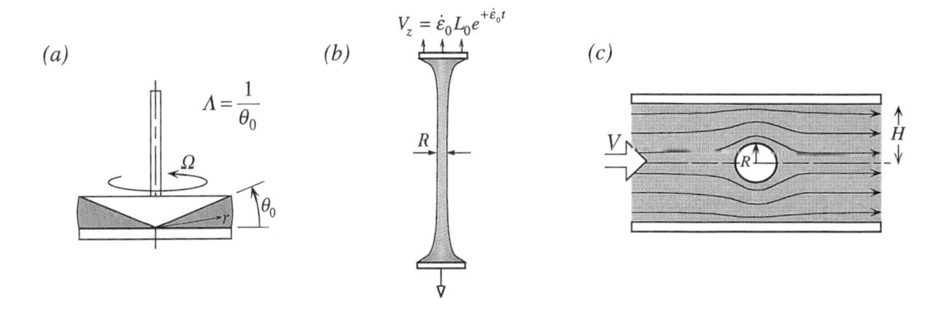

Fig. 4. Some representative geometries that involve viscoelastic flows with curved streamlines and exhibit purely elastic instabilities; (a) steady torsional shear flow in a cone-and-plate rheometer; (b) uniaxial elongational flow of a thin elastic fluid filament pinned between two disks; (c) flow past a confined cylinder; a "mixed" flow with regions exhibiting both shearing and extensional kinematics.

Ample experimental, numerical and analytical evidence now exists to indicate that in simple shear flows of viscoelastic fluids, the purely elastic instabilities — such as the Taylor-Couette instability or the Bernoulli spiral instability [20] that develops in the cone-and-plate geometry of Fig. 4(a) — arise from a coupling between (i) the curvature of fluid streamlines and (ii) the elastic normal stresses that give rise to a tension along each streamline [15]. In more complex flows, each of these quantities may vary throughout the domain of interest. However, the coupling between these terms still affords the possibility of the onset of elastic flow instabilities. We have recently shown that the kinematic and dynamic conditions corresponding to the curvature of the flow and the tensile stress along the streamlines can be combined into a single dimensionless criterion with a magnitude M [38]. There is a critical value of this dimensionless group that must be exceeded for the onset of purely elastic instabilities. This criterion can be represented in the general form

$$M \equiv \left[\frac{\lambda_1 U}{\mathcal{R}} \cdot \frac{\tau_{11}}{(\eta_0 \dot{\gamma})}\right]^{1/2} \geq M_{crit} \tag{4}$$

for onset of elastic instability, where U is a characteristic measure of the fluid velocity, $\mathcal{R}$ is a characteristic radius of the curvature of the fluid streamline ($\kappa = \mathcal{R}^{-1}$ is the curvature), τ_{11} is the tensile stress in the "1" or flow direction, $\eta_0 = \eta_p/(1-\beta)$ is the zero-shear-rate viscosity of the fluid, and $\dot{\gamma}$ is a characteristic value of the local deformation rate in the flow.

Physically, the dimensionless criterion given by Eq. (4) captures both the kinematic curvature of the streamlines in the flow and the dynamical influence of viscoelastic normal stresses. This dimensionless grouping does not reduce trivially to either of the familiar dimensionless groups (the Weissenberg number or the Deborah number) characterising viscoelastic effects in a flow. However, we have shown how a connection to these dimensionless groups can be made for simple geometries. In [8], we interpret the length scale given by $\ell \sim \lambda_1 U$ as the characteristic distance over which perturbations to the base viscoelastic stress and velocity fields relax, and the ratio $(\ell/\mathcal{R})$ is thus a dimensionless measure of the relative distance over which disturbances are advected compared to the local curvature of the flow. The relative magnitude of the coupling of these perturbations to the elastic stresses in the viscoelastic base flow is then characterised by the second dimensionless term on the left-hand side of Eq. (4). A more rigorous justification of these conclusions based on the governing linear stability equations is discussed elsewhere [38].

Of course, almost all viscoelastic polymer solutions exhibit shear-thinning in the elastic normal stresses (cf. Fig. 1). Both experiments and analyses have shown that this shear-thinning can dramatically affect the locus of the stability boundaries for purely elastic flow transitions. The dimensionless criterion given by Eq. (4) can be generalised to incorporate such variations by recognising that the characteristic fluid relaxation time is a rate-dependent quantity that can be estimated from the variations in the viscosity and first normal stress coefficient using an equation of the general form

$$\lambda(\dot{\gamma}) = \frac{\Psi_1(\dot{\gamma})}{2\eta_p(\dot{\gamma})} = \frac{\Psi_1(\dot{\gamma})}{2[\eta(\dot{\gamma}) - \eta_s]} . \tag{5}$$

Substituting this definition into Eq. (4) and rearranging leads to a generalised stability criterion of the form

$$\left[\frac{\lambda_1 U}{\mathcal{R}} \cdot (\lambda_1 \dot{\gamma}) \cdot \left(\frac{\Psi_1(\dot{\gamma})}{\Psi_{10}}\right) \cdot \left(\frac{\lambda(\dot{\gamma})}{\lambda_1}\right)\right]^{1/2} \geq \frac{M_{crit}}{\sqrt{2(1-\beta)}} . \tag{6}$$

The selection of appropriate measures for the variation in the streamline curvature $\mathcal{R}(\Lambda)$ as the geometric aspect ratio Λ is varied, coupled with either independent measurement of the shear-rate-dependent viscometric functions, $\{\eta_p(\dot{\gamma}), \Psi_1(\dot{\gamma})\}$, or the predictions of molecular theories such as the FENE-PM model discussed in §2, allows Eq. (6) to be rearranged in such a way that measurements of the critical velocity U_{crit} for the onset of elastic instability can be used to determine M_{crit}. The utility of this general expression has recently been demonstrated by comparison to experimental observations and numerical predictions of elastic instabilities in numerous geometries [38]. We illustrate this in Fig. 5 by showing the critical conditions for the onset of the spiral cone-and-plate instability predicted by a single-mode Giesekus model with various values of the nonlinear model parameter α. The stability boundaries are predicted to be sensitive functions of the Deborah number (De), the geometry (characterised by $\Lambda \equiv \theta_0^{-1}$) and the nonlinear parameter (α) governing the fluid rheology (α).

Of course, a dimensionless criterion such as Eq. (6) can only provide guidance about the functional form of the predicted stability boundary. The value of M_{crit} for each geometry must be determined by either regression to measurements of critical flow conditions or by comparison to numerical analysis. The value of M_{crit} for the cone-and-plate geometry has been determined analytically by Olagunju [39] to be $M_{crit} = 21.193\ldots$, and the single-mode Giesekus model clearly captures the general form of the data in Fig. 5. A more elaborate multi-mode analysis permits an almost quantitative description of

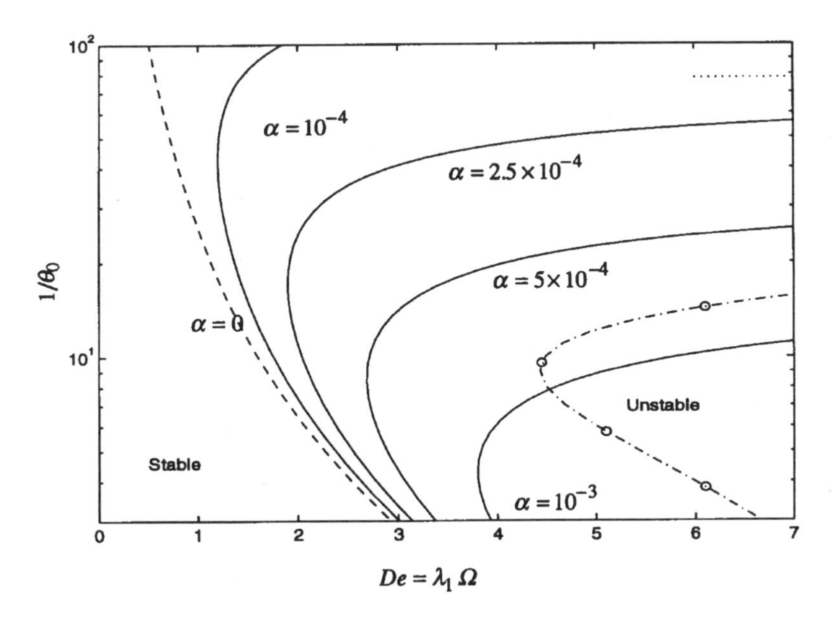

Fig. 5. Predicted shape of the neutral stability loci from the scaling estimate in Eq. (6) with $M_{crit} = 21.193$ for $\alpha = 10^{-3}$, 5×10^{-4}, 2.5×10^{-4} and 10^{-4} respectively. The experimental data (o) are for a 0.31 wt% PIB Boger fluid described in [20]. The asymptotic estimate (- - -) shown for $\alpha = 2.5 \times 10^{-4}$ is given by Eq. (10).

the stability boundaries which is consistent with both the steady shear and transient elongational rheology of the experimental test fluid [6].

Since the radius of the curvature and the streamwise elastic stress in Eq. (4) are functions of spatial position $\boldsymbol{x}$ in a complex geometry, the value of the dimensionless quantity M will also vary spatially. This scalar field, $M(\boldsymbol{x})$, can be computed during post-processing of numerical simulations and may be useful in identifying (and possibly eliminating through geometric modifications or partial re-design) critical regions of a complex flow that are prone to the onset of elastic instabilities [40].

One general conclusion to be drawn from the form of the criteria in Eq. (4) or Eq. (6) is that similar mechanisms for purely elastic instabilities may be expected in shear flows, extensional flows or "mixed" two-dimensional flows. Viscoelastic fluids with curved free surfaces of viscoelastic fluids may also be prone to purely elastic instabilities, and this can be of importance in processing operations such as coating flows [41]. An elastic free surface instability has recently been observed in measurements with Boger fluids in a filament stretching rheometer of the type discussed earlier and sketched in Fig. 4(b).

Three different views of the instability are shown in Fig. 6. The plan view of the fluid near the endplate shows that the smooth cylindrical surface of the elongating and narrowing filament becomes unstable due to an azimuthally-periodic disturbance. This results in the formation of a number of thin fibrils

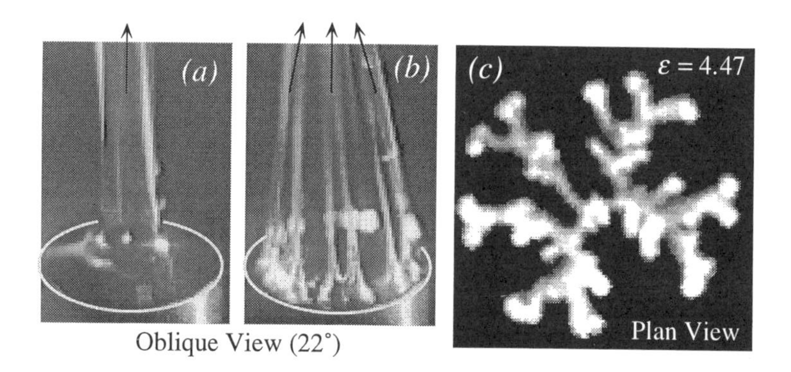

Fig. 6. Curved streamlines near the rigid end plates and large elastic tensile stresses lead to the onset of an elastic instability during uniaxial elongation of the PS Boger fluid in a filament stretching extensional rheometer: (a) oblique view, $De = 5$, $\varepsilon = 3.9$ before the onset; (b) oblique view, $De = 5$, $\varepsilon = 4.48$ following the onset; (c) plan view from underneath a specially constructed endplate at the same conditions as (b).

that connect the main column to the endplate. These fibrils are under tension and subdivide and elongate as they propagate radially outwards. At very large strains, the filament decoheres completely from the endplate [17] as a result of the elastic instability — but does not directly "fracture" or "rupture" as observed in experiments and analyses of polymer melts at high strains [42].

In this flow, the perturbations will be driven by elastic tensile stresses, $\tau_{11}(De)$, and will be dampened by the interfacial tension of the curved fluid air or fluid-fluid free surface. The dimensionless group appearing in Eq. (4) will thus be modified to balance elastic stresses and stress gradients across curved streamlines with the interfacial stress arising from surface tension. An important dimensionless group differentiating melts and polymer solutions is the ratio of surface tension stresses, $(\sigma/\mathcal{R})$, to the elastic modulus G of the material $\sigma/G\mathcal{R} \equiv De/[(1-\beta)\ Ca]$. In [17], we show how such arguments can be used to construct the appropriate form of a "stability diagram" describing the parameter space for stable operation of filament stretching rheometers. However, no detailed stability analysis on the onset or propagation of such elastic instabilities on curved free surfaces has yet been performed.

Interrelation of Material Properties in Shear and Extension

The experimental observations and numerical calculations of purely elastic flow instabilities discussed in the preceding section also hint at a connection between the viscometric properties of a fluid measured in steady shear flow and the magnitude of the increase in the extensional viscosity in uniaxial elongation. In particular, linear stability calculations with nonlinear differential constitutive single-mode models such as the Chilcott-Rallison and Giesekus models [20], as well as multi-mode formulations of the Giesekus model [6], show that even in a base state consisting of a unidirectional shear flow, the stability boundaries are sensitive functions of the nonlinear model parameter (L^2 or α respectively) governing the shear-thinning in the polymer contribution to the viscosity, $\eta_p(\dot{\gamma})$, and/or the first normal stress coefficient, $\Psi_1(\dot{\gamma})$.

This can be demonstrated more clearly by evaluating the asymptotic forms of the dimensional stability criterion, M^2, at large Wi. For the Giesekus model, using the asymptotic results for the viscometric properties given in [43], one obtains [38]

$$\lim_{\lambda_1\dot{\gamma}\to\infty} M^2(\dot{\gamma}) \to \frac{(1-\beta)\theta_0}{\alpha}. \tag{7}$$

Similarly, for the Chilcott-Rallison FENE-CR model, one finds

$$\lim_{\lambda_1\dot{\gamma}\to\infty} M^2(\dot{\gamma}) \to (1-\beta)\,(L^2-3)\,\theta_0\,. \tag{8}$$

Rearranging these expressions will yield lower bounds for the size of the cone-angle for which elastic spiral instabilities will be observed, or

$$\left(\frac{1}{\theta_0}\right)_{CR} \leq \frac{(1-\beta)\,(L^2-3)}{{M_{crit}}^2} \tag{9}$$

and

$$\left(\frac{1}{\theta_0}\right)_{Giesekus} \leq \frac{(1-\beta)}{\alpha}\frac{1}{{M_{crit}}^2}\,. \tag{10}$$

This latter asymptotic limit is indicated in Fig. 5 using the critical value of ${M_{crit}}^2 = 21.193\ldots$ determined from a perturbation analysis [39]. Surprisingly, the dependence of these asymptotic limits on the relevant nonlinear constitutive parameter are very similar to the limiting values for the planar elongational Trouton ratio, $\bar{\eta}_p/\eta_0$, predicted by each model. It is well-known that the invariants, I_1, I_2 of the finite strain tensors are identical for planar elongation and steady simple shear flow [43]. However, to our knowledge, no connection has to date been made between the material functions, $\eta_p(\dot{\gamma})$, and $\Psi_p(\dot{\gamma})$ in shear and $\bar{\eta}_p(\dot{\varepsilon})$ in planar elongation.

Tracing the definition of the shear-rate-dependent relaxation time, $\lambda(\dot{\gamma})$, to Eq. (5), it becomes clear that this asymptotic connection between the model predictions in uniaxial elongation and steady shear can best be expressed in the form

$$\lim_{\lambda_1\dot{\varepsilon}_0\gg\frac{1}{2}} ((\tau_{zz}-\tau_{rr})_{ext}) \to (\bar{\eta}_{p,\infty}\dot{\varepsilon}_0) \leftarrow \lim_{\lambda_1\dot{\gamma}\to\infty} \left((\tau_{xx}-\tau_{yy})_{shear}{}^2/\tau_{xy}\right) \tag{11}$$

or in dimensionless form by

$$\lim_{\lambda_1\dot{\varepsilon}_0\gg\frac{1}{2}} \left(\frac{\bar{\eta}_p(\dot{\varepsilon})}{\eta_p}\right) \to \frac{\bar{\eta}_{p,\infty}}{\eta_p} \leftarrow \lim_{\lambda_1\dot{\gamma}\to\infty} \left(\frac{(\Psi_1(\dot{\gamma}))^2\,\dot{\gamma}^2}{\eta_p(\dot{\gamma})\,\eta_p}\right), \tag{12}$$

where $\bar{\eta}_{p,\infty}$ is the asymptotic or limiting value of the polymeric contribution to the planar elongational viscosity. For the FENE-P model discussed in connection with Eq. (1), this is given by $\bar{\eta}_{p,\infty} \equiv (\Delta\tau_0/\dot{\varepsilon}_0) = 2\eta_p L^2[(1-3/L^2) - (2De)^{-1}\ldots]$.

Furthermore, the asymptotic values of the polymer contribution to the planar elongational viscosity, $\bar{\eta}_{p,\infty} = (\bar{\eta}_\infty - 4\eta_s)$, are identical to the contribution to the uniaxial elongation viscosity for many simple models, including the Giesekus and FENE-P type models [44]. To demonstrate the validity of this

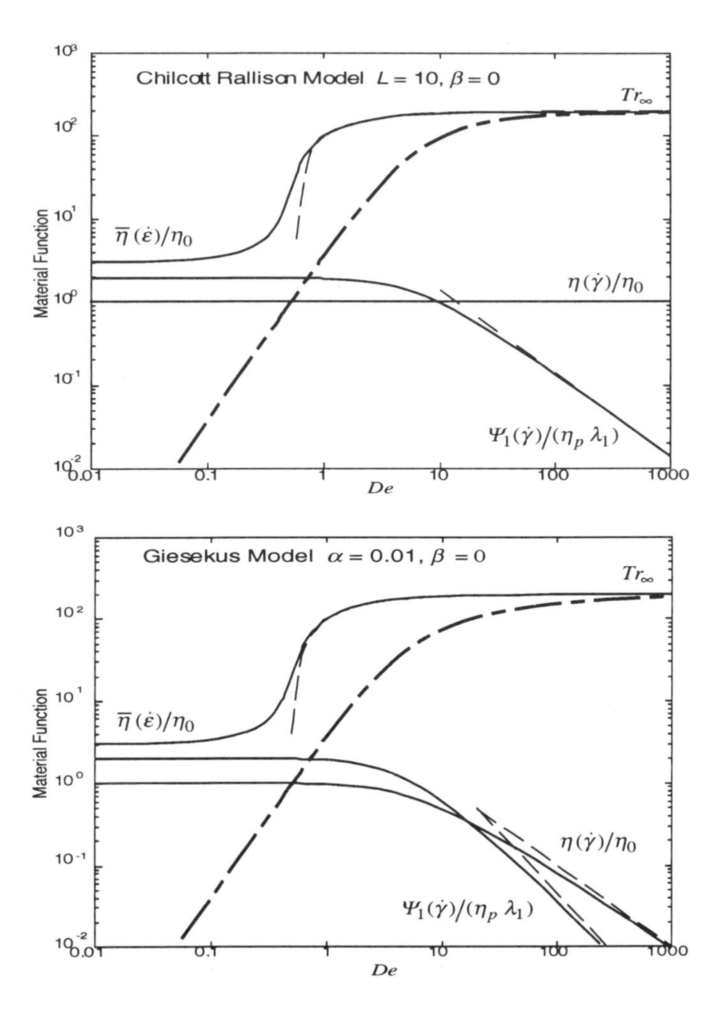

Fig. 7. Asymptotic connections between the form of the viscometric material functions and the ultimate plateau value of the planar elongational viscosity for (a) the FENE-CR dumbbell model of Chilcott and Rallison with $L = 10$; (b) the Giesekus model with $\alpha = 0.01$. Solid lines show the material functions and dashed lines indicate appropriate asymptotic estimates at high ***ADe***. The correlation given in Eq. (12) is shown by the broken line (— - —).

asymptotic connection, we show in Fig. 7 a comparison of the material functions $\eta_p(\dot{\gamma})/\eta_p$, $\Psi_1(\dot{\gamma})/(\eta_p\lambda_1)$ and $\bar{\eta}_p(\dot{\varepsilon}_0)/\eta_p$ with the predictions of the product given on the right-hand side of Eq. (12) for the Chilcott-Rallison and Giesekus models. This estimate clearly provides a good prediction of the planar and/or uniaxial elongational viscosity at large De. Unfortunately, the convergence is somewhat slow; however, the agreement is excellent for $De \gg 10$. A similar agreement is also found in other simple dumbbell-like models (such as FENE-P) and in the Phan-Thien-Tanner model (with the linear form of the friction term).

Since the stresses in multi-mode models are combined linearly, it is straightforward to define a shear-rate-dependent relaxation time for each mode as $\lambda_i = \Psi_{1i}(\dot{\gamma})/2\eta_i(\dot{\gamma})$ and extend this estimate of the asymptotic extensional viscosity to multiple relaxation modes. For the spectrum of relaxation times, $\{\eta_i, \lambda_i\}$, and the values of the nonlinear parameters, $\{L_i^2, \alpha_i\}$, obtained for typical elastic fluids (see for example, [14, 45]) this value is heavily weighted toward the mode with the largest (smallest) value of L_i^2 (or α_i), respectively.

This unexpected correlation can be understood, at least qualitatively, by considering the mechanisms by which stress is transmitted by a simple dumbbell model [46]. In a shear flow, the principal axes of stress are initially oriented at $\chi = 45°$ to the flow direction at low De (that is, in the direction associated with the principal stretch) and become progressively aligned with the flow direction as De increases ($\tan 2\chi = \tau_{12}/(\tau_{11} - \tau_{22})_{shear} \sim 1/De$). In contrast, in a planar extensional flow, the molecules are aligned with the flow and the stress is given by $(\tau_{11} - \tau_{22})_{ext}$. Resolving the elastic normal stress difference experienced in steady shear flow into the direction of the planar extensional flow then leads to the balance in Eq. (11).

This result may be useful in dilute polymer solutions as a heuristic rule (analogous to the Cox-Merz rule or Laun's rule [43]) for estimating the limiting magnitude of the increase in the uniaxial elongational viscosity which is expected in experimental devices such as the filament stretching apparatus. The application of Eq. (12), however, requires a careful separation of the polymeric and solvent contributions to the total viscosity since $\eta_p(\dot{\gamma}) = \eta(\dot{\gamma}) - \eta_s$ appears in the denominator of Eq. (12). Although the correlation is valid for many simple single-mode models in which the elastic stress is carried by a single (nonlinear) spring, other models, such as the exponential form of the PTT non-affine network model often employed for polymer melts [7], do not lead to a limiting result of the form in Eq. (12). This is because the extensional viscosity is predicted to pass through a maximum and then subsequently decreases, whereas the ratio of the normal stress difference and shear stress in steady simple shear flow increases monotonically.

Conclusions

In this chapter, we have attempted to review some of the recent areas of discussion related to the viscoelasticity of elastic polymer solutions that have taken place at the Isaac Newton Institute as part of the programme on the "Dynamics of Complex Fluids". The viscoelasticity of polymeric systems is, of course, a vast area and only a few topics related directly to the rheology and dynamics of "ideal elastic fluids" have been covered in this essay. These Boger fluids have played important roles as model systems for detailed comparisons between the numerical calculations and experimental measurements of many complex flows [11]. Indeed, they continue to be of use in providing convenient experimental pathways for studying phenomena such as the relative contributions of entropic elasticity and internal conformational dynamics to the total stress carried by an unravelling polymer chain discussed briefly on page 12.

The level of internal consistency that can be obtained between the measurements of: (i) the viscometric material functions; (ii) the transient uniaxial elongational properties of these fluids; and (iii) the hydrodynamic stability of many flows with mixed shearing and extensional kinematics is extremely encouraging. The future challenge is to extend this quantitative predictive capability to less ideal systems such as concentrated polydisperse solutions and linear or branched melts.

Acknowledgements

The research discussed in this article is supported by grants from the National Science Foundation (through the PFF programme, grant no. CTS9553216) and from NASA (NAG3-1793) to GHM. I would also like to thank Drs. S. H. Spiegelberg, P. Pakdel and A. Öztekin for their contributions and insight. My participation in the DCF Programme at the Isaac Newton Institute for the Mathematical Sciences was made possible by a fellowship from the Gabriella and Paul Rosenbaum Fund.

Discussion

D. James How bad is the discrepancy between the measured rheology and the Rouse model?

G. H. McKinley There are two important changes. First, the coefficient of proportionality relating the average relaxation time in steady shear flow to the longest relaxation time (i.e. $\lambda_s \equiv \Psi_{10}/2\eta_p = k\lambda_{longest}$ and k is the coefficient of proportionality) is modified due to the values of the zeta functions arising in the infinite (or finite) sum over individual modes. Second, and most importantly, however, the scaling of the slopes observed in the linear viscoelastic spectra changes systematically. For example, the frequency dependence of the elastic rigidity, $2\eta''(\omega)/\omega$, changes from ω^{-2} for a dumbbell model to $\omega^{-3/2}$ for the Rouse model (with no hydrodynamics) to $\omega^{-4/3}$ for the Zimm model. The data shown in Fig. 1 and for many other Boger fluids composed of dilute polymer chains in very viscous solvents appear to be more consistent with the dominant hydrodynamics of the Zimm model.

R. Keunings How does the FENE model without the closure approximation fare in predicting the extensional rheology?

G. H. McKinley Qualitatively, it is very similar to the change in the form of the response of the FENE-PM model when compared to the FENE-P model. There is a somewhat faster increase in the stress at short times followed by a more gradual approach to steady state. This qualitative similarity, however, arises form different physical mechanisms. In the FENE-PM model, the shorter time constants of the higher modes result in a faster linear viscoelastic response at small strains, whereas in the FENE dumbbell the configuration distribution becomes skewed and non-Gaussian.

R. Tanner What sort of models would we use in finite element simulations that incorporate these instantaneous viscous stresses?

G. H. McKinley Both John Rallison and John Hinch have proposed simple constitutive models which have a dumbbell-like convection-diffusion equation for the configuration tensor $\boldsymbol{A}$, and a Kramers-like expression for the stress with the presence of an additional term which is explicitly dissipative and scales with the deformation rate and $tr\boldsymbol{A}$ to some power. Oliver Harlen has recently pointed out, however, that these models have somewhat unphysical responses in transient shear flows such as flow past a sphere accelerating from rest in an elastic fluid.

J. R. A. Pearson The viscoelastic free surface instability looks reminiscent of Taylor's work on peeling of thin films. Have you contrasted your observations with this work?

G. H. McKinley Yes there are many similarities plus some important differences. In the instability we observed, the peeling angle is not small and the relevant balance is of elasticity (which tends to elongate the fluid column and decrease the radius of curvature near the endplates of the filament device) and surface tension which resists this deformation. However, in both cases, the onset of instability appears to be governed by a critical value of the negative pressure gradient across the fluid interface. A linear stability analysis of this elastic free surface instability is very challenging but should prove very interesting.

References

1. Petrie, C. J. S. & Denn, M. M. (1976) Instabilities in polymer processing. *A. I. Ch. E Journal* **22**, 209–235
2. Pearson, J. R. A. (1976) Instability in non-Newtonian flow. *Ann. Rev. Fluid Mech.* **8**, 163–181
3. Larson, R. G. (1992) Instabilities in viscoelastic flows. *Rheol. Acta* **31**, 213–263
4. Boger, D. V. (1977/78) A highly elastic constant-viscosity fluid. *J. Non-Newtonian Fluid Mech.* **3**, 87–91
5. Larson, R. G., Muller, S. J. & Shaqfeh, E. S. G. (1994) The effect of fluid rheology on the elastic Taylor-Couette flow instability. *J. Non-Newtonian Fluid Mech.* **51**, 195–225
6. Öztekin, A., Brown, R. A. & McKinley, G. H. (1994) Quantitative prediction of the viscoelastic instability in cone-and-plate flow of a Boger fluid using a multi-mode giesekus model, *J. Non-Newtonian Fluid Mech.* **54**, 351–379
7. Larson, R. G. (1988) *Constitutive Equations for Polymer Melts and Solutions.* Boston: Butterworths
8. Pakdel, P. & McKinley, G. H. (1996) Elastic instability and curved streamlines. *Phys. Rev. Lett.* **77**, 2459–2462
9. Boger, D. V. & Nguyêñ, H. (1978) A model viscoelastic fluid. *Polym. Eng. & Science* **18**, 1037–1043
10. Boger, D. V. & Mackay, M. E. (1991) Continuum and molecular interpretation of ideal elastic fluids. *J. Non-Newtonian Fluid Mech.* **41**, 133–150
11. Boger, D. V. & Yeow, Y. L. (1992) The impact of ideal elastic liquids in the development of non-Newtonian fluids. *Exp. Thermal and Fluid Sci.* **5**, 633–640
12. Magda, J. J., Lou, J., Baek, S.-G. & DeVries, K. L. (1991) Second normal stress difference of a Boger fluid. *Polymer* **32**, 2000–2009
13. Prilutski, G., Gupta, R. K., Sridhar, T. & Ryan, M. E. (1983) Model viscoelastic liquids. *J. Non-Newtonian Fluid Mech.* **12**, 233–241
14. Quinzani, L. M., McKinley, G. H., Brown, R. A. & Armstrong, R. C. (1990) Modelling the rheology of polyisobutylene solutions. *J. Rheol.* **34**, 705–748
15. Shaqfeh, E. S. G. (1996) Purely elastic instabilities in viscometric flows. *Ann. Rev. Fluid Mech.* **28**, 129–185

16. James (1996) Constitutive Relations and their Applications. Cambridge: Isaac Newton Institute
17. Spiegelberg, S. H. & McKinley, G. H. (1996) Stress relaxation and elastic decohesion of viscoelastic polymer solutions in extensional flow. *J. Non-Newtonian Fluid Mech.* **67**, 49–76
18. Bird, R. B., Curtiss, C. F., Armstrong, R. C. & Hassager, O. (1987) *Dynamics of Polymeric Liquids. Volume 2: Kinetic Theory.* (New York: Wiley Interscience)
19. Satrape, J. V. & Crochet, M. J. (1994) Numerical simulation of the motion of a sphere in a Boger fluid. *J. Non-Newtonian Fluid Mech.* **55**, 91–111
20. McKinley, G. H., Öztekin, A., Byars, J. A. & Brown, R. A. (1994) Self-similar spiral instabilities in elastic flows between a cone and a plate. *J. Fluid Mech.* **285**, 123–164
21. Rajagopalan, D. R., Arigo, M. T. & McKinley, G. H. (1996) Sedimentation of a sphere through an elastic fluid. Part II: Transient Motion. *J. Non-Newtonian Fluid Mech.* **65**, 17–46
22. Harlen, O. G., Rallison, J. M. & Szabo, P. (1995) A split Eulerian-Lagrangian method for simulating transient viscoelastic flows. *J. Non-Newtonian Fluid Mech.* **60**, 81–104
23. Entov, V. M. & Hinch, E. J. (1997) Effect of a spectrum of relaxation times on the capillary thinning of a filament of elastic liquid. *J. Non-Newt. Fluid Mech.* **72**, 31–53
24. Wedgewood, L. E., Ostrov, D. N. & Bird, R. B. (1991) A finitely extensible bead-spring chain model for dilute polymer solutions. *J. Non-Newtonian Fluid Mech.* **40**, 119–139
25. Keunings, R. (1997) On the Peterlin approximation for finitely extensible dumbbells. *J. Non-Newtonian Fluid Mech.* **68**, 85–100
26. Herrchen, M. & Öttinger, H. C. (1997) A detailed comparison of various FENE dumbbell models. *J. Non-Newtonian Fluid Mech.* **68**, 17–43
27. Tirtaatmadja, V. & Sridhar, T. (1993) A filament stretching device for measurement of extensional viscosity. *J. Rheol.* **37**, 1081–1102
28. Ooi, Y. W. & Sridhar, T. (1994) Extensional rheometry of fluid S1. *J. Non-Newtonian Fluid Mech.* **52**, 153–162
29. Solomon, M. J. & Muller, S. J. (1996) The transient extensional behaviour of polystyrene-based Boger fluids of varying solvent quality and molecular weight. *J. Rheol.* **40**, 837–856
30. Spiegelberg, S. H., Ables, D. C. & McKinley, G. H. (1996) The role of end effects on measurements of extensional viscosity in viscoelastic polymer solutions with a filament stretching rheometer. *J. Non-Newtonian Fluid Mech.* **64**, 229–267
31. Sridhar, T. & McKinley, G. H. (1998) Filament Stretching Rheometers and Extensional Rheometry of Polymer Solutions. In *Advances in the Flow and Rheology of Non-Newtonian Fluids*, ed. D. Siginer, R. P. Chhabra & D. DeKee. New York: Elsevier
32. Larson, R. G., Perkins, T. T., Smith, D. E. & Chu, S. (1997) The hydrodynamics of a DNA molecule in a flow field. *Phys. Rev. E* **55**, 1794–1797

33. Orr, N. & Sridhar, T. (1996) Stress relaxation in uniaxial extension. *J. Non-Newtonian Fluid Mech.* **67**, 77–104
34. van Nieuwkoop, J. & von Czernicki, M. M. O. M. (1996) Elongation and subsequent relaxation measurements on dilute polyisobutylene solutions. *J. Non-Newtonian Fluid Mech.* **67**, 105–124
35. Rallison, J. M. (1997) Dissipative stresses in dilute polymer solutions. *J. Non-Newtonian Fluid Mech.* **68**, 61–83
36. Doyle, P., Shaqfeh, E. S. G. & Gast, A. P. (1997) Dynamic simulation of freely draining, flexible polymers in linear flows. *J. Fluid Mech.* **334**, 251–291
37. Larson, R. G., Shaqfeh, E. S. G. & Muller, S. J. (1990) A purely elastic instability in Taylor-Couette Flow. *J. Fluid Mech.* **218**, 573–600
38. McKinley, G. H., Pakdel, P. & Öztekin, A. (1996) Geometric and rheological scaling of purely elastic elow instabilities. *J. Non-Newtonian Fluid Mech.* **67**, 19–48
39. Olagunju, D. O. (1995) Elastic instabilities in cone-and-plate flow: small gap theory. *Z. A. M. P.* **46**, 946–959
40. Öztekin, A., Alakus, A. & McKinley, G. H. (1997) Elastic instability in planar stagnation flow. *J. Non-Newtonian Fluid Mech.* **72**, 1–29
41. Aidun, C. K. (1991) Principles of hydrodynamic instability: application in coating systems. Part 2. Examples of flow instability. *Tappi J.* **74**, 213–220
42. Malkin, A. Y. & Petrie, C. J. S. (1997) Some conditions for rupture of polymer liquids in extension. *J. Rheol.* **41**, 1–25
43. Bird, R. B., Armstrong, R. C. & Hassager, O. (1987) *Dynamics of Polymeric Liquids. Volume 1: Fluid Mechanics.* (New York: Wiley Interscience)
44. Petrie, C. J. S. (1990) Some asymptotic results for planar extension. *J. Non-Newtonian Fluid Mech.* **34**, 37–62
45. James, D. F. & Sridhar, T. (1995) Molecular conformation during steady-state measurements of extensional viscosity. *J. Rheol.* **39**, 713–724
46. Renardy, M. (1996) Personal communication
47. Doyle, P. S., Shaqfeh, E. S. G., McKinley, G. H., & Spiegelberg, S. H. (1998) Relaxation of Dilute Polymers Solutions following Extensional Flow. *J. Non-Newtonian Fluid Mech.* In press

Dynamics of Complex Fluids, pp. 30–37
ed. M. J. Adams, R. A. Mashelkar, J. R. A. Pearson & A. R. Rennie
Imperial College Press–The Royal Society, 1988

Chapter 2

Surface Instabilities During Extrusion of Linear Low Density Polyethylene

R. P. G. RUTGERS AND M. R. MACKLEY*
Department of Chemical Engineering, University of Cambridge, UK

D. GILBERT
BP Chemicals, Meyrin, Switzerland
**E-mail: mrm1@cheng.cam.ac.uk*

The objective of this research is to understand the mechanisms underlying the extrudate surface defect that occurs during the extrusion of Linear Low Density Polyethylene (LLDPE) under certain process conditions. The characterisation of the flow field with optical stress birefringence is carried out simultaneously with the study of the extrudate surface. The qualitative and quantitative characterisation of the defect as a function of process conditions is achieved via scanning electron microscopy (SEM) and surface profile analysis respectively. The physical characteristics of the defect suggest a melt failure at the die exit under the influence of high tensile stresses, resulting in random crack formation on the extrudate.

Background

During extrusion, LLDPE shows a variety of instabilities. We are interested in the mechanism underlying the onset of the first instability to occur with increasing flow rate, which affects only the surface of the extrudate. A large amount of experimental work has been carried out in this field.

The mechanisms proposed range from the theories based on wall slip [1] and stick-slip effects [2, 3] to rupture at the exit due to exit stresses [5] and cavitation at the wall due to acceleration at the die exit [4]. Others [6, 7] have also researched the effects of polymer chain architecture on the instability. No widespread agreement on the mechanism, or even on the physical characteristics and critical process conditions of the defect, has yet been reached. We aim to study the qualitative and quantitative characteristics of the distortion as a function of extrusion conditions and stress fields in the melt.

Extrusion

We use a Betol 25 mm screw extruder equipped with an abrupt entry slit die which allows rheo-optical measurements. The stainless steel slit die used is 1 mm wide, 8 mm long and 15 mm deep with a contraction ratio of 15:1. The process conditions studied cover shear rates from 7 s^{-1} to 200 s^{-1} and shear

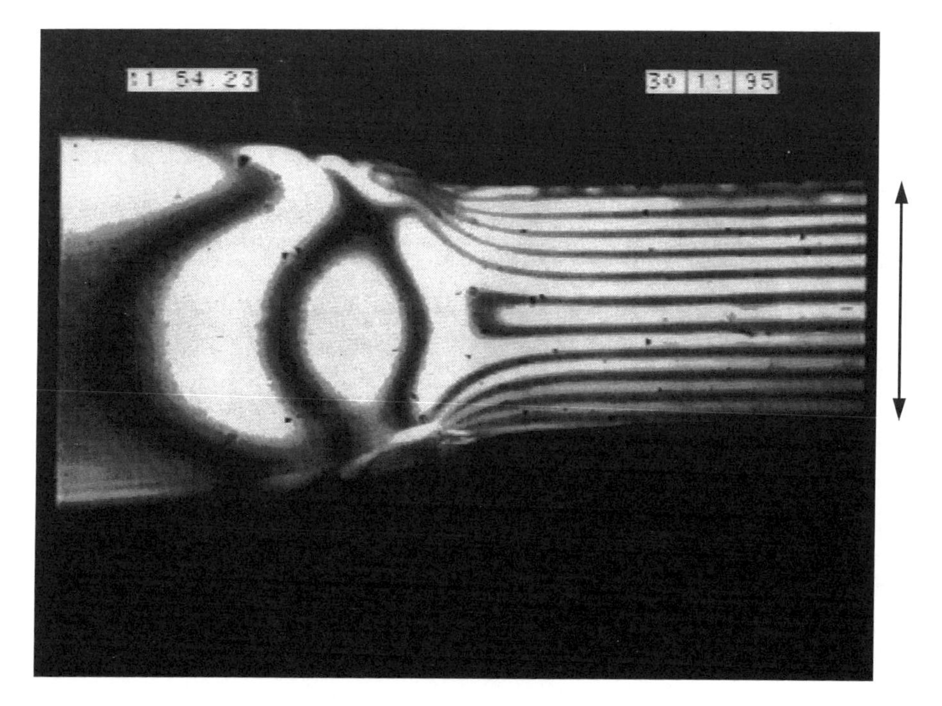

Fig. 1. Flow birefringence pattern at die exit at temperature T = 191–196°C and flow rate Q = 1.5–2.3 g/min. Arrow = 1 mm.

stresses from 0.03 MPa to 0.4 MPa. The widely reported critical shear stress for the onset of surface defects is circa 0.1 to 0.15 MPa.

Optical Stress Birefringence

Flow induced birefringence measurements provide a measure of the level of stress in the die exit region under stable and unstable flow conditions. The experimental set-up used in this work corresponds to the set-up used in previous studies carried out din this laboratory on the steady abrupt constriction flow of various polyethylenes [8]. Figures 1 and 2 show the fringes in the exit region under conditions of low and high stress levels, respectively. The flow in both figures is from right to left. The fringes at the exit show large concentrations of stress as a result of the extensional stresses involved in the transition from a constrained flow at the die wall to a no-friction condition at the free surface of the extrudate. The stress field is steady for all process conditions studied

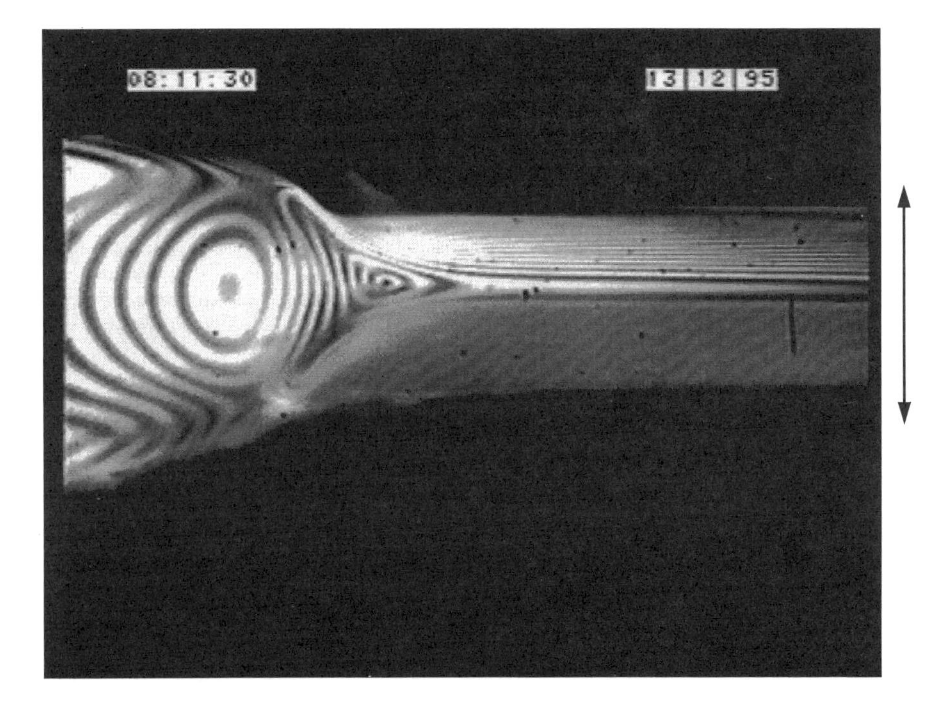

Fig. 2. Flow birefringence pattern at die exit at T = 174–178°C and Q = 18.8–22.5 g/min. Arrow = 1 mm. Flow from right to left.

apart from the lowest temperature range (135–139°C), where fluctuations of the stresses cause blurring and unsteadiness of the fringes that represent the average stress level over the depth of the slit. The stress levels increase as expected with increasing flow rate and falling temperature. At the highest flow rates, shear heating causes deflection of the light bundle near the wall.

Scanning Electron Microscopy

Scanning electron microscope photographs were made from gold sputtered samples of the extruded tape using a Leica SEM Stereoscan 430. The samples used were taken after the frost-line but before contact was made with the haul-off rollers. Figures 3 and 4 show SEM photographs of the extrudate surface under the same process conditions as Figs. 1 and 2.

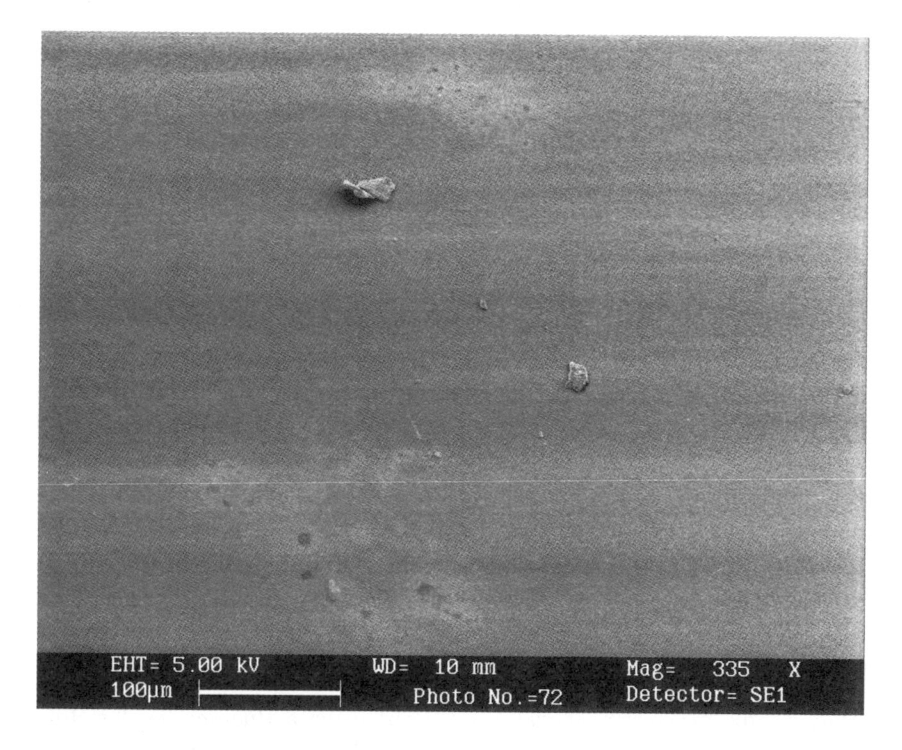

Fig. 3. SEM photo of extrudate surface at melt temperature T = 174–178° C and flow rate Q = 1.5–2.3 g/min. Flow from right to left.

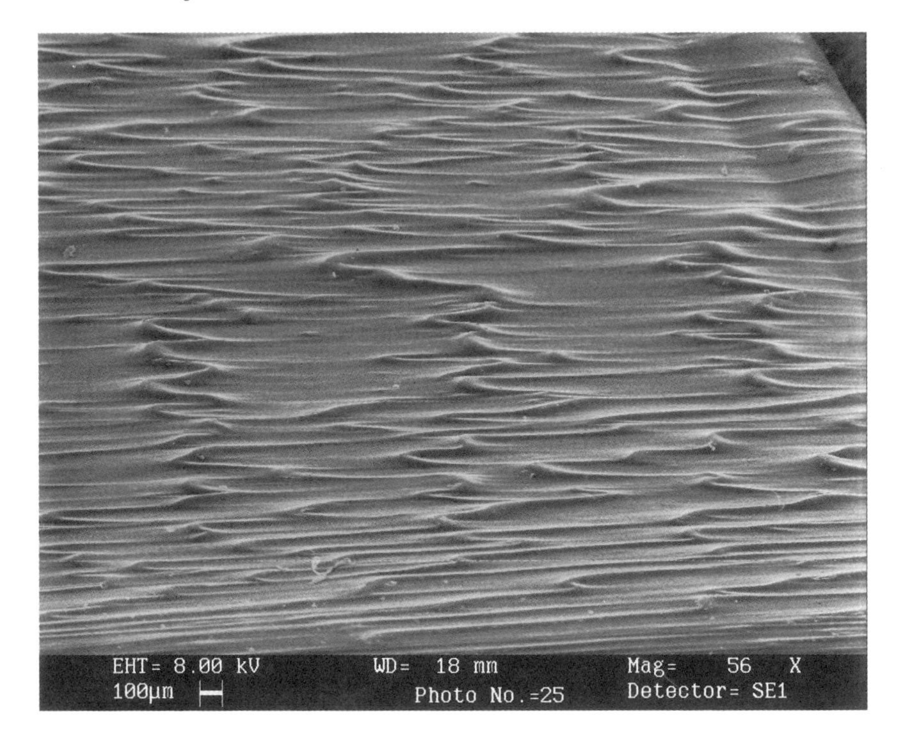

Fig. 4. SEM photo of extrudate surface at melt temperature T = 174–178° C and flow rate Q = 18.8–22.5 g/min. Flow from right to left.

The surface is smooth at low flow rates but shows a distinct wave-like pattern at higher flow rates. The waves appear to have a sharp front in the direction of the flow. The typical spacing of the random pattern is of order 100 to 500 μm, and the typical height of the distortion is estimated at one to 50 μm. At the low temperature range of 135–139°C, the surface shows striping in the flow direction which develops into a severe undulation of the extrudate at higher flow rates.

Surface Profile Analysis

Surface profile measurement was carried out using a Pneumo Taylor Hobson Talisurf system. A stylus is traversed across the surface parallel to the flow direction over a distance of approximately 10 mm. A filter eliminating all fluctuations the height of the surface with a wavelength greater than 2.5 mm is applied to the data. The sampling length is five times the filter length. A typical profile is given in Fig. 5. It can be seen that the profile consists of shallow valleys with relatively sharp peaks which are sharper at the front in the direction of flow. The average absolute deviation of the mean height along the sampling length is the average roughness, (Ra). The Ra value of this type

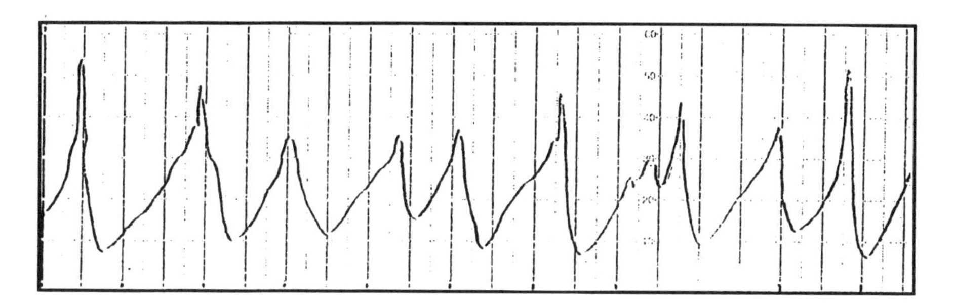

Fig. 5. Extrudate profile at T = 174–178°C and Q = 18.8–22.5 g/min. Flow from left to right. Horizontal div. = 1 mm, vertical div. = 10 μm.

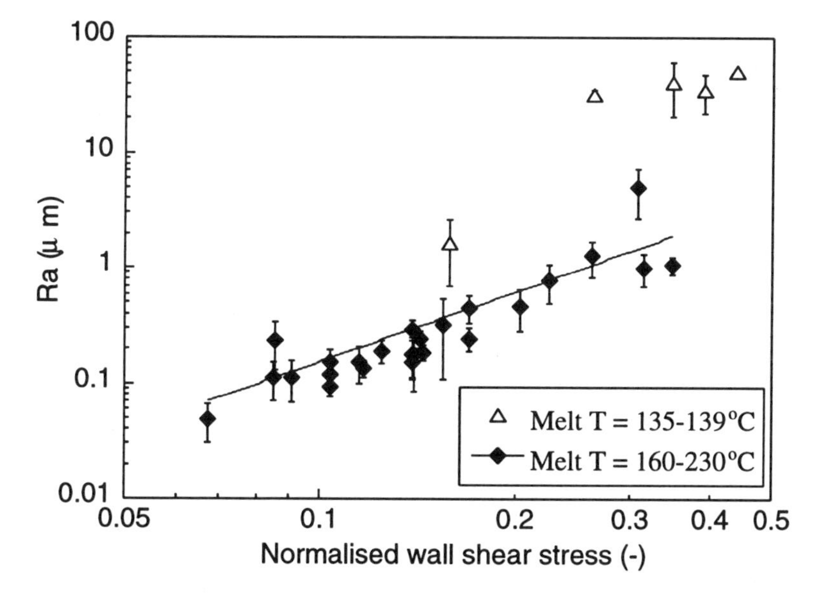

Fig. 6. Ra versus wall shear stress (normalised to meet BP Chemicals confidentiality requirements).

of the peak to valley height. Figure 6 gives Ra as a function of shear stress, where the symbols represent the average values of three to eight repeated measurements on different specimens and the error bars represent the standard deviation. The curve represents an exponential fit to the data. No critical shear stress for the onset is evident from this data. The instability develops gradually with increasing shear stress and the severity is independent of temperature. The roughness of the samples obtained at the low range of temperatures (around 137°C) is far greater. It is believed that this is the effect of a different type of instability that affects the bulk of the melt.

Conclusions

LLDPE extrudate exhibits a wave-like surface defect with an amplitude ranging from 0.5 to 50 μm. The waves exhibit a sharp front in the direction of the flow. The amplitude of the distortion increases exponentially with increasing exit stress level independent of temperature. When this defect occurs, no time dependence of the stress field is observed inside the die. The unstable stress field at lower temperatures is believed to be related to a bulk-type instability which has a very different appearance and order of magnitude. It is suggested that the extrudate surface distortion is a result of elongational stress concentrations at the exit of the die, which result in random crack formation along the width of the exit. As LLDPE has a weak melt strength, it may not withstand the elongational stresses which are particularly high at the surface of the extrudate due to die swell. This could explain why LLDPE is much more susceptible to surface instabilities than LDPE which is strain hardening and does not show this effect.

References

1. Denn, M. M. (1992) Surface-induced effects in polymer melt flow. In *XI International Congress on Rheology*, Vol. 1, ed. P. Molenaers & R. Keunings, p. 45, Elsevier, Brussels
2. Kurtz, S. J. (1992) The dynamics of sharkskin melt fracture: effect of die geometry. In *XI International Congress on Rheology*, Vol. 1, ed. P. Molenaers & R. Keunings, Brussels, Elsevier
3. Kurtz, S. J. (1995) Sharkskin melt fracture — experimental observations. In *Material Instabilities*, ed. J. Goddard & B. Coleman, California: IMM, Davis
4. Tremblay, B. (1991) Sharkskin defects of polymer melts: the role of cohesion and adhesion. *J. Rheology* **35** (6), 985–998

5. El Kissi, N., Leger, L., Piau, J-M. & Mezghani, A. (1994) Effect of surface properties on polymer melt slip and extrusion defects. *J. Non-Newtonian Fluid Mech.* **52**, 249–261
6. Karbashevski, E., Rudin, A., Kale, L., Tchir, W. J. & Schreiber, J. P. (1991) Effects of polymer structure on the onset of processing defects in linear low density polyethylene. *Pol. Eng. Sci.* **31** (22), 1581
7. Ajji, A., Varennes, S., Schreiber, H. P. & Duchesne, D. (1993) Flow defects in linear low density polyethylene processing: instrumental detection and molecular weight dependence. *Pol. Eng. Sci.* **33** (23), 1524–1531
8. Ahmed, R. & Mackley, M. R. (1995) Experimental centreline planar extension of polyethylene melt flowing into a slit die. *J. Non-Newtonian Fluid Mech.* **56**, 127–149

Dynamics of Complex Fluids, pp. 38–46
ed. M. J. Adams, R. A. Mashelkar, J. R. A. Pearson & A. R. Rennie
Imperial College Press–The Royal Society, 1998

Chapter 3

High Weissenberg Number Boundary Layers and Corner Singularities in Viscoelastic Flows

M. RENARDY

Department of Mathematics, Virginia Tech,
Blacksburg, VA 24061-0123, USA
E-mail: renardym@math.v.t.edu

Viscoelastic fluids can form sharp stress boundary layers on walls when the shear rate is varying and the Weissenberg number is high. We shall explain the reasons for this phenomenon and derive equations to describe it. These boundary layers occur, for example, in flow between eccentric cylinders or flow past an obstacle. They are also a prominent feature in the behaviour of viscoelastic fluids near re-entrant corners. Recent progress on the corner behaviour of the upper convected Maxwell (UCM) and Phan-Thien Tanner (PTT) fluids will be summarised.

Introduction

Stress boundary layers at high Weissenberg number (W) are a common occurrence in non-Newtonian flows. In numerical simulations, they have been observed in flow between eccentric cylinders [5] and flow past a sphere [6]. They also play a significant role in flow near a re-entrant corner [7, 9].

The reason for the formation of these boundary layers is as follows. Constitutive relations of the Maxwell type contain a convected derivative term which is important when the Weissenberg number is high and the shear rate is variable. However, the convected derivative vanishes on the wall where the velocity is zero. Consequently, the dominant balance of terms in the equations away from the wall is fundamentally different from the balance at the wall where viscometric flow applies. As the Weissenberg number is increased, the transition region becomes thin. In contrast to the well-known high Reynolds number boundary layers in Newtonian fluids, the high Weissenberg number boundary layers have nothing to do with satisfying boundary conditions. The only boundary condition at the wall is the no-slip condition, and it is satisfied by the outer solution which applies outside the boundary layer. Indeed, the high Weissenberg number boundary layers occur even if the constitutive equation is integrated in a prescribed Newtonian velocity field.

Re-entrant corner behaviour has been a particularly troublesome issue in the numerical simulation of viscoelastic fluids. The corner behaviour of Newtonian fluids is well understood [2]. There is a fundamental difference between the non–re-entrant case, where velocity gradients and stresses vanish at the corner, and the re-entrant case, where they are infinite. In the non-Newtonian case, non-reentrant corners can be analysed by the usual regular perturbation series which applies at low Weissenberg numbers. In contrast, re-entrant corners are an infinite Weissenberg number limit, and hence the behaviour is highly nonlinear and dependent on the constitutive law. Recent results [1, 4, 7–9] have shed considerable light on the behaviour of the UCM fluid at re-entrant corners. Boundary layers occur on the upstream and downstream walls while the solution in the core region can be described by a potential flow. Another important feature is a strong downstream instability which can lead to large spurious stresses in numerical integrations. In contrast, the PTT model leads to less singular stresses, much less pronounced boundary layers, and no spurious stresses downstream [11].

Stress Boundary Layers at High Weissenberg Number

We consider the equations for steady creeping plane flow of the UCM fluid:

$$
\begin{gathered}
\operatorname{div}\mathbf{T} - \nabla p = 0\,, \\
\operatorname{div}\mathbf{v} = 0\,, \\
W[(\mathbf{v}\cdot\nabla)\mathbf{T} - (\nabla\mathbf{v})\mathbf{T} - \mathbf{T}(\nabla\mathbf{v})^T] + \mathbf{T} = \nabla\mathbf{v} + (\nabla\mathbf{v})^T\,. \qquad (1)
\end{gathered}
$$

For the derivation of boundary layer equations, it is advantageous to express the stress components on a basis aligned with the stream lines. We have the planar velocity vector $\mathbf{v} = (u, v)$ and we define a perpendicular vector by

$$\mathbf{w} = \frac{1}{u^2 + v^2}(-v, u)\,. \tag{2}$$

We then set

$$\mathbf{T} = -\frac{1}{W}\mathbf{I} + \lambda \mathbf{v}\mathbf{v}^T + \mu(\mathbf{v}\mathbf{w}^T + \mathbf{w}\mathbf{v}^T) + \nu \mathbf{w}\mathbf{w}^T\,. \tag{3}$$

The constitutive equation then takes the form

$$\begin{aligned} W(\mathbf{v}\cdot\nabla)\lambda + \lambda + 2W\mu\,\mathrm{div}\,\mathbf{w} &= \frac{1}{W|\mathbf{v}|^2}\,, \\ W(\mathbf{v}\cdot\nabla)\mu + \mu + W\nu\,\mathrm{div}\,\mathbf{w} &= 0\,, \\ W(\mathbf{v}\cdot\nabla)\nu + \nu &= \frac{|\mathbf{v}|^2}{W}\,. \end{aligned} \tag{4}$$

We are interested in the situation where W is large, and solutions vary rapidly across a thin layer (of order $1/W$) near a wall. In general, the wall may be curved. The natural coordinates are the stream function, ψ, and a second coordinate, x, which measures the arc length along the wall and has level curves perpendicular to the stream lines. A self-consistent scaling is obtained if one assumes that the derivatives perpendicular to the wall scale with a factor, W, relative to tangential derivatives. In addition, we scale the velocity with a factor, W^{-1}, the stream function with W^{-2}, λ with W^3 and ν with W^{-3}. With these rescalings, Eq. (4) can be shown to turn into the following set of equations, at leading order in W (see [10]):

$$\begin{aligned} u\lambda_x + \lambda - \frac{2\mu u_\psi}{u} &= 0\,, \\ u\mu_x + \mu - \frac{\nu u_\psi}{u} &= 0\,, \\ u\nu_x + \nu &= u^2\,. \end{aligned} \tag{5}$$

Here u denotes the (rescaled) speed of the fluid. In the context of the boundary layer approximation, the motion is primarily tangent to the wall. The momentum balance equation,

$$\mathrm{curl}\,\mathrm{div}\,\mathbf{T} = 0\,, \tag{6}$$

becomes, at leading order,

$$[u(\lambda u)_x + u\mu_\psi]_\psi = 0\,. \tag{7}$$

We comment briefly on the approximations contained in Eqs. (5) and (7) relative to the full equations. First, the right-hand side of the first equation in Eq. (4) has been dropped. Second, the term $\operatorname{div}\mathbf{w}$ in Eq. (4) can be split into two contributions: one associated with the variation of the fluid speed and the other associated with the variation of the direction of the velocity. In the boundary layer approximation, the second contribution is neglected. These are the only differences between Eqs. (4) and (5). The momentum equation, on the other hand, is much more severely truncated. Essentially, Eq. (7) is a lubrication approximation based on the thinness of the layer.

Solutions of the Boundary Layer Equations

The task at hand now is to find the solutions to Eqs. (5) and (7) along with the boundary condition

$$u(x,0) = 0\,. \tag{8}$$

Since we are still dealing with a nonlinear system of partial differential equations, this is by no means easy. However, there are a number of situations which lead to considerable simplification. First, we can look for similarity solutions. The similarity variable is $\xi = x^{-\alpha}\psi$ and we make the ansatz

$$u = x\tilde{u}(\xi), \quad \lambda = x^{2-2\alpha}\tilde{\lambda}(\xi), \quad \mu = x^{2-\alpha}\tilde{\mu}(\xi), \quad \nu = x^2\tilde{\nu}(\xi)\,. \tag{9}$$

With this, the equations are reduced to a set of ordinary differential equations. There are still no obvious explicit solutions. However, the ordinary differential equations can be integrated numerically. The solutions found in this manner arose in the study of the re-entrant corner flow [9]. For a 270° corner, the relevant value of α is 7/3.

Another situation for which explicit solutions can be found is the linearisation at parallel shear flow. Actually, this is a special case of the linearisation of the equations for the UCM model about plane Couette flow. An explicit solution for that problem is found in [3]. In the context of Eqs. (5) and (7), parallel shear flow is the solution $u = \psi^{1/2}$, $\nu = \psi$, $\mu = 1/2$ and $\lambda = 1/(2\psi)$. We are looking at small periodic perturbations

$$u(x,\psi) = \psi^{1/2}(1+\tilde{u}(\psi)e^{ix}),$$
$$\nu(x,\psi) = \psi + \tilde{\nu}(\psi)e^{ix},$$
$$\mu(x,\psi) = \frac{1}{2} + \tilde{\mu}(\psi)e^{ix},$$
$$\lambda(x,\psi) = \frac{1}{2\psi} + \tilde{\lambda}(\psi)e^{ix}. \tag{10}$$

Such solutions would be of relevance in periodic geometries, such as flow between eccentric cylinders and flow through corrugated pipes. The linearised equations can be combined into the single equation,

$$\frac{d}{dy}\left[\frac{y\tilde{u}'}{2(iy+1)}\right] + i\tilde{u} + \tilde{u}' = C, \tag{11}$$

where $y = \psi^{1/2}$, the prime denotes differentiation with respect to y and C is an integration constant. Equation (11) has the remarkably simple solution, $\tilde{u} = C/i$, and it can be shown [10] that this is the only solution which behaves in a physically reasonable manner as $\psi \to 0$ and as $\psi \to \infty$. Since $\tilde{u}$ does not depend on ψ, the velocity for this linearised solution has no boundary layer at all. However, there are boundary layers in the stresses; we have $\tilde{\nu} \sim \psi$, $\tilde{\mu} \sim 1$, $\tilde{\lambda} \sim 1/\psi$ as $\psi \to 0$ and, on the other hand, $\tilde{\nu} \sim \psi^{1/2}$, $\tilde{\mu} \sim 1/\psi$ and $\tilde{\lambda} \sim \psi^{-5/2}$ as $\psi \to \infty$. The solution constructed here has a spatially periodic shear rate, and the stresses have boundary layers. Nonlinear effects would, presumably, lead to a boundary layer of the velocity as well. Further analysis shows that even in the full nonlinear problem, the velocity can be expected to behave like $\psi^{1/2}$ for both small and large ψ, but with different coefficients [10].

Re-entrant Corner Behaviour of Viscoelastic Fluids

The flow through a contraction has been a benchmark problem for the evaluation of numerical codes. Numerical simulations with the UCM model have consistently run into trouble at the re-entrant 270° corner, while other models which limit stresses at high deformation rates, such as the PTT and Chilcott-Rallison models, have allowed successful simulations. We shall now review recent work which explains both the analytical nature of the corner singularity and the source of the numerical difficulties.

Away from the walls, the dominant terms in the constitutive relation, Eq. (1), are the quadratic ones. This means that the leading order balance is

$$(\mathbf{v}\cdot\nabla)\mathbf{T}-(\nabla\mathbf{v})\mathbf{T}-\mathbf{T}(\nabla\mathbf{v})^T=0\,. \tag{12}$$

A simple solution of this equation is

$$\mathbf{T}=g(\psi)\mathbf{v}\mathbf{v}^T\,, \tag{13}$$

where g is an arbitrary function of the stream function. If Eq. (13) is inserted into the momentum equation, it is found that $\mathbf{u}=g^{1/2}\mathbf{v}$ is a solution of the Euler equation [4, 9]

$$\begin{aligned}(\mathbf{u}\cdot\nabla)\mathbf{u}-\nabla p&=0\,,\\ \operatorname{div}\mathbf{u}&=0\,.\end{aligned} \tag{14}$$

We can then obtain a simple solution for $\mathbf{u}$ by potential flow. Moreover, a matching argument with the viscometric flow at the upstream wall leads to the conclusion that $g(\psi)$ is proportional to $\psi^{-8/7}$ [4]. Consequently, the stresses behave like $r^{-2/3}$ while the velocity behaves like $r^{5/9}$.

The potential flow solution breaks down in a boundary layer near each wall, where it needs to be matched to one of the similarity solutions of the boundary layer equations discussed above. For the upstream wall, this is done in [9], where the boundary layer equations are integrated numerically starting from the wall. Integration starting from the wall is an ill-posed problem for the downstream boundary, and hence the nonlinear ODEs governing the boundary layer have to be treated as a boundary value problem rather than an initial value problem. This remains to be carried out.

The integration of stresses for the UCM model has a strong instability when the downstream wall is approached. This was demonstrated in [7], where the stresses were integrated in a given (Newtonian) velocity field. The source of the instability is best understood in terms of the transformed Eq. (4). As long as one remains outside the viscometric boundary layer, the dominant balance in these equations is

$$\begin{aligned}W(\mathbf{v}\cdot\nabla)\lambda+2W\mu\operatorname{div}\mathbf{w}&=0\,,\\ W(\mathbf{v}\cdot\nabla)\mu+W\nu\operatorname{div}\mathbf{w}&=0\,,\\ W(\mathbf{v}\cdot\nabla)\nu&=0\,.\end{aligned} \tag{15}$$

Near the downstream wall, $\operatorname{div}\mathbf{w}$ becomes very large because $\mathbf{w}$ is singular at the downstream wall. Moreover, ν is multiplied by the factor $\mathbf{w}\mathbf{w}^T$ to recover the actual stress, and this factor is also large. These two effects combine to lead to a growth of stresses as the downstream wall is approached. As far as the actual stresses are concerned, this growth is compensated by the upstream dynamics, which ensure that the λ-component dominates the stress in the core region while μ and ν are small. Discretisation errors, however, will continually feed some of the dominant λ-component back into the other components. As the downstream wall is approached, these errors are magnified and can lead to large spurious stresses. These spurious stresses were avoided in [8] simply by basing the discretisation on the transformed Eq. (4) instead of the original constitutive Eq. (1). I believe that the downstream magnification of discretisation errors is the main reason for the failure of numerical simulations near re-entrant corners.

It is also instructive to contrast the UCM behaviour with another constitutive model such as the PTT fluid. The stresses for both models were integrated in the prescribed Newtonian velocity field near a 270° corner [7, 11]. We can observe the following contrasts:

1. The stress singularity in the core region (away from walls) is $r^{-0.74}$ for the Maxwell model and $r^{-0.33}$ for the PTT model. Note that the elastic stress in the PTT case is actually less singular than the Newtonian stress.
2. The boundary layer at the wall is given by $\theta = \mathrm{O}(r^{1-\alpha})$ in the UCM case and $\theta = \mathrm{O}(r^{(1-\alpha)/3})$ in the PTT case, where $\alpha = 0.5445$ is the exponent for the Newtonian velocity field.
3. The downstream instability in the stress integration is still present for the PTT model, but much less severe. In particular, it is not sufficiently strong to magnify discretisation errors to a troublesome level. However, there is actually a genuine stress maximum on each streamline as the downstream wall is approached.

In summary, the PTT model leads to less singular stresses, but it also makes the numerical simulation easier in more subtle and probably more crucial ways (less sharp boundary layers and a milder downstream instability).

Acknowledgement

This research was supported by the Office of Naval Research under Grant N00014-92-J-1664 and by the National Science Foundation under Grant DMS-9306635. A substantial part of it was carried out when the author visited the Isaac Newton Institute at the University of Cambridge.

Discussion

J. Goddard Could we have a picture? Do we have flow around the corner? And a sketch would help with these cusp-shaped boundary layers.

M. Renardy Yes, we have flow around the corner and, on each wall, we have a boundary layer which is shaped like a cusp.

J. Goddard At the corner, where the boundary layers vanish, we have an elastically dominated region where you deduced this form and where the stress is the velocity dyad. What is peculiar about that form is that it suggests that it is all primary normal stress. I might expect an elastic response associated with a sudden step strain, in which case I would expect to get a more complicated kind of stress pattern. In the balance that you did in the equations, when you looked at the quadratic terms, I would have thought that we just take stress rate equal to strain rate and integrate that up. That does not necessarily give us stresses which are dominated by the primary normal stresses. Is this the only solution?

M. Renardy No, it is not the only solution. It is the solution which has the right upstream behaviour. Because you start with conditions upstream where the primary normal stress is clearly the dominant stress component, and you just follow that around.

M. Mackley Aren't you asking for trouble when you are ask a material like the UCM fluid to flow around a corner like that?

M. Renardy Yes, of course you are asking for trouble. And that was part of the point of my lecture; the PTT model behaves much more nicely.

M. Mackley Absolutely, but does a material exist that obeys UCM characteristics in a flow like this?

M. Renardy Probably not.

K. Walters It is good problem for mathematicians.

M. Denn The steps that are required to get the boundary layer equation sounded like there is an explicit or implicit assumption of an infinite radius of curvature. It is a lubrication approximation near the wall, which is perfectly flat. Is this not causing problems near the corner?

M. Renardy The boundary layer Eqs. (5) are in a transformed coordinate system, and in this form they are also valid for a curved wall.

M. Denn Even at a corner?

M. Renardy No, not at a corner. It must be a smooth boundary.

M. Denn My concern is that as you approach the corner from both sides, you are using a set of equations that must break down at the corner, and that is the point you are interested in.

J. Goddard That is where the cusp is.

M. Renardy Yes, that is where the cusp comes in. You are quite right, you cannot have a boundary layer going around the corner.

References

1. Davies, A. R. & Devlin, J. (1993) On corner flows of Oldroyd-B fluids. *J. Non-Newtonian Fluid Mech.* **50**, 173–191
2. Dean, W. R. & Montagnon, P. E. (1949) On the steady motion of viscous liquid in a corner. *Proc. Cambridge Philos. Soc.* **45**, 389–394
3. Gorodtsov, V. A. & Leonov, A. I. (1967) On a linear instability of a plane parallel Couette flow of viscoelastic fluid. *J. Appl. Math. Mech. (PMM)* **31**, 310–319
4. Hinch, E. J. (1993) The flow of an Oldroyd fluid around a sharp corner. *J. Non-Newtonian Fluid Mech.* **50**, 161–171
5. King, R. C., Apelian, M. R., Armstrong, R. C. & Brown, R. A. (1988) Numerically stable finite element techniques for viscoelastic calculations in smooth and singular geometries. *J. Non-Newtonian Fluid Mech.* **29**, 147–216
6. Lunsmann, W. J., Genieser, L., Armstrong, R. C. & Brown, R. A. (1993) Finite element analysis of steady viscoelastic flow around a sphere in a tube: calculations with constant viscosity models. *J. Non-Newtonian Fluid Mech.* **48**, 63–99
7. Renardy, M. (1993) The stresses of an upper convected Maxwell fluid in a Newtonian velocity field near a re-entrant corner. *J. Non-Newtonian Fluid Mech.* **50**, 127–134
8. Renardy, M. (1994) How to integrate the upper convected Maxwell (UCM) stresses near a singularity (and maybe elsewhere, too). *J. Non-Newtonian Fluid Mech.* **52**, 91–95
9. Renardy, M. (1995) A matched solution for corner flow of the upper convected Maxwell fluid. *J. Non-Newtonian Fluid Mech.* **58**, 83–89
10. Renardy, M. (1997) High Weissenberg number boundary layers for the upper convected Maxwell fluid. *J. Non-Newtonian Fluid Mech.* **68**, 125–132
11. Renardy, M. (1997) Re-entrant corner behaviour of the PTT fluid. *J. Non-Newtonian Fluid Mech.* **69**, 99–104

Dynamics of Complex Fluids, pp. 47–60
ed. M. J. Adams, R. A. Mashelkar, J. R. A. Pearson & A. R. Rennie
Imperial College Press–The Royal Society, 1998

Chapter 4

Computing with High Viscoelasticity

R. I. TANNER
Department of Mechanical & Mechatronic Engineering,
The University of Sydney,
Sydney, 2006, Australia
E-mail: rit@mech.eng.usyd.edu.au

Some suggestions for confronting the large computing tasks posed by three-dimensional highly viscoelastic flows at large Weissenberg numbers are given. Progress with the use of structured and unstructured finite volume methods are described and some remarks on recent progress on the flow around spheres are included. It is believed that we are approaching the stage where computing problems are becoming less significant than the problems arising from inadequate rheological description.

Introduction

In the last 40 years of computing viscoelastic problems, the speed-up of computers has been phenomenal (of order 10^6) and the increase in rheological knowledge has also been considerable. However, the practical application of computing to highly nonlinear, three-dimensional viscoelastic flows is still far from routine. Here we describe progress with some finite-volume techniques and we also report on the test problem of flow around a sphere. Much of the work described has been done at the University of Sydney by a group led by Nhan Phan-Thien, with significant inputs from Rong Zheng, Shicheng Xue, Xiaofang Huang, Junsuo Sun and Hao Jin.

We are concerned with three-dimensional incompressible, isothermal viscoelastic flows. The relevant equations are:

$$\nabla \cdot \boldsymbol{u} = 0 \tag{1}$$

and

$$\rho \left(\frac{\partial \boldsymbol{u}}{\partial t} + \boldsymbol{u} \cdot \nabla \boldsymbol{u} \right) = \nabla \cdot \boldsymbol{\sigma}\,, \tag{2}$$

where ρ is the density of the fluid, t is time and $\boldsymbol{u} = (u, v, w)$ the velocity vector. The total stress tensor, $\boldsymbol{\sigma}$, is also given by

$$\boldsymbol{\sigma} = -p\mathbf{1} + 2\eta_N \boldsymbol{D} + \boldsymbol{\tau}\,, \tag{3}$$

where p is the hydrostatic pressure, 1 the unit tensor, $2\eta_N \boldsymbol{D}$ is the Newtonian contribution to the stress tensor with η_N being the solvent viscosity and $\boldsymbol{D}$ the strain rate tensor, and $\boldsymbol{\tau}$ is the non-Newtonian stress tensor which can be connected with kinematic quantities by constitutive equations.

We will mostly omit gravity, non-isothermal behaviour and inertia as these factors do not present significant extra difficulties.

Constitutive Equations

For viscoelastic fluids, the stress $\boldsymbol{\tau}$ is assumed to satisfy the MPTT model [1]

$$g\boldsymbol{\tau} + \lambda \left(\frac{\partial \boldsymbol{\tau}}{\partial t} + \boldsymbol{u} \cdot \nabla \boldsymbol{\tau} - \boldsymbol{L}\boldsymbol{\tau} - \boldsymbol{\tau}\boldsymbol{L}^T \right) = 2\eta_m \boldsymbol{D} \tag{4}$$

with

$$g = 1 + \frac{\lambda \epsilon}{\eta_{mo}} \mathrm{tr}\boldsymbol{\tau}\,; \quad \boldsymbol{L} = \boldsymbol{L} - \xi \boldsymbol{D}\,, \tag{5}$$

where $\boldsymbol{L} = \nabla \boldsymbol{u}^T$ is the velocity gradient tensor, ^{T} denotes the transpose operation, λ the (single) relaxation time, ξ and ϵ are material parameters, η_{mo} is the zero shear rate molecular-contributed viscosity, and

$$\eta_m = \eta_{mo} \frac{1 + \xi(2-\xi)\lambda^2 \dot{\gamma}^2}{(1 + \Gamma^2 \dot{\gamma}^2)^{(1-n)/2}}\,, \tag{6}$$

where $\dot{\gamma} = \sqrt{2\mathrm{tr}\boldsymbol{D}^2}$ is the generalised shear rate, n is the power-law index and Γ is a time parameter. In this paper, we assume that $\Gamma = \lambda$ but this is not essential.

The zero-shear rate viscosity of the fluid is $\eta_0 = \eta_N + \eta_{mo}$. By defining $\beta = \eta_{mo}/\eta_0$ as the retardation ratio, we have $\eta_N = (1-\beta)\eta_0$, and

$$\boldsymbol{\sigma} = -p\mathbf{1} + 2(1-\beta)\eta_0 \boldsymbol{D} + \boldsymbol{\tau}\,, \tag{7}$$

$$\lambda\left(\frac{\partial \boldsymbol{\tau}}{\partial t} + \nabla\cdot(\boldsymbol{u}\boldsymbol{\tau})\right) = 2\beta\eta_0\mu\boldsymbol{D} + \lambda(\boldsymbol{L}\boldsymbol{\tau} + \boldsymbol{\tau}\boldsymbol{L}^T) - g\boldsymbol{\tau}\,, \tag{8}$$

and

$$\mu = \frac{1+\xi(2-\xi)\lambda^2\dot{\gamma}^2}{(1+\Gamma^2\dot{\gamma}^2)^{(1-n)/2}}\,. \tag{9}$$

This covers the following six special cases of constitutive models:

(i) the Newtonian fluid model ($\lambda = 0$, $\epsilon = 0$, $\xi = 0$, $\beta = 0$ and $\eta_0 = \eta_N$);
(ii) the upper-convected Maxwell (UCM) model ($\epsilon = 0$, $\xi = 0$, $\beta = 1$ and $\eta_0 = \eta_m = \eta_{mo}$);
(iii) the Oldroyd-B model ($\epsilon = 0$, $\xi = 0$, $0 < \beta < 1$ and $\eta_0 = \eta_m = \eta_{mo}$);
(iv) the simplified PTT (SPTT) model ($\xi = 0$, $\beta = 1$ and $\eta_0 = \eta_m = \eta_{mo}$);
(v) the PTT model ($\beta = 1$ and $\eta_0 = \eta_m = \eta_{mo}$);
(vi) the MPTT model ($\beta = 1$)

We shall mostly use the UCM, Oldroyd-B and SPTT models in examples since they present the maximum difficulty. Only one relaxation time is used, but realism in simulation will eventually need more.

The goal is to solve the above governing equations with appropriate boundary conditions to predict the velocity and stress fields of incompressible viscoelastic fluid flow. We considered some problems of the flow in non-circular ducts with and without singular stress points. The incompressible, isothermal and steady-state fully developed flow in a non-circular pipe is not generally rectilinear due to normal-stress effects.

Since [1], for the PTT and the MPTT models, the second normal stress coefficient is not proportional to the viscosity function, the flow cannot be rectilinear in pipes with non-circular cross sections, and some secondary flow is to be expected.

The problem of flow in the rectangular duct was solved [2] using a finite volume method, without difficulty, on a workstation. The details of the work are given in the above paper [2], including the new SIMPLEST algorithm.

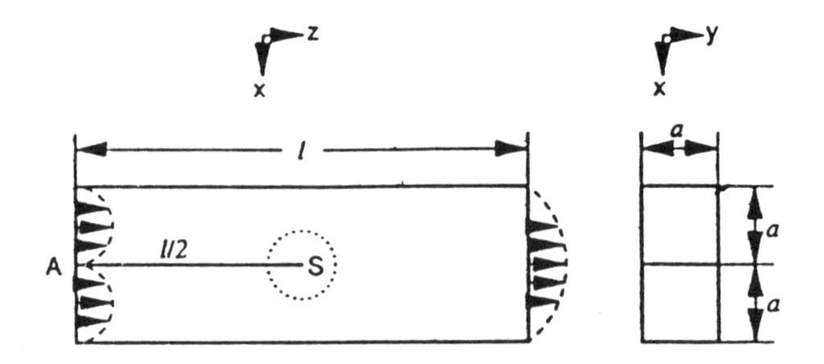

Fig. 1. Schematic view of stick-slip flow geometry.

This flow is smooth, and in order to see whether the method could tackle less smooth flows (that is, problems with singularities), the SPTT model was used to solve a stick-slip 3-D problem (Fig. 1) where AS is a thin blade with no-slip boundary conditions placed in a rectangular channel. This problem is challenging in view of the stress singularity at S.

We scaled length with a, velocity with $Q/(2a^2)$, where Q is the flow rate, and pressure and stress with $\eta Q/(2a^3)$. In the calculations, we chose $\ell = 6a$. Flows at different Weissenberg numbers ranging from 0 to 3.3 were modelled. Here the Weissenberg number (a dimensionless measure of the fluid's elasticity) is defined by $Wi = \lambda\dot{\gamma}_w$, where $\dot{\gamma}_w$ is the maximum wall shear rate in the fully developed upstream region. Computations for higher Wi numbers are still possible, but the length of the downstream domain needs to be extended as judged by velocity and stress contours.

No-slip boundary conditions are applied on walls and the AS plane. At the inlet and outlet, we impose a pressure difference, ΔP, and set $u = v = 0$. By assuming that the flow in the upstream region is fully developed, the stress components at the inlet can be expressed in terms of velocity gradients only. Thus, the stress boundary conditions can be calculated in each iteration.

Figures 2(a) and 2(b) show the shear stresses, τ_{xz}, and the first normal stress difference, $N_1 = \tau_{zz} - \tau_{xx}$, along the centreline of the rectangular duct. In Fig. 2(a), both Newtonian and UCM fluids show a sudden change in shear stress in the vicinity of the point S. In the case of the UCM fluid, the shear stress is also found to change its sign in a small region just downstream of the singular point. The normal stress differences (Fig. 2(b)) are seen to rise dramatically near the singular point, with the peak values growing with Wi. The maximum value of N_1 is 20.07 for the Newtonian case, 84.15 for $Wi = 2.5$ and 112.11 for

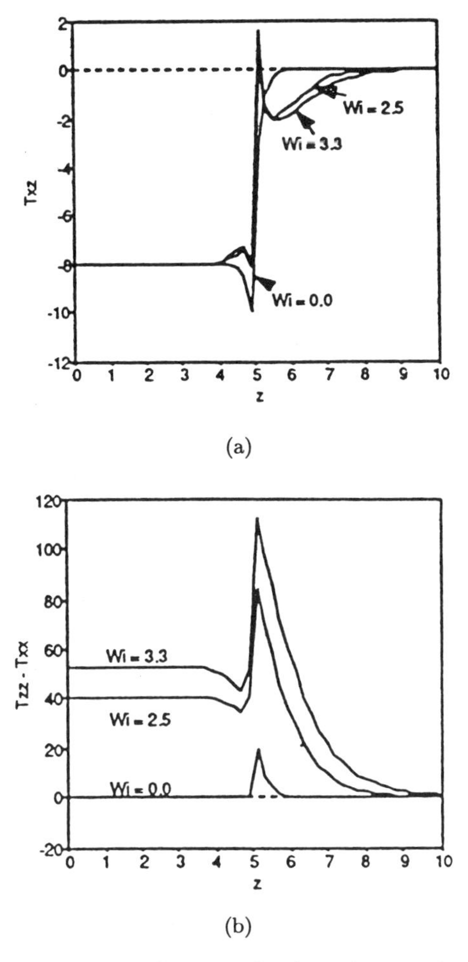

Fig. 2. (a) The shear stress, τ_{xz}, as a function of z along the centreline. (b) The first normal stress difference as a function of z along the centreline.

$Wi = 3.3$. Local asymptotic analysis for the UCM fluid in a 2-D stick-slip flow given by Tanner and Huang [3] shows that the normal stress near the singularity point behaves almost like $\tau \sim f(\theta)r^{-1}$, where (r, θ) represent a polar co-ordinate system with its origin at the singularity. Similar behaviour is found in the 3-D flow case. Further developments of the method to solve the $(4 \times 4) : (1 \times 1)$ three-dimensional square, square contraction flow have been reported [4] and these computations used about 700,000 unknowns. This is an order of magnitude larger than previous viscoelastic flow computations, and they were again done on workstations.

Non-Rectangular Elements

The above computations use rectangular or brick-shaped finite volume elements and are somewhat restrictive geometrically. We have therefore explored how to use random mesh triangular volumes with a view to using tetrahedral elements in 3-D. Since this involves jettisoning the staggered grid representation, a key article of faith in FEV technology, we record our experience here. A more detailed exposition is given by Huang *et al.* [5].

To begin, we use the split form of the PTT model due to Perera and Walters [6], so that the extra stress is written in a form slightly different from Eq. (3) above: we set

$$\boldsymbol{\tau} = \boldsymbol{\tau}_1 + \boldsymbol{\tau}_2 \,, \tag{10}$$

where $\boldsymbol{\tau}_2 = \eta_0(\nabla \boldsymbol{u} + \nabla \boldsymbol{u}^T)$, a viscous portion.

The problem to be considered here is the annular flow between the two eccentric cylinders of radii R_i (inner cylinder) and R_o (outer cylinder), offset by a centre-to-centre distance of e. Distances will be made dimensionless with respect to R_i. We define the dimensionless gap as

$$\mu = (R_o - R_i)/R_i \,, \tag{11}$$

and the eccentricity ratio as

$$\epsilon = e/(R_o - R_i) \,. \tag{12}$$

The minimum and maximum dimensionless gaps are therefore $\mu_m = \mu(1-\epsilon)$, and $\mu_M = \mu(1+\epsilon)$ respectively. Different types of boundary conditions can be prescribed at the cylinder walls; here we are concerned with Dirichlet boundary conditions, where both velocity components are given. The inner cylinder is rotating with an angular velocity of ω and the outer cylinder is held stationary. There is an exact Stokes solution to this problem, and therefore the accuracy of the numerical solution can at least be verified in this limiting case. In the viscoelastic case, there are good solutions by the finite element and by the spectral/finite element (SFEM) and perturbation methods, and our numerical solution can be benchmarked against these [7].

In the iteration process, the same pseudo-diffusion terms are added to both sides of the constitutive Eq. (4), the effects of which vanish as the algorithm converges to a steady state solution. They are, however, treated differently: the terms on the right-hand side of the equation are considered to be sources, and terms on the left-hand side participate in the finite volume discretisation process. All the governing equations are written as a general transport equation in the form [5] of

$$\nabla \cdot \boldsymbol{J} = S_p \Phi + S_c \equiv S\,, \tag{13}$$

where $\boldsymbol{J}$ is the "flux" of the variable, Φ, and S is the source term. The flux $\boldsymbol{J}$ is of the form

$$\boldsymbol{J} = \Lambda \boldsymbol{u} \Phi - \Gamma \nabla \Phi\,. \tag{14}$$

Table 1 defines the portions of the fluxes and sources; Φ consists of the (unknown) nodal variables.

Table 1. Definition of variables.

Equation	Φ	Λ	Γ	S
Continuity	1	ρ	0	0
Momentum	$\boldsymbol{u}$	ρ	η_0	$-\nabla p + \nabla \cdot \boldsymbol{\tau}_1$
Constitutive	$\boldsymbol{\tau}_1$	λ	0	$-\lambda(\boldsymbol{u} \cdot \nabla \boldsymbol{\tau}_2 - \nabla \boldsymbol{u}^T \cdot \boldsymbol{\tau}_2 - \boldsymbol{\tau}_2 \cdot \nabla \boldsymbol{u})$ $-\boldsymbol{\tau}_1 + \lambda(\nabla \boldsymbol{u}^T \cdot \boldsymbol{\tau}_1 + \boldsymbol{\tau}_1 \cdot \nabla \boldsymbol{u})$

Numerical Method

Unstructured Control Volume Method (*UCV*)

The SIMPLER algorithm is by far the most popular finite volume method to solve conservation equations of the form Eq. (13), especially for Newtonian flows at moderate to high Reynolds numbers. The method and its modifications are becoming attractive in numerical viscoelastic flows [2]. With the SIMPLER method, a staggered mesh is invariably used. Thus, the x-direction velocity u is calculated at the faces that are normal to the x-direction. The same rule is applied to v and w. One ends up having four different types of control volumes, three for the velocity components, and one for the pressure. The staggered grid is designed to eliminate checkerboard patterns in the pressure field. However, no counterpart of this staggering is needed in the finite element method because of the manner in which weak solutions are generated; here we use the SIMPLER algorithm without a staggered mesh.

No extra boundary conditions are applied with the artificial diffusion terms, which implies the application of no-flux boundary conditions on the quantities involved. The new terms will not affect the balance of the governing equations, and we refer to this as the self-consistent diffusion technique. However, if the terms are treated differently, depending on which sides of the equation they belong to, then the discretised constitutive equations can be made elliptic. The effect is similar to that of artificial stress diffusion introduced by various upwinding schemes and also to under-relaxation. For one-dimensional model problems, we find that the magnitude of the stress diffusivity should be of the order of the mesh Weissenberg number. In practice, we keep the stress diffusivity at a fraction of the global Weissenberg number, starting from a value of about 1% and gradually reducing to about 0.001% of the global Weissenberg number in the convergent solutions. We find that the quality of the solution is not affected by the inclusion of a small amount of diffusion introduced on both sides of the constitutive equations. In general, the larger the stress diffusivity, the more stable the numerical solution becomes. However, more iterations are needed to obtain an accurate solution. For example, in calculating the solutions at an eccentricity ratio of 0.8, a small gap of 0.1 and a Deborah number of 2, we have used a range of stress diffusivity of 0.0001–0.1 times the Deborah number. Starting from the same initial value, after 60 iterations the differences in the stresses are less than 10^{-9} in all cases.

A lack of convergence in some numerical schemes can also be due to steep stress boundary layers. We thus endeavour to supply boundary stresses as accurately as possible, whenever they can be derived from the constitutive equations and the specialised form of the kinematics at the boundary. In the eccentric cylinder problems (and some others), the velocities on the boundary and their derivatives along the boundary direction are known. In this case, the constitutive equations are simplified into ordinary differential equations along the boundary, and these can be solved as part of the solution procedure at every iteration to provide the stress components on the boundary.

A comparison between the FVM results and Beris's [7] results (SFEM) for the total load component is given in Fig. 3. The FVM continued to converge at high Weissenberg numbers.

The Sphere Test Problem

We have not yet studied the sphere-in-tube (sphere radius equal to 0.5 tube radius) problem using FVM, but this difficult problem has shown us that

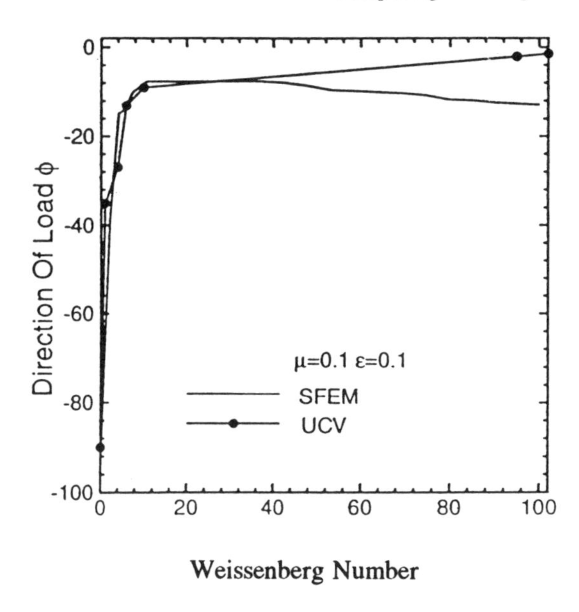

Fig. 3. The inner cylinder direction of the load.

progress is being made in the world of viscoelastic computation. Some recent results for the drag coefficient, $(F/6\pi\eta R_s U)$, where F is the drag force, η is the zero-shear fluid viscosity, R_s is the sphere radius, and U is the speed of the sphere relative to the tube are shown in Fig. 4. Good agreement for the drag force between various methods of computation is now evident up to a Weissenberg number Wi (defined as $\lambda U/R_s$) of about 1.5. Recent results up to about $Wi = 2.2$ show some divergence. Clearly, more investigation of the divergences are needed, although the work of Fan and Crochet [8] and our work [9] are reasonably close. The work by Sun [9] uses a variation of the EVSS method, called AVSS. In the Automatic Viscous Stress Splitting (AVSS) method, a variable viscosity splitting is used to achieve diagonal matrix dominance; the split is done within the programme.

The results extending beyond $Wi = 1.6$ indicate, I believe, that there is no barrier, except mesh refinement, to getting results in this region despite earlier opinions on its solvability for $Wi > 1.6$. Whether we shall see an upturn in the drag coefficient for the UCM model remains to be seen.

Table 2. Characteristics of various numerial methods with differential constitutive equations of the UCM type.

Creators	Name	Stability	h-Conv	Flexi-bility	Speed	Comments
Marchal/ Crochet (1987)	4×4 SUPG	Good	$\surd$	Good	Slow	Coupled scheme fails with non-smooth probl.
Marchal/ Crochet (1987)	4×4 SU	Excellent	Poor	Good	Slow	Coupled — Ok on non-smooth; not a true method of weighted residuals
Brown *et al.* (1988)	EEME	V Good	$\surd$	Poor	Slow	Coupled; very difficult with free boundary; problems with non-smooth flows
Sugeng *et al.* (1986)	BEM	Med.	$\surd$	V Good	Slow	Uncoupled; hard to do highly non-lin. probl.
Beris (1987)	Spectral SFEM	Excellent	Good	Poor	Slow	Coupled; fails for non-smooth probl.
Xue, Phan-Thien *et al.* (1995); Yoo (1987)	FVM (2D & 3D)	Excellent	$\surd$	Med. (rect. elem.)	Quick	Uncoupled (tridag solver; Ok for non-smooth)
Brown *et al.* (1990)	EVSS	V Good	$\surd$	Good	Slow	Coupled; ? on non-smooth flows
Fortin (1995)	Splitting/ Disc. Gal.	V Good	$\surd$	Good	Ok	GMRES-decoupled; Ok for non-smooth
Sun (1986)	AVSS	V Good	$\surd$	$\surd$	Quick	Ok for non-smooth

- Stability refers to ability to reach high Wi on smooth problems with UCM equation (for example, sphere/tube).
- h-conv varies with smoothness.
- Flexibility refers to geometry.

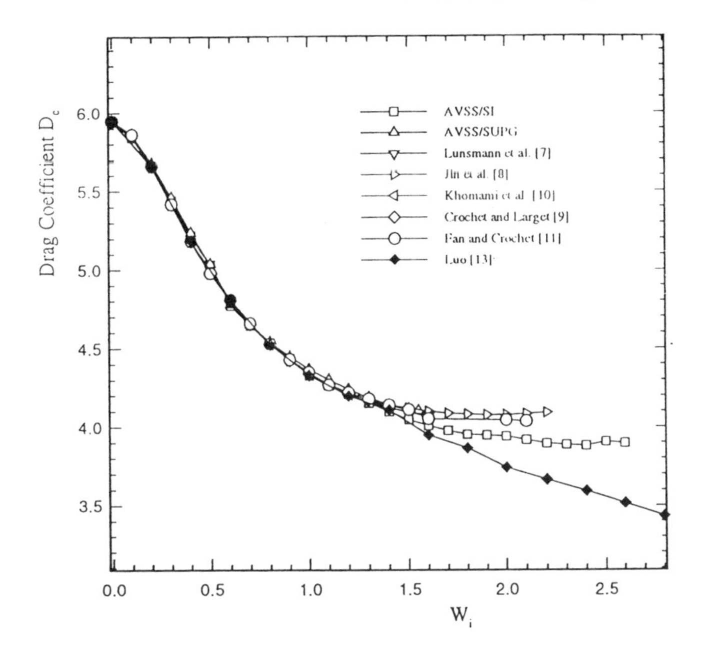

Fig. 4. The predicted drag coefficient versus the Weissenberg number, Wi.

Conclusion

Table 2 gives a summary of where I believe we are now. Problems regarding the form of the solutions near singularities (compare point S in Fig. 1) are now being resolved both analytically and computationally. The physical laws of real materials in these regions remain unknown, but it is clear that literally infinite stresses will not occur in practice. Numerical diffusion is often present near these points; we note the MUCM model of Armstrong *et al.* which deliberately made the behaviour near singular points viscous in nature, thereby easing the calculation.

Acknowledgement

This work was supported by the Australian Research Council.

Discussion

M. M. Denn Could you express your optimism about our ability to deal with singular problems?

R. I. Tanner Let us consider creeping Newtonian flow in such problems — for example, extrudate swell. We know that the stress components, for example, σ_{zz}, behave like $1/\sqrt{r}$ near the singular points, where r is the distance from S. Except at S, a good solution is obtained, and we should aim to get our viscoelastic solutions to this standard. In spite of the fact that the analytical solution is usually lacking in the viscoelastic case, I believe we can now get reasonable viscoelastic solutions by various methods.

R. A. Armstrong One needs to refine a great deal near a singularity so that the solution becomes mesh-independent. What do you sacrifice without mesh refinement?

M. M. Denn Is the problem well-posed? I am sceptical about viscoelastic singular problems.

M. Renardy I will tell you the answer to the form of the singularity tomorrow; I can't tell you if it is a well-posed problem.

R. I. Tanner Suppose we could compute the results near singularities as well as we wish. It is clear that the physics of the (Maxwell) model will fail near S. There is a clear precedent here in the propagation of an elastic crack — a plasticity model takes over at the tip of the crack.

R. A. Armstrong If one alters the constitutive relation near the singularity S, then one can get a converged solution near S for the 4:1 contraction.

R. I. Tanner I recommend doing the stick-slip problem instead of the 4:1 contraction.

R. Keunings At high Weissenberg numbers, mesh refinement makes life much tougher and we are in dangerous waters. First, one must be able to get numbers. Second, we must get accurate numbers. Third, are the computations related to reality? What about the boundary layers?

R. I. Tanner In the eccentric cylinder (bearing) problem, the boundary layers are of $O(1/Wi)$ and cannot be resolved at $Wi \approx 100$ with the present mesh. I believe more elements would enable this. Unlike previous solutions, which appear to be on the edge of stability at $Wi \approx 100$, the finite volume solution is quite stable at this value of Wi.

R. Keunings I am concerned about numerical limits; there is no exact solution even for 4:1 contraction.

Unknown voice I am concerned about the discretisation of the left-hand sides.

R. I. Tanner If the scheme converges, I believe the cancellation of the diffusive terms is alright; we are doing hundreds of iterations here.

A. R. Davies However, the diffusive terms on either side are treated differently, so they may not cancel.

R. I. Tanner A numerical analysis would be useful here. However, both terms arise from volume integrals, so some averaging does take place.

A. R. Davies You have chosen an easy problem — have you tried large eccentricity?

R. I. Tanner Yes, we covered the same range as Beris did, up to 0.8, without problems.

A. R. Davies What is the effect of the diffusion terms?

R. I. Tanner A very large amount of diffusion swamps the calculation and slows down convergence. Typically, we choose the alpha parameter to be (0.01 De) or lower. As the calculation continues, it is made smaller.

R. Keunings What about the sphere problem?

R. I. Tanner It is my highest priority. Does the drag curve go up as Wi increases? We need an asymptotic solution here.

K. Walters Experimentally it does seem to.

G. McKinley Experiments are not done with Maxwell fluids.

R. A. Armstrong What about boundary conditions?

R. I. Tanner No extra boundary conditions are imposed, implying $d\tau/dn = 0$ at the boundary.

K. Walters I would like to ask where the field will be in five years' time.

R. I. Tanner I think computing will soon be more routine, and consulting firms will be able to offer a service. I believe we must address the inadequacy of our constitutive relations, and further experiments are needed. Three-dimensional flows are still a challenge, and we must use multiple-mode models since we know a single time constant is inadequate.

M. M. Denn At the 1972 Engineering Foundation Conference at Asilomar, we wondered what we would be doing professionally in five years, as all problems were on the verge of solution. What happened?

R. I. Tanner I left the Laboratory and went into computing — maybe in error!

References

1. Tanner, R. I. (1988) *Engineering Rheology.* Oxford University Press
2. Xue, S.-C. *et al.* (1995) Numerical study of secondary flows of viscoelastic fluid in straight pipes by an implicit finite volume method. *J. Non-Newtonian Fluid Mech.* **59**, 191–213
3. Tanner, R. I. & Huang, X. (1993) Stress singularities in non-Newtonian stick-slip and edge flows. *J. Non-Newtonian Fluid Mech.* **50**, 135–160
4. Tanner, R. I. (1996) Finite volume computational methods for polymer processing. *Proc. XIX Int. Congr. Theor. and Applied Mech.*, pp. 367–368. Kyoto, Japan
5. Huang, X. *et al.* (1996) Viscoelastic flow between eccentric rotating cylinders: unstructured control volume method. *J. Non-Newtonian Fluid Mech.* **64**, 71–92
6. Perera, M. G. N. & Walters, K. (1977) Long range memory effects in flows involving abrupt changes in geometry. 1. Flows associated with L-shaped and T-shaped geometries. *J. Non-Newtonian Fluid Mech.* **2**, 49–82
7. Beris, A. N. *et al.* (1987) Spectral/finite element calculations of the flow of a Maxwell fluid between eccentric rotating cylinders. *J. Non-Newtonian Fluid Mech.* **22**, 129–167
8. Fan, Y. & Crochet, M. J. (1995) High-order finite element methods for steady viscoelastic flows. *J. Non-Newtonian Fluid Mech.* **57**, 283–311
9. Sun, J. *et al.* (1996) An adaptive viscoelastic stress splitting scheme and its applications: AVSS/SI and AVSS/SUPG. *J. Non-Newtonian Fluid Mech.* **65**, 75–92

Dynamics of Complex Fluids, pp. 61–72
ed. M. J. Adams, R. A. Mashelkar, J. R. A. Pearson & A. R. Rennie
Imperial College Press–The Royal Society, 1998

Chapter 5

Brownian Dynamics Simulations in Polymer Physics

H. C. ÖTTINGER

ETH Zürich, Department of Materials, Institute of Polymers,
and Swiss F.I.T. Rheocenter, CH-8092 Zürich, Switzerland
E-mail: hco@ifp.mat.ethz.ch

An equivalence between diffusion equations in polymer configuration space (which are ubiquitous in polymer kinetic theory) and stochastic differential equations for the polymer configurations is the basis for very general stochastic simulation techniques for polymer dynamics in solutions and melts. These simulation techniques, the most well-known of which are Brownian dynamics simulations, play an important role in the investigation of polymer dynamics. We sketch the state of the art in the simulation of polymer dynamics. Recent developments include higher-order schemes, in particular derivative-free higher-order schemes, semi-implicit methods and variance reduction methods for the numerical integration of stochastic differential equations. We summarise various variance reduction schemes, based either on important sampling strategies or on the use of control variables.

Introduction

Conventionally, the kinetic theory models for the dynamics of polymers in solutions as well as in melts are formulated in terms of diffusion equations for suitable configurational distribution functions [1]. The dynamics described by these diffusion equations can be related to stochastic processes which are governed by stochastic differential equations, often referred to as Langevin equations.

In kinetic theory, one is usually not interested in the motion of a single polymer molecule, but rather in an average effect of a large ensemble of molecules. An important task is to compute ensemble averages such as a kinetic theory expression for the stress tensor in a given flow situation, or quantities describing the conformational properties of the polymer molecules. Two different ways to compute the averages lead to identical results: either the diffusion equation is solved for the configurational distribution function and the averages are evaluated by computing an integral weighted with the distribution function, or the averages are estimated from an ensemble of trajectories constructed by simulating the corresponding stochastic differential equation. Stochastic simulation techniques have been established as a powerful tool for investigating various kinetic theory models. The most popular simulations are Brownian dynamics simulations for polymers in a dilute solution. Nevertheless, stochastic simulations for reptation models and other systems have also been carried out.

Simulations of polymer dynamics have gained additional importance because they can be combined with the finite element method to calculate the complex flows of polymeric liquids. For example, with the CONNFFESSIT idea (Calculation of Non-Newtonian Flow: Finite Elements & Stochastic Simulation Techniques), the flows are calculated without using a constitutive equation. Like in real liquids, the stresses are obtained directly from the configurations of a large number of polymer molecules in the system.

We summarise the state of the art in the simulation of polymer dynamics. Recent developments include higher-order schemes, in particular derivative-free higher-order schemes, semi-implicit methods, and variance reduction methods for the numerical integration of stochastic differential equations. More details and extensive lists of references can be found in a recent book (henceforth HCO) [2].

Stochastic Formulation of Polymer Dynamics

In view of the immense number of degrees of freedom and the wide range of time scales involved in polymer problems, the derivation of tractable kinetic theory models requires some coarse-grained or trace description of such problems (for example, by mechanical bead-rod-spring rather than atomistic models). The effects of the rapidly fluctuating degrees of freedom associated with very short length scales are usually taken into account through random forces which perturb the time evolution of the slower degrees of freedom. For that reason, the basic equations for most kinetic theory models are stochastic in nature. From a more intuitive point of view, thermal noise turns the equations of motion into stochastic differential equations.

Most kinetic theory models developed to describe the rheological behaviour of polymeric liquids are specified in terms of diffusion or Fokker-Planck equations for the dynamics of the polymers [1, 3]. The existence of an equivalence between diffusion equations and stochastic differential equations makes it possible to rewrite directly the familiar models in a form which is suitable for computer simulations. More precisely, the diffusion equation,

$$\frac{\partial}{\partial t}p(t,\boldsymbol{x}) = -\frac{\partial}{\partial \boldsymbol{x}} \cdot [\boldsymbol{A}(t,\boldsymbol{x})p(t,\boldsymbol{x})] + \frac{1}{2}\frac{\partial}{\partial \boldsymbol{x}}\frac{\partial}{\partial \boldsymbol{x}} : [\boldsymbol{D}(t,\boldsymbol{x})p(t,\boldsymbol{x})]\,, \tag{1}$$

and the stochastic differential equation,

$$\mathrm{d}\boldsymbol{X}_t = \boldsymbol{A}(t,\boldsymbol{X}_t)\mathrm{d}t + \boldsymbol{B}(t,\boldsymbol{X}_t)\cdot \mathrm{d}\boldsymbol{W}_t\,, \tag{2}$$

with $\boldsymbol{D}(t,\boldsymbol{x}) = \boldsymbol{B}(t,\boldsymbol{x})\cdot\boldsymbol{B}^T(t,\boldsymbol{x})$ are equivalent. This means that all the transition probability densities of the Markovian solution of the stochastic differential Eq. (2) are obtained by solving the Fokker-Planck Eq. (1) under suitable initial conditions. In these equations, $\boldsymbol{x}$ is the configuration vector, t is time, $p(t,\boldsymbol{x})$ is the probability density in configuration space, $\boldsymbol{A}$ is the drift vector, and $\boldsymbol{D}$ is the diffusion matrix. The symbols $\boldsymbol{W}_t$ and $\boldsymbol{X}_t$ denote stochastic processes, namely the Wiener process and the solution of the stochastic differential equation.

In computer simulations based on the numerical integration of stochastic differential equations, we construct stochastic trajectories. It is therefore crucial to give a precise meaning to these random objects, the trajectories of stochastic processes, and not only to introduce their probability distribution, as is usually done in the applied sciences and engineering. This is particularly important in constructing sophisticated integration schemes, that is, for developing efficient simulation algorithms. The *art of designing efficient algorithms*, and the required mathematical background, are the subject of HCO [2]. Moreover, the investigation of stochastic differential equations of motion for the polymer configurations has additional advantages compared to the usual consideration of partial differential equations for the time-evolution of probability densities in configuration space: we obtain a more direct understanding of the polymer dynamics and a better feeling of the degree of complexity for the types of models.

Brownian dynamics simulations have been applied in the field of polymer dynamics since the late 1970s. However, the great progress made with numerical methods for stochastic differential equations has not really been exploited yet in simulating polymer dynamics. If state-of-the-art techniques are used (such as higher-order integration schemes, predictor-corrector schemes, Runge-Kutta ideas, implicit methods, time-step extrapolation, non-Gaussian random numbers, and variance reduction), one is then in a position to attack the most important and exciting problems in polymer dynamics, and some progress has already been made in this direction. On the one hand, one can work out the *universal dynamic properties* of long chain molecules. This expected universality, which is related to the famous self-similarity and scaling laws for high-molecular-weight polymers, may also serve as a justification for the simple mechanical models used in polymer kinetic theory. On the other hand, one can simulate extremely large ensembles of molecules and, in combination with the finite element method, one can then directly *solve complex flow problems* with kinetic theory models. The stresses required in the finite element calculation of the flow field are read off from the molecular configurations. These exciting perspectives are the motivation for the stochastic approach described in this presentation.

The advantages of using stochastic equations rather than Fokker-Planck equations seem to violate an empirical conservation law which states that solving a given problem by any reasonable approach requires roughly the same amount of effort. This paradox can be resolved by noting that solution of the Fokker-Planck equation for a high-dimensional problem yields full information about the distribution. The numerical integration of a stochastic differential equation yields an ensemble of trajectories from which one can evaluate averages, but one cannot obtain the distribution of the solution in a high-dimensional space. One might try to obtain information about the distribution by dividing the configuration space into cells and counting the number of trajectories which assume values in each cell. In other words, one could try to construct a histogram approximating the distribution. However, in a high-dimensional configuration space, for any reasonable approximation to the distribution function, the number of cells would be so large that most cells would be unoccupied. In conclusion, high-dimensional distribution functions can neither be obtained from Fokker-Planck equations nor from stochastic differential equations. However, averages or contracted distribution functions can be obtained directly by numerical integration of stochastic differential equations, whereas the Fokker-Planck equation approach fails when the full distribution cannot be determined. This is the origin of the *first kind of miracle* to be emphasised in this presentation.

Recipes for Simulating Stochastic Differential Equations

We have seen how one can rewrite the polymer dynamics that are usually described by diffusion or Fokker-Planck equations in terms of stochastic differential equations of motion. What can we actually gain by such a reformulation? Is the reformulated problem any more tractable than the original one?

Like deterministic differential equations, their stochastic counterparts can be solved only in very special cases. Solvable stochastic differential equations are essentially the linear ones and those which can be reduced to linear equations by suitable nonlinear transformations. The stochastic differential equations associated with all the interesting nonlinear models of polymer kinetic theory cannot be solved analytically. However, powerful numerical integration schemes have been developed for solving stochastic differential equations. Since random numbers are involved, such integration schemes may be regarded as simulation algorithms. The significance of the relationship between Fokker-Planck equations and stochastic differential equations can hence be described as follows: for a high-dimensional configuration space, it is hopeless to solve a Fokker-Planck equation by the usual numerical methods used in treating partial differential equations; the associated stochastic differential equations, however, can be integrated numerically even for a large number of degrees of freedom, and averages or other properties of the distribution of the solution can be evaluated from an ensemble of trajectories.

Since the mid-1970s, many papers on sophisticated numerical integration schemes for stochastic differential equations have appeared in the mathematical literature. Twenty years later, this is still a field of very active research. A comprehensive description of efficient numerical integration methods can be found in a recent book by P. E. Kloeden and E. Platen [4]. The simulation techniques that are most useful for simulating polymer dynamics are explained in great detail in HCO.

Here we only summarise the various considerations that are required, or at least helpful, before a simulation program for a given problem is developed. The first step is to decide whether the underlying stochastic differential equation is of the Itô or Stratonovich type [2]. There is a straightforward procedure for going back and forth between Itô and Stratonovich equations, and if one starts from a Fokker-Planck equation or diffusion equation, an unambiguous formulation is possible. If the starting point is a stochastic differential equation or Langevin equation, such an equation should also be accompanied by an interpretation rule, which has to be found from a deeper understanding of the physical system. The second step is to decide whether a strong or weak solution is needed, that is, whether one is interested in individual trajectories or only in ensemble averages. In the weak case, it is much simpler to construct good approximation schemes, and the efficiency can be further improved by replacing Gaussian random variables with more readily generated ones. Finally, the nature and complexity of the problem determine the order of strong or weak convergence of the approximation scheme to be selected, the necessity of eliminating derivatives of the coefficient functions (Runge-Kutta schemes, multi-step schemes, predictor-corrector schemes), and the requirements for obtaining numerical stability (implicit schemes, semi-implicit schemes). The details of these methods and their various applications and further references can be found in HCO.

Time-step extrapolation can be regarded as an alternative method to increase the order of convergence. For example if the first-order Euler scheme is used for two different time steps and the results are extrapolated linearly, the extrapolated result is of the second order in the time-step width. Even higher-order convergence can be achieved by working with three (or more) different time steps and using quadratic (or higher-order) extrapolation. Of course, one can also improve the order of convergence for higher-order schemes by using extrapolation methods. Time-step extrapolation is therefore a very simple and valuable tool for improving simulations, and it should always be employed when a weak approximation scheme is used to integrate a stochastic differential equation.

It may be worth pointing out a *second miracle* here: many physicists have denied the existence of subtle differences between deterministic and stochastic differential equations. While there is nothing to be gained by denying these differences, the generalisation of the familiar integration methods for deterministic differential equations certainly requires deeper insight into the peculiarities of stochastic differential equations.

Variance Reduction

The idea of variance reduction methods is to transform a given stochastic differential equation or to modify its solution so that the transformed equation yields the same averages but with a smaller variance for a certain quantity of interest. Then, when a given number of trajectories is simulated, a correspondingly smaller statistical error bar is obtained. The possibility of finding a miraculous transformation reducing the error bars in a simulation seems too good to be true. Hence we refer to variance reduction as a *third kind of miracle.* Indeed, there is a catch: the construction of the transformation requires an *a priori* understanding of the simulated system. Typically, an approximate solution is needed in order to construct a variance reduced integration scheme, and the better the approximation, the more the variance is reduced. The possibility of constructing considerably more efficient integration schemes can thus be a motivation for developing good approximations.

The first variance reduced calculations for polymer problems were based on the idea of importance sampling. In the most elementary realisation of importance sampling for stochastic differential equations, a biased final configuration is first selected. Alternatively, one can produce a biased final distribution by modifying the drift and initial conditions such that the final configurations are distributed according to the desired probability density. The proper modification of the drift and the determination of the corresponding correction factor can be based on a suitably constructed Girsanov transformation, which is a profound result of stochastic calculus. Details on this idea, and its successful implementation for Hookean dumbbells with hydrodynamic interaction, can be found in [5].

Very recently, a variance reduction method not based on importance sampling has been adapted to polymer kinetic theory [6]. The basic idea is to estimate the fluctuations in the quantity of interest as obtained in an unbiased standard simulation. By subtracting an approximate expression for the fluctuations (the so-called control variable), one can reduce the fluctuations, that is, the variance, without changing averages. This approach is suited ideally when the fluctuations can be estimated from a Gaussian approximation and, even better, when the quantity of interest is a quadratic function of configuration. Under these conditions, the variance for the viscometric functions and the mean-square size of Hookean dumbbells with hydro/dynamic interaction at the start-up of shear flow has been reduced by two orders of magnitude, uniformly in time and at little extra cost [6]. If the quantity of interest is not a quadratic function of configuration, the variance can still be reduced by subtracting the results of two simulations, one for the exact stochastic differential equation and the other for an approximate differential equation for which the averages are known. The trajectories of the two simulations must be constructed with the same sequence of random numbers so that the fluctuations are approximately cancelled.

Two particularly important applications of the idea of control variables are mentioned here:

(i) In order to obtain the linear viscoelastic behaviour of a kinetic theory model, one can simulate the model at a very small shear rate. A very efficient control variable is then obtained from a parallel equilibrium simulation, with the same initial ensemble and the same sequence of random numbers. For decreasing shear rates, the variance reduction becomes more efficient, so that the decreasing shear stress signal can actually be detected without any loss in accuracy [7].

(ii) The new idea of configuration fields in flow calculations based on stochastic simulations introduces strong correlations in the stress fluctuations at different points in space [8]. When forming gradients of the stress tensor, the fluctuations cancel, and very substantial variance reduction results. The relationship between configuration fields and variance reduction by control variables is explained in greater detail in [9].

Conclusion

We have described three miracles associated with stochastic simulation techniques in polymer physics: (i) the enormous advantages of reformulating diffusion equations as stochastic differential equations; (ii) the long persistence of the fruitless and counterproductive denial of the subtle differences between deterministic and stochastic differential equations; and (iii) the power of variance reduction ideas. We hope that, in view of these miracles, the following statement becomes more widely accepted: What Monte Carlo simulations are to the theory of static polymer properties, *Brownian dynamics simulations* and other simulations based on the *numerical integration of stochastic differential equations* are to the theory of polymer dynamics. Both simulation methods are unique tools for obtaining exact results in the fields of equilibrium and non-equilibrium thermodynamics, respectively.

Discussion

M. M. Denn I don't want to minimise the miracle of the first kind, but isn't there a long history of taking advantage of the dualism between diffusion-like equations and stochastic processes?

H. C. Öttinger There certainly is a long history of Brownian dynamics for polymer problems, starting with the work of Fixman, say 20 years ago. But what exactly are you referring to as "a long history?"

M. M. Denn The idea actually goes back to Einstein — the whole notion that there are times, if you want to solve a diffusion-type equation, when it is simply computationally more efficient to solve the equivalent stochastic problem.

H. C. Öttinger This is not an example of the miracle effect of the first kind. For Einstein's one- or three-dimensional problem, if you would do all the calculations on a computer, both approaches would require roughly the same order of computer time. The miracle that stochastic simulations are significantly more efficient than numerical solutions of diffusion equations happens only when many degrees of freedom are involved, which is not the case for Einstein's problem. For low-dimensional problems, I would usually prefer deterministic methods.

M. M. Denn I recall that, when I was editing the AIChE Journal, there was at least one paper demonstrating that for a diffusion-reaction problem in one dimension, for example, it was computationally more efficient to solve the equivalent stochastic problem.

H. C. Öttinger At best, marginally more efficient — certainly no miracle! A major advantage, however, is the generally much simpler programming of stochastic methods.

R. C. Armstrong What happens to the efficiency of stochastic methods when they are applied to concentrated solutions or melts, when one is interested in intermolecular interactions? Do you still think it's more efficient to do the stochastic simulations, or would the diffusion-equation approach be more appropriate?

H. C. Öttinger The diffusion-equation approach is ideal in reducing problems formally, but not to solve them explicitly. For example, you would use the

Liouville equation and the BBGKY hierarchy to derive simplified or reduced equations. However, you would clearly prefer Newton's or Hamilton's equations of motion for developing computer simulation algorithms (molecular dynamics). In the kinetic theory of polymer melts, the diffusion-equation approach is a very useful tool for deriving closed equations for a probe chain or even a single chain segment. Once a reduced diffusion-equation for a concentrated solution or melt has been derived, and it usually involves not too many degrees of freedom or explicit interactions, one can solve this equation by stochastic methods.

S. F. Edwards May I ask you a more general question? There are no beads and springs — they are just a convenient model. You could say that the polymer is a completely continuous chain, and you can spend your time trying to construct eigenfunctions for the polymer, and not doing simulations of the kind you made. Why is it that you and other investigators like these Hookean dumbbells so much when you know that, in the end, they cannot come into the answer? The polymer itself is just like a piece of elastic: it's like a rubber band. You don't treat it as a continuous object that has its own eigenfunctions, but you prefer to put these points all the way along.

H. C. Öttinger For the simulations, I want a finite number of degrees of freedom ...

S. F. Edwards ... yes, but you could have a finite set of eigenfunctions. But I am an analyst, and I don't have the slightest idea how to use the computer ...

H. C. Öttinger I do not see a fundamental difference between having a finite number of beads or a finite number of eigenfunctions. In both cases, you can only achieve a certain spatial resolution, and you neglect more local details. Now, you usually want to treat interactions and you need to determine the forces between different chain segments — and that is much easier if you work with the positions of the chain segments rather than with eigenmodes. In particular, if you have highly nonlinear interactions, working with modes may be very unpleasant.

S. Ramaswamy This is an extension or clarification of the previous point. You can take the dynamics of a chain moving along as a partial differential equation in the chain parameter and time. Now what Sam Edwards is asking, I guess, is this: you often obtain numerical solutions of partial differential equations in terms of Fourier modes. Isn't it profitable in some cases to do that for the polymer case as well, numerically? (Analytically, of course, much is done in precisely that way.)

H. C. Öttinger A highly nonlinear problem is, in general, very difficult to analyse in that way. And the question remains: how much of the physics can you capture with just a few modes. Once again, I do not see a fundamental difference between discretising by means of beads or modes.

S. F. Edwards The answer is: he likes it that way — you the other.

R. Keunings I think Hans Christian is being too modest here. He has just written a very nice book in which there is a whole chapter on the description of the use of stochastic techniques for polymer melts, including reptation models. So, stochastic techniques can obviously be used for systems other than just simple dumbbells.

J. D. Goddard Can't one use path-integral techniques (obtained from local propagators or Green's functions) to get efficient computations of distributions $p(t, x)$ and, hence, of averages?

H. C. Öttinger Path integrals certainly are another possibility of solving stochastic differential equations — quite complementary to simulation techniques. The obtained solutions are rather formal and certainly extremely difficult to evaluate for nonlinear problems. However, path-integral techniques may be very useful for developing approximations in general, and perturbation theories in particular. A very nice reference on what can be done with path integrals in this field is found in the book by J. Honerkamp [10] (Chapters 11 and 12).

References

1. Bird, R. B., Curtiss, C. F., Armstrong, R. C. & Hassager, O. (1987) *Dynamics of Polymeric Liquids, Vol. 2, Kinetic Theory*, 2nd edn. New York: Wiley-Interscience
2. Öttinger, H. C. (1996) *Stochastic Processes in Polymeric Fluids: Tools and Examples for Developing Simulation Algorithms.* Berlin: Springer
3. Doi, M. & Edwards, S. F. (1986) *The Theory of Polymer Dynamics.* Oxford: Clarendon
4. Kloeden, P. E. & Platen, E. (1992) *Numerical Solution of Stochastic Differential Equations.* Berlin: Springer
5. Melchior, M. & Öttinger, H. C. (1995) Variance reduced simulations of stochastic differential equations. *J. Chem Phys.* **103**, 9506–9509

6. Melchior, M & Öttinger, H. C. (1996) Variance reduced simulations of polymer dynamics. *J. Chem. Phys.* **105**, 3316–3331
7. Wagner, N. J. & Öttinger, H. C. (1997) Accurate simulation of linear viscoelastic properties by variance reduction through the use of control variates. *J. Rheol.* **41**, 757–768
8. Hulsen, M. A., van den Brule, B. H. A. A. & van Heel, A. P. G. (1998) Simulation of the flow of an Oldroyd-B fluid using Brownian configuration fields. In *Dynamics of Complex Fluids*, ed. M. I. Adams, R. A. Mashelkar, J. R. A. Pearson & A. R. Rennie, pp. 106–123. Singapore: World Scientific. See also "Simulation of viscoelastic flows using Brownian configuration fields" (1997) *J. Non-Newtonian Fluid Mech.* **70**, 79–101
9. Öttinger, H. C., van den Brule, B. H. A. A. & Hulsen, M. A. (1997) Brownian configuration fields and variance reduced CONNFFESSIT. *J. Non-Newtonian Fluid Mech.* **70**, 255–261
10. Honerkamp, J. (1993) *Stochastic Dynamical Systems.* New York: VCH

Dynamics of Complex Fluids, pp. 73–87
ed. M. J. Adams, R. A. Mashelkar, J. R. A. Pearson & A. R. Rennie
Imperial College Press–The Royal Society, 1998

Chapter 6

Two-Dimensional, Time-Dependent Viscoelastic Flow Calculations Using Molecular Models

M. LASO
Department of Chemical Engineering, ETSII,
José Gutiérrez Abascal, 2,
E-28006 Madrid (Spain)
E-mail: laso@iris.digi.upm.es

Viscoelastic flow calculations in complex geometries (journal bearing, abrupt contraction) for a variety of molecular models are presented. These calculations are based on CONNFFESSIT (Calculation of Non-Newtonian Flows: Finite Element and Stochastic Simulation Technique). The flow is considered to be incompressible and isothermal. The momentum conservation equation is integrated using a time-marching procedure in which local ensembles of dumbbells act as stress calculators. The CONNFFESSIT calculations combine deterministic (finite elements) and stochastic techniques to advance in time the velocity and stress fields. The ability of CONNFFESSIT to treat models for which no closed-form constitutive equation can be derived is illustrated by performing calculations using FENE dumbbells and using a Brownian Dynamics model for a concentrated colloidal dispersion. Additionally, experience gathered from parallel and vector versions of CONNFFESSIT is reported.

Introduction

Most current numerical methods for the calculation of viscoelastic flows are based on a continuum-mechanical description of the polymer solution or of the melt. They therefore yield a full description of the field variables at the macroscopic level typical of discretisation methods (for example, finite elements, volumes and differences). They are, however, unable to provide any information at the molecular level (molecular in the sense of the relatively simple mechanical models, such as dumbbells and networks, which underlie some macroscopic constitutive equations (CE's) [2, 3, 10]; the terms "molecule" and "molecular" will be used in this coarse-grained sense throughout).

Consequently, and in spite of the substantial progress achieved in flow calculations, most current numerical methods are inherently unable to offer insights into molecular behaviour: in traditional isothermal and incompressible calculations, the fluid is entirely and exclusively characterised by the relationship between stress and strain history expressed as a differential or integral constitutive equation. Even for those CE's which allow a microscopic interpretation in terms of simple molecular models, the averaging procedure required to arrive at a macroscopic expression for the stress erases all microscopic (molecular) information. The primary variables involved in classical flow calculations are only velocity, pressure and stress; there is no way to retrieve any details of the lost molecular picture.

Partly motivated by the idea of extracting molecular information from viscoelastic flow calculations, an alternative computational method has recently been introduced that combines stochastic and traditional discretisation methods [4, 7, 8, 9]. In the CONNFFESSIT (Calculation Of Non-Newtonian Flows: Finite Elements and Stochastic Simulation Technique), the primary variables in isothermal, incompressible calculations are velocity, pressure and molecular configurations (degrees of freedom), with stress having only a subordinate function.

As in most viscoelastic flow calculations, the objective of a CONNFFESSIT calculation is to solve the equations of mass and momentum conservation for an incompressible, isothermal non-Newtonian fluid:

$$(\nabla \cdot \boldsymbol{\nu}) = 0$$

$$\rho \frac{\partial}{\partial t}\boldsymbol{\nu} + [\nabla \cdot \rho \boldsymbol{\nu}\boldsymbol{\nu}] + [\nabla \cdot \boldsymbol{\pi}] = 0\,,$$

where $\boldsymbol{\pi}$ is the total momentum-flux or total stress tensor which can be split in the following way:

$$\boldsymbol{\pi} = p\boldsymbol{\delta} + \boldsymbol{\tau}\,.$$

This set of partial differential equations must be closed by a third equation (the CE) relating $\boldsymbol{\tau}$ to the history of the flow. The form of this closure is responsible to a large extent for the mathematical complexity and the considerable numerical difficulties often encountered in the calculation of viscoelastic flows.

The most characteristic feature of CONNFFESSIT is the replacement of the traditional, closed-form CE by a direct stochastic simulation of the molecular model itself. As an illustrative example, the dynamics of Hookean dumbbells is given by

$$d\boldsymbol{Q} = \left([\boldsymbol{\kappa} \cdot \boldsymbol{Q}] - \frac{2H}{\zeta}\boldsymbol{Q}\right) dt + \sqrt{\frac{4kT}{\zeta}} d\boldsymbol{w} \,,$$

where each of the terms on the right-hand side has a clear physical interpretation (effect of velocity gradients, restoring spring force and Brownian motion).

A great many molecules (the so-called global ensemble), such as be rigid or elastic dumbbells and bead-spring chains, act as stress calculators in each element. All molecules in a given element (the so-called local ensemble) contribute to the average value of the stress in that element.

The polymer contribution to the extra stress is treated as a constant, right-hand side body force in a classical finite element time-marching procedure. The molecules are entrained by the macroscopic flow of the fluid as represented by the velocity field. Stored in their configurations, they carry the deformation history of the flow and therefore automatically perform the same function as tracking and integration in the integral formulation of CE's [5]. In addition to their macroscopic degrees of freedom (spatial co-ordinates), they possess internal degrees of freedom, the time evolution of which is governed by stochastic differential equations. The internal degrees of freedom fully describe the molecules. Simple molecular models, like Hookean or FENE dumbbells, are described by few degrees of freedom. It suffices to specify the three components of the vector joining the two masses that make up the dumbbell. In other models, the molecular motions are described by reptation processes [3, 10]. Standard numerical techniques are used to integrate both internal and external degrees of freedom.

A typical CONNFFESSIT flow calculation is time-dependent, although it need not be so [4]. It starts from a given velocity field and a given set of molecules, usually distributed uniformly over the integration domain. In this case of uniform polymer concentration, each element contains a number of molecules proportional to its area or volume.

In addition to the traditional initial conditions, it is necessary to assume a flow history in some way ([5], Sec. 9.2.6). The configurations of the molecules that initially fill the integration domain must reflect what has happened to them during the infinite time interval before $t = 0$. Initial conditions

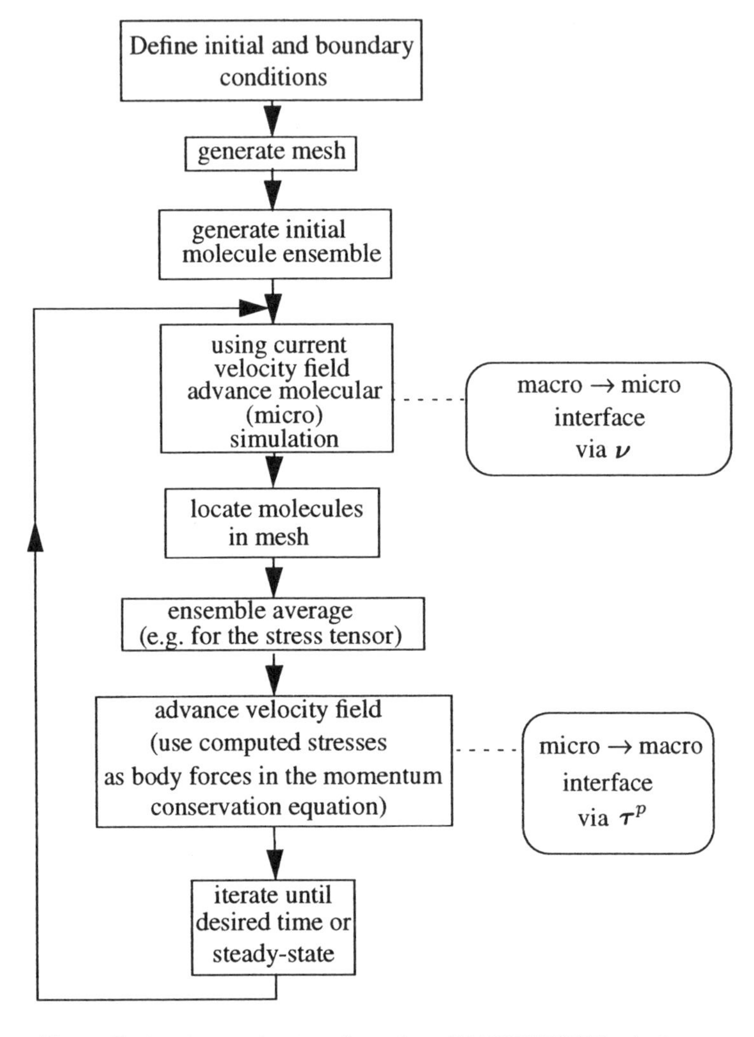

Fig. 1. Basic scheme of a time-dependent CONNFFESSIT calculation.

correspond typically (but not necessarily) to zero velocity and to a flow history of zero strain throughout the domain. The initial configurations of the dumbbells are then drawn from a suitable equilibrium configurational distribution function (see, for example, Table 11.5–1 of [2]). In addition, if open boundaries exist and a non-zero strain history is specified, it is first necessary to determine, analytically or numerically, the molecular configurations at the inlet of the integration domain.

The stochastic (micro) and finite element (macro) parts of CONNFFESSIT are interfaced through the velocity and stress fields as indicated in Fig. 1. In the terminology of computational rheology, CONNFFESSIT is a fully decoupled method. The calculation of $\boldsymbol{\nu}$ and $\boldsymbol{\tau}^p$ occur in separate sequential steps, with each variable being a constant.

CONNFFESSIT has already been used to solve one- and two-dimensional steady-state and time-dependent viscoelastic flow problems. Within the error bars due to its stochastic nature, CONNFFESSIT yields results in complete agreement with traditional techniques. In addition, CONNFFESSIT has been used to perform flow calculations with molecular models for which there is no equivalent closed-form constitutive equation (for example, the Finitely Extensible Nonlinear Elastic or FENE dumbbell [12]), or which are too complex [2, 3] for traditional numerical methods.

Examples

In the next sub-sections, we present some results of two-dimensional CONNFFESSIT calculations. For more details, the reader is referred to the original publications mentioned therein.

Journal Bearing

The first 2-d, time-dependent CONNFFESSIT calculations were performed in the journal bearing geometry [8] for the FENE and the FENE-P fluids.

In spite of the fluctuations present in the stochastic solution, CONNFFESSIT is perfectly adequate for resolving the different behaviour of the true FENE and the linearised FENE-P model. The smooth continuous line in Fig. 2 represents the solution obtained with a finite element code using the analytical FENE-P constitutive equation [11]. No such calculation is possible for the FENE fluid.

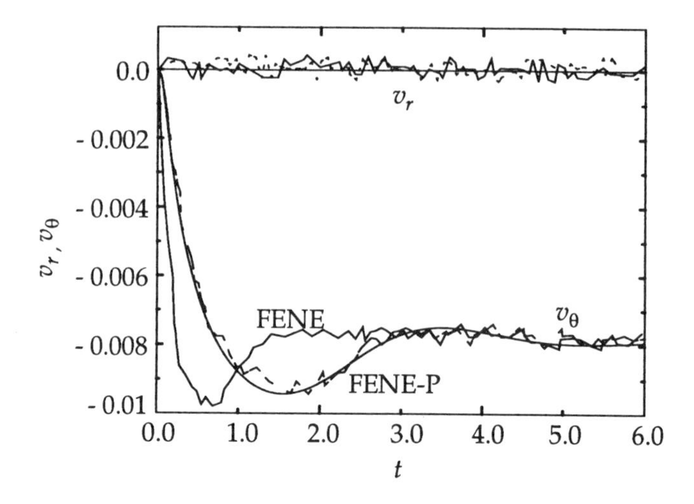

Fig. 2. Temporal evolution of radial and azimuthal velocities in the journal bearing at low gap and eccentricity. The CONNFFESSIT calculation can clearly distinguish between the true FENE and the linearised FENE-P models [8].

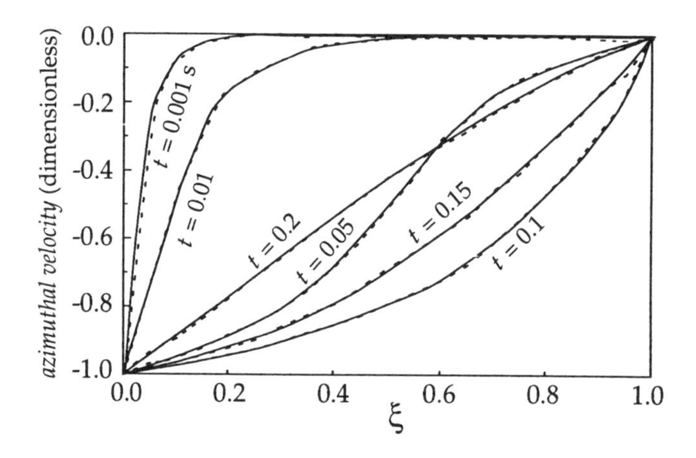

Fig. 3. Start-up of concentric journal bearing flow for an Oldroyd-B model on a mesh of 1600 elements (20 in the radial direction and 80 in the azimuthal) at low dimensionless gap and zero eccentricity (0.1). Solid line: exact solution; dashed line: 2D CONNFFESSIT calculation [8].

Since an exact solution for the start-up of Couette flow (zero eccentricity) for the Oldroyd-B model in the concentric journal bearing geometry is known, it is possible to use this analytical solution to assess the reliability of

CONNFFESSIT. Figure 3 shows the analytical (solid lines) and CONNFFESSIT (dashed lines) solutions to the start-up problem (azimuthal velocity as a function of time and dimensionless radial position).

The very minor discrepancies between both are due to finite ensemble size (2^{19} Hookean dumbbells) and to limited spatial resolution of the mesh. The very low noise level in the solution could be obtained by exploiting the azimuthal symmetry of the problem. When no such symmetry is present, the intensity of fluctuations for the 2D problem (non-zero eccentricity) using a similar global ensemble size and mesh refinement is of course higher (Fig. 2).

Contraction Flow

The suitability of CONNFFESSIT for performing realistic 2D calculations in complex geometries with open boundaries has been demonstrated recently [9]. A viscoelastic classic, the die entry, has been rerun for the FENE fluid using CONNFFESSIT on a mesh of 1920 elements.

Typical streamlines are given in Fig. 4. Besides this kind of continuum-mechanical picture, the CONNFFESSIT calculation also yields molecular information such as molecular extension.

Plot (a) in Fig. 5 shows the configurations of a few hundred dumbbells that have left the domain at the upper tenth of the narrow channel width, while Fig. 5(b) corresponds to the region closest to the axis of symmetry. Each dot in

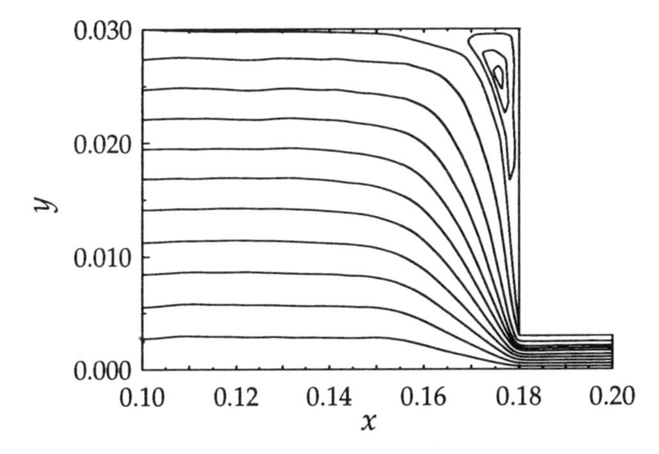

Fig. 4. Streamline pattern in the neighbourhood of the re-entrant corner of a planar 10:1 contraction for the FENE fluid at $De = 4.1$ [9].

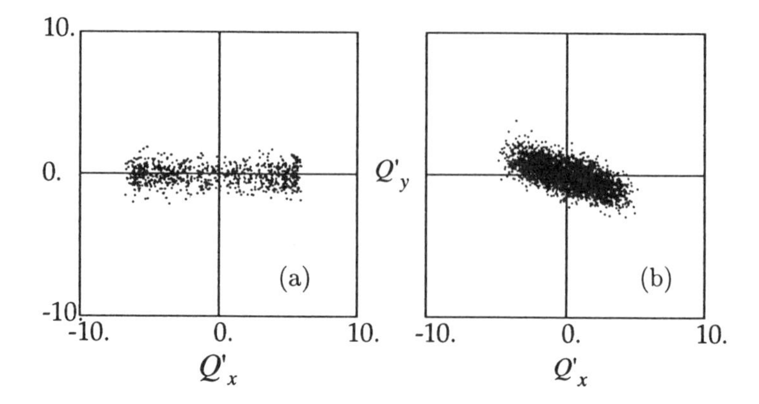

Fig. 5. x- and y-components of FENE dumbbells at the outlet of the planar 10:1 contraction for the FENE fluid at $De = 4.1$: (a) closest to the channel wall; (b) closest to the middle of the channel [9].

these plots is the tip of a vector with its origin at (0, 0) and the components (Q'_x, Q'_y) representing an individual dumbbell configuration. The very high degree of orientation of the dumbbells closest to the wall is apparent. Even molecules closest to the centre of the channel show a strong departure from isotropicity (Fig. 5(b)). The prediction of resulting anisotropies in macroscopic properties (for example, mechanical and optical) from known molecular configurations is a fascinating possibility to which access can be gained through a CONNFFESSIT calculation.

Colloidal Dispersion

The idea of coupling microscopic simulations with macroscopic finite elements is not restricted to molecular models of non-interacting particles. It is possible to model complex flows of strongly interacting systems such as colloidal dispersions. The main difficulty of this calculation (besides the sheer amount of number crunching) is that a multi-particle system requires some kind of spatially replicated confining cell in order to avoid edge effects. This cell must be deformed in a way that is consistent with the macroscopic flow. The consequence is that for large strains, this cell becomes excessively deformed, thus rendering the microscopic calculation unfeasible.

It is possible to use Lees-Edwards boundary conditions in the very restricted case of plane Couette flow which allow for arbitrarily large strain. Although

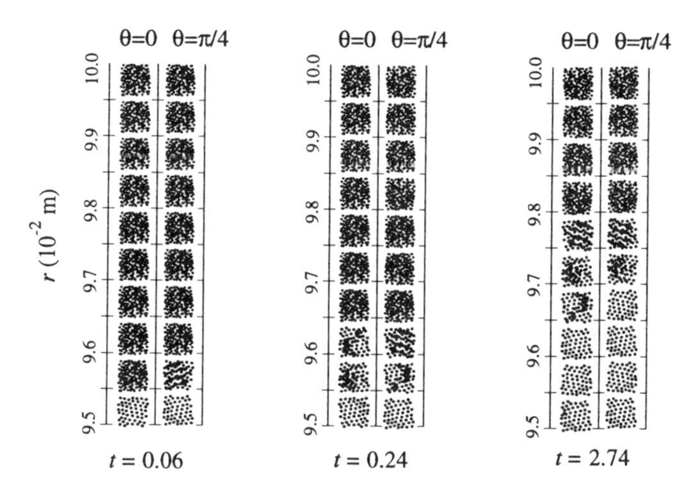

Fig. 6. Projection of colloidal particles on the vorticity-gradient plane. Comparison at two azimuthal positions differing by $\theta = \pi/4$ at three different times (in seconds) [6].

there are ways to partially alleviate this problem [6], they can only delay the inevitable. However, if the process we are interested in takes place in a time scale during which the simulation cell does not have time to become too skewed, useful results can be obtained.

Figure 6 is such an application: it represents the appearance and growth of an ordered domain during start-up of Couette flow of a dense, sterically stabilised colloidal dispersion. It is sheared in the gap between two concentric cylinders (it is known both experimentally and from simulations that such colloids can develop ordered phases under shear). In this radial section, the lower part of the plot corresponds to the inner cylinder, which is impulsively set in motion. Each of the small many-particle plots are projections of all the colloidal spheres in a given simulation cell on the vorticity-gradient plane. Each cell corresponds to one finite element of the micro-micro calculation. The use of the Rectangular Cell algorithm [6] allows us to follow the start-up for a sufficiently long time (approximately three times longer than using a standard Parrinello-Rahman technique [1]) in order to observe the growth of the ordered domain. The appearance of the *hss* (hexagonal shear string) phase is evident from the plot.

Furthermore, the advance of the boundary between the ordered and the disordered colloidal dispersion is not linear but follows the curve depicted in

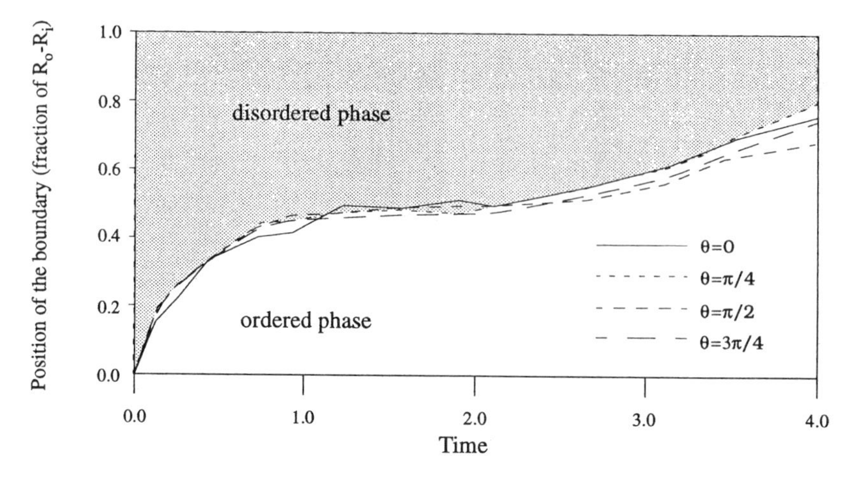

Fig. 7. Dependence of the location of the order-disorder boundary on the azimuthal position.

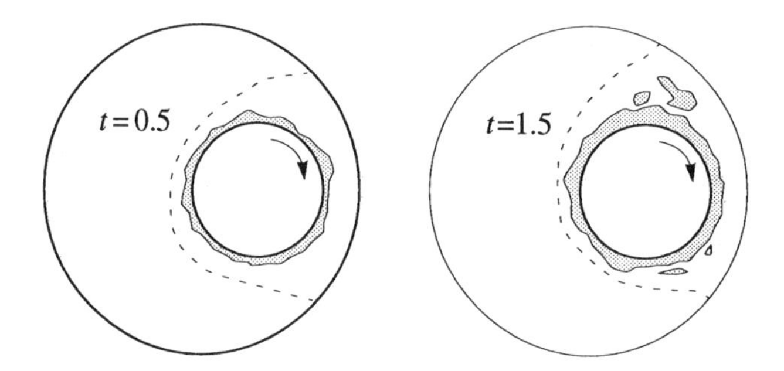

Fig. 8. Domain of existence of ordered phases in a colloidal dispersion in the journal bearing at large gap and eccentricity at $t = 0.5$ and $t = 1.5$ [6].

Fig. 7. After a period of rapid growth lasting about 1 s, the boundary stalls for a similar period before finally resuming its progress at a slower pace.

The plots in Fig. 8 represent a journal bearing at high eccentricity and gap in which the same colloidal dispersion is sheared starting from rest. The shaded area in Fig. 8 represents the region where the sheared dispersion has undergone the disorder-order transition and has a hexagonal shear string structure. The ordered domain grows from the inner cylinder outwards and eventually fills the whole area to the right of the separation line.

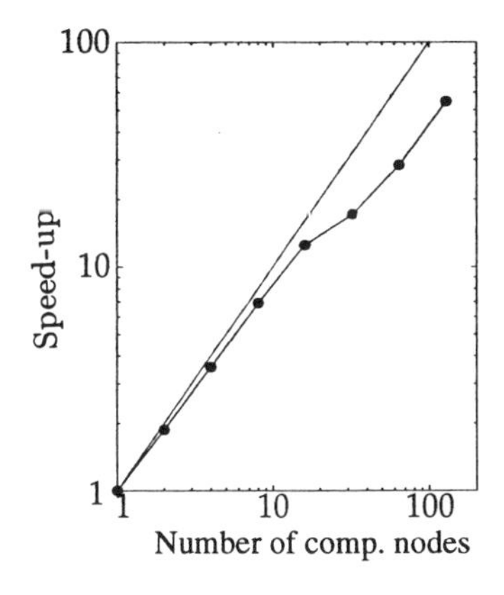

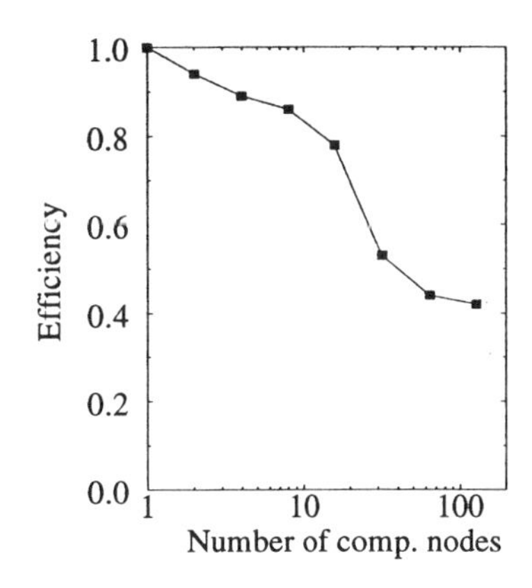

Fig. 9. Speed-up and efficiency as a function of the number of computational nodes.

Parallelisation and Vectorisation

Fortunately, the vast number of computations required in a CONNFFESSIT calculation lend themselves very well to implementation on modern computer architectures. The use of efficient data structures allows vectorised CONNFFESSIT codes to reach sustained performance of 65% peak and average vector length of 97.2% of the hardware vector length on standard supercomputers [8].

On the other hand, since the simulation of the molecules is performed at the element level, it is only natural to split the task over several processors and assign sub-sets of the global ensemble to each of them. The finite element calculation can run in scalar mode or be parallelised as well, depending on the fraction of time spent in it. Whereas efficient parallelisation and load balancing of finite element codes are a delicate task, the molecular part of CONNFFESSIT belongs to the class of embarrassingly parallel problems. Figure 9 shows the good parallelisation characteristics of a typical CONNFFESSIT code. In spite of being programmed by a very inexperienced parallel programme, a sizable 40% efficiency for 128 computational nodes can be reached.

Numerical and Algorithmic Aspects

The algorithmic details of 2D CONNFFESSIT calculations have been presented elsewhere [6, 8]. A distinctive feature of CONNFFESSIT calculations is that a large fraction of the actual computation time is not devoted to the kernel of finite element routines, but to routines that move, locate and sort molecules. This is a natural consequence of the very large number of molecules required in a CONNFFESSIT calculation. In the following except from a profile listing, bold-face names correspond to routines which manipulate molecules (FENE dumbbells in this case) and are completely separated from the finite element part. Names in italics are finite element routines from a public domain kernel. The two remaining routines are random number generators. Routines are sorted according to the fraction of time they consume. It may come as a surprise that 43% of CPU time is spent sorting and relocating molecules while only about 10% is devoted to the traditional finite element solver.

This unusual distribution of computational effort should be borne in mind when scaling CONNFFESSIT calculations to large-scale problems. Of the above routines, only two are model-dependent (dostepfene and stressfene). Together, they take up less than 10% of the total CPU time. The rest of the time is spent either in finite elements or in molecule manipulations which have nothing to do with the physics of the molecular model. This is excellent news: we can increase the complexity of the polymer dynamics by a factor of 10 and it will result in only a doubling of the overall effort. The stochastic simulation of some of the most complex models is only slightly more complex than that of FENE dumbbells. By significantly improving polymer dynamics (which is one of the main goals), it will not result in a significant increase in computation. In that respect, a CONNFFESSIT calculation for a FENE fluid contains all the sorting and relocation overheads and comparatively little

Time (%)	Procedure	Time (%)	Procedure
18.0	**sortdmb**	4.5	** **stressfene** **
11.0	**bitsort**	4.2	vcarry
8.8	**relocate 2**	4.1	** **dostepfene** **
6.3	**dumbvels**	2.6	gaussian
6.2	**move**	2.3	*slvasi*
6.2	*vematr*	2.0	**stressavg**
5.0	*ddot*	1.5	**shft_l**
5.0	**btest**	0.9	*4shap*

physics. As the molecular models become more sophisticated, CONNFFESSIT will become increasingly efficient and competitive.

Final Comments

In summary, it can be said that CONNFFESSIT represent a viable alternative to traditional continuum-mechanical numerical methods for the calculation of viscoelastic flows in complex geometries. It is based on the use of ensembles of molecular models to compute the stresses. CONNFFESSIT solutions are therefore inherently noisy.

When the viscoelastic fluid can be represented by a closed-form constitutive equation and the goal is to obtain velocity and stress fields, traditional calculations are certainly superior in speed and precision. However, CONNFFESSIT is currently the method of choice if, in addition to the macroscopic quantities, molecular information is desired as well or the molecular model does not allow a closed-form CE. Other interesting phenomena, such as chemical reaction (degradation, polymerisation) during flow, velocity gradient-induced migration of polymer molecules across streamlines and molecule-wall interactions can be incorporated in a CONNFFESSIT calculation with ease. Even molecular systems of interacting particles can be used (with limitations on the achievable strain).

The main drawback of CONNFFESSIT nowadays is the heavy computational burden. Tens of hours of supercomputer time, as well as a large memory, are required in order to reduce the strength of the fluctuations in the velocity to the 1% level. Fortunately, CONNFFESSIT computations can both be vectorised and parallelised with ease, and will run optimally on a new generation of vector-parallel machines. If a 10% noise level is acceptable, and due to the square-root dependence of fluctuation on ensemble size, satisfactory solutions can then be obtained on inexpensive workstations.

As the power of computational algorithms and hardware grows, the methods available to solve complex flow problems using a combination of microscopic and macroscopic techniques will play an increasingly important role in the arsenal of those interested in the dynamics of complex fluids.

Notation

De	Deborah number	(-)
H	spring constant	(N/m)

k Boltzmann's constant (J/K)
Q connector vector (m)
maximum value of connector vector (m)
T temperature (K)
t time (s)
$\boldsymbol{\nu}$ velocity (m/s)
$\boldsymbol{w}$ three-dimensional Wiener process (-)
$\bar{z}$ penalty parameter
η_s viscosity of Newtonian fluid (kg/m/s)
$\boldsymbol{\kappa}$ transposed velocity gradient $(\nabla\boldsymbol{\nu})^T$ (s^{-1})
λ time constant of FENE dumbbells (s)
$\boldsymbol{\pi}$ total stress tensor (Pa)
ρ density (kg/m^3)
$\boldsymbol{\tau}_p$ polymer contribution to the extra stress (Pa)
ζ friction coefficient (N.s./m)
ξ normalised radial coordinate (-)

References

1. Allen, M. P. & Tildesley, D. J. (1987) *Computer Simulation of Liquids.* Clarendon Press
2. Bird, R. B., Curtiss, C. F., Armstrong, R. C. & Hassager, O. (1987) *Dynamics of Polymeric Liquids: Kinetic Theory Vol. II.* John Wiley
3. Doi, M. & Edwards, S. F. (1992) *The Theory of Polymer Dynamics.* Oxford University Press
4. Feigl, K., Laso, M. & Öttinger, H. C. (1995) The CONNFFESSIT approach for solving a two-dimensional viscoelastic fluid problems. *Macromolecules* **28** (9), 3261
5. Keunings, R. (1989) *Fundamentals of Computer Modelling for Polymer Processing*, ed. C. L. Tucker, pp. 402. Carl Hanser Verlag
6. Laso, M. (1995) Habilitationsschrift Micro/macro computational methods for complex fluids. *ETH Zürich*
7. Laso, M. & Öttinger, H. C. (1993) Calculation of viscoelastic flow using molecular models: the CONNFFESSIT approach. *J. Non-Newtonian Fluid Mech.* **47** (1)
8. Laso, M., Picasso, M. & Öttinger, H. C. (1996) Two-dimensional, time-dependent viscoelastic flow calculations using CONNFFESSIT. *AlChEJ*, submitted
9. Laso, M., Picasso, M. & Öttinger, H. C. (1997) Calculation of flows with large elongational components: CONNFFESSIT calculation of the flow of a FENE fluid in a planar 10:1 contraction. In *Flexible Polymer Chains in Elongational Flow*, ed. H. H. Kausch & Q. T. Nguyen. Springer

10. Öttinger, H. C. (1996) *Stochastic Processes in Polymeric Fluids.* Springer
11. Purnode, B. (1996) *Vortices and Change of Type in Contraction Flows of Viscoelastic Fluids.* PhD Thesis, Université Catholique de Louvain
12. Warner, H. R. (1972) Kinetic theory and rheology of dilute suspensions of finitely extensible dumbbells. *Ind. Eng. Chem. Fundtls.* **11**, 379

Dynamics of Complex Fluids, pp. 88–105
ed. M. J. Adams, R. A. Mashelkar, J. R. A. Pearson and A. R. Rennie
Imperial College Press–The Royal Society, 1998

Chapter 7

Macroscopic and Mesoscale Approaches to the Computer Simulation of Viscoelastic Flows

R. KEUNINGS* AND P. HALIN

Division of Applied Mechanics, Université Catholique de Louvain,
B-1348 Louvain-la-Neuve, Belgium
**E-mail: rk@mema.ucl.ac.be*

We review the field of computational rheology with special emphasis on the numerical simulation of viscoelastic flows in complex geometries. The conventional approach, based on a macroscopic description of the underlying physics, has met with steady and significant progress over the last 15 years. A new approach, which combines a macroscopic formulation of the conservation laws with a mesoscale kinetic theory model of the flowing polymer, is emerging as a useful complementary tool. We discuss the main issues involved in micro-macro simulations, and describe some of our preliminary results obtained with nonlinear dumbbell models for dilute polymer solutions.

Introduction

Computer simulation techniques for viscoelastic liquids have been under active development over the last 20 years [1–4]. They allow, in principle, a detailed analysis of the flow of polymeric liquids in complex geometries that are relevant to laboratory or processing work. The objective of the scientist or engineer is to understand and possibly control the nonlinear coupling between the fluid's rheological behaviour, the flow-induced evolution of the micro-structure, the operating flow conditions, and the final mechanical properties [5–7]. In the case of polymeric liquids, major challenges include a huge number of micro-structural degrees of freedom, and a broad range of time scales (10^{-15} s $\rightarrow 10^{3}$ s) and length scales (10^{-10} m $\rightarrow$ 1 m). In view of the computer resources currently available, three basic modelling approaches can be considered for performing flow analysis of polymers in complex geometries. The first approach is that of atomistic modelling, with non-equilibrium molecular dynamics being the associated computational framework. Although an active research topic in fundamental polymer rheology studies, atomistic modelling has not yet been exploited in the simulation of complex flows. A second approach is based on a continuum-mechanical description of the underlying physics, namely the conservation principles and the fluid's rheological behaviour. The associated numerical framework is that of grid-based computational techniques for solving partial differential equations. The continuum approach has been under active development over the last 20 years and is reviewed in the next section. Finally, a third approach uses a coarse-grain description of the flow-induced molecular configurations which effectively separates very small time and length scales from the longer ones that are of interest in complex flows. The associated computational tools are stochastic numerical techniques (or Brownian dynamics). Over the last four years, complex flow simulations have been made feasible by combining such a microscopic description of the fluid's rheology with a conventional macroscopic formulation of the conservation principles. We review the micro-macro approach on pages 6 to 7, and describe our own preliminary results for dilute polymeric solutions on pages 7 to 8. Directions for future work are identified in the final section.

Macroscopic Computational Techniques

Until the mid-1970s, much of the research focus in theoretical rheology has been centred on the development of a rigorous mathematical framework for the *continuum mechanical* description of the rheology of polymeric liquids [5]. The objective is to formulate, by means of suitable *constitutive equations*, the intricate relationship between deformation and internal stresses which is responsible for many flow phenomena not seen in classical fluid mechanics [8]. A variety of macroscopic constitutive equations have thus been proposed which can be used together with the classical conservation laws (momentum, mass, energy) to perform flow analyses [9].

Since 1975 or so, a number of researchers worldwide realised the potential of computer simulation techniques in predicting the flow of polymers in complex geometries. They embarked upon the development of specialised numerical techniques for solving the nonlinear set of partial differential equations describing the macroscopic flow of viscoelastic fluids. These early works met with significant numerical and modelling difficulties, as reviewed in the monograph by Crochet, Davies and Walters [1]. The progress made in this field over the last 15 years has been steady and very impressive indeed [2–4]. The 1989 review by Keunings [3] details the numerical challenges associated with macroscopic viscoelastic simulations: (i) the constitutive models currently in use lead to governing equations of the mixed mathematical type (elliptic-hyperbolic) with possible local changes in type; (ii) stress boundary layers develop in many flow fields where the corresponding purely viscous Newtonian fluid mechanical problem is smooth (a major question is the relevance of these phenomena); (iii) stress singularities (for example, at die exit lips or re-entrant corners) are much stronger than in the Newtonian case, to the point where they are suspected of being physically unrealistic; and (iv) the nonlinear qualitative behaviour is very rich (for example, multiplicity of solutions, bifurcations) and it can be affected by the discretisation process.

As shown by an inspection of consecutive review papers [2–4], progress has been steady since the early days (circa 1978) along the path of *getting numbers* (not a trivial task in a nonlinear context!), assessing their *numerical accuracy* (with respect to mostly unknown exact solutions of the governing equations) and *physical relevance* (through detailed comparisons with experimental observations). There are now a number of vastly different numerical techniques that have provided mesh-converged solutions in benchmark flows. For example, the computational benchmark of the steady flow of an upper-convected Maxwell fluid past a sphere is now solved with great accuracy by a number of techniques (for example, finite elements, spectral, Lagrangian) up to a Weissenberg number of order 2 [10]. Beyond this limit, accurate numerical results are very difficult to obtain in view of the intense stress boundary layer close to the sphere surface [11]. The most successful macroscopic viscoelastic techniques are those that take into specific account the purely hyperbolic nature of the constitutive equations and the elliptic character of the momentum equations. They include mixed finite element formulations [12–16], high-order adaptive techniques [17–19] and Lagrangian methods [20–22]. For a complete description of the state of the art, the reader is referred to the review by Baaijens [4].

Computational work has been partly responsible for a better understanding of the dynamics of viscoelastic fluids in complex flows. Rheological issues of great interest can now be addressed, such as the evaluation in complex flows of constitutive models, the development of *computational rheometry* to aid data interpretation, and applications to polymer processing [23–31]. The computational community has spent a lot of effort on polymer solutions (the so-called Boger fluids), with good reasons. Much remains to be done there, and also on concentrated solutions and melts. For example, improved techniques for 3D flows, moving boundary problems or integral constitutive equations remain to be developed.

In summary, the progress made over the last 15 years in the field of macroscopic viscoelastic flow simulations has been very impressive indeed, to the extent that numerical techniques now provide a useful analysis tool in fundamental rheology or polymer processing applications. Although there is still much room for further numerical and algorithmic developments in this field, its very success reveals that improved *modelling* of the rheological behaviour is now required. Further progress will not come from continuum mechanical arguments alone. Here, the input of kinetic theory (or coarse-grain molecular modelling) appears essential, as discussed in the next section.

Micro-Macro Approach

As stated in the introduction, the direct simulation of complex polymer flows by means of molecular dynamics techniques is likely to remain out of reach for many years to come. The use of the simpler coarse-grain models of kinetic theory [6] is becoming feasible, with the availability of powerful parallel computers. For illustrative purposes, let us consider the case of polymer solutions. Kinetic models are based on mechanical contrivances (for example, bead-rod-spring models) that describe intra-molecular interactions. The effects of rapidly fluctuating degrees of freedom associated with very short length scales are taken into account through *random forces* which perturb the temporal evolution of the slower configurational degrees of freedom. (This is similar to Langevin's analysis of Brownian motion.) Since molecular velocities fluctuate much more rapidly than molecular conformations, one can assume equilibration in momentum space. Stochastic differential equations (SDE) are thus obtained that describe the evolution of the coarse-grained polymer conformations. Alternatively, one can write a *deterministic* partial differential equation (the Fokker-Planck equation or FPE) for the temporal evolution of the probability density in conformation space. Kinetic models have been proposed for polymer solutions and melts [6].

The continuum mechanical and kinetic approaches are best viewed as *complementary.* Ideally, one would wish to use in complex flow simulations a macroscopic constitutive equation derived from kinetic theory. Unfortunately, only the very crudest kinetic models yield equivalent macroscopic constitutive equations that could be used in flow simulations. In some cases, such as the Doi-Edwards theory [32] based on de Gennes' reptation concept for melts, constitutive equations can be derived from the underlying kinetic model using appropriate (closure) mathematical approximations. How much of the physics is lost in the approximation process remains, however, an open issue, especially in complex flows. Our recent work [33] using kinetic models for dilute polymer solutions (FENE dumbbells) shows that mathematical approximations necessary to derive macroscopic constitutive equation can have a significant impact on the statistical and rheological properties of the resulting model.

If further progress is to come from kinetic models, one should thus find a way of incorporating such models in flow simulation codes *without* any further mathematical simplifications. A possible approach is to solve, numerically, the FPE for the configurational probability density associated with the kinetic model. Relevant macroscopic variables (such as the polymer contribution to the stress tensor) are then computed as statistical averages of the polymer configurations. To date, this approach is limited mostly to simple flows and kinetic models with few degrees of freedom. Brownian dynamics or stochastic simulation techniques provide a powerful alternative [34]. The latter draw on the mathematical equivalence between FPE's and SDE's. For each FPE derived for a particular kinetic model, one can associate a SDE whose solution, i.e., a stochastic process, has a transition probability that satisfies the corresponding FPE [34]. Thus, instead of solving the deterministic FPE for the distribution function, which can be a formidable if not impossible task, one can compute a number of realisations of the solution to the SDE by means of suitable numerical techniques [35], which can be considerably simpler. Macroscopic fields of interest are then obtained by averaging over an ensemble of realisations of the stochastic process.

Over the last few years, the stochastic approach has been used successfully in the analysis of simple flows of polymer solutions and melts [34]. The analysis of complex flows can be performed using a *micro-macro* approach that combines classical macroscopic techniques with stochastic simulations. A first step in that direction has been taken recently by Laso and Öttinger [36]. The basic idea is to solve the macroscopic conservation laws by means of appropriate grid-based numerical techniques (for example, finite elements) with the polymer stress being computed through a stochastic simulation. The work of Öttinger and his collaborators [36–38] is very promising indeed. Much research remains to be done, however, before the micro-macro approach to polymer flow simulations becomes an established tool for fundamental and applied rheological studies.

FENE Dumbbell Model for Dilute Polymer Solutions

We describe below our preliminary results on stochastic simulations of the flow of dilute polymer solutions in a complex geometry. We focus on the problem of computing the polymer contribution to the stress tensor for specified flow kinematics. The selected kinetic model is known as the Warner Finitely Extensible Nonlinear Elastic (FENE) dumbbell model [6], where the polymer solution is represented as a suspension of non-interacting dumbbells in a Newtonian solvent. Each dumbbell consists of two beads connected by a nonlinear spring that models intra-molecular interactions. As they move through the solvent, the beads experience thermal Brownian motion, Stokes drag and the spring force. For FENE dumbbells, the spring force reads

$$\boldsymbol{F}^c = \frac{H}{1 - \boldsymbol{Q}^2/Q_0^2}\,\boldsymbol{Q}\,, \tag{1}$$

where H is a spring constant, $\boldsymbol{Q}$ is the vector connecting the two beads, and Q_0 is the maximum spring length. It is only in the limit of a linear Hookean spring, ($Q_0 \to \infty$), that it is possible to derive a constitutive equation for the polymer stress *without* any closure approximation. For finite values of Q_0, however, it is impossible to obtain a constitutive equation that is mathematically equivalent to the FENE kinetic theory. The use of the following closure approximation, due to Peterlin,

$$\boldsymbol{F}^c = \frac{H}{1 - \langle \boldsymbol{Q}^2\rangle/Q_0^2}\,\boldsymbol{Q}\,, \tag{2}$$

yields the FENE-P constitutive equation [6]. Here, the angular brackets denote the configuration space average, $\langle\cdot\rangle = \int \cdot\psi(\boldsymbol{Q},t)\;d\boldsymbol{Q}$, where ψ is the distribution function.

It is usually believed that the FENE-P constitutive model is a good approximation of the FENE kinetic theory, particularly in elongational flows where the distribution of dumbbell configurations can be highly localised. The

Peterlin approximation is indeed accurate in *steady-state* extension. In a recent paper [33], we have shown by means of stochastic simulations of *transient* elongational flows that the Peterlin approximation has a significant impact on the statistical and rheological properties of the model. The new results described below reveal that the same conclusion holds true for complex kinematics, which by nature are always transient in the *Lagrangian* sense.

The FENE model involves a relaxation time, $\lambda_H = \zeta/4H$, and a dimensionless finite extensibility parameter, $b = HQ_0^2/kT$, where ζ is the friction coefficient, k is Boltzmann's constant and T is the absolute temperature. In this work, we specify $b = 50$. All subsequent equations and computational results are given in dimensionless form. The connector vector $\boldsymbol{Q}$, the time t and the velocity gradient $\boldsymbol{\kappa}$ are made dimensionless with $\sqrt{kT/H}$, λ_H and λ_H^{-1}, respectively. The magnitude of the dimensionless velocity gradient $\boldsymbol{\kappa}$ can thus be viewed as a Weissenberg number. The polymer stress $\boldsymbol{\tau}_p$ is made dimensionless with nkT, where n is the dumbbell number density. Finally, we define, for the sake of convenience, the notation $h(\kappa) = 1/(1 - \kappa/b)$.

The FPE that describes the evolution of the configuration distribution function, $\psi(\boldsymbol{Q}, t)$, for FENE dumbbells reads

$$\frac{\partial \psi}{\partial t} = -\frac{\partial}{\partial \boldsymbol{Q}} \cdot \left\{ \left[\boldsymbol{\kappa} \cdot \boldsymbol{Q} - \frac{1}{2} \boldsymbol{F}^c(\boldsymbol{Q}) \right] \psi \right\} + \frac{1}{2} \frac{\partial}{\partial \boldsymbol{Q}} \cdot \frac{\partial}{\partial \boldsymbol{Q}} \psi , \tag{3}$$

with $\boldsymbol{F}^c(\boldsymbol{Q}) = h(\boldsymbol{Q}^2)\, \boldsymbol{Q}$. The FENE polymer stress is provided by Kramer's expression

$$\boldsymbol{\tau}_p = \langle \boldsymbol{Q} \boldsymbol{F}^c \rangle - \boldsymbol{\delta} , \tag{4}$$

where $\boldsymbol{\delta}$ is the unit tensor; the stress, $\boldsymbol{\tau}_p$, vanishes at equilibrium (no macroscopic flow).

Solving the FPE (3) is mathematically equivalent to solving the following SDE:

$$d\boldsymbol{Q}_t = \left\{ \boldsymbol{\kappa} \cdot \boldsymbol{Q}_t - \frac{1}{2} \boldsymbol{F}^c(\boldsymbol{Q}_t) \right\} dt + d\boldsymbol{W}_t , \tag{5}$$

where $\boldsymbol{W}_t$ is the three-dimensional Wiener process, that is, an idealisation of Brownian motion. Indeed, Eq. (5) is an evolution equation for the Markov process $\boldsymbol{Q}_t$ whose probability density, ψ, is solution to the FPE (3).

In the stochastic simulation approach, one solves Eq. (5) numerically along flow trajectories for a large number N_d of dumbbells, namely for N_d individual realisations $\boldsymbol{Q}_t^{(i)}$ of the stochastic process. Macroscopic observables of interest are then approximated as ensemble averages. For example, the polymer stress $\boldsymbol{\tau}_p$ is given by

$$\boldsymbol{\tau}_p(t) \approx \frac{1}{N_d} \sum_{i=1}^{N_d} \boldsymbol{Q}_t^{(i)} \boldsymbol{F}^c \left(\boldsymbol{Q}_t^{(i)} \right) - \boldsymbol{\delta} \,. \tag{6}$$

The statistical error implied in Eq. (6) is of order $N_d^{-1/2}$.

For FENE-P dumbbells, Eqs. (3) to (6) remain valid but with a consistent pre-averaging of the connector force: $\boldsymbol{F}^c(\boldsymbol{Q}) = h\left(\langle \boldsymbol{Q}^2 \rangle\right) \boldsymbol{Q}$. It is then possible to derive an evolution equation for the covariance tensor $\boldsymbol{A} = \langle A\boldsymbol{QQ} \rangle$:

$$\frac{d\boldsymbol{A}}{dt} - \boldsymbol{\kappa} \cdot \boldsymbol{A} - \boldsymbol{A} \cdot \boldsymbol{\kappa}^\dagger = \boldsymbol{\delta} - h(tr(\boldsymbol{A}))\boldsymbol{A} \,, \tag{7}$$

whose solution $\boldsymbol{A}$ yields the FENE-P polymer stress through $\boldsymbol{\tau}_p = h(tr(\boldsymbol{A}))\boldsymbol{A} - \boldsymbol{\delta}$. Equation (7) thus gives a macroscopic constitutive equation for the FENE-P model.

In the next section, we compare FENE and FENE-P polymer stresses obtained for specified kinematics $\boldsymbol{\kappa}$ corresponding to the steady flow between eccentric rotating cylinders. For FENE dumbbells, we solve the SDE (5) numerically along the flow trajectories by means of the semi-implicit predictor-corrector scheme due to Öttinger [34]. Tracking of the dumbbells along the streamlines is performed by means of a fourth-order Runge-Kutta scheme. For FENE-P dumbbells, we either solve the associated SDE, as for the FENE model, using the Euler-Maruyama scheme [34] or, equivalently, in the limit of vanishing discretisation errors, we solve the macroscopic evolution Eq. (7) for $\boldsymbol{A}$ by means of the Streamline Upwinding 4×4 finite element method. More information on the numerical techniques can be found in Halin's PhD thesis [39].

Numerical Results

The flow kinematics are illustrated in Fig. 1. We consider the steady-state planar flow between two eccentric cylinders: the inner cylinder (of radius R_i) is rotating at constant angular speed while the outer cylinder (of radius R_o) is fixed. The dimensionless eccentricity measure, $\varepsilon = e/(R_o - R_i)$, is equal to 0.67, where e is the eccentricity. Figure 1 illustrates a number of flow trajectories obtained from a separate finite element solution of the Stokes flow equations. There is a large re-circulation zone where the fluid particles are moving relatively slowly, while the flow near the rotating cylinder is nearly viscometric. In

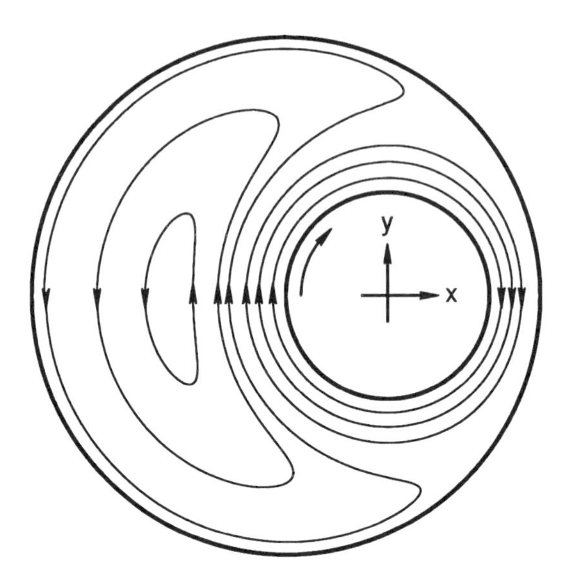

Fig. 1. Sample flow trajectories for steady-state Stokes flow between eccentric cylinders.

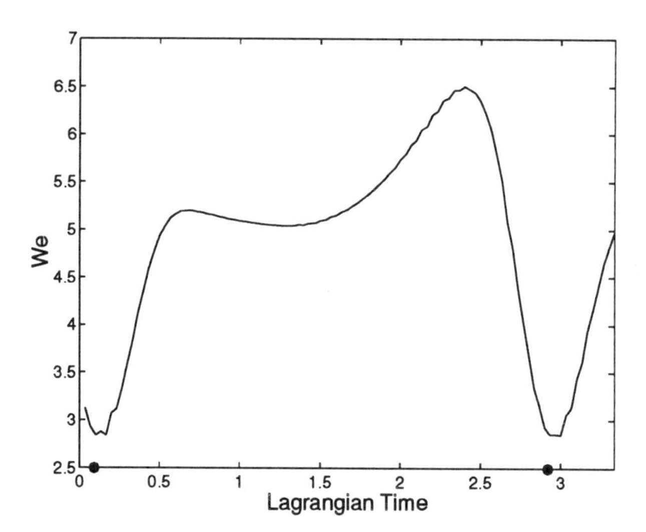

Fig. 2. Local Weissenberg number as a function of travel time along a flow trajectory located near the rotating (inner) cylinder. The highest value corresponds to the location of the thinnest gap between the two cylinders. A full rotation is achieved in about three dimensionless time units.

the sequel, we consider results computed along a trajectory that is close to the rotating cylinder. Figure 2 shows the evolution of the local Weissenberg number We along that trajectory as a function of the fluid particles Lagrangian (travel) time. Clearly, the fluid elements experience a transient deformation process as they travel along the (Eulerian) steady-state flow lines. As a check on the numerical accuracy of the stochastic approach, we compare in Fig. 3 FENE-P stochastic results with those obtained by means of the macroscopic finite element calculations. The stochastic simulation uses $N_d = 10,000$ dumbbells and assumes an equilibrium distribution (zero stress) initially. An inspection of Fig. 3 reveals an excellent agreement between the stochastic and macroscopic results once the Eulerian steady-state is achieved in the stochastic simulation.

Finally, we focus on the impact of the Peterlin closure approximation. Figure 4 shows the computed mean square polymer extension, $tr\langle \boldsymbol{AQQ} \rangle$, obtained with the FENE-P and FENE models. One should note that the maximum possible mean square extension is $b = 50$. We find that both the dynamics and the level of extension are significantly affected by the closure approximation. A similar comment applies to the polymer stress results shown in Fig. 5.

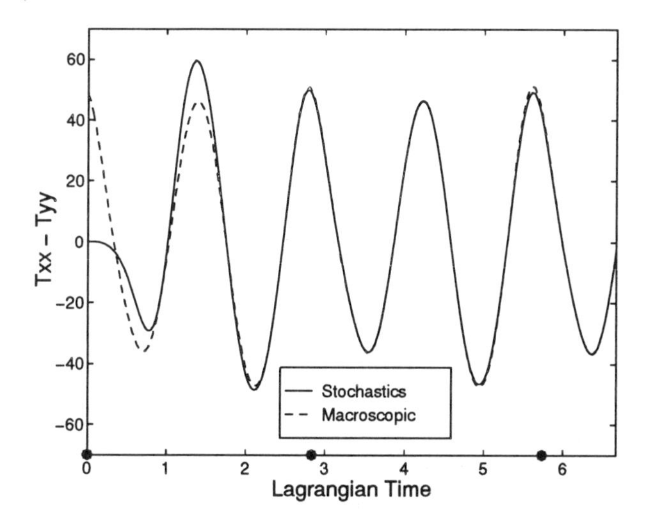

Fig. 3. FENE-P dumbbells: Normal stress difference as a function of travel time along a flow trajectory located near the rotating (inner) cylinder. Comparison of results obtained with stochastic simulation and macroscopic finite element technique.

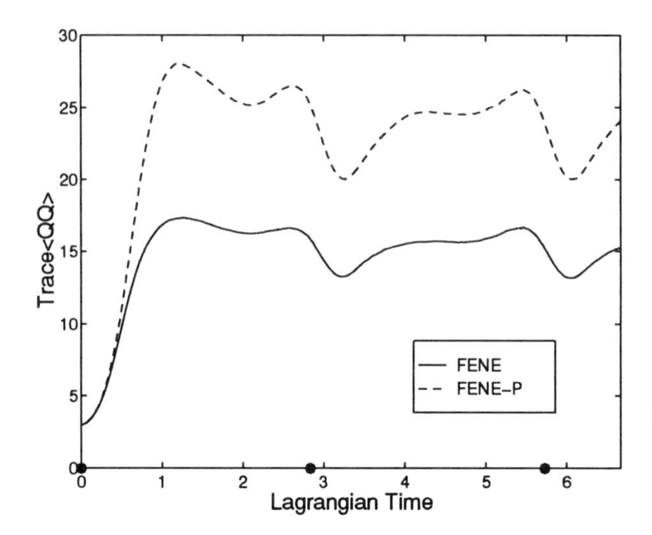

Fig. 4. Average square polymer extension as a function of travel time along a flow trajectory located near the rotating (inner) cylinder. Comparison of results obtained for FENE and FENE-P models.

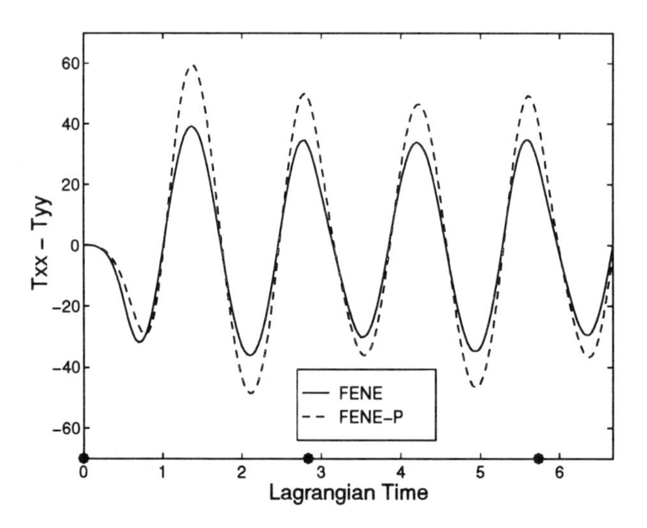

Fig. 5. Normal stress difference as a function of travel time along a flow trajectory located near the rotating (inner) cylinder. Comparison of results obtained for FENE and FENE-P models.

Additional insights into the Peterlin approximation can be gained from Fig. 6, where we compare the radial distribution (that is, the distribution of dumbbell length) for the FENE and FENE-P models. These results are computed at a particular point in space close to the rotating cylinder, in the form of normalised histograms of the N_d realisations of $\sqrt{\boldsymbol{Q}_t^2}$. The equilibrium radial distribution is also shown for comparison. In the case of the FENE model, the flow kinematics deform the equilibrium distribution, that is, the dumbbells get extended but their length is always below the upper limit $\sqrt{b} \approx 7$. The FENE-P dumbbells behave in a very different manner. Indeed, many *individual* dumbbells do cross the $\sqrt{b}$ limit as they are deformed by the flow field. As a further check on the FENE-P stochastic results, we have computed the radial distribution in a different way by taking into account the fact that the FENE-P distribution function is Gaussian with zero mean. Therefore, it is fully determined once its second moment $\boldsymbol{A}$ is known. Thus, from the finite element results for $\boldsymbol{A}$, we can determine the distribution function which is then contracted to yield the radial distribution [33]. An excellent agreement with the stochastic result is obtained.

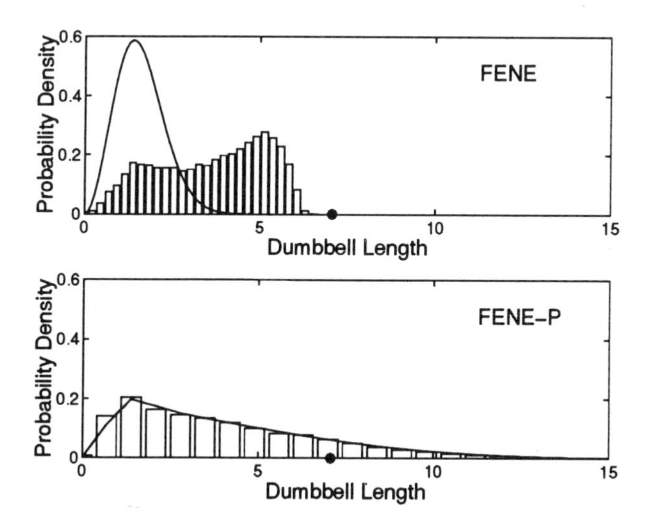

Fig. 6. Radial distribution computed near the rotating (inner) cylinder at the point of thinnest gap. Comparison of FENE and FENE-P models. The dot on the horizontal axis marks the $\sqrt{b}$ FENE upper limit. The upper figure also shows the equilibrium distribution (continuous curve) while the lower figure includes the radial FENE-P distribution computed using the covariance tensor $\boldsymbol{A}$ (continuous curve).

The above results are an extension of those obtained by Keunings [33] for specified homogeneous kinematics. We conclude that in a complex flow field, where the polymer molecules experience a time-dependent velocity gradient (at least in the Lagrangian sense) as they are convected along the flow trajectories, drastic differences are obtained in simulations using the FENE kinetic theory and its approximate FENE-P macroscopic version. A similar conclusion holds in coupled simulations, where the polymer stress is allowed to alter the kinematics through the momentum equations [36, 38–39]. Clearly, evaluation and further improvement of kinetic models should avoid the use of questionable closure approximations. Stochastic simulations are a useful tool for this purpose.

Conclusions

The progress made over the years in the field of viscoelastic flow computations, which we have briefly reviewed in the present paper, is indeed impressive. Although computing viscoelastic flows remains a formidable challenge, macroscopic numerical techniques have become a useful tool for fundamental or applied rheological studies. Further work on macroscopic techniques is needed in a number of directions, including accurate yet feasible methods for complex 3D flows, improved techniques for integral constitutive equations, schemes for moving boundary problems, and tools for the temporal stability analysis of complex flows.

The very progress of macroscopic viscoelastic simulations has revealed an equally important need for improved modelling of the rheology of polymeric fluids. Here, the input of kinetic theory appears essential. We have seen that stochastic techniques are a mathematically sound and accurate tool that allows the direct use of kinetic theory models in flow simulations, thereby avoiding the need for questionable closure approximations. As for macroscopic simulations, work on stochastic modelling has mostly been done for dilute solutions using relatively simple dumbbell models. More realistic theories (for solutions and melts) are now being considered, as well as detailed micro-physics phenomena (for example, wall rheology) [4].

Finally, micro-macro simulation techniques combining a macroscopic description of the conservation laws with a kinetic theory for the coarse-grain polymer configurations is an emerging methodology that is being developed by a number of research groups worldwide. The computer resources involved in micro-macro simulations are even more significant than with conventional macroscopic methods. The exploitation of parallel computing architectures [40] should alleviate this issue. As far as modelling is concerned, the phenomenology is now at the level of the kinetic models rather than that of the constitutive equation. It is hoped, but remains to be shown, that coarse-grain modelling will increase our understanding of polymer dynamics in complex flows.

Acknowledgements

The work of Roland Keunings was performed while on sabbatical leave at the University of Cambridge (UK), as visitor at the Department of Applied Mathematics and Theoretical Physics, and as a participant in the research programme *Dynamics of Complex Fluids* at the Isaac Newton Institute for Mathematical Sciences. RK wishes to thank his hosts, E. J. Hinch and J. M. Rallison (DAMTP), and J. R. A. Pearson, K. Walters and T. C. B. McLeish (INIMS), for their kind hospitality and a very stimulating environment. We also wish to acknowledge partial support of the BRITE/EURAM project MPFLOW CT96-0145.

Discussion

M. Mackley I do not agree with the view that the "job is done" in relation to continuum numerical solutions of processing problems.

R. Keunings I fully agree with your statement, and hope that my message regarding the state of the art of macroscopic simulations has not been misunderstood. As stated explicitly during my talk, many difficult challenges remain in the continuum numerical framework. It is also true, however, that the very success of available techniques shows the need for improved modelling. To this end, kinetic theory, as implemented via stochastic simulations, has great potential.

M. Mackley It is entirely valid to make the transition from continuum to Brownian structures. It is *a* way forward but not necessarily *the* way forward, because of meso-structures between these length scales. Many complex fluids have micro-structures on an intermediate length scale.

R. Keunings Here again, I fully agree with your statement. To my knowledge, intermediate length scales have not been described yet in micro-macro simulations. Innovative work is needed in that direction as well.

R. Baranger Is somebody working on adaptive techniques for macroscopic viscoelastic simulations?

R. Keunings I know indeed of the work by my UCL colleague Dr. V. Legat with his PhD student V. Warichet [11, 19]. They have developed a so-called *h-p* finite element algorithm for differential viscoelastic models. Their high-Weissenberg number results for the sphere benchmark are, I believe, of impressive numerical accuracy.

R. Baranger I wish to comment that very little is known on the mathematical analysis of continuum viscoelastic equations (well-posedness, weak solutions, stability, boundary layers), as well as on the numerical analysis of available techniques.

R. Keunings I should agree with your comment. I refer the audience to published reviews [3–4] where such theoretical work is referred to.

D. Peric In relation to the present state of finite element development, it should be emphasised that there still exists a number of issues which require substantial research input and clarification, even for the so-called "simple fluids", not to mention the complex ones. As far as adaptivity is concerned, significant advances have recently been made that include sharp error estimates for nonlinear problems, such as plasticity and viscoelasticity, and also associated numerical procedures. Important advances have also been made in the development of efficient equation solvers which rely on domain decomposition and/or multi-grid techniques, and their implementation to solutions of complex nonlinear problems in both fluid and solid mechanics. The list may be made longer. However, it illustrates the fact that there is still scope for substantial advancement in finite element methodology.

R. Keunings On the basis of my previous work on finite element techniques, I could not agree more with your comment. As far as adaptivity and error estimates are concerned, I refer you again to Legat's work. My research group has also been active on parallel decomposition techniques applied to computational mechanics [40–41]. Yes, there is indeed a future for researchers on finite elements!

References

1. Crochet, M. J., Davies, A. R. & Walters, K. (1984) *Numerical Simulation of Non-Newtonian Flow*. Amsterdam: Elsevier

2. Crochet, M. J. (1989) Numerical simulation of viscoelastic flow: a review. *Rubber Chem. Techn.* **62** (3), 426–455
3. Keunings, R. (1989) Simulation of viscoelastic flow. In *Computer Modelling for Polymer Processing*, ed. C. L. Tucker, pp. 404–469, Munich: Hanser
4. Baaijens, F. P. T. (1998) Review paper in preparation, to appear in *J. Non-Newtonian Fluid Mech.*
5. Bird, R. B., Armstrong, R. C. & Hassager, O. (1987) *Dynamics of Polymeric Liquids. Vol. 1: Fluid mechanics*, 2nd edn. New York: John Wiley & Sons
6. Bird, R. B., Curtiss, C. F., Armstrong, R. C. & Hassager, O. (1987) *Dynamics of Polymeric Liquids. Vol. 2: Kinetic theory*, 2nd edn. New York: John Wiley & Sons
7. Gelin, B. R. (1994) *Molecular Modelling of Polymer Structures and Properties.* Munich: Carl Hanser Verlag
8. Boger, D. V. & Walters, K. (1993) *Rheological Phenomena in Focus.* Amsterdam: Elsevier
9. Larson, R. G. (1988) *Constitutive Equations for Polymer Melts and Solutions.* Boston: Butterworth
10. Brown, R. A. & McKinley, G. H. (1994) Report on the VIII international workshop on numerical methods in non-Newtonian flow. *J. Non-Newtonian Fluid Mech.* **52**, 407–413
11. Warichet, V. & Legat, V. (1997) Adaptive *hp* finite element prediction of the drag correction factor for a Maxwell-B fluid. *J. Non-Newtonian Fluid Mech.* **73**, 95–114
12. Baaijens, F. P. T. (1994) Application of low-order discontinuous Galerkin methods to the analysis of viscoelastic flows. *J. Non-Newtonian Fluid Mech.* **52**, 37–57
13. Debae, F., Legat, V. & Crochet, M. J. (1994) Practical evaluation of four mixed finite element methods for viscoelastic flow. *J. Rheol.* **38** (2), 421–441
14. Guénette, R. & Fortin, M. (1995) A new mixed finite element method for computing viscoelastic flows. *J. Non-Newtonian Fluid Mech.* **60**, 27–52
15. Brown, R. A., Szady, M. J., Northey, P. J. & Armstrong, R. C. (1993) On the numerical stability of mixed finite element methods for viscoelastic flows governed by differential constitutive equations. *Theor. and Comp. Fluid Dynamics* **5**, 77–106
16. Szady, M. J. *et al.* (1995) A new mixed finite element method for viscoelastic flows governed by differential constitutive equations. *J. Non-Newtonian Fluid Mech.* **59**, 215–243
17. Fan, Y.-R. & Crochet, M. J. (1995) High-order finite element methods for steady viscoelastic flows. *J. Non-Newtonian Fluid Mech.* **57**, 283–311
18. Talwar, K. K., Khomani, B. & Ganpule, H. K. (1994) A comparative study of higher- and lower-order finite element techniques for computation of viscoelastic flows. *J. Rheol.* **38** (2), 255–289
19. Warichet, V. & Legat, V. (1996) Adaptive *hp*-finite element viscoelastic flow calculations. *Comput. Methods Appl. Mech. Engng.* **136**, 93–110

20. Harlen, O. G., Rallison, J. M. & Szabo, P. (1995) A split Lagrangian-Eulerian method for simulating transient viscoelastic flows. *J. Non-Newtonian Fluid Mech.* **60**, 81–104
21. Rasmussen, H. K. & Hassager, O. (1995) Simulation of transient viscoelastic flow with second-order time integration. *J. Non-Newtonian Fluid Mech.* **56**, 65–84
22. Yuan, X. F., Ball, R. C. & Edwards, S. F. (1994) Dynamical modelling of viscoelastic extrusion flows. *J. Non-Newtonian Fluid Mech.* **54**, 423–435
23. Ahmed, R., Liang, R. & Mackley, M. R. (1995) The experimental observation and numerical prediction of planar entry flow and die swell for molten polyethylene. *J. Non-Newtonian Fluid Mech.* **59**, 129–153
24. Hartt, W. H. & Baird, D. G. (1996) The confined flow of polyethylene melts past a cylinder in a planar channel. *J. Non-Newtonian Fluid Mech.* **65**, 247–268
25. Baaijens, J. H. W., Peters, G. W. M., Baaijens, F. P. T. & Meijer, H. E. H. (1995) Viscoelastic flow past a confined cylinder of a polyisobutylene solution. *J. Rheol.* **39**, 1243–1277
26. Quinzani, L. M., Armstrong, R. C. & Brown, R. A. (1995) Use of coupled birefringence and LDV studies of flow through a planar contraction to test constitutive equations for concentrated polymer solutions. *J. Rheol.* **39** (6), 1201–1227
27. Purnode, B. & Crochet, M. J. (1996) Flows of polymer solutions through contractions. Part 1: flows of polyacrylamide solutions through planar contractions. *J. Non-Newtonian Fluid Mech.* **65**, 269–289
28. Crochet, M. J., Debbaut, B., Keunings, R. & Marchal, J. M. (1992) Polyflow: a multi-purpose finite element program for continuous polymer flows. In *Computer Modelling for Extrusion and Other Continuous Polymer Processes*, ed. K. T. O'Brien, pp. 25–50, Munich: Carl Hanser Verlag
29. Kiriakidis, D. G. *et al* (1993) A study of stress distribution in contraction flows of an LLDPE melt. *J. Non-Newtonian Fluid Mech.* **47**, 339–356
30. Barakos, G. & Mitsoulis, E. (1995) Numerical simulation of extrusion through orifice dies and prediction of Bagley correction for an IUPAC-LDPE melt. *J. Rheol.* **39** (1), 193–209
31. Goublomme, A. & Crochet, M. J. (1993) Numerical prediction of extrudate swell of a high-density polyethylene: further results. *J. Non-Newtonian Fluid Mech.* **47**, 281–287
32. Doi, M. & Edwards, S. F. (1986) *The Theory of Polymer Dynamics.* Oxford: Clarendon Press
33. Keunings, R. (1997) On the Peterlin approximation for finitely extensible dumbbells. *J. Non-Newtonian Fluid Mech.*, **68**, 85–100
34. Öttinger, H. C. (1996) *Stochastic Processes in Polymeric Fluids.* Berlin: Springer-Verlag
35. Kloeden, P. E. & Platen, E. (1992) *Numerical Solution of Stochastic Differential Equations.* Berlin: Springer-Verlag
36. Laso, M. & Öttinger, H. C. (1993) Calculation of viscoelastic flow using molecular models: the CONNFFESSIT approach. *J. Non-Newtonian Fluid Mech.* **47**, 1–20

37. Feigl, K., Laso, M. & Öttinger, H. C. (1995) CONNFFESSIT approach for solving a two-dimensional viscoelastic fluid problem. *Macromolec.* **28**, 3261–3274
38. Laso, M. (1995) *Micro/Macro Computational Methods for Complex Fluids.* E. T. H. Zürich, Switzerland: Habilitationsschrift für die Privat-Dozentur
39. Halin, P. PhD Thesis, Université catholique de Louvain, in preparation
40. Keunings, R. (1995) Parallel finite element algorithms applied to computational rheology. *Computers Chem. Engng.* **19** (6/7), 647–669
41. Zone, O., Vanderstraeten, D., & Keunings, R. (1995) A parallel finite element solver based on automatic domain decomposition applied to viscoelastic flows. In *Parallel Computational Fluid Dynamics: New Algorithms and Applications*, ed. N. Satofuka, J. Periaux & A. Ecer, pp. 297–304. Amsterdam: Elsevier

Dynamics of Complex Fluids, pp. 106–123
ed. M. J. Adams, R. A. Mashelkar, J. R. A. Pearson & A. R. Rennie
Imperial College Press–The Royal Society, 1998

Chapter 8

Simulation of the Flow of an Oldroyd-B Fluid Using Brownian Configuration Fields

M. A. HULSEN, A. P. G. VAN HEEL AND B. H. A. A. VAN DEN BRULE*

Delft University of Technology,
Laboratory for Aero-and Hydrodynamics,
Rotterdamseweg 145, 2628 AL Delft, The Netherlands
**E-mail: b.h.a.vandenbrule@siep.shell.com*

We present a new approach for calculating viscoelastic flows. The polymer stress is not determined from a closed-form constitutive equation, but from a microscopic model. In this description, we replace the collection of individual polymer molecules by an ensemble of configuration fields representing the internal degrees of freedom of the polymers. Similar to the motion of real molecules, these configuration fields are convected and deformed by the flow and are subjected to Brownian motion. We incorporate this field description in a finite element calculation. An important advantage of our approach is that the difficulties associated with particle tracking of individual molecules are circumvented. In order to validate our approach and to demonstrate its robustness, we present the results for the start-up of planar flow of an Oldroyd-B fluid past a cylinder between two parallel plates. The results are very promising. We find excellent agreement between the results of the configuration field formulation and those obtained using a closed-form constitutive equation. Moreover, the microscopic method appears to be more robust than the conventional macroscopic technique.

Introduction

The field of viscoelastic fluid research has recently shown some exciting developments which offer opportunities to study systems that are well beyond the limitations inherent to the conventional macroscopic approach. These limitations were due essentially to the need for a closed-form constitutive equation (CE) to complement the conservation laws of mass and momentum. This CE can be founded either on a macroscopic approach, that is, within the framework of continuum mechanics, or on a microscopic description using kinetic theory. Unfortunately, however, many of the intuitively appealing microscopic models cannot be cast into a closed form. A comprehensive review of both approaches can be found in [1, 2].

Recently, a promising technique has been introduced to avoid the limitations described above, by simply *bypassing* the need for a CE. The essential idea of this so-called CONNFFESSIT approach is to combine traditional finite element techniques and Brownian dynamics simulations. In contrast to a conventional finite element approach, however, the polymer contribution to the stress, needed in the finite element calculation, is not calculated from a CE, but instead from the configuration of a large ensemble of model polymers [3–5]. The time evolution of this ensemble is calculated using Brownian dynamics.

This approach has been shown to be able to reproduce the known analytical results for start-up flow of a Maxwell fluid for 1-D plane Couette flow [3] and start-up of 2-D flow between two concentric cylinders [6]. For problems for which no analytic solution is at hand, the method has been tested by comparing its results with those obtained using a purely macroscopic finite element calculation. The latter approach was applied to the problem of steady flow of an Oldroyd-B fluid in an abrupt 4:1 axisymmetric contraction [7]. In this case, the stress was either calculated from an ensemble of Hookean dumbbells or from the equivalent macroscopic Oldroyd-B equation.

The most promising feature of the new approach, however, is its ability to calculate fluid flows using models for which no closed-form CE is known, such as, for example, a FENE model [3, 6]. Additional advantages are that it is very easy to change models and that, in principle, it is possible to incorporate effects such as chemical reactions, polydispersity and polymer migration.

In this paper, we present a new approach to solve time-dependent viscoelastic flows based on an ensemble of *Brownian configuration fields.* The basic idea of this description is to replace the collection of individual molecules by an ensemble of configuration fields representing the internal degrees of freedom (configuration) of the polymer molecules. An important feature of the configuration fields, as will be explained in the next section, is that in this approach the configurations are defined at *every* point of the domain.

Analogous to centre-of-mass convection and evolution of the configuration of a real polymer molecule, the fields in this approach are convected by the flow and are subjected to Brownian motion. Moreover, given the way the Brownian motion is implemented, the configuration fields are continuous. As we shall see on pages 8 to 11, these properties render the method particularly suitable for implementation in a finite element formulation of the problem. A major advantage of the field description is the ease with which convection is incorporated into the simulation, thus avoiding the difficulties associated with the tracking of individual particles.

In order to validate the configuration field method and to demonstrate its robustness, we consider, on pages 11 to 17 the start-up of planar flow of an Oldroyd-B fluid past a cylinder between two parallel plates. We solved this problem using two different methods: first by an entirely macroscopic calculation and then by using the configuration field formulation. In the first case, the polymer contribution to the stress is calculated from the CE for an Oldroyd-B fluid. In the second case, this stress is obtained from an ensemble of Hookean dumbbell configuration fields. A summary with conclusions is given in the last section of this paper.

Governing Equations

The isothermal flow of an incompressible fluid with density ρ is governed by the momentum balance and the continuity equation,

$$\rho\frac{\partial \boldsymbol{u}}{\partial t} + \rho\boldsymbol{u}\cdot\nabla\boldsymbol{u} = -\nabla p + \nabla\cdot\boldsymbol{\tau}\,, \tag{1}$$

$$\nabla\cdot\boldsymbol{u} = 0\,. \tag{2}$$

In these equations, p and $\boldsymbol{u}$ denote the pressure and the velocity field, and $\boldsymbol{\tau}$ is the extra stress generated in the fluid due to its motion. It is standard practice to write the extra stress of a polymer solution as the sum of a solvent contribution, $\boldsymbol{\tau}_s$, and a contribution due to the presence of the polymer, $\boldsymbol{\tau}_p$. For the solvent contribution, we take the constitutive equation of a Newtonian fluid, that is, $\boldsymbol{\tau}_s = 2\eta_s\boldsymbol{d}$, where $\boldsymbol{d} = (\nabla\boldsymbol{u} + (\nabla\boldsymbol{u})^T)/2$ is the rate-of-deformation tensor and η_s is the solvent viscosity. The momentum balance can thus be written as

$$\rho\frac{\partial \boldsymbol{u}}{\partial t} + \rho\boldsymbol{u}\cdot\nabla\boldsymbol{u} = -\nabla p + \nabla\cdot(2\eta_s\boldsymbol{d} + \boldsymbol{\tau}_p)\,. \tag{3}$$

The polymer contribution to the stress can be calculated using either a constitutive equation or from a Brownian dynamics simulation. In this paper, we will use the so-called Oldroyd-B model. This model finds its origin in continuum mechanics but it can also be derived from a microscopic model. The Oldroyd-B constitutive equation reads

$$\boldsymbol{\tau}_p + \lambda \overset{\triangledown}{\boldsymbol{\tau}}_p = 2\eta_p \boldsymbol{d}\,. \tag{4}$$

In this equation, λ is the relaxation time of the fluid, η_p is the polymer contribution to the viscosity, and $\overset{\triangledown}{\boldsymbol{\tau}}_p$ is the upper convected derivative of the stress defined by $\overset{\triangledown}{\boldsymbol{\tau}}_p = \partial\boldsymbol{\tau}_p/\partial t + \boldsymbol{u}\cdot\nabla\boldsymbol{\tau}_p - \boldsymbol{\kappa}\cdot\boldsymbol{\tau}_p - \boldsymbol{\tau}_p\cdot\boldsymbol{\kappa}^T$, where $\boldsymbol{\kappa}$ is the transpose of the velocity gradient, $\boldsymbol{\kappa} = (\nabla\boldsymbol{u})^T$.

From a microscopic point of view, the Oldroyd-B equation is the result of the Hookean dumbbell model. In this model, a polymer solution is considered as a suspension of non-interacting elastic dumbbells consisting of two Brownian beads with the friction coefficient, ζ, connected by a linear spring. The configuration of a dumbbell, that is, the length and orientation of the spring connecting the two beads, is indicated by a vector, $\boldsymbol{Q}$. The spring force can thus be written as $\boldsymbol{F}^{(c)} = H\boldsymbol{Q}$, where H is the spring constant. The relaxation time and viscosity are related to microscopic parameters by $\lambda = \zeta/4H$ and $\eta_p = nkT\lambda$.

Once the configuration distribution function $\psi(\boldsymbol{Q}, t)$, which gives the probability of finding a dumbbell with a configuration $\boldsymbol{Q}$, is known, it is possible to calculate the stress using the Kramers expression which for this model reads

$$\boldsymbol{\tau}_p = -nkT\mathbf{1} + nH\langle\boldsymbol{Q}\boldsymbol{Q}\rangle\,, \tag{5}$$

where n is the number density of dumbbells and the angular brackets denote an ensemble average.

In the usual analytical approach to polymer kinetic theory [2], the distribution function is derived from the Fokker-Planck equation. In this paper, however, we generate the configuration distribution using a stochastic differential equation. For details on the formal relationship between the Fokker-Planck equation and stochastic differential equations, the reader is referred to [5]. For the Hookean dumbbell model considered here, the stochastic equation reads

$$\mathrm{d}\boldsymbol{Q}(t) = \left(\boldsymbol{\kappa}(t)\cdot\boldsymbol{Q}(t) - \frac{2H}{\zeta}\boldsymbol{Q}(t)\right)\,\mathrm{d}t + \sqrt{\frac{4kT}{\zeta}}\mathrm{d}\boldsymbol{W}(t)\,, \tag{6}$$

where $\boldsymbol{W}(t)$ is a Wiener process which accounts for the random displacements of the beads due to thermal motion. The Wiener process is a Gaussian process with zero mean and covariance $\langle\boldsymbol{W}(t)\boldsymbol{W}(t')\rangle = \min(t,t')\mathbf{1}$. In a Brownian dynamics simulation, this equation is integrated for a large number of dumbbells. A typical algorithm, based upon an explicit Euler integration, reads

$$\boldsymbol{Q}(t+\Delta t) = \boldsymbol{Q}(t) + \left(\boldsymbol{\kappa}(t)\cdot\boldsymbol{Q}(t) - \frac{2H}{\zeta}\boldsymbol{Q}(t)\right)\Delta t + \sqrt{\frac{4kT}{\zeta}}\Delta\boldsymbol{W}(t)\,, \tag{7}$$

where the components of the random vector, $\Delta\boldsymbol{W}(t)$, are independent Gaussian variables with zero mean and variance Δt. Once the configurations are known, the stress can be estimated by

$$\boldsymbol{\tau}_p \approx -nkT\mathbf{1} + nH\frac{1}{N_d}\sum_{i=1}^{N_d}\boldsymbol{Q}_i\boldsymbol{Q}_i\,, \tag{8}$$

where N_d is the number of dumbbells.

In order to solve a flow problem, it is necessary to find an expression for the stress at a specified position $\boldsymbol{x}$ at time t. From a microscopic point of view, this means that we have to convect a sufficiently large number of molecules through the flow domain until they arrive at $\boldsymbol{x}$ at time t. Neglecting centre-of-mass diffusion, all the molecules experienced the same deformation history but were subjected to different, and independent, stochastic processes.

However simple in theory, a number of problems need to be addressed in practice. For instance, if we disperse a large number of dumbbells into the flow domain, we not only have to calculate all their individual trajectories, but the local value of the stress as well. At every time step, we must also sort all the dumbbells into cells (or elements). Another problem that arises is due to the fact that the tracking occurs with a finite accuracy so that we have to check for dumbbells leaving the flow domain through the boundaries. Once these problems are solved, it is possible to construct a transient code to simulate a non-trivial flow problem. This has in fact been done by Laso *et al.* [6] for the start-up of flow in a journal bearing.

We used a different approach to this problem which overcomes the problems associated with particle tracking. Instead of convecting discrete particles specified by their configuration vector $\boldsymbol{Q}_i$, an ensemble of N_f continuous configuration fields, $\boldsymbol{Q}_i(\boldsymbol{x},t)$, is introduced. Initially, the configuration fields are spatially uniform and their values are independently sampled from the equilibrium distribution function of the Hookean dumbbell model. After start-up of the flow field, the configuration fields are convected by the flow and are deformed by the action of the velocity gradient, elastic retraction and Brownian motion in exactly the same way as a discrete dumbbell. The evolution of a configuration field is thus governed by

$$\mathrm{d}\boldsymbol{Q}(\boldsymbol{x},t) = \left(-\boldsymbol{u}(\boldsymbol{x},t)\cdot\nabla\boldsymbol{Q}(\boldsymbol{x},t) + \boldsymbol{\kappa}(\boldsymbol{x},t)\cdot\boldsymbol{Q}(\boldsymbol{x},t) - \frac{2H}{\zeta}\boldsymbol{Q}(\boldsymbol{x},t)\right)\mathrm{d}t$$
$$+\sqrt{\frac{4kT}{\zeta}}\mathrm{d}\boldsymbol{W}(t)\,. \tag{9}$$

The first term on the RHS of Eq. (9) accounts for the convection of the configuration field by the flow. It should be noted that $\mathrm{d}\boldsymbol{W}(t)$ depends only on time and hence it affects the configuration fields in a spatially uniform way. For this reason, the gradients of the configuration fields are well defined and smooth functions of the spatial coordinates. Of course, the stochastic processes acting on different fields are uncorrelated.

From the point of view of the stress calculation, this procedure is completely equivalent to the tracking of individual dumbbells: an ensemble of configuration vectors, $\{\boldsymbol{Q}_i\}$ with $i = 1, N_f$, is generated at $(\boldsymbol{x}, t)$ all of whose members went through the same kinematical history but experienced different stochastic processes. This is precisely what is required in order to determine the local value of the stress.

In the remainder of this paper, we prefer to scale the length of the configuration vector with $\sqrt{kT/H}$, which is one-third of the equilibrium length of a dumbbell. The relevant equations thus become

$$\boldsymbol{\tau}_p = nkT(-\mathbf{1} + \boldsymbol{b}) = \frac{\eta_p}{\lambda}(-\mathbf{1} + \boldsymbol{b})\,, \tag{10}$$

where $\boldsymbol{b} = \langle\tilde{\boldsymbol{Q}}\tilde{\boldsymbol{Q}}\rangle$ is the conformation tensor, which is dimensionless and reduces to the unity tensor at equilibrium. The Oldroyd-B equation can now be written as

$$\boldsymbol{b} + \lambda \overset{\nabla}{\boldsymbol{b}} = \mathbf{1}\,. \tag{11}$$

The equation for the evolution of the configuration fields becomes

$$\mathrm{d}\tilde{\boldsymbol{Q}}(\boldsymbol{x},t) = \left(-\boldsymbol{u}(\boldsymbol{x},t)\cdot\nabla\tilde{\boldsymbol{Q}}(\boldsymbol{x},t) + \boldsymbol{\kappa}(\boldsymbol{x},t)\cdot\tilde{\boldsymbol{Q}}(\boldsymbol{x},t) - \frac{1}{2\lambda}\tilde{\boldsymbol{Q}}(\boldsymbol{x},t)\right)\mathrm{d}t$$
$$+\sqrt{\frac{1}{\lambda}}\mathrm{d}\boldsymbol{W}(t)\,. \tag{12}$$

Finally, the conformation tensor field follows from

$$\boldsymbol{b}(\boldsymbol{x},t) = \frac{1}{N_f}\sum_{i=1}^{N_f} \tilde{\boldsymbol{Q}}(\boldsymbol{x},t)\tilde{\boldsymbol{Q}}(\boldsymbol{x},t)\,. \tag{13}$$

From now on, the tildes are dropped and it is understood that $\boldsymbol{Q}$ is a dimensionless quantity.

Numerical Methods

Preliminaries

For the spatial discretisation of the system of equations, we will use the finite element method. In order to obtain better stability and extend the possible stress space, we use the Discrete Elastic-Viscous Split Stress (DEVSS) formulation of Guénette & Fortin [8] for the discretisation of the linear momentum balance and the continuity equation. The discontinuous Galerkin (DG) formulation will be used to discretise the constitutive equation in the macroscopic case and the equation for the configuration fields in the microscopic case. In the DG formulation, the interpolation functions are discontinuous across elements, leading to minimal coupling between elements. This means that in our time-stepping scheme, the stresses and the configuration variables at the next time step can be computed at element level. In this way, we avoid solving a large number of coupled equations which, in particular, would be an almost impossible task for configuration fields. The combination of the DEVSS formulation with DG has been introduced by Baaijens *et al.* [9] and it leads to a remarkably stable method.

Momentum Balance and Continuity Equation

For the DEVSS formulation [8], we introduce an extra variable, $\boldsymbol{e} = 2\eta_p\boldsymbol{d}$: the viscous polymer stress. The weak form of the momentum balance Eq. (1) and the continuity Eq. (2) becomes: find $(\boldsymbol{u}, \mathrm{p}, \boldsymbol{e}) \in U \times P \times E$ such that for all $(\boldsymbol{v}, \mathrm{q}, \boldsymbol{f}) \in U \times P \times E$, we have

$$\left(\boldsymbol{v}, \rho\frac{\partial \boldsymbol{u}}{\partial t} + \rho\boldsymbol{u}\cdot\nabla\boldsymbol{u}\right) - (\nabla\cdot\boldsymbol{v}, p) + (\nabla\boldsymbol{v}, 2\eta\boldsymbol{d}(\boldsymbol{u}) - \boldsymbol{e} + \boldsymbol{\tau}_p) = (\boldsymbol{v}, \boldsymbol{\sigma})_\Gamma\,, \tag{14}$$

$$-(\boldsymbol{q}, \nabla\cdot\boldsymbol{u}) = 0\,, \tag{15}$$

$$-(\boldsymbol{f}, \nabla\boldsymbol{u}) + \frac{1}{2\eta_p}(\boldsymbol{f}, \boldsymbol{e}) = 0\,, \tag{16}$$

where $2\boldsymbol{d}(\boldsymbol{u}) = \nabla\boldsymbol{u} + (\nabla\boldsymbol{u})^T$, and $(\cdot,\cdot)$ and $(\cdot,\cdot)_\Gamma$ are proper L^2 inner products on the domain Ω and on the boundary Γ respectively. The viscosity η is the zero-shear-rate viscosity, $\eta_s + \eta_p$, and $\boldsymbol{\sigma}$ is the traction vector on the boundary. The system for $(\boldsymbol{u}, \mathrm{p}, \boldsymbol{e})$ is symmetrical except for the $\rho\boldsymbol{u}\cdot\nabla\boldsymbol{u}$ term.

The discrete form of the equations is obtained by requiring that the weak form is valid on approximating subspaces $U_h \times P_h \times E_h$ which consist of piecewise polynomial spaces. In this work, we use quadrilateral elements with continuous bi-quadratic polynomials, (Q_2), for the velocity space, U_h, discontinuous linear polynomials, (P_1), for the pressure space, P_h, and continuous bilinear polynomials, (Q_1), for viscous polymer stress space, E_h.

Constitutive Equation

For the discretisation of the constitutive equation, we use the discontinuous Galerkin method [10] which is best formulated on a single element. The weak formulation of the CE becomes: find $b \in T$ on all elements e_i, such that for all $s \in T$, we have

$$\left(\boldsymbol{s}, \frac{\partial \boldsymbol{b}}{\partial t} + \boldsymbol{u}\cdot\nabla\boldsymbol{b} - \boldsymbol{\kappa}\cdot\boldsymbol{b} - \boldsymbol{b}\cdot\boldsymbol{\kappa}^T + \frac{1}{\lambda}(\boldsymbol{b} - \mathbf{1})\right)_{e_i} + (\boldsymbol{s}, \boldsymbol{n}\cdot\boldsymbol{u}(\boldsymbol{b}^+ - \boldsymbol{b}))_{\gamma_i^{\text{in}}} = 0\,, \tag{17}$$

where the functional space for $\boldsymbol{b}$ is denoted by T, $(\cdot,\cdot)_{e_i}$ denotes an L^2 inner product on element e_i only, $(\cdot,\cdot)_{\gamma_i^{\text{in}}}$ denotes an integral on the part of the element boundary where $\boldsymbol{u}\cdot\boldsymbol{n} < 0$ with $\boldsymbol{n}$ being the outside normal. Furthermore, $\boldsymbol{b}^+$ is either the value of $\boldsymbol{b}$ in the upstream neighbour element or the imposed value at the inflow boundary part of Γ. In this work, we will use discontinuous bilinear polynomials, (Q_1), for the space, T_h.

Configuration Fields

The equation for solving the fields $\boldsymbol{Q}_j$, $j = 1, \ldots, N_f$ given by Eq. (12) looks very similar to the constitutive equation. We discretise this equation by the DG method and obtain the following weak formulation: find $\boldsymbol{Q}_j \in Q$ in all elements, e_i, such that for all $\boldsymbol{R} \in Q$, we have

$$\left(\boldsymbol{R}, \mathrm{d}\boldsymbol{Q}_j + (\boldsymbol{u}\cdot\nabla\boldsymbol{Q}_j - \boldsymbol{\kappa}\cdot\boldsymbol{Q}_j + \frac{1}{2\lambda}\boldsymbol{Q}_j)\,\mathrm{d}t - \sqrt{\frac{1}{\lambda}}\mathrm{d}\boldsymbol{W}\right) + (\boldsymbol{R}, \boldsymbol{n}\cdot\boldsymbol{u}(\boldsymbol{Q}_j^+ - \boldsymbol{Q}_j)\,\mathrm{d}t)_{\gamma_i^{\text{in}}} = 0\,, \tag{18}$$

where Q is the functional space of $\boldsymbol{Q}_j$ and $\boldsymbol{Q}_j^+$ is the value of $\boldsymbol{Q}_j$ in the upstream neighbour element or the imposed value at the inflow boundary part of Γ. For the approximating space Q_h of Q, we use discontinuous bilinear polynomials, (Q_1). The conformation tensor $\boldsymbol{b}$ is found by projection: find $\boldsymbol{b} \in T$ such that for all all $\boldsymbol{s} \in T$, we have

$$\left(\boldsymbol{s}, \boldsymbol{b} - \frac{1}{N_f} \sum_{j=1}^{N_f} \boldsymbol{Q}_j \boldsymbol{Q}_j \right) = 0\,, \tag{19}$$

For the approximation space, T_h, we again use discontinuous bilinear polynomials, (Q_1).

Time Discretisation

For the time discretisation of the constitutive equation Eq. (17) and the equation for the configuration fields Eq. (18), we use an explicit Euler scheme. At each step, we find $\boldsymbol{b}^{n+1}$ by solving either Eq. (17) case or Eq. (18) and subsequently Eq. (19) for the microscopic case. All equations can be solved at element level.

Next, we substitute $\boldsymbol{\tau}_p^{n+1} = \eta_p/\lambda(-\mathbf{1} + \boldsymbol{b}^{n+1})$ into Eq. (14) and apply a semi-implicit Euler scheme, where all terms are taken implicitly except for the nonlinear inertia term. The system matrix for solving $(\boldsymbol{u}_{n+1}, p_{n+1}, \boldsymbol{e}_{n+1})$ is symmetrical and LU decomposition is performed at the first time step. Since this matrix is constant in time, solutions at later time steps can be found by back substitution only.

Results for Flow Past a Cylinder

Problem Description

We consider the planar flow of an Oldroyd-B fluid past a cylinder of radius a positioned between two flat plates separated by a distance $2H$. The ratio a/H is equal to two and the total length of flow domain is $30a$. In the following, we will use an (x, y) co-ordinate system with the origin positioned at the centre of the cylinder.

Rather than specifying inflow and outflow boundary conditions, we take the flow to be periodic. This means that we extend the flow domain periodically such that cylinders are positioned $30a$ apart. The flow is generated by specifying a flow rate Q that is constant in time. The required pressure gradient is computed at each instant in time. The direction of the flow is from left to right. We assume no-slip boundary conditions on the cylinder and the walls of the channel. Since the problem is assumed to be symmetrical, we consider only half of the domain and use symmetry conditions on the centre line, that is, zero tangential traction.

The dimensionless parameters governing the problem are the Reynolds number, $Re = \rho U a/\eta$, the Deborah number, $De = \lambda U/a$, and the viscosity ratio, η_s/η, where $U = Q/2H$ is the average velocity and η is the viscosity of the fluid given by $\eta_s + \eta_p$. In this paper, we take $Re = 0.01$ and $\eta_s/\eta = 1/9$, which are the values used by Bodart and Crochet [11] for the falling sphere problem. In the following, we will only use dimensionless quantities: the time variable has been made dimensionless with the characteristic time scale of the flow a/U, velocities with U and macroscopic lengths with a. We will define the drag coefficient by $C_d = F_x/\eta U$, where F_x is the drag force per unit length on the cylinder. Since the objective of this paper is to compare two methods, we will focus on the start-up flow for a single Deborah number.

To solve the problem numerically, we use five meshes denoted by mesh1 to mesh5. Each mesh is derived from the previous one by a uniform refinement which approximately doubles the number of elements. The numerical parameters of the meshes are summarised in Table 1.

Table 1. Numerical parameters.

	mesh1	mesh2	mesh3	mesh4	mesh5
Number of elements N_{el}	256	508	1024	2032	4096
Number of nodal points	1113	2155	4273	8373	16737
Smallest radial element size	0.0747	0.0504	0.0380	0.0254	0.0191
Time step	0.01	0.01	0.007	0.007	0.005

Table 2. Drag coefficient C_d. For the Oldroyd-B fluid, the value after start-up at $De = 0.6$ and $t = 7$ is given.

	mesh1	mesh2	mesh3	mesh4	mesh5
C_d Newtonian	132.4058	132.3878	132.3612	132.3588	132.3584
C_d Oldroyd-B	99.791	98.653	98.222	98.156	98.124

Macroscopic Calculations

We first compute the Newtonian flow for which the drag coefficient C_d for the various meshes is given in Table 2. We see a fast convergence.

The value of the drag coefficient at $t = 7$ for various meshes is given in Table 2. Though not as fast as in the Newtonian case, convergence is evident. We checked, by using smaller time steps, that time discretisation errors are negligible here. The drag reduction compared to the Newtonian flow is almost 26%.

To show convergence with mesh refinement, we have plotted the value of b_{xx} along the centre line in Fig. 1 and the cylinder surface as a function of x. We see a good convergence, except maybe close to the maximum, where we still need a more refined mesh. The convergence of b_{xy} and b_{yy} (not shown) is even better.

Microscopic Calculations

For the microscopic computations of an Oldroyd-B fluid, the drag coefficient until $t = 7$ is shown in Fig. 2 for mesh1. We have a good agreement with the

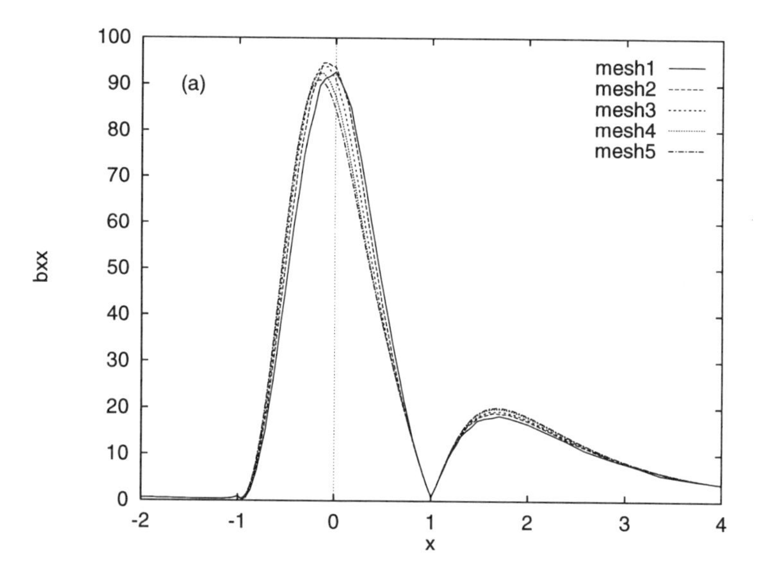

Fig. 1. The b_{xx}-component of $\boldsymbol{b}$ on the centre line and the wall of the cylinder for $De = 0.6$ as a function of x. Time = 7. Note that $-1 \leq x \leq 1$ is on the surface of the cylinder.

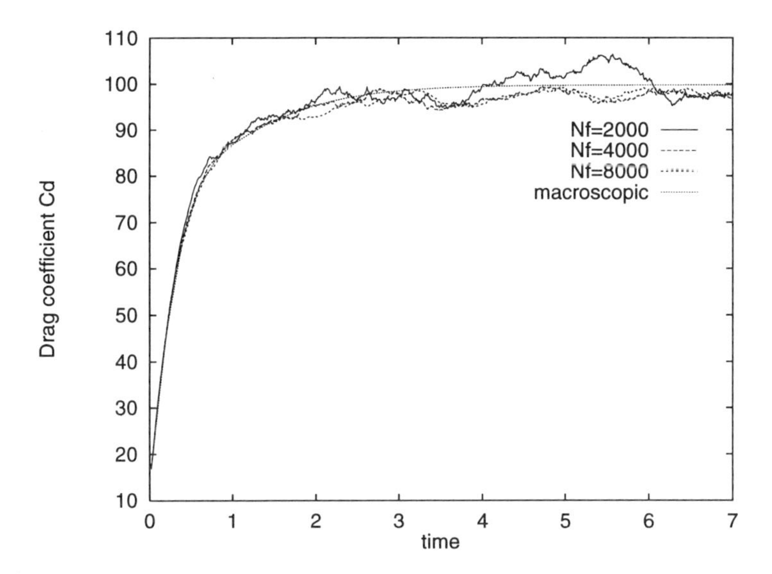

Fig. 2. The drag on the cylinder $De = 0.6$ as a function of time for the configuration field method. Mesh is mesh1.

macroscopic method on the same mesh. Increasing the number of fields, N_f, reduces the fluctuations and the drag coefficient seems to converge to a curve slightly below the macroscopic curve.

To show convergence upon mesh refinement, we plotted the value of b_{xx} along the centre line in Fig. 3 and the cylinder surface as a function of x for three meshes. We see that the results for the configuration field method compare very well with the macroscopic method on the same mesh. Convergence is very similar for both methods, except for the wake in mesh1 where the error compared to the converged solution (see Fig. 1) is more than twice as large. This difference, however, disappears for the finer meshes. For the components b_{xy} and b_{yy} (not shown), the results are even better. Error bars are given only on the local maxima. The relative error in the stress appears to be approximately constant throughout the domain and is $\approx 3\%$ for $N_f = 2000$. As expected the error decreases in proportion to $1/\sqrt{N_f}$ with an increase in the number of fields, N_f.

To show the stability of the configuration field method, we performed calculations up to $t = 30$ (50 relaxation times) for $N_f = 4000$ on mesh1. In Fig. 4, the drag coefficient is given and in Fig. 5, the minimum value of det $\boldsymbol{b}$ as

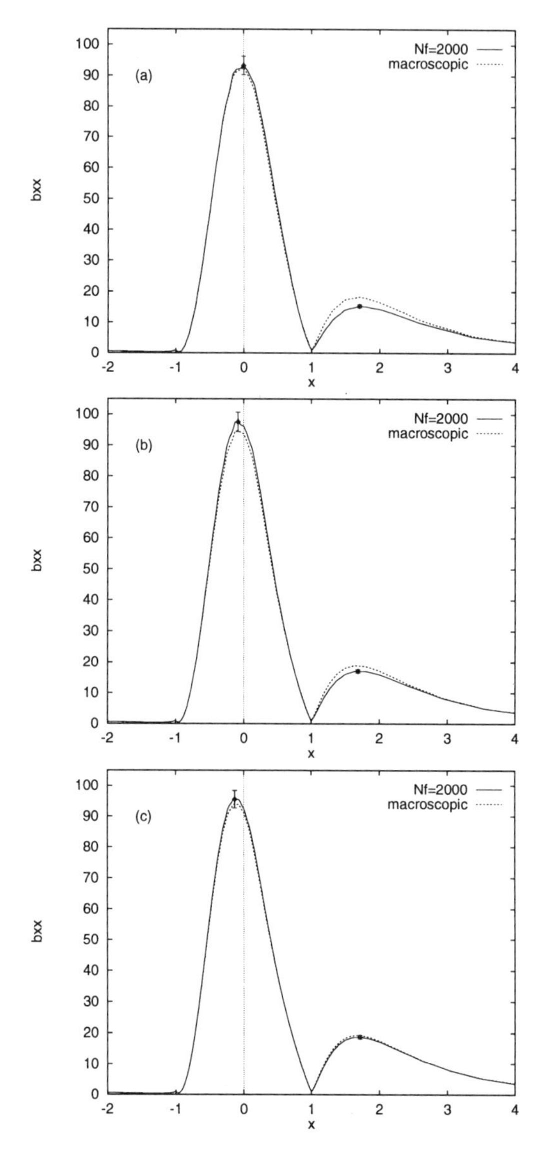

Fig. 3. The b_{xx} component of $\boldsymbol{b}$ on the centre line and the surface of the cylinder for $De = 0.6$ as a function of x. Time = 7. Note that $-1 \leq x \leq 1$ is on the surface of the cylinder. Error bars given on local maxima: $\pm$ standard deviation. (a) mesh1, (b) mesh2, (c) mesh3.

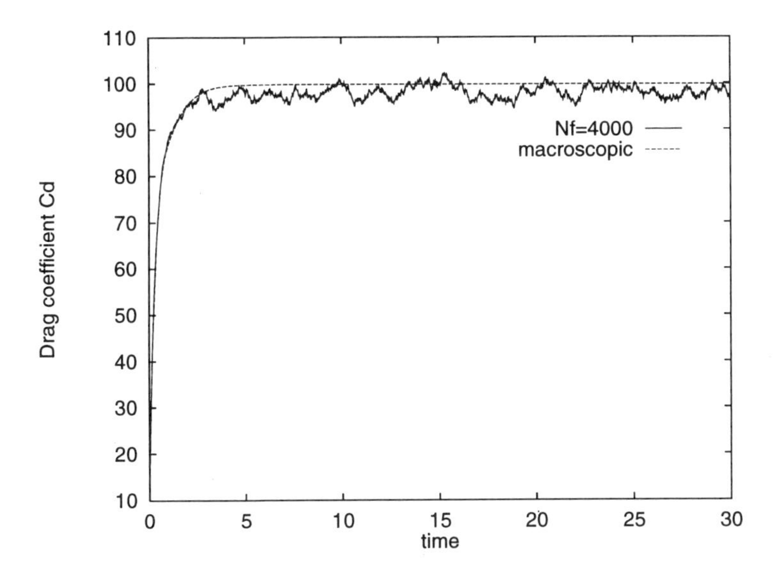

Fig. 4. The drag on the cylinder $De = 0.6$ as a function of time. Mesh is mesh1.

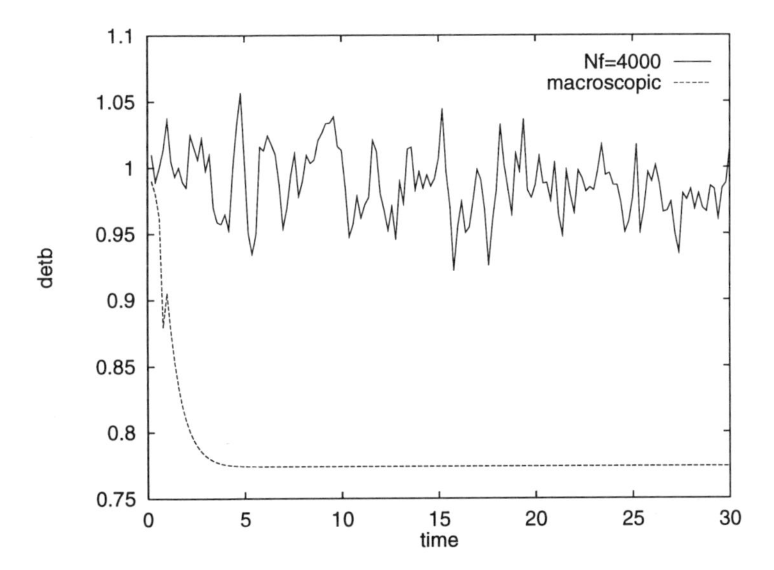

Fig. 5. The minimum value in the Gauss integration points of det $\boldsymbol{b}$ for $De = 0.6$ as a function of time. Mesh is mesh1.

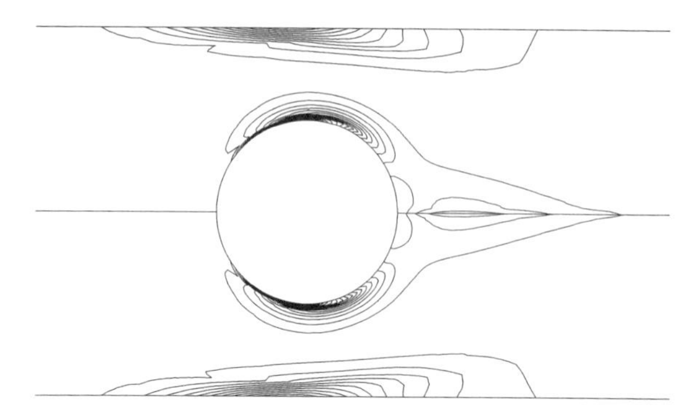

Fig. 6. Contours of b_{xx} for $De = 0.6$. Mesh is mesh3. Time = 7. Upper half: microscopic $N_f = 20000$. Lower half: macroscopic.

a function of time. The theoretical minimum of det $\boldsymbol{b}$ for an Oldroyd-B fluid is 1 [12]. We see that both the macroscopic and the microscopic methods are stable for this Deborah number. There is, however, an important difference that is typical for these computations: in the microscopic method, the minimum value of det $\boldsymbol{b}$ fluctuates close to the theoretical minimum 1 whereas the macroscopic method shows a minimum that is below 1 for coarse meshes.

To show that the configuration field method leads to smooth functions in space, we have shown a plot of the contours of b_{xx} in Fig. 6 for $N_f = 2000$ and mesh3. For the purpose of comparison, we have also included the results of the macroscopic computations on the same mesh. We see that the agreement is excellent.

Higher Deborah Numbers

We have not explored the limits of the configuration field method yet. However, in order to show that the method is more robust than the macroscopic method, we performed a calculation for the start-up at $De = 1.2$ on mesh1. In Fig. 7, we show the results for the minimum value of det $\boldsymbol{b}$ in the Gauss points. For this Deborah number, det $\boldsymbol{b}$ becomes negative for the macroscopic method and eventually blows up. The microscopic method remains stable with det $\boldsymbol{b}$ fluctuating close to the theoretical minimum of 1. This stability or robustness of the configuration field method is somewhat unexpected and seems to be related to the inherently positive formulation of the conformation tensor $\boldsymbol{b}$.

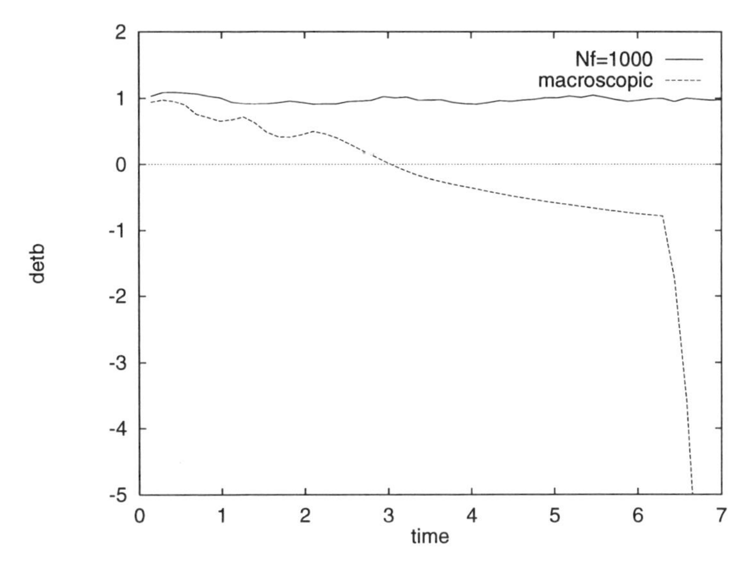

Fig. 7. The minimum value in the Gauss integration points of det $\boldsymbol{b}$ for $De = 1.2$ as a function of time. Mesh is mesh1.

Conclusions

In this paper, we have shown that the idea of using Brownian configuration fields, instead of following discrete particles, works very well in practice. The implementation, using the discontinuous Galerkin method, is rather straightforward and does not require complicated particle tracking and sorting procedures. Since the operations on the configuration fields involve only simple do-loops, full vectorisation (and probably also parallelisation) can be achieved without much "code tuning". Another advantage of the configuration field method is that the statistical error at *any* point in the flow is controlled by the single parameter, N_f, independent of the mesh. This is in contrast to a discrete particle method where, in order to maintain a given statistical error, mesh refinement necessarily implies increasing the number of dumbbells, Furthermore, since small elements generally contain the smallest number of dumbbells, the statistical errors will be largest at precisely those locations where we strive for accuracy.

The computations of the flow past a cylinder show that the accuracy of the configuration field method is similar to the macroscopic method on the same mesh. The microscopic method, however, appears to be more stable or robust than the macroscopic method. This is probably because the conformation tensor $\boldsymbol{b}$ is positive definite by design. Furthermore, it turns out that $\boldsymbol{b}$ not only remains positive definite, but in the computations done so far the minimum value of det $\boldsymbol{b}$ remains close to the theoretical minimum for even very coarse meshes. The method leads to stress fields that are smooth in space and stochastic fluctuations, due to a finite number of fields, only show up as fluctuations in time, for which the statistical errors are approximately proportional to the local stress value (and, of course, to $1/\sqrt{N_f}$).

However, a price has to be paid for all of these: the computer requirements are much more demanding than for macroscopic methods, both in CPU time and memory. Therefore, in order for the method to become "competitive", future research need to show that it can meet the following two requirements:

1. The promise of bypassing the closure problem leads to significantly better prediction of flows in industrial environments or experimental setups in research environments.

2. Methods to reduce the CPU and memory requirements will become available in the near future.

Both aspects are part of our current research programme.

Discussion

Entov You start with a set of uniformly distibuted initial variables. Is it clear that this produces the same results as a purely stochastic process?

B. H. A. A. *van den Brule* The fields are uncorrelated. The stochastic processes within each field are spatially correlated — it is the same everywhere. The fields are independent. Since we take the averages over all fields, I don't think there is a problem.

Entov If I understood it properly, you have all the variables in one realisation oriented in exactly the same way throughout the whole domain.

B. H. A. A. *van den Brule* That is one way of thinking about it. We actually prefer not to think about individual molecules. It is just a continuous field that is deformed and convected by the flow and updated in a random fashion in such a way that if you look at a specific point in the flow domain, it generates the right set of vector values. This set is exactly identical to the set that would have been generated by tracking individual particles. Mathematically, it is the same. Of course, it is hard to imagine having molecules all wiggling in the same way.

J. R. A. *Pearson* When you say that you have spatial correlation within each field, that confused me.

B. H. A. A. *van den Brule* Initially, each field is spatially uniform. Each individual field is then convected and distorted by the flow and on top of that, we have the change of each field through a random process. Only this stochastic process within each individual field is correlated. After a while, of course, each field is no longer uniform due to the action of the flow process. Uniformity of the vector field is only there at the beginning.

References

1. Bird, R. B., Armstrong, R. C. & Hassager, O. (1987) *Dynamics of Polymer Liquids, Vol. 1*, 2nd edn. New York: John Wiley
2. Bird, R. B., Curtiss, C. F., Armstrong, R. C. & Hassager, O. (1987) *Dynamics of Polymer Liquids, Vol. 2*, 2nd edn. New York: John Wiley
3. Laso, M. & Öttinger, H. C. (1993) Calculation of viscoelastic flow using molecular models: the CONNFFESSIT approach. *J. Non-Newtonian Fluid Mech.* **47**, 1–20
4. Öttinger, H. C. & Laso, M. (1994) Bridging the gap between molecular models and viscoelastic flow calculations. In *Lectures on Thermodynamics and Statistical Mechanics*, pp. 139–153. Singapore: World Scientific
5. Öttinger, H. C. (1996) *Stochastic Processes in Polymeric Fluids.* Berlin: Springer Verlag
6. Laso, M., Picasso, M. & Öttinger, H. C. (1996) Two-dimensional time-dependent viscoelastic flow calculations using CONNFFESSIT. *AIChE Journal* **43**, 877–892
7. Feigl, K., Laso, M. & Öttinger, H. C. (1995) CONNFFESSIT approach for solving a two-dimensional viscoelastic fluid problem. *Macromolecules* **28**, 3261–3274
8. Guénette, R. & Fortin, M. (1995) A new mixed finite element method for computing viscoelastic flows. *J. Non-Newtonian Fluid Mech.* **60**, 27–52
9. Baaijens, F. P. T., Baaijens, H. P. W., Selen, J. H. A., Peters, G. W. N. & Meijer, H. E. H. (1996) Viscoelastic flow past a confined cylinder of a LDPE melt. *J. Non-Newtonian Fluid Mech.* **68**, 173–203
10. Fortin, M. & Fortin, A. (1989) A new approach for the fem simulation of viscoelastic flows. *J. Non-Newtonian Fluid Mech.* **32**, 295–310
11. Bodart, C. & Crochet, M. J. (1994) The time-dependent flow of a viscoelastic fluids around a sphere. *J. Non-Newtonian Fluid Mech.* **54**, 303–329
12. Hulsen, M. A. (1988) Some properties and analytical expressions for plane flow of Leonov and Giesekus models. *J. Non-Newtonian Fluid Mech.* **30**, 85–92

Part Two
Polymeric and Self-Assembled Systems

Introduction 127

Chapter 9. Role of Energetic Interaction in the Dynamics of Polymer Networks: Some New Suggestions 131
A. K. Lele and R. A. Mashelkar

Chapter 10. The Prediction of Universal Viscometric Functions for Dilute Polymer Solutions Under Theta Conditions 155
J. Ravi Prakash

Chapter 11. Polymeric Liquids at High Shear Rates 176
G. Marrucci and G. Ianniruberto

Chapter 12. Deuterium NMR Investigations of Liquid Crystals During Shear Flow 188
C. Schmidt, S. Müller and H. Siebert

Chapter 13. Formation of Polymer Brushes 193
J. Wittmer and A. Johner

Chapter 14. Dynamics of Adsorbed Polymer Layers 199
J. L. Harden and M. E. Cates

Chapter 15. The Molecular Dynamics Simulations of Boundary-Layer Lubrication: Atomistic and Coarse-Grained Simulations 213
D. J. Tildesley

Chapter 16. Direct Simulation of Surfactant Solution Phase Behaviour 230
R. G. Larson

Introduction

This section considers molecular-based descriptions of the phenomenological behaviour of a wide range of polymeric and lower molecular weight amphiphiles. The theoretical advances are strongly based on early concepts, such as those developed by Rouse, Zimm and Flory, and more recent ideas such as those introduced by Doi and Edwards. Atomistic and coarse grained theoretical and numerical modelling have made enormous progress over the last few years and this is reflected in the papers presented here. The current work is distinguished by the level of complexity of the systems which these approaches attempt to describe. Significant developments have also been made in the measurement of micro-structural evolution under flow and a paper representing this progress is included in the section.

Many experimental results have demonstrated that concentrated solutions and gels of associating polymers exhibit relaxation times that are orders of magnitude longer than those that may be accounted for using conventional models such as the Upper Convected Maxwell and Oldroyd-B. Lele and Mashelkar start this section by introducing the notion of a transient network model to describe the hydrogen bonding interaction kinetics which incorporates Flory theory to obtain the average cross-link density. Their key results are: (a) the local viscosity is proportional to the local cross-link density distribution; and (b) the timescales of bond reformation are dominated by co-operative network mobility and orientation constraints and, consequently, are much longer than those associated with the entanglement–disentanglement dynamics. The model is successfully used to describe a number of experimental observations, such as shear and first normal stress overshoots in polyacrylamide solutions and an apparent slip in capillary flow which continues to be a controversial area.

Unlike the temporary network systems described in the paper by Lele and Mashelkar, the details of the chemical structure do not greatly influence the viscometric behaviour of dilute polymer solutions which contributed to the early successes of bead-spring coarse grain models. In the second paper, Prakash reviews the parameters used to describe the hydrodynamic interactions which, together with the number of beads, are essentially arbitary and yet critically determine the trends predicted by such models. He considers a twofold normal approximation in order to improve predictions of the shear viscosity, and also the first and second normal stress differences as a function of shear rate for large molecular weight polymers in theta solvents. The values of these viscometric functions are extrapolated to an infinite chain length when the results become parameter-free or "universal". Their accuracy was verified against Brownian Dynamics simulations.

Experimental data have shown that the non-linear behaviour of polymer melts at high shear rates follow the Cox-Merz rule rather than the instabilities expected in the Doi-Edwards theory. In the next paper, by Marrucci and Lanniruberto, they point out that thermally driven entanglement renewal is effectively "frozen" at high shear rates, and convection becomes the dominant mechanism although this is not sufficient to completely account for the measured flow behaviour. They show that it is also necessary to consider the validity of the Doi-Edwards assumption that the tube diameter remains constant. In particular, by combining the effects of a convective constraint mechanism and a reduction in the tube diameter, due to an increasing entanglement renewal frequency, they were able to obtain agreement with the Coz-Merz rule. However, in the case of a step strain perturbation, the steady state diameter is recovered following instantaneous contraction.

The fourth paper by Schmidt *et al.* describes the application of deuterium NMR in examining the influence of flow on the orientation of liquid crystals in both cone-and-plate and Couette geometries. The power of quadrupole coupling of the deuterium nucleus as a probe in this context is illustrated for a shear-aligning polysiloxane, which tends to a stable director orientation at high shear rates, compared to a poly-methacrylate in nematic solution which does not shear-align and shows a continuous increase in this angle with increasing shear rate. The results are also given for lyomesophases of surfactants which again demonstrate the value of rheo-NMR in studying mesoscopic structures.

Wittmer and Johner continue the theme in the fifth paper with a consideration of associating polymeric systems. In particular, they develop a theoretical model of the dissociation/association process in systems such as block copolymer mesophases where association arises from the self-assembly of the poor soluble blocks. Like Lele and Mashelkar, they conclude that the flow of such systems is driven by local factors. This involves a thermally activated desorption process with local stretching of the chain determining the extraction process. The main result of this work is that the friction associated with chain extraction is independent of molecular weight.

The following paper by Harden and Cates is also concerned with adsorption/desorption of polymer brush structures, in this case on solid surfaces formed either by strongly adsorbing head groups or insoluble blocks in a copolymer. This is a critically important area in the rheology of sterically stabilised colloidal suspensions and polymeric lubricant systems. They develop a semi-phenomenological theory to describe the non-uniform deformation of the grafted chains in strong solvent flows. A major advance of this work is to account for the solvent flow inside a brush. The paper includes a consideration of chain desorption and fractionation.

Over the last five years, there has been a major advance in understanding boundary lubrication through the use of lateral force devices capable of measuring gaps in the nanometer range. This has led to the realisation that the performance of these systems often depends upon the transient structures developed by the sliding motion of the solid surfaces. Spectroscopic tools (for example, sum frequency spectroscopy) are now available to characterise these structures at the molecular scale. At the same time, there has been significant developments in full atomistic and coarse-grained simulation techniques, which should complement the experimental developments and provide a detailed understanding of the mechanisms involved. Tildesley gives a summary of the current state of these simulation procedures by using amphiphiles as a specific example. Clearly, time and length scales remain a restriction but the results provide a useful framework for questioning the validity of any proposed mechanisms. He finds that the coefficient of friction decreases with increasing normal force, which is in accord with the interfacial rheological data of Briscoe and Tabor, and increases with increasing shearing velocity. The latter finding is not always observed experimentally due to the retardation in compression mechanism identified by Briscoe. It requires that the compressive, as well as the shear, relaxation time be considered for real (rough) surfaces due to the cyclic compressive field at sliding point contacts such as those formed by asperities.

The final paper by Larson describes a simple Monte Carlo lattice model for predicting composition-temperature phase diagrams of surfactant solutions. The basis of this work is that surfactants with a wide range of chemical structures, including non-ionic, anionic, cationic and zwitterionic amphiphiles, exhibit remarkably similar phase behaviour. He concludes that lyotropic phase transitions are driven by the following universal factors: volume filling constraints, entropies, energies of mixing, and entropies of surfactant chain conformation. The lattice model successfully predicts the main features of the phase behaviour and supports Israelachvili's contention that the major governing factor is the ratio of the effective head to tail sizes, which controls the preferred curvature of the locus-connecting junctions between head and tail groups.

Dynamics of Complex Fluids, pp. 131–154
ed. M. J. Adams, R. A. Mashelkar, J. R. A. Pearson & A. R. Rennie
Imperial College Press–The Royal Society, 1998

Chapter 9

Role of Energetic Interactions in the Dynamics of Polymer Networks: Some New Suggestions

A. K. LELE

National Chemical Laboratory, Pune-411 008, India
E-mail: lele@dalton.ncl.res.in

R. A. MASHELKAR*

Anusandhan Bhavan, 2, Rafi Marg, New Delhi, India 110 001
**E-mail: dgcsir@csirhq.ren.nic.in*

Concentrated solutions and gels of polar polymers, such as polyacrylamide or poly(acrylic acid), show unusually large relaxation times in certain experiments. These time scales are orders of magnitude larger than the molecular relaxation times estimated from standard rheological measurements. We propose that such systems form transient networks which are crosslinked through energetic interactions and chain entanglements. Examples of energetic interactions include hydrogen bonding and hydrophobic interactions. Since the energy of H-bonds is significantly lower than that of a covalent bond, the interpolymer H-bonds in a network can be easily broken by separating the participating sites beyond a minimum distance by the action of an imposed deformation. However, the reformation of such bonds would require not only that the sites come close to each other, but also that they approach with a proper orientation. This involves a co-operative motion of the network, which is very slow. Hence large restoration times would be needed to achieve the original undeformed state of the network. Based on this proposition, we present in this paper some interesting dynamic phenomena in such networks and provide new suggestions for certain hitherto unexplained anomalous flow behaviour.

Introduction

Several independent experimental results on concentrated polymer systems show unusually long relaxation times, which are orders of magnitude longer than molecular relaxation time scales. For example, we have shown previously [1] that the swelling capacity of a physically sheared crosslinked gel does not recover its original unsheared value even seven hours after the shear had stopped. Another example is that of the simple experiment of a sphere settling in a polymer solution. Bisgaard [2], Cho *et al.* [3] and Ambeskar and Mashelkar [4] have shown that a waiting period of several minutes is required before successively dropped spheres can have the same settling velocities. If the spheres are dropped during this time, the second sphere has a higher velocity than the first. This phenomenon is independent of the elasticity of the solution. Recently, Gheissary and van den Brule [5] have shown that if the solution is stirred after the first ball has settled, the next ball would have the same settling velocity as the first. They have also shown that for a poly(acrylic acid) solution, the next ball has an unusally high settling velocity after a 12-hour waiting period. Jones *et al.* [6] have demonstrated that the settling velocity of a sphere undergoes a transient overshoot before reaching terminal velocity, and that the transients occur over a time scale that is much longer than any molecular relaxation times. These transient effects in the falling ball problem have remained an unresolved problem for a long time.

There are other transient effects that have remained unresolved. We refer to the unusual stress growth experiments on a polyacrylamide solution in ethylene glycol by Kulicke *et al.* [7, 8]. The authors observed that above a critical shear rate, both the shear stress and the first normal stress show large double overshoots. However, on reshearing the same solution after a waiting period of up to 60 minutes, the second overshoot was not seen. The authors argued that the second overshoot was due to energetic associations which, once formed, require a long time to dissociate.

None of the well-known rheological models, such as the Upper Convected Maxwell model or the Oldroyd-B model, can predict the large relaxation times mentioned above. Transient network theories are best suited to explain the dynamics of concentrated polymeric systems. The development of such theories began with Lodge's [9] theory and its subsequent modifications [10]. During the model's formulation, the network was conceived to be formed due to the dynamics of continuous chain entanglement-disentanglement. These theories can predict standard rheological phenomena such as shear thinning, stress overshoot and stress relaxation. However, the predicted time scales are related to the time scale over which the entanglements remain effective, which is again much smaller than the long restoration times mentioned above. Recently, several network models that incorporate the different molecular level dynamics of network breakage have been formulated [11–15]. However, none of these theories incorporate the dynamics of network formation. We believe that the long relaxation or "healing" times described above are associated with the dynamics of network formation through energetic interactions.

It is well-known that the H-bonding energies (3–5 kcal/mol) are at least of an order of magnitude lower than those of covalent bonds, while the distance of separation between the participating donor-acceptor sites is roughly similar (2–4 A°). Therefore, the force required to separate the sites to break the H-bond can be much lower than that required to break a covalent bond. Even a relatively weak deformation field can be expected to break such energetic interactions. In the absence of a deforming force, the formation of H-bonds not only requires that the participating sites are at a minimum distance of separation corresponding to a minima in the H-bonding energy, but also that the sites approach with a particular orientation. This is shown schematically in Fig. 1. The polymer chain is represented as a bead-rod model, while the H-bonding site is shown as a "sticker bead" attached to the main chain. Since the sticker beads have rotational freedom, the probability of reforming the H-bond is reduced if the sticker beads do not maintain a proper orientation

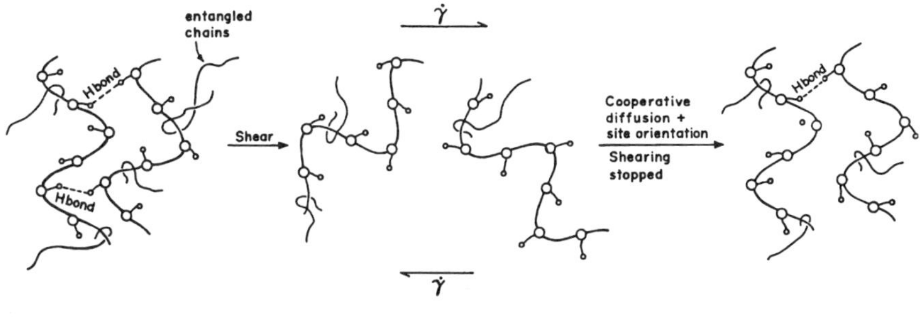

Fig. 1. Scheme of shear-induced breakage of energetic interactions followed by the reformation process through network diffusion and site orientation on stopping the shear: (a) shows the chains of the network formed through entanglements as well as energetic interactions; (b) shows a deformed network and breakage of the interactions due to the imposed shearing deformation; (c) shows that upon the cessation of the shear, not all energetic interactions can be reformed due to improper site orientations.

with respect to each other as the two main chains diffuse towards each other. The latter process involves a co-operative diffusion of the network, which is very slow. Thus, on the whole, the reformation of energetic interactions can be a slow process.

In this paper, we will propose an outline of a transient network model which incorporates the breakage-reformation kinetics of energetic interactions. The real polymeric systems which are pertinent to our study include polar polymers such as polyacrylamide and poly(acrylic acid). We use the simple model to predict the time-dependent swelling of a microgel and describe some additional interesting experimental results for this system. The microgel studied here consists of a network formed mainly through H-bonding and chain entanglements. It contains a very small amount of covalent crosslinks. Next, we proceed to study concentrated polymer solutions, which can be conceived as a network formed from only chain entanglements and H-bonding. We show that the simple model outlined here can qualitatively predict the anomalous transient behaviour discussed above as well as some steady state flow anomalies such as the apparent slip behaviour in pipe flow and the lower film thicknesses observed in wire withdrawal experiments. We show that the model does not require the assumptions of "depleted region" or a "slip velocity". We are in the process of developing a more quantitative theory which will incorporate the molecular level physics of formation of a transient network through energetic interactions.

Theory

In a simplified treatment, the kinetics of breakage and formation of chain entanglements and H-bonds can be written as a simple kinetic equation,

$$E_i \rightleftarrows 2S_i \,, \tag{1}$$

where $\langle E_i \rangle$ and $\langle S_i \rangle$ denote the volume averaged concentrations of the crosslinks and the uncrosslinked sites of type i, respectively. The type i of the crosslinks considered here are entanglements and energetic interactions. k_1^i and k_2^i are rate constants for breakage and formation of crosslinks of type i. In general, k_1^i would depend on the local shear rate and the chain length between crosslinks, whereas k_2^i, for energetic interactions in particular, will depend on the co-operative mobility of the network and orientation of participating sites. In the preliminary model discussed here, we will assume that

k_1^i is proportional to the local shear rate and that k_2^i is a constant which is significantly greater than k_1^i.

If $\langle S_i \rangle_T$ is the volume averaged concentration of the total number of sites of type i available for crosslinking, the evolution of crosslinks with time can be calculated as

$$\langle E_i \rangle = \frac{\alpha^i{}_+ - \alpha^i - A_i \exp(-B_i t)}{1 - A_i \exp(-B_i t)}, \tag{2}$$

where the various symbols are defined as

$$\begin{aligned}
\alpha^i{}_\pm &= \frac{b_i \pm (b_i{}^2 - 4a_i c_i)^{1/2}}{2a_i}, \\
a_i &= 2k_1^i, \\
b_i &= k_1^i + 2k_2^i [S_i]_T, \\
c_i &= (k_2^i/2)[S_i]_T^2, \\
A_i &= \frac{2\langle S_i \rangle_T - \alpha^i{}_+}{2\langle S_i \rangle_T - \alpha^i{}_-}, \\
B_i &= a_i(\alpha^i{}_+ - \alpha^i{}_-).
\end{aligned} \tag{3}$$

The total volume averaged crosslink density, which is a sum of the covalent crosslinks and physical crosslinks, determines the swelling of a microgel system. According to the Flory theory, the equilibrium swelling pressure of a gel is given by

$$\frac{\Pi}{RT} = \ln(1-\phi) + \phi + \chi\phi^2 + \nu_s \left(\frac{\nu_c + \nu_p}{V_0}\right)\left(\phi^{1/3} - \frac{\phi}{2}\right) = 0, \tag{4}$$

where ν_c and ν_p are covalent and physical crosslinks, the latter being given by Eqs. (2) and (3) provided the kinetics of breakage occur faster than the swelling of the gel. The swelling capacity, which is a macroscopic property of the gel defined as the ratio of the volume of swollen gel to that of the dry gel, can be calculated as

$$q = 1/\phi. \tag{5}$$

Equations (2) to (5) predict the time-dependent swelling of a microgel gel system.

A concentrated polymer solution is also a transient network. It differs from the microgel system in that the solution does not contain covalent crosslinks. We consider such a network in which the crosslinks are in the form of chain entanglements and energetic interactions. Under the action of an imposed stress or deformation, the crosslinks can break and reform in a dynamic process. We assume: (i) that at a microscopic level, the flow is homogeneous; (ii) that there is a limiting number of sites, S_∞, which can be formed at very large stresses; and (iii) that the viscosity of the solution is directly proportional to the instantaneous crosslink density of the network. The last assumption is shown to be valid for "strong" networks by Wang [11]. If $[E_i]$ and $[S_i]$ represent the distribution functions of crosslinked segments and uncrosslinked segments of type i at any point in the solution, the conservation equation for the sites of the network can be written as

$$\frac{\partial[E_i]}{\partial t} + \bar{\nu} \cdot \nabla_Q [E_i] = 2k_1[E_i] - k_2([S_i] - [S_i]_\infty)^2\,, \tag{6}$$

where ∇_Q represents the gradient along the vector $\bar{Q}$, which is the end-to-end vector of chains. The similarity between the polymer solution and the microgel system is seen from a comparison of Eqs. (1) to (3) and Eq. (6). Since we are interested in the macroscopic properties of the microgel, such as its swelling capacity, we consider the volume average quantities for the gel whereas in the case of solutions undergoing deformation, the local network properties are pertinent.

The velocity field $\bar{\nu}$ in Eq. (6) is given by the momentum equation

$$\frac{\partial \bar{\nu}}{\partial t} + \bar{\nu} \cdot \nabla \bar{\nu} = \nabla \cdot \bar{\tau} - \nabla p + \bar{g}\,. \tag{7}$$

Equations (4) and (5) are coupled by the second assumption given by

$$\bar{\tau} = H\langle QQ \rangle\,, \tag{8}$$

where H is a spring constant and the averaging is done over the distribution function of crosslinked sements, $[E_i]$. At steady state and in the strong network limit assumption [14], Eq. (8) yields

$$\eta = K[E]_{ss}\,, \tag{9}$$

where K is a constant of proportionality and $[E]_{ss}$ is given by $[E]_{ss} = \alpha_+$ of Eqs. (2) and (3). Equation (9) is the main equation in the paper which suggests that the local viscosity is proportional to the local steady state distribution of crosslink density of the network.

If the deformation is stopped after reaching a dynamic steady state, the network can reform back according to the kinetics,

$$E = E_{ss} + \frac{2k_2(E_0 - E_{ss})^2 \Delta t}{1 + 2k_2(E_0 - E_{ss})\Delta t}, \tag{10}$$

where E_0 is the density of crosslinks in the network before shearing it, E_{ss} is the density of crosslinks at the dynamic steady state, and Δt is the time interval after the deformation has been stopped. The driving force for this reformation could be thermodynamic in origin.

In summary, we propose that transient networks are formed simultaneously from chain entanglement-disentanglement dynamics and dynamics of energetic interactions. These two physical processes have inherently different time scales. Whereas the lifetime of entanglements are of the same order of magnitude as the molecular relaxation times that are typically measured from rheological experiments, the time scales of reformation of energetic interactions could be significantly larger due to co-operative network mobility and orientational constraints.

Results and Discussions

Many of the experiments mentioned in the introduction which show the unusually long relaxation time of transient networks are transient in nature. We begin by qualitatively explaining some of these experiments using the transient network model described above. Specifically, we discuss the microgel system, the falling ball experiment and the stress growth experiments of Kulicke *et al.* [7]. We then show that the same model can also explain some flow anomalies in steady state flows, in particular the pipe flow of solutions and wire coating from solutions.

Transient Swelling of Sheared Microgels

We begin by looking at a microgel which has very few covalent crosslinks and a large number of transient crosslinks. The microgel is formed from starch-*g*-hydrolysed polyacrylonitrile (HSPAN). The chemical structure of the microgel and a scheme of its physical structure is shown in Fig. 2. The microgel was

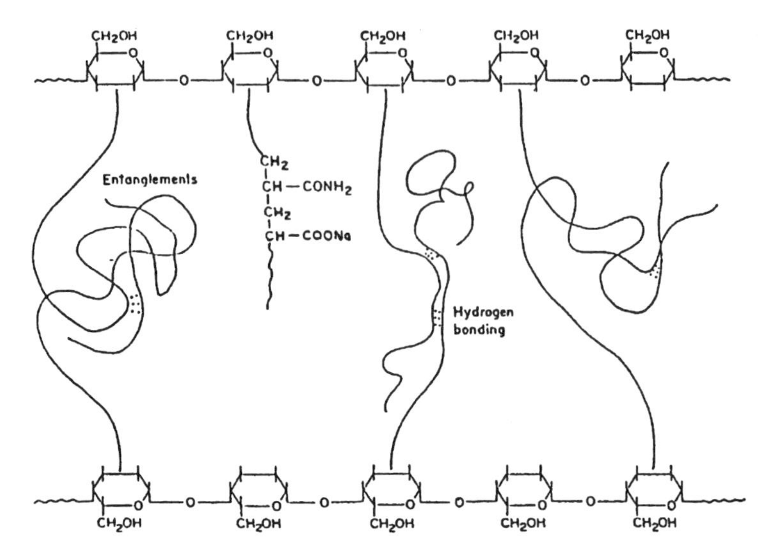

Fig. 2. Scheme of the HSPAN microgel. The grafted PAN chains form transient crosslinks through H-bonding as well as entanglements.

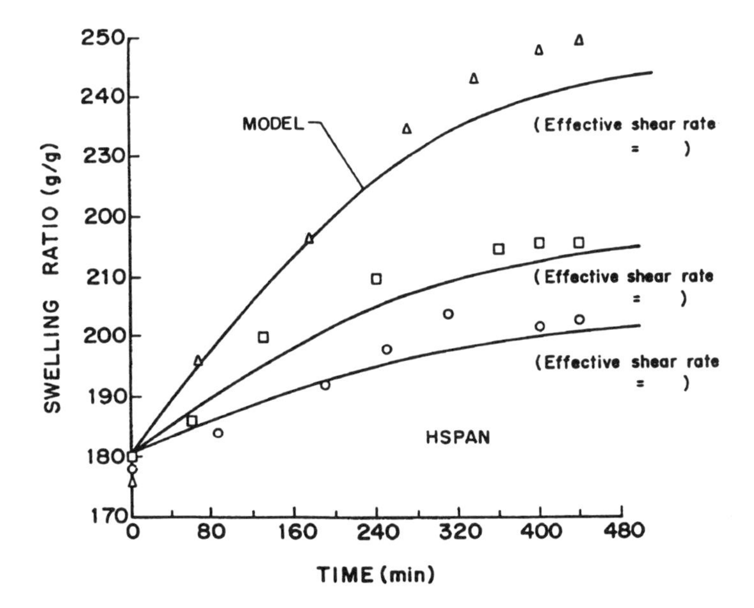

Fig. 3. Time-dependent swelling of the HSPAN gel while being sheared in a couette device. Points show experimental data, lines show theoretical calculations.

sheared in a specially constructed Couette device in which *in situ* swelling measurements can be done during the shearing process [1]. Figure 3 shows the time-dependent swelling of the gel at three different rotation speeds of the inner cylinder. The swelling capacity increases with time until a saturation value is reached, which is higher for greater rotation speeds. The lines in Fig. 3 are predictions using Eqs. (2) and (3) of the simple theory outlined above. It is interesting to note that by fitting the theory to the data for one rotation speed, all parameters of the model can be obtained and the data for the other two rotation speeds become completely predictive.

There are many other interesting observations in the microgel system. On stopping the shear after attaining a saturation swelling capacity, the microgel de-swells extremely slowly as indicated in Fig. 4. It appears that the gel might not completely attain its original swelling capacity. This indicates that the reformation is not only a very slow process, but it can also remain incomplete thus indicating a "partial memory" in the system. Our simple theory predicts a complete memory instead of a partial memory for the microgel. Further, the

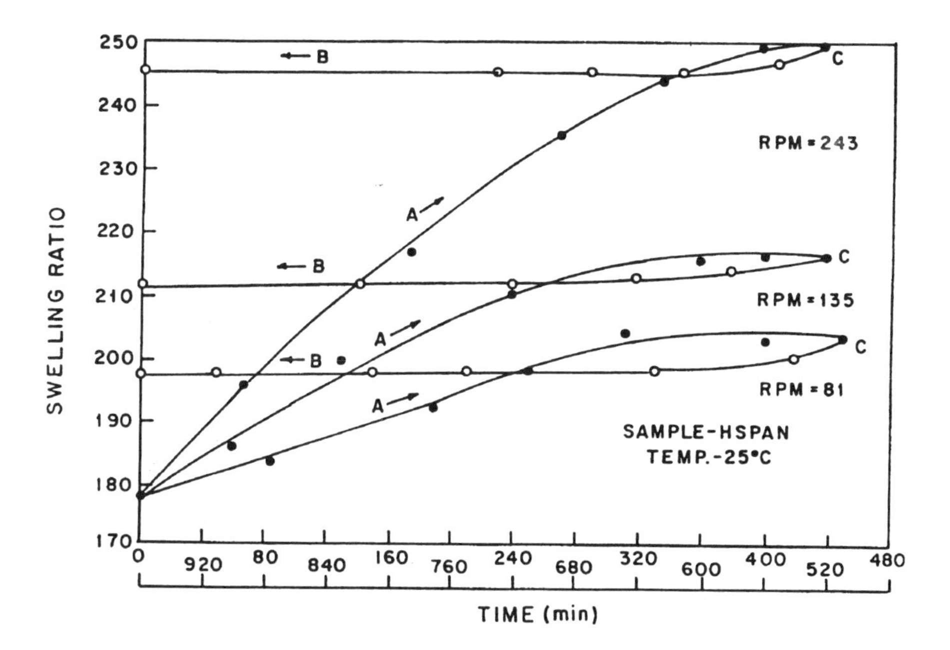

Fig. 4. Slow and partial de-swelling of the HSPAN microgel upon the cessation of the shear. The lines through the data are drawn only for better visual clarity.

time scale associated with the reformation process is greater than any molecular relaxation time scale by several orders of magnitude. This indicates that the reformation would involve a co-operative mobility of the network rather than local chain motions. As will be shown below, these features are also seen in concentrated polymer solutions, which indicate that the underlying mechanism of network breakage and formation through co-operative motion is common to physically crosslinked systems.

Stress-Growth Experiments

Kulicke *et al.* [7, 8] have shown that when a solution of polyacrylamide in ethylene glycol is sheared at shear rates above 525 s^{-1}, the shear stress and the first normal stress exhibit a double overshoot, which is time- and history-dependent (Fig. 5). A gel fracture type response was typically observed after the second overshoot and, in fact, the solution was seen to gel up in the gap of the rheometer. The authors believe that the first overshoot is, us usual, associated with the chain entanglement-disentanglement dynamics while the second overshoot is due specifically to energetic associations. On attaining a

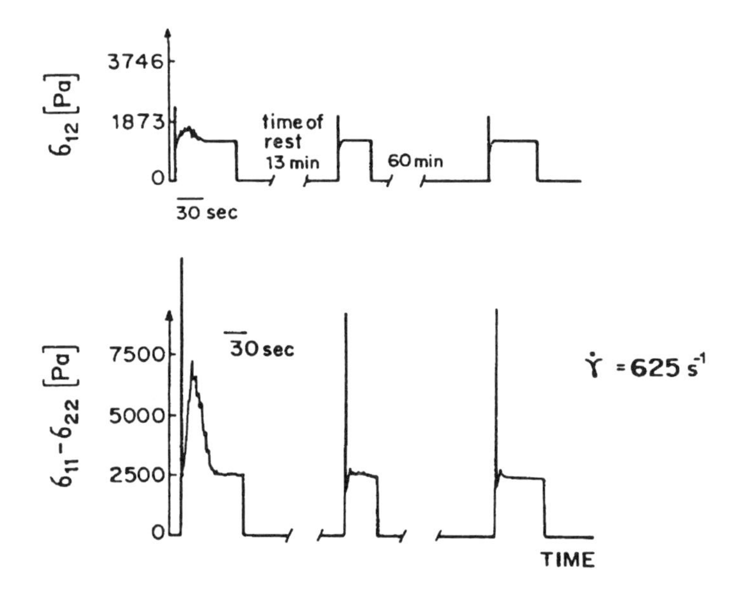

Fig. 5. Experimental data from Kulicke *et al.* [7] showing double overshoots of shear stress and first normal stress in a 2% polyacrylamide solution in ethylene glycol.

steady state stress after the initial double overshoot, when the experiment was stopped for as long as 60 minutes and then re-started, the second overshoot was not observed. This experiment clearly demonstrates: (i) that energetic interactions can become important under certain experimental conditions; (ii) that the time scales associated with the dynamics of the two types of associations are very different; and (iii) that the large time required for reformation is associated with some kind of co-operative motion of the network. Somewhat similar observations were reported by Peterlin and Turner earlier [16]. The large times scales are similar to those observed in the microgel system, indicating that the mechanism of slow relaxation of an energetically associated network is common to both systems.

There are two possible ways by which a double stress overshoot could be observed. If there is more than one relaxation mode for the polymer in solution, such as the different time scales of entanglement dynamics and energetic interactions, each mode will manifest a separate overshoot after a certain shear rate at which the experimental time scale is lower than those of both modes. On the other hand, it is also possible that the polymer molecules associate through energetic interactions above a certain shear rate at which the chain extension and orientation favour intermolecular association. This sudden energetic association can also cause the second overshoot phenomenon. By writing the full equations for components of stress tensor from Eqs. (6) to (8), it is possible to predict the double overshoot phenomenon. Our preliminary results in this attempt have been successful and details of this work will be part of a forthcoming publication.

Time-Dependent Settling of Spheres

We next consider the case of spheres settling in a polymer solution, such as polyacrylamide, which can exhibit energetic interactions. It has been shown by Ambeskar and Mashelkar [4] that spheres drop one after another along the same path in a polyacrylamide solution, but with a fixed intermittent time gap showing increasing terminal settling velocities until a saturation velocity is reached (Fig. 6). The increase in the settling velocity reduces with an increase in the time gap and the waiting time required before the next sphere reaches the same settling velocity as the one before it is of the order of minutes. Thus, it appears that the first sphere thins the liquid in its path and that the liquid retains the memory of this reduced apparent viscosity for a long time, allowing the next ball to travel in a thinner fluid at a higher velocity. Joseph *et al.* [17]

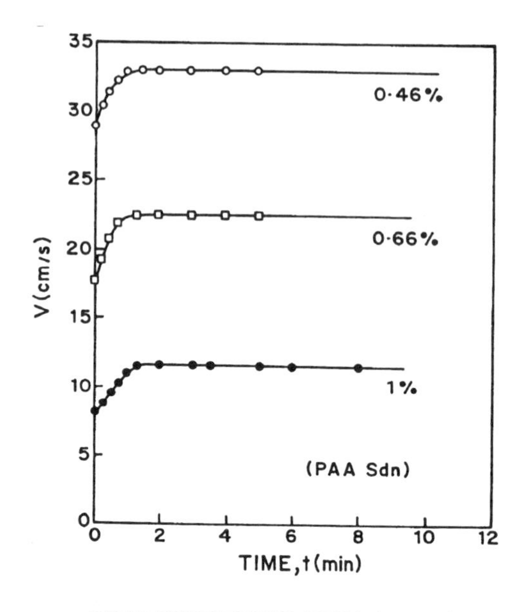

Fig. 6. Time-dependent terminal velocities of spheres settling consecutively in polyacrylamide solution: effect of intermediate time gap. Data from Ambeskar and Mashelkar [4].

have suggested that in order to describe this phenomenon, it appears that the polymer solutions should have a "memory of shear thinning". It is interesting to note that the long waiting time needed before the second sphere reaches the same terminal velocity as the first indicates the similarity between the polymer solution and the microgel. It is suggestive that the underlying mechanism for these effects might be similar.

One of the theories used to explain the time-dependent terminal settling velocities of spheres is that of a "depleted region" [18]. It is argued that the polymer molecules migrate away from regions of high shear near the surface of the sphere, thus causing a depleted region with lower viscosity. The subsequent sphere therefore experiences a lower viscosity and hence a higher settling velocity. However, Ambeskar and Mashelkar [4] have in fact shown experimentally that no measurable depletion of polymer was seen. Recently, Ianniruberto and Marucci [19] and Gheissary and van den Brule [5] have expressed doubts about the accuracy of the sample withdrawal procedure used by Ambeskar and

Mashelkar. However, we believe that, first, the cylindrical syringe which was inserted in the path traversed by the sphere would withdraw mostly solution from that path. Second, the accuracy of measurements on the concentration of polyacrylamide was 20 ppm, whereas the observed increase in the settling velocities would have required a depletion of at least 200 ppm.

We argue that the simple transient network model proposed here is able to explain, at least qualitatively, the above phenomenon. The settling of the first sphere sets up a shear rate gradient which varies from a maximum at

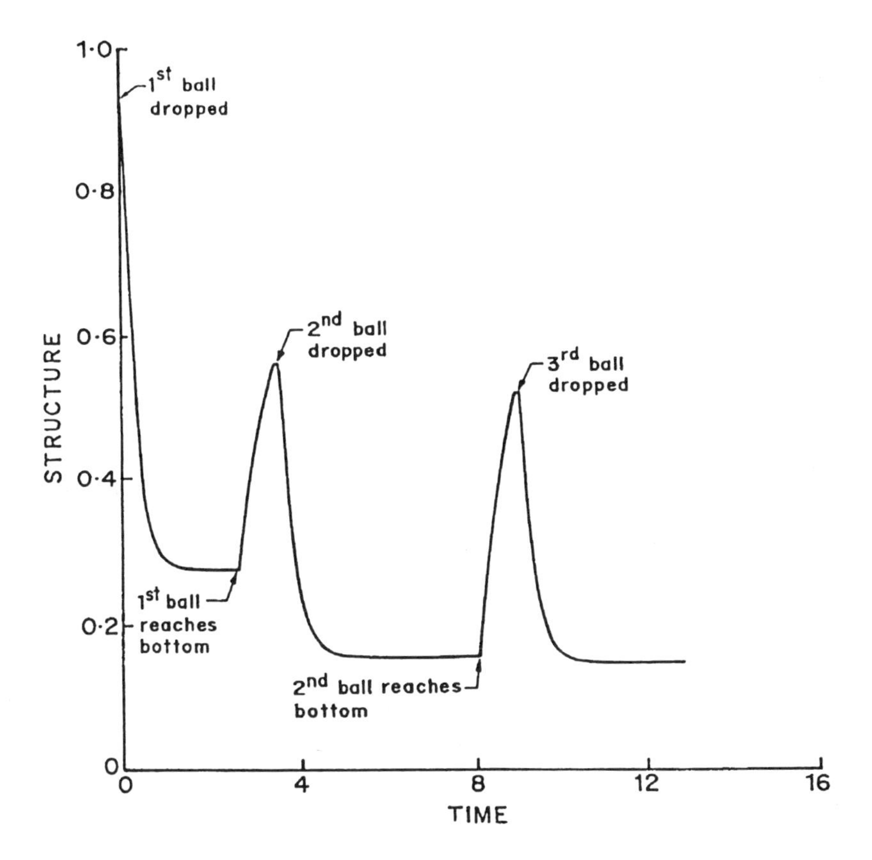

Fig. 7. Qualitative predictions of the effective resistance to the settling of consecutive spheres dropped in a polymer solution having energetic interactions with entanglements. A decrease in the resistance occurs due to the network breakage by the falling ball. The network recovers partially in the intermediate time gap through co-operative diffusivity and site orientations. The second ball is dropped before complete network recovery.

the sphere surface to almost zero within a short distance that scales approximately as $1/r^2$, with r being the sphere radius. This shear rate gradient causes a gradient in the crosslink density of the network in much the same way as the pipe flow. At steady state, the decrease in the crosslink density corresponds to the local shear rate which depends on the settling velocity. Thus, the average crosslink density in the path of the sphere is lower than that outside the path. As the first sphere settles down, the network starts reforming during the interval before the next sphere is dropped. Thus, the path of the first sphere can be said to have a "memory", due to which the viscosity (in general, the resistance

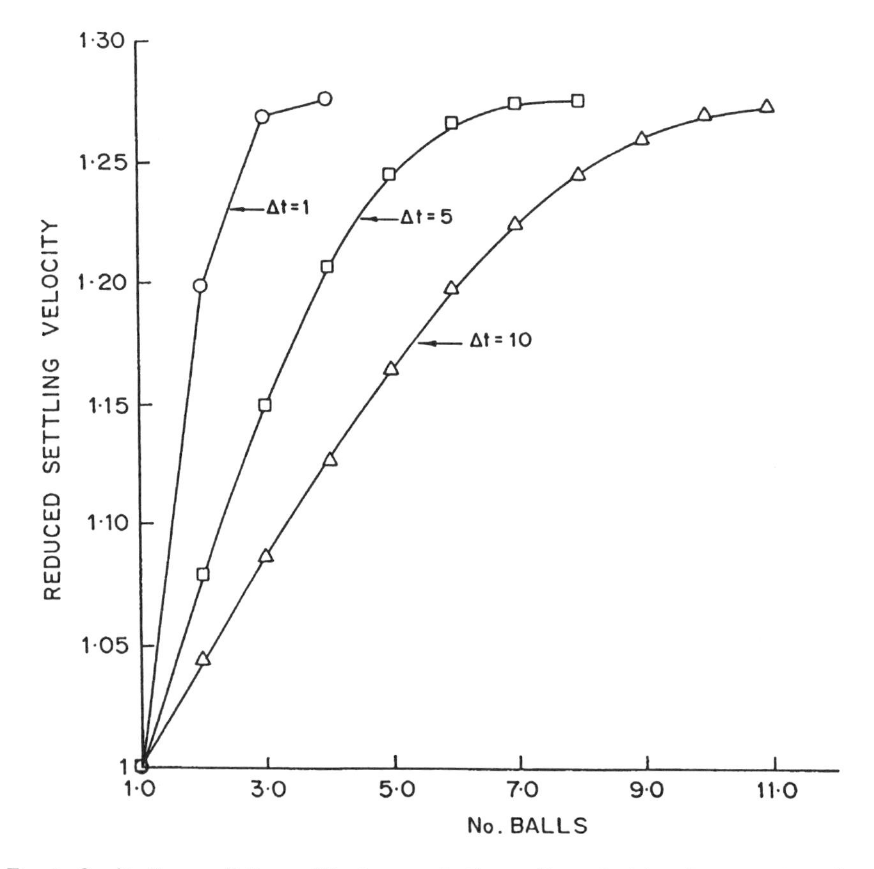

Fig. 8. Qualitative predictions of the increase in the settling velocities of consecutive spheres dropped in a polymer solution. The effect of increasing intermediate time gap is to decrease the difference in the settling velocities of consecutive spheres.

to flow) builds up over time. If the next sphere is dropped within the time required for complete reformation, it will experience a lower effective viscosity compared to the first sphere as it falls. Since the mass of the sphere is the same as the first sphere and the effective drag force is now lower, the sphere starts to settle with a higher velocity. This in turn imposes a higher shear rate and therefore a larger breakage rate of the network and a lower apparant viscosity. These lead to a higher terminal velocity of the second sphere. This continues for subsequent spheres until the network breaks to a state of E_∞ after which no further breakage is possible. The subsequent spheres then have the same terminal velocities, which is the saturation velocity seen experimentally. Figure 7 shows the average apparant "viscosity" of the solution in the path traversed by consecutive spheres. Figure 8 shows the increasing terminal velocities of settling spheres. It is also seen that as the intermediate time gap between two spheres is increased, the increase in the settling velocities is reduced. This is because the network is allowed a greater time to reform to a larger extent. The calculations presented in Figs. 7 and 8 are very preliminary in nature and are only used to demonstrate the qualitative predictions of the theory. We are in the process of solving the more complete and involved problem of the flow of such a transient network around a sphere using 3-D FEM calculations. It might be added that such time-dependent viscosity changes are not seen in a cone and plate steady shear experiment. This is because either a constant shear rate or shear stress is imposed on the solution independent of the resistance to flow, whereas in the falling ball viscometer the shear rate (due to the settling velocity) depends on the resistance to flow. Thus, in the former the constant shear rate would always correspond to the same steady state network density (and hence the effective resistance to flow), while in the latter the decreased flow resistance increases the shear rate through higher terminal velocity.

The three cases discussed above, namely the sheared microgel, the double overshoot and time-dependent settling velocities, are cases where network reformation governs time-dependent behaviour. We now look at some steady state experiments and show that the same model discussed above can in fact provide an alternative explanation for certain flow anomalies observed in steady state experiments. We consider the cases of steady flow in a pipe and sheet coating from a polymer solution.

Apparant Slip in Pipe Flow

We consider first the flow of polyacrylamide solutions in a pipe. It is well-known that these solutions exhibit apparant slip after a critical shear rate is

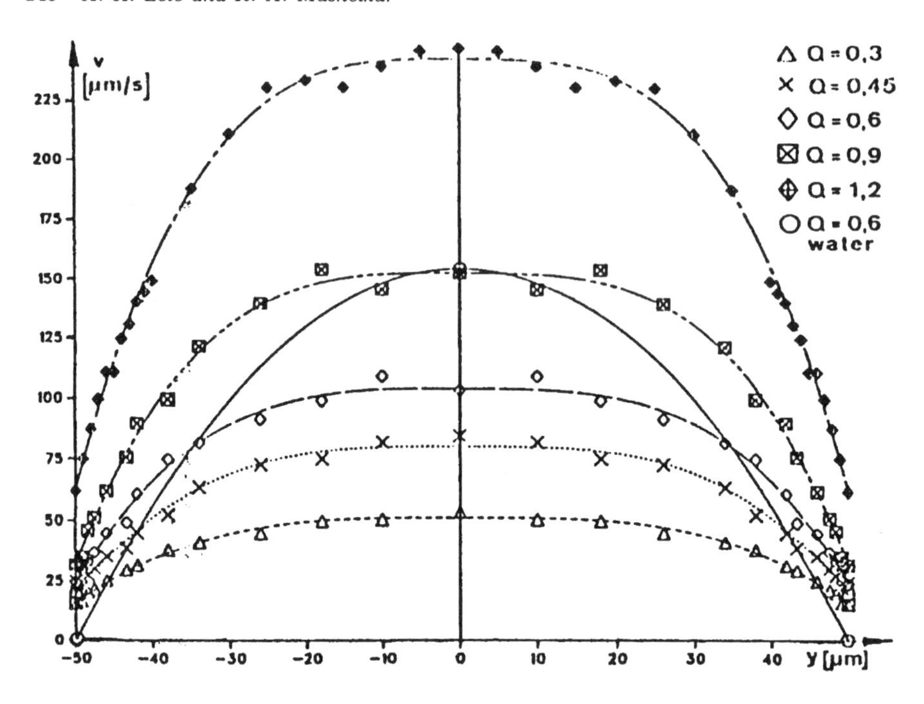

Fig. 9. Experimental data of Müller-Mohnssen [21] showing the apparent slip flow at the walls of a capillary for a polyacrylamide solution.

exceeded [20, 21]. For example, Fig. 9 shows the data of Müller-Mohnssen [21] giving the experimentally measured apparant slip of polyacrylamide solution flowing in a slit. The slip causes an increase in the net flow rate, which increases with a decrease in the pipe diameter. The increase in the flow rate has been modelled by assuming a finite slip velocity at the pipe walls instead of a no-slip condition. However, it is not yet known with certainity whether the no-slip condition is indeed violated at a segmental scale near the wall. It has been suggested that a depleted region of lower viscosity is formed near the walls of the pipe due to migration, and this causes the apparant slip to occur. However, a number of contradictory predictions of polymer migration theory exist in the literature which have been summarised by Agarwal *et al.* [18].

We propose an alternative simple mechanism in which the breakage rate of the network is proportional to the local shear rate. This is so that the effective crosslink density of the network is minimum (E_∞) near the pipe walls where the shear rate is maximum, and maximum (E_0) at the centre line where the shear rate is zero. Since the apparant viscosity of the solution is proportional to the effective crosslink density, a viscosity gradient is set up which in turn further increases the local velocity of flow near the walls. This coupling is evident in Eqs. (4) to (6) in the theory. The combined effect results in a steep velocity gradient near the wall which resembles the apparant slip. It is important to note that we have not invoked the apparant slip assumption. The boundary conditions are assumed to be

$$\nu = 0 \quad \text{and} \quad E = E_\infty \quad \text{at} \quad r = R\,,$$

$$\frac{\partial \nu}{\partial r} = 0 \quad \text{and} \quad E = E_0 \quad \text{at} \quad r = 0\,. \tag{11}$$

Figure 10 shows the velocity profile for the flow of a solution exhibiting shear rate dependent physical crosslinking compared to that of a "Newtonian" fluid

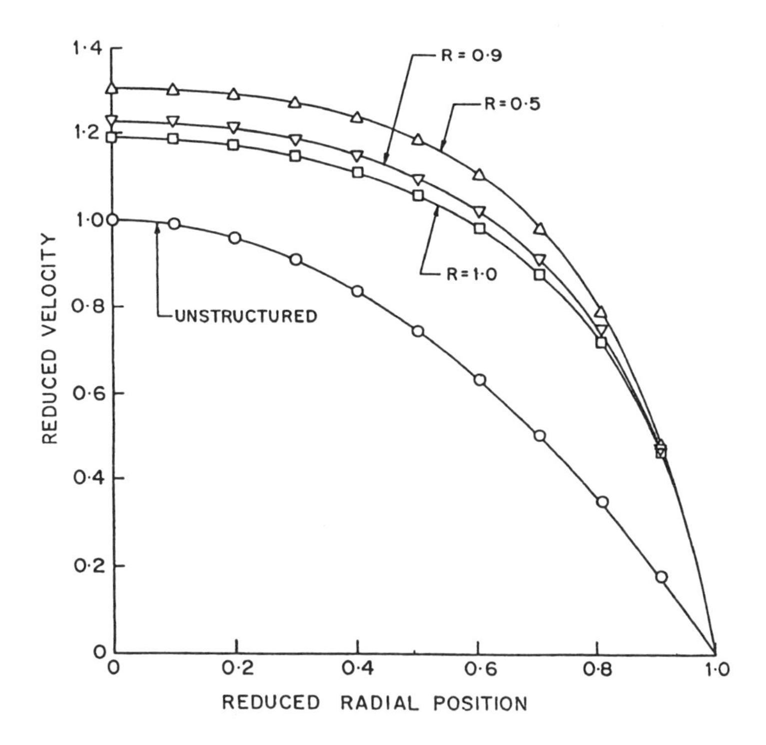

Fig. 10. Predictions of the velocity profile for a polyacrylamide solution by the transient network theory. The sharp gradients at the wall are similar to those observed experimentally as shown in Fig. 9.

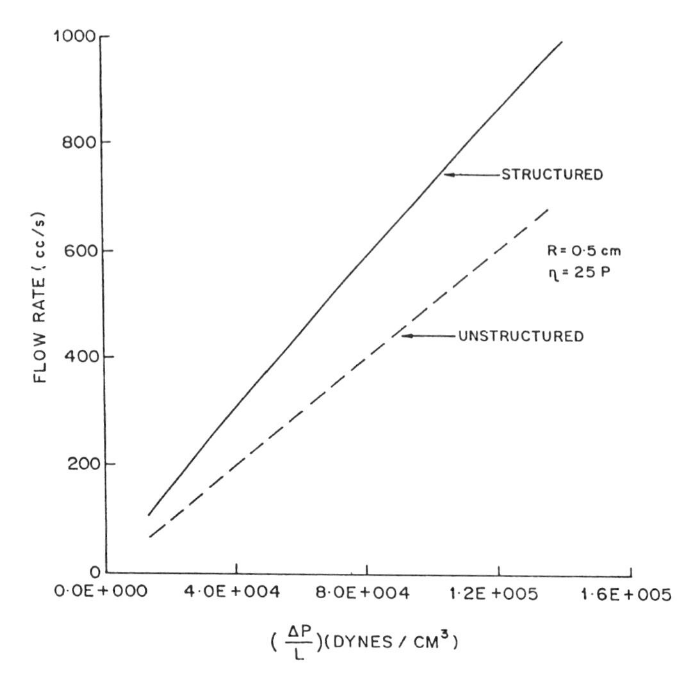

Fig. 11. Predictions of the increase in flow rate of polyacrylamide solution flowing through a capillary due to network breakage at the capillary walls.

having a constant viscosity. Similarly, Fig. 11 shows an increase in the net flow rate of the solution compared to a constant viscosity fluid. Our hypothesis is in some ways similar to the model of Brochard and de Gennes [22] who proposed that after a critical shear rate, the polymer chains adsorbed on the walls undergo a coil-stretch transition which becomes more effective due to the entanglements with the flowing bulk chains. This transition causes a sudden disentanglement very close to the wall and is the cause of the apparant "slip" seen in experiments. Our theory does not propose a coil-stretch transition, but rather proposes a continuous breakage of the network with increasing shear rates.

Film Coating

The final example is that of the coating of a polymer film on a sheet which is withdrawn from a bath of polymer solution. It is known that the Tallmadge model predicts the experimentally measured film thicknesses very well for many polymer soultions. However, polyacrylamide and Carbopol solutions are known

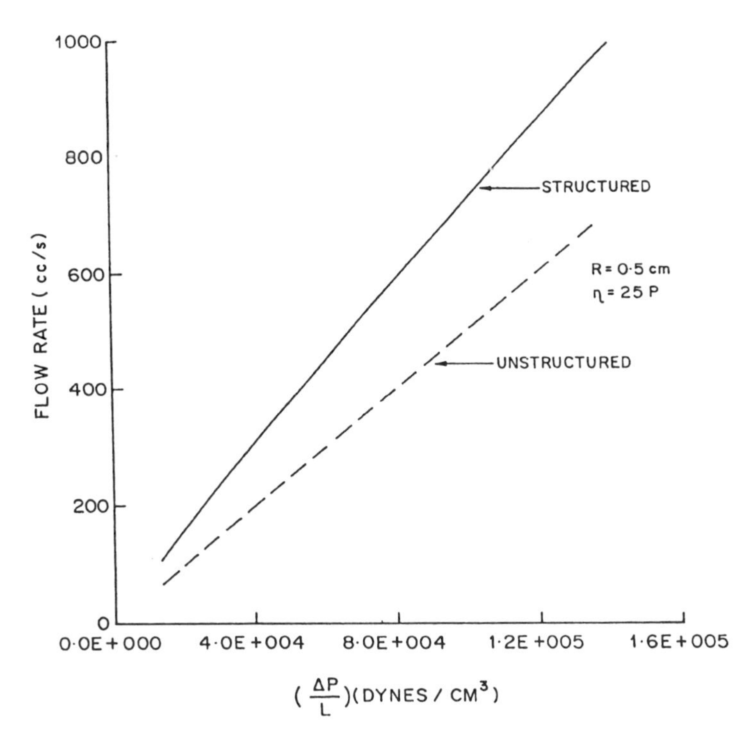

Fig. 12. Predictions of the decrease in the film's thickness for a polyacrylamide solution during the wire withdrawal experiment.

to form thinner films than those predicted by the Tallmadge relation. Dutta and Mashelkar [23] have provided a simple explanation for this behaviour by invoking the slip hypothesis, where it is shown that by assuming a finite slip velocity at the wire surface in the velocity profile of the Tallmadge model, a smaller film thickness is easily predicted. We propose that the transient network model can provide an alternative explanation for the observed effects without invoking the slip assumption. The shear rates in the coating experiment range from a maximum at the sheet surface to zero at the air interface. Consequently, the network will have the lowest crosslink density and therefore the lowest viscosity at the sheet surface and the maximum values at the air interface. For a Newtonian liquid, the film thickness is given by the relation between the dimensionless film thickness, T_∞, and the capillary number, Ca, as

$$\frac{T_\infty}{(1 - T_\infty^2)^{2/3}} = 0.944\ Ca^{1/6}\,. \tag{12}$$

Here,

$$T_\infty = H_\infty \left(\frac{\rho g}{\eta U}\right)^{1/2} \quad \text{and} \quad Ca = \frac{\eta U}{\sigma}, \tag{13}$$

where H_∞ is the ultimate film thickness, U is the withdrawal velocity and σ is the surface tension.

Thus, the simple transient network model, which couples the dynamics of transient networks with the conservation equations, can successfully predict a number of "anomalous" flow behaviours of polymer solutions. These anomalies are explained without making the assumptions of apparent "slip velocities" or polymer migration. However, the predictions presented here are very preliminary and therefore only qualitative in nature. A generalised theory which incorporates the salient features of energetic interactions and used for obtaining detailed solutions to the flow problems will be published separately.

Conclusions

We have provided some new suggestions on the important role played by energetic interactions in the dynamics of transient networks such as microgels and concentrated polymer solutions. The energetic interactions, such as hydrogen bonding, are prevalent in many "benchmark" fluids such as polyacrylamide and Carbopol solutions, which are routinely used for rheological measurements. The fact that such systems form networks through chain entanglements and energetic interactions has been substantiated by experiments such as the stress growth measurements by Kulicke *et al.* [8–9] and the time-dependent swelling of microgels by [3]. The salient features of these interactions include the long recovery times need for the removal of stress and different time scales for breakage-reformation dynamics compared to the chain entanglement-disentanglement dynamics.

We have proposed a simple transient network model by taking into consideration the existence of energetic interactions. By coupling the kinetics of volume averaged density of network segments with the quasi-equilibrium swelling pressure, we could quantitatively predict the time-dependent macroscopic swelling of a sheared microgel. The same model can also explain some anomalous flow behaviours of polymer solutions. We have shown that the same model can qualitatively explain transient flow problems, such as the time-dependent settling velocities of spheres and double overshoots in stress growth experiments. In the case of polymer solutions undergoing steady state flow, when the conservation equation for network segments is coupled with the momentum equation, we could qualitatively predict the flow enhancements observed in capillary flows without assuming either slip or a depleted region at the walls. Similarly, the lower thickness of polymer coatings during wire withdrawal experiments could be explained.

The predictions presented in this paper indicate our preliminary attempts in this direction. The detailed solutions to these problems are under investigation. We are also investigating the nature of the long restoration times by using a rheo-NMR device which can probe both molecular as well as segmental relaxations.

Discussions

M. Mackley Is there a second level of micro-structure with your swelling polymers when subject to shear? In order to maintain fluid properties, presumably the swollen gel must form a micro-structure to sustain large strain deformation.

A. Lele The microgel system indeed has a second level of micro-structure which is formed by the transient crosslinks such as hydrogen bonds. The microgel can be viewed as a suspension of covalently crosslinked microgel beads held to each other by transient crosslinks. The entire system (that is, the microgel beads and the interconnecting transient crosslinks) contain large quantities of absorbed water. On imposing shear, it is the transient crosslinks which break and allow the microgel system to sustain large strain deformations.

G. Marucci How does the model reconcile the drop in viscosity in the wake of the falling ball to the shearing viscosity of a normal viscosity measurement?

A. Lele The viscosity measured in a steady shear experiment is predicted by the model as the viscosity associated with a steady-state "structure" in the solution given by $\eta = K[E]_{ss}$. The steady state structure is formed by the dynamic balance between the rates of breakage and formation of the transient crosslinks, which in turn depend only on the shear rates. Thus, as long as the shear rate is held constant, the model will predict a certain steady viscosity. However, in the falling ball experiment, the second ball falling along the same path as the first ball experiences a lower initial viscosity due to the partial recovery of the structure broken by the first ball. This causes a rise in the settling velocity of the second ball, which in turn increases the shear rate on the solution and therefore a lower viscosity. Thus, dropping a second ball which is identical to the first ball in terms of its mass and size does not imply imposing the same shear rates by both balls during their settling.

V. Kumaran The important parameter which determines the rate at which the micro-structure relaxes to the original state after a perturbation is the difference in free energy for hydrogen bonds. Should not the activation energy be an important parameter, and will it not be affected by co-operative motions of the polymers?

A. Lele One of the driving forces for the relaxation of the micro-structure should indeed be the free energy of hydrogen bonds. One can think of other kinds of free energies, such as those associated with hydrophobic interactions, in case such interactions are prevalent in the solution. Thus, the activation energy for the formation of such energetic interactions should indeed be an important parameter. However, there are perhaps other "entropic" parameters, such as the appropriate orientations of the interacting groups which can influence the reformation of energetic interactions. Both the activation energy, as well as the entropic parameters, will indeed be affected by the co-operative mobility of the polymer molecules. None of the existing transient network models have incorporated any details of the kinetics of reformation of transient crosslinks. We believe that such kinetics should incorporate the above-mentioned features.

V. Entov What is the lowest value of concentration at which the falling ball paradox is observed? What is the value of the molecular mass of PAA? Have you visualised the flow in the falling sphere experiment, and how long does it take for it to decay after the sphere reaches the bottom?

A. Lele All the concentrations which we have looked at are above the c^* value. We have not really examined dilute solutions. However, within the concentration range studied, we find that the percentage increase in the settling velocities of consecutively dropped balls decreases with decreasing concentration. Thus, in dilute solutions, the changes in the settling velocities are expected to be small for consecutively settling balls. The molecular mass of Separan used was about 4×10^6. The wake created by the falling ball visually disappears within a minute for the most concentrated solutions we have used. It is difficult and probably not reliable to relate the visual relaxation of the wake to the restoration times of the settling velocities.

V. Entov Of course, it is nothing strange in long after effects if gels are partially destroyed by falling balls. However, it would be important to make sure that the flow decays before the next sphere is dropped.

M. Denn What is the effect of pH on the PAA experiments?

A. Lele All the experiments reported here are done at neutral pH. We have used distilled de-ionised water for preparing solutions. It may be worth carrying out the experiments at acidic and alkaline pH to determine the role of the charges created on the polymer backbones. However, the effects of enhanced viscosity due to polyelectrolyte effects will have to be properly accounted for.

G. Goddard Have the authors considered using the technique of parallel superposition to investigate the effects of entanglements and gel formations?

A. Lele We have not looked at the rheological response on a Separan solution subjected to oscillations imposed on steady shear flow. This experiment should be quite interesting as it might separate the responses of the physical entanglements from those of a gel formed through energetic interactions.

B. Costello (TA Instruments) Such an experiment has been done for various industrial gels and a sharp gel/entanglement transition is usually observed occuring over less than a decade of shear rate.

References

1. Badiger, M. V., Lele, A. K., Kulkarni, M. G. & Mashelkar, R. A. (1994) Swelling and phase transitions in deforming polymeric gels. *Ind. Eng. Chem.* **33**, 2426
2. Bisgaard, C. (1983) Velocity fields around spheres and bubbles investigated by laser doppler anemometry. *J. Non-Newtonian Fluid Mech.* **12**, 283
3. Cho, Y. I., Hartnett, J. P. & Lee, W. Y. (1984) Non-Newtonian viscosity measurements in the intermediate shear rate range with falling ball viscometer. *J. Non-Newtonian Fluid Mech.* **15**, 61
4. Ambeskar, V. D. & Mashelkar, R. A. (1990) On the role of stress-induced migration on time-dependent terminal velocities of falling spheres. *Rheol. Acta* **29**, 182
5. Gheissary, G. & van den Brule, B. H. A. A. (1996) Unexpected phenomena observed in particle settling in non-Newtonian media. *J. Non-Newtonian Fluid Mech.* **67**, 1

6. Jones, W. M., Price, A. H. & Walters, K. (1994) The motion of a sphere falling under gravity in a constant viscosity elastic liquid. *J. Non-Newtonian Fluid Mech.* **53**, 175
7. Kulicke, W. M., Kniewske, R. & Klien, J. (1982) Preparation, characterisation, solution properties and rheological behaviour of polyacrylamide. *Prog. Polym. Sci.* **8**, 373
8. Kulicke, W. M., Klien, J. & Porter, R. S. (1979) Viscoelastic properties of linear chain macro-molecular gels. Growth of stress at the onset of steady shear flow. *Die Angew Makromol Chemie* **76/77**, 151
9. Lodge, A. S. (1956) A network theory of flow birefringence and stress in concentrated polymer solution. *Trans. Faraday Soc.* **52**, 120
10. Bird, R. B., Armstrong, R. C. & Hassager, O. (1977) *Dynamics of Polymeric Liquids, Vols. 1 & 2*, New York: John Wiley & Sons
11. Wang, S.-Q. (1992) Transient network theory for shear thickening fluids and physically crosslinked systems. *Macromoecules* **25**, 7003
12. Marucci, G., Bhargava, S. & Cooper, S. L. (1993) Models for shear-thickening behaviour in physically crosslinked networks. *Macromolecules* **26**, 6483
13. Tanaka, F. & Edwards, S. F. (1992) Viscoelastic properties of physically crosslinked networks. Transient network theory. *Macromolecules* **25**, 1516
14. Vrahopoulou, E. P. & McHugh, A. J. (1987) Theory with non-Gaussian chain segments. *J. Rheol.* **31**, 371
15. Ahn, K. H. & Osaki, K. (1995) Mechanism of shear-thickening investigated by a network model. *J. Non-Newtonian Fluid Mech.* **56**, 267
16. Peterlin, A. & Turner, D. T. (1965) Temporary network formation in shearing solutions of polymethyl methacrylate in aroclor. *Polym. Lett.* **3**, 517
17. Joseph, D. D., Liu, Y. J., Polette, M. & Feng, J. (1994) Aggregation and dispersion of spheres falling in viscoelastic liquids. *J. Non-Newtonian Fluid Mech.* **54**, 45
18. Agarwal, U. S., Dutta, A. & Mashelkar, R. A. (1994) Migration of macromolecules under flow: the physical origin and engineering implications, *Chem. Eng. Sci.* **49**, 1693
19. Ianniruberto, G. & Marucci, G. (1994) Falling spheres in polymeric solutions. Limiting results of the two fluid theory of migration. *J. Non-Newtonian Fluid Mech.* **54**, 231
20. Cohen, Y. & Metzner, A. B. (1982) Slip phenomena in polymeric solutions flowing through small channels. *A.I.Ch.E. Symp. Ser.* No. 212, **78**, 77
21. Muller-Mohnssen, H., Lobl, H. P. & Schauerte, W. (1987) Direct determination of apparent slip for a ducted flow of polyacrylamide solutions. *J. Rheol.* **31**, 323
22. Brochard, F. & de Gennes, P. G. (1992) Shear-dependent slippage at a polymer-solid interface. *Langmuir* **8**, 3033
23. Dutta, A. & Mashelkar, R. A. (1982) On slip effect in free coating of non-Newtonian fluids. *Rheol. Acta.* **21**, 52

Dynamics of Complex Fluids, pp. 155–175
ed. M. J. Adams, R. A. Mashelkar, J. R. A. Pearson & A. R. Rennie
Imperial College Press–The Royal Society, 1998

Chapter 10

The Prediction of Universal Viscometric Functions for Dilute Polymer Solutions Under Theta Conditions

J. RAVI PRAKASH
Department of Chemical Engineering,
Indian Institute of Technology, Madras, India 600 036
E-mail: rprakash@acer.iitm.ernet.in

The universal behaviour of different polymer systems, independent of detailed chemical structure, is used to justify the introduction of coarse-grained bead-spring models to represent real polymer molecules in many theories for the rheological properties of dilute polymer solutions. However, in all the current theories which incorporate the effect of hydrodynamic interaction, there occur two parameters apart from a basic length and time scale, namely, the number of beads N, and the strength of the hydrodynamic interaction h^*. The ultimate choice of values for these parameters is essentially arbitrary, and this choice crucially influences the model's predictions. In the limit of N going to infinity, however, these predictions can be shown to become parameter-free and, consequently, independent of the details of the mechanical model. In this paper, the traditional treatment of hydrodynamic interaction is briefly reviewed and the recently introduced twofold normal approximation, which permits an evaluation of the viscosity and the first and second normal stress differences in simple shear flow as a function of shear rate for very large values of N, is discussed. The universal predictions of these viscometric functions, obtained by the extrapolation of the long chain results to the limit $N \to \infty$, are also displayed and discussed.

Introduction

Polymer kinetic theory represents one of the means by which constitutive equations for complex polymeric fluids may be derived. The polymer molecule is typically represented by a coarse-grained model, and the rapidly fluctuating motions of the solvent molecules surrounding the polymer molecule are replaced by a randomly varying force field. The polymer model then interacts with itself and with the force field leading to the macroscopically observed dynamic behaviour. The usefulness of a constitutive equation developed in this manner is verified by comparing the prediction of viscometric functions in certain simple flows, such as simple shear flows or elongational flows, with experimental results.

The most significant observations of dilute polymer solution behaviour in steady simple shear flow are (i) the non-vanishing of the first and second normal stress differences and (ii) the dependence of all the viscometric functions on the shear rate. Strikingly, these viscometric functions exhibit universal behaviour when the experimental data on shear rate dependence for various polymer systems are plotted in terms of appropriately normalised co-ordinates. Such universal behaviour is believed to arise because the macroscopic properties of the polymer solution are determined by a few large-scale properties of the polymer molecule and not by the fine details of the chemical structure. Dilute solutions that differ from each other with regard to, for example, the chemical structure, the molecular weight of the dissolved polymer and the temperature behave similarly so long as a few parameters specifying the molecular characteristics are the same.

Molecular theories usually cite this universal behaviour in order to justify the introduction of crude mechanical models to represent real polymer molecules. For instance, a frequently used model is a linear chain of identical beads connected by Hookean springs. Such models do indeed represent certain large-scale features of the polymer molecule, such as its stretching and orientation by the solvent flow field. However, it turns out that the predictions of these models are by no means universal since, apart from a basic length and time scale, there occur other parameters that need to be specified, such as the number of beads, N, in the chain and the strength of hydrodynamic interaction, h^*. Hydrodynamic interaction is a long-range interaction between the beads of a bead-spring chain model which arises because of the solvent's capacity to propagate one bead's motion to another through disturbances in its velocity field. Its inclusion, as discussed in greater detail below, leads to a significant improvement in the prediction of dilute polymer solution rheological properties.

This paper places in context the recent introduction of an approximate treatment of hydrodynamic interaction which makes it possible to overcome the limitations of the mechanical polymer model and to verify the universal predictions of viscometric functions in simple shear flow.

The next section is a review, in roughly chronological order, of attempts to incorporate hydrodynamic interaction under theta conditions (that is, neglecting excluded volume effects) into molecular theories for dilute polymer solution behaviour. The diagonalise and decouple procedure by which long chain rheological behaviour may be explored is then discussed. A representative sample of universal properties in steady simple shear flow, obtained by Prakash and Öttinger [1] who extrapolate the finite long chain results of the recently introduced twofold normal approximation to the infinite chain limit, concludes this paper.

Kinetic Theory with Hydrodynamic Interaction

Basic Equations

The kinetic theory of the bead-spring chain model, suspended in a Newtonian solvent with a homogeneous velocity field and with hydrodynamic interaction taken into account, requires that $\psi(\boldsymbol{Q}_1, \boldsymbol{Q}_2, \ldots \boldsymbol{Q}_{N-1})$, the configurational distribution function for the $N-1$ bead connector vectors $\boldsymbol{Q}_j$, satisfies the "diffusion" equation [2]

$$\frac{\partial \psi}{\partial t} = -\sum_{j=1}^{N-1} \frac{\partial}{\partial \boldsymbol{Q}_j} \cdot \left\{ \left(\boldsymbol{\kappa} \cdot \boldsymbol{Q}_j - \frac{H}{\zeta} \sum_{k=1}^{N-1} \widetilde{\mathbf{A}}_{jk} \cdot \boldsymbol{Q}_k \right) \psi \right\} + \frac{k_B T}{\zeta} \sum_{j,k=1}^{N-1} \frac{\partial}{\partial \boldsymbol{Q}_j} \cdot \widetilde{\mathbf{A}}_{jk} \cdot \frac{\partial}{\partial \boldsymbol{Q}_k} \psi \,, \tag{1}$$

where $\boldsymbol{\kappa}(t)$ is a traceless tensor that determines the homogeneous velocity field, H is the spring constant, ζ is the bead friction coefficient, k_B is Boltzmann's constant, and T is the absolute temperature. The tensor, $\widetilde{\mathbf{A}}_{jk}$, accounts for the presence of hydrodynamic interaction and is defined by

$$\widetilde{\mathbf{A}}_{jk} = A_{jk} \mathbf{1} + \zeta \left(\boldsymbol{\Omega}_{j,k} + \boldsymbol{\Omega}_{j+1,k+1} - \boldsymbol{\Omega}_{j,k+1} - \boldsymbol{\Omega}_{j+1,k} \right) \,, \tag{2}$$

where $\mathbf{1}$ is the unit tensor and A_{jk} is the Rouse matrix given by

$$A_{jk} = \begin{cases} 2 & \text{for } |j-k| = 0\,, \\ -1 & \text{for } |j-k| = 1\,, \\ 0 & \text{otherwise} \end{cases} \tag{3}$$

and the tensor, $\boldsymbol{\Omega}_{\mu\nu}$, which describes the hydrodynamic interaction between beads μ and ν, is here assumed to be the given by the Oseen-Burgers expression:

$$\boldsymbol{\Omega}_{\mu\nu} = \begin{cases} \dfrac{1}{8\pi\eta_s r_{\mu\nu}} \left(\mathbf{1} + \dfrac{\boldsymbol{r}_{\mu\nu}\boldsymbol{r}_{\mu\nu}}{r_{\mu\nu}^2} \right), \quad \boldsymbol{r}_{\mu\nu} = \boldsymbol{r}_\mu - \boldsymbol{r}_\nu & \text{for } \mu \neq \nu \\ 0 & \text{for } \mu = \nu \end{cases}. \tag{4}$$

The polymer contribution to the stress tensor, $\boldsymbol{\tau}^p$, is given by the Kramers expression,

$$\boldsymbol{\tau}^p = -nH \sum_{j=1}^{N-1} \langle \boldsymbol{Q}_j \boldsymbol{Q}_j \rangle + (N-1) n k_B T\, \mathbf{1}\,, \tag{5}$$

where n is the number density of polymers and the angular brackets represent averaging with respect to the configurational distribution function, ψ.

The second moments, $\langle \boldsymbol{Q}_j \boldsymbol{Q}_j \rangle$, are calculated from the following evolution equation:

$$\frac{d}{dt} \langle \boldsymbol{Q}_j \boldsymbol{Q}_k \rangle = \boldsymbol{\kappa} \cdot \langle \boldsymbol{Q}_j \boldsymbol{Q}_k \rangle + \langle \boldsymbol{Q}_j \boldsymbol{Q}_k \rangle \cdot \boldsymbol{\kappa}^T + \frac{2k_BT}{\zeta} \langle \widetilde{\mathbf{A}}_{jk} \rangle$$
$$- \frac{H}{\zeta} \sum_{m=1}^{N-1} \left[\left\langle \boldsymbol{Q}_j \boldsymbol{Q}_m \cdot \widetilde{\mathbf{A}}_{mk} \right\rangle + \left\langle \widetilde{\mathbf{A}}_{jm} \cdot \boldsymbol{Q}_m \boldsymbol{Q}_k \right\rangle \right]. \tag{6}$$

The central problem in attempting to predict the rheological properties of dilute polymer solutions is to find the solution to Eq. (6). It is not a closed equation for the second moments since it involves more complicated moments on the right-hand side. The various treatments of hydrodynamic interaction are essentially distinguished by the closure approximation that each proposes.

Mean Field Theories

An early attempt at obtaining a constitutive equation with kinetic theory was the model proposed by Rouse [3], which ignored the presence of hydrodynamic interaction altogether. This model leads to a constitutive equation that is a multi-mode generalisation of the convected Jeffreys or Oldroyd-B model. It is well-known that while this constitutive equation accounts for the presence of viscoelasticity through the prediction of a non-zero first normal stress difference in simple shear flow, it does not predict several other features of the behaviour of dilute solutions such as the non-vanishing of the second normal stress difference, the shear rate dependence of the viscometric functions, the correct molecular weight dependence of the viscosity, the first normal stress difference, and the diffusion coefficient.

Significant progress was made when Zimm [4] incorporated the effect of hydrodynamic interaction in a *pre-averaged* form by replacing $\widetilde{\mathbf{A}}_{jk}$ in the diffusion Eq. (1), and in the second moment Eq. (6) with its equilibrium average, $\widetilde{A}_{jk}$, which is called the modified Rouse matrix and is given by

$$\widetilde{A}_{jk} = A_{jk} + \sqrt{2}\, h^* \left(\frac{2}{\sqrt{|j-k|}} - \frac{1}{\sqrt{|j-k-1|}} - \frac{1}{\sqrt{|j-k+1|}} \right) . \tag{7}$$

Here, $h^* = a(H/\pi k_B T)^{\frac{1}{2}}$ is the hydrodynamic interaction parameter with a being the bead radius.

The predictions of the Zimm theory of the molecular weight dependence of the viscometric functions, the diffusion coefficient, the relaxation time, and the small amplitude oscillatory shear flow material functions are in excellent agreement with experiment. However, similar to Rouse, it too fails to predict the occurrence of a non-vanishing second normal stress difference and the shear rate dependence of the viscometric functions.

A non-zero *positive* second normal stress difference and shear rate dependent viscometric functions were predicted when Öttinger [5, 6] accounted for the presence of hydrodynamic interaction through a self-consistent averaging procedure. The self-consistent averaging method replaces the tensor, $\widetilde{\mathbf{A}}_{jk}$, in Eqs. (1) and (6) with $\overline{\mathbf{A}}_{jk}$, an average carried out with the non-equilibrium distribution function. This linearising procedure renders the diffusion equation solvable, and the solution ψ is indeed a Gaussian distribution similar to Rouse and Zimm. The $(N-1) \times (N-1)$ matrix with the tensor components, $\overline{\mathbf{A}}_{jk}$, is given by

$$\overline{\mathbf{A}}_{jk} = A_{jk}\, \mathbf{1} + \sqrt{2}\, h^* \times \left[\frac{\mathbf{H}(\hat{\boldsymbol{\sigma}}_{j,k}) + \mathbf{H}(\hat{\boldsymbol{\sigma}}_{j+1,k+1})}{\sqrt{|j-k|}} - \frac{\mathbf{H}(\hat{\boldsymbol{\sigma}}_{j,k+1})}{\sqrt{|j-k-1|}} - \frac{\mathbf{H}(\hat{\boldsymbol{\sigma}}_{j+1,k})}{\sqrt{|j-k+1|}} \right] . \tag{8}$$

Here, the function $\mathbf{H}(\boldsymbol{\sigma})$ is given by

$$\mathbf{H}(\boldsymbol{\sigma}) = \frac{3}{2(2\pi)^{3/2}} \int d\boldsymbol{k} \, \frac{1}{k^2} \left(\mathbf{1} - \frac{\boldsymbol{k}\boldsymbol{k}}{k^2} \right) \exp\left(-\frac{1}{2} \boldsymbol{k} \cdot \boldsymbol{\sigma} \cdot \boldsymbol{k} \right) \tag{9}$$

and the tensors, $\hat{\boldsymbol{\sigma}}_{\mu\nu}$, are defined by

$$\hat{\boldsymbol{\sigma}}_{\mu\nu} = \hat{\boldsymbol{\sigma}}_{\mu\nu}^T = \hat{\boldsymbol{\sigma}}_{\nu\mu} = \frac{1}{|\mu - \nu|} \frac{H}{k_B T} \sum_{j,k = \min(\mu,\nu)}^{\max(\mu,\nu)-1} \boldsymbol{\sigma}_{jk} \, , \tag{10}$$

where

$$\boldsymbol{\sigma}_{jk} \equiv \langle \boldsymbol{Q}_j \boldsymbol{Q}_k \rangle \, . \tag{11}$$

The tensors, $\boldsymbol{\sigma}_{jk}$, are not symmetric, but satisfy the relation

$$\boldsymbol{\sigma}_{jk} = \boldsymbol{\sigma}_{kj}^T \, . \tag{12}$$

It must be noted that the sign of the second normal stress difference has not been experimentally conclusively established. However, an exact solution obtained by Brownian dynamics simulations leads, at low shear rates, to a negative value for the second normal stress difference. Furthermore, while the predictions of the shear rate dependence of the viscosity and the first normal stress difference using the self-consistent averaging procedure are in qualitative agreement with Brownian dynamics simulations, they do not agree quantitatively.

The Incorporation of Fluctuations

The introduction of the Gaussian approximation for the hydrodynamic interaction [7–10] has led to predictions of viscometric functions that are in close agreement with the results of Brownian dynamics simulations. While the pre-averaging assumption of Zimm and the self-consistent averaging method of Öttinger solve the closure problem by replacing the tensor, $\widetilde{\mathbf{A}}_{jk}$, with an average, the Gaussian approximation makes no assumption with regard to the hydrodynamic interaction. Thus, fluctuations in the hydrodynamic interaction are not neglected. The Gaussian approximation assumes that the solution to the diffusion Eq. (1) may be approximated by a Gaussian distribution. As a result, all the complicated averages on the right-hand side of Eq. (6) can be reduced to the functions of the second moment, which gives rise to the closed second moment equation

$$\frac{d}{dt}\boldsymbol{\sigma}_{jk} = \boldsymbol{\kappa}\cdot\boldsymbol{\sigma}_{jk} + \boldsymbol{\sigma}_{jk}\cdot\boldsymbol{\kappa}^T + \frac{2k_BT}{\zeta}\,\overline{\mathbf{A}}_{jk} - \frac{H}{\zeta}\sum_{m=1}^{N-1}\left[\boldsymbol{\sigma}_{jm}\cdot\overline{\mathbf{A}}_{mk} + \overline{\mathbf{A}}_{jm}\cdot\boldsymbol{\sigma}_{mk}\right]$$

$$-\frac{H}{\zeta}\frac{H}{k_BT}\sum_{m,l,p=1}^{N-1}\left[\boldsymbol{\sigma}_{jl}\cdot\boldsymbol{\Gamma}_{lp,mk} : \boldsymbol{\sigma}_{pm} + \boldsymbol{\sigma}_{mp} : \boldsymbol{\Gamma}_{lp,jm}\cdot\boldsymbol{\sigma}_{lk}\right]\,. \tag{13}$$

Here, the $(N-1)^2\times(N-1)^2$ matrix with the fourth ranked tensor components, $\boldsymbol{\Gamma}_{lp,jk}$, is defined by

$$\boldsymbol{\Gamma}_{lp,jk} = \frac{3\sqrt{2}\,h^*}{4}\left[\frac{\theta(j,l,p,k)\,\mathbf{K}(\hat{\boldsymbol{\sigma}}_{j,k}) + \theta(j+1,l,p,k+1)\,\mathbf{K}(\hat{\boldsymbol{\sigma}}_{j+1,k+1})}{\sqrt{|j-k|^3}}\right.$$

$$\left. - \frac{\theta(j,l,p,k+1)\,\mathbf{K}(\hat{\boldsymbol{\sigma}}_{j,k+1})}{\sqrt{|j-k-1|^3}} - \frac{\theta(j+1,l,p,k)\,\mathbf{K}(\hat{\boldsymbol{\sigma}}_{j+1,k})}{\sqrt{|j-k+1|^3}}\right], \tag{14}$$

where the function, $\theta(j,l,p,k)$, is unity if l and p lie between j and k, and zero otherwise

$$\theta(j,l,p,k) = \begin{cases} 1 & \text{if } j \le l,p < k \quad \text{or} \quad k \le l,p < j \\ 0 & \text{otherwise} \end{cases}\,. \tag{15}$$

The function of the second moments, $\mathbf{K}(\boldsymbol{\sigma})$, is given by

$$\mathbf{K}(\boldsymbol{\sigma}) = \frac{-2}{(2\pi)^{3/2}}\int d\boldsymbol{k}\,\frac{1}{k^2}\boldsymbol{k}\left(\mathbf{1} - \frac{\boldsymbol{k}\boldsymbol{k}}{k^2}\right)\boldsymbol{k}\,\exp\left(-\frac{1}{2}\boldsymbol{k}\cdot\boldsymbol{\sigma}\cdot\boldsymbol{k}\right)\,. \tag{16}$$

Note that the conventions, $\mathbf{H}(\hat{\boldsymbol{\sigma}}_{jj})/0 = 0$, and $\mathbf{K}(\hat{\boldsymbol{\sigma}}_{jj})/0 = 0$ have been adopted in Eqs. (8) and (14). Both the hydrodynamic interaction functions, $\mathbf{H}(\boldsymbol{\sigma})$, and $\mathbf{K}(\boldsymbol{\sigma})$ can be evaluated analytically in terms of elliptic integrals [6, 10–12].

Öttinger [7] has examined small amplitude oscillatory shear flows and steady shear flow in the limit of zero shear rate with the Gaussian approximation for chains with $N \le 30$ beads, while Zylka [10] has obtained the material functions in steady shear flow for chains with $N \le 15$ beads and compared his results with those of Brownian dynamics simulations. As mentioned above, of all the approximate treatments of hydrodynamic interaction introduced so far, the Gaussian approximation compares best with simulation results. However, for both the flow situations just discussed, the examination of chains with larger values of N becomes increasingly difficult due to computational intensity. As a consequence, it is impossible to obtain results for very long chains.

Long Chain Behaviour

It is important to examine long chain behaviour for two reasons. First, the representation of more degrees of freedom makes it possible to explore more length (and time) scales. Consequently, aspects of the behaviour polymer solutions which are hidden when only short chains are considered might be revealed. Magda, Larson and Mackay [13], and Kishbaugh and McHugh [14] cite this as the reason for introducing a *decoupling approximation* which is much more computationally efficient (and which consequently enables them to examine long chain behaviour) than the self-consistent averaging procedure while at the same time retaining its accuracy. The second and more important reason for considering long polymer chains is that in the limit $N \to \infty$, parameter-free results are obtained. This has been shown to be true analytically by Öttinger in both the generalised Zimm model [5] and in the zero shear rate limit of the Gaussian approximation [7]. Prakash and Öttinger have recently verified this result numerically for arbitrary shear rates in the twofold normal approximation [1]. As a consequence, predictions of rheological properties in this limit are independent of the details of the mechanical model and represent the general consequences of the assumptions on hydrodynamic interaction.

The key to obtaining approximate treatments of hydrodynamic interaction, which are not excessively computationally intensive, lies in decoupling the connector vectors which are coupled to one another in the diffusion Eq. (1) and in the second moment Eq. (6). This is achieved rigorously in the Rouse and Zimm models, and approximately in the decoupling approximation of Magda, Larson and Mackay [13], and Kishbaugh and McHugh [14] by mapping the connector vectors, $\boldsymbol{Q}_1, \ldots, \boldsymbol{Q}_{N-1}$, to a new set of "normal" co-ordinates, $\boldsymbol{Q}'_1, \ldots, \boldsymbol{Q}'_{N-1}$, with the transformation

$$\boldsymbol{Q}_j = \sum_{k=1}^{N-1} \Pi_{jk} \boldsymbol{Q}'_k \,, \tag{17}$$

where Π_{jk} are the elements of an orthogonal matrix with the property

$$(\Pi^{-1})_{jk} = \Pi_{kj} \tag{18}$$

such that

$$\sum_{m=1}^{N-1} \Pi_{mj} \Pi_{mk} = \delta_{jk} \,. \tag{19}$$

Here, δ_{jk} is the Kronecker delta. In the Rouse model, the orthogonal matrix, Π_{jk}, diagonalises the Rouse matrix, A_{jk} [see Eq. (3)]. On the other hand, in the Zimm model, the orthogonal matrix, Π_{jk}, diagonalises the modified Rouse matrix, $\widetilde{A}_{jk}$ [see Eq. (7)]. In the decoupling approximation [13, 14], it is *assumed* that the time invariant Rouse (or Zimm) orthogonal matrix, Π_{jk}, diagonalises the matrix of *tensor* components, $\overline{\mathbf{A}}_{jk}$ [see Eq. (8)]. As a result of this diagonalisation, the diffusion equation becomes uncoupled in all these models and only the $(N-1)$ variances, $\boldsymbol{\sigma}'_j \equiv \langle \boldsymbol{Q}'_j \boldsymbol{Q}'_j \rangle$, remain non-zero. All the rheological properties are then obtained by solving the evolution equations that govern the variances, $\boldsymbol{\sigma}'_j$.

The decoupling approximation overcomes the computational intensity of the consistent averaging approximation and is therefore capable of predicting the behaviour of long chains. However, as mentioned earlier, the results obtained with the consistent averaging approximation are not in agreement with Brownian dynamics simulations. The *twofold normal approximation* introduced recently [1], though not as computationally intensive as the Gaussian approximation, is just as accurate. In other words, material functions can now be obtained that agree closely with Brownian dynamics simulations with considerably less computational effort.

The twofold normal approximation assumes that the configurational distribution function, ψ, is a normal distribution as in the Gaussian approximation. Again, as in the Gaussian approximation, the hydrodynamic interaction tensor, $\widetilde{\mathbf{A}}_{jk}$, is not replaced by an average, with the result that fluctuations in the hydrodynamic interaction are accounted for. Furthermore, the Rouse or Zimm orthogonal matrix, Π_{jk}, is used to map the bead connector vectors, $\boldsymbol{Q}_j$, to a new set of "normal" co-ordinates, $\boldsymbol{Q}'_j$, as in Eq. (17). The key assumption, however, is that Π_{jk} is assumed to diagonalise the covariance matrix, $\boldsymbol{\sigma}_{jk}$, that is,

$$\sum_{j,k=1}^{N-1} \Pi_{jp} \, \boldsymbol{\sigma}_{jk} \, \Pi_{kq} = \langle \boldsymbol{Q}'_p \boldsymbol{Q}'_q \rangle = \boldsymbol{\sigma}'_p \, \delta_{pq} \,. \tag{20}$$

The time evolution equations for the variances, $\boldsymbol{\sigma}'_j$, can then be shown to have the following form [1]:

$$\frac{d}{dt}\boldsymbol{\sigma}'_j = \boldsymbol{\kappa}\cdot\boldsymbol{\sigma}'_j + \boldsymbol{\sigma}'_j\cdot\boldsymbol{\kappa}^T + \frac{2k_BT}{\zeta}\,\boldsymbol{\Lambda}_j - \frac{H}{\zeta}\,\left[\boldsymbol{\sigma}'_j\cdot\boldsymbol{\Lambda}_j + \boldsymbol{\Lambda}_j\cdot\boldsymbol{\sigma}'_j\right]$$

$$- \frac{H}{\zeta}\,\frac{H}{k_BT}\sum_{k=1}^{N-1}\left[\boldsymbol{\sigma}'_j\cdot\boldsymbol{\Delta}_{jk} : \boldsymbol{\sigma}'_k + \boldsymbol{\sigma}'_k : \boldsymbol{\Delta}_{jk}\cdot\boldsymbol{\sigma}'_j\right], \tag{21}$$

where $\boldsymbol{\Lambda}_j \equiv \widetilde{\boldsymbol{\Lambda}}_{jj}$ are the diagonal tensor components of the matrix, $\widetilde{\boldsymbol{\Lambda}}_{jk}$, given by

$$\widetilde{\boldsymbol{\Lambda}}_{jk} = \sum_{l,p=1}^{N-1} \Pi_{lj}\,\overline{\mathbf{A}}_{lp}\,\Pi_{pk} \tag{22}$$

and the matrix, $\boldsymbol{\Delta}_{jk}$, is given by

$$\boldsymbol{\Delta}_{jk} = \sum_{l,m,n,p=1}^{N-1} \Pi_{lj}\,\Pi_{pk}\,\boldsymbol{\Gamma}_{lp,mn}\,\Pi_{mj}\,\Pi_{nk}\,. \tag{23}$$

In Eqs. (22) and (23), the tensors, $\overline{\mathbf{A}}_{jk}$, and $\boldsymbol{\Gamma}_{lp,mn}$ are given by Eqs. (8) and (14), respectively. The argument of the hydrodynamic interaction functions, $\hat{\boldsymbol{\sigma}}_{\mu\nu}$, is, however, now given by

$$\hat{\boldsymbol{\sigma}}_{\mu\nu} = \frac{1}{|\mu-\nu|}\,\frac{H}{k_BT}\sum_{j,k=\min(\mu,\nu)}^{\max(\mu,\nu)-1}\;\sum_{m=1}^{N-1}\Pi_{jm}\,\Pi_{km}\,\boldsymbol{\sigma}'_m\,. \tag{24}$$

Equation (21) is a set of $(N-1)$ equations for the variances, $\boldsymbol{\sigma}'_j$. The "twofold normal approximation" to the stress tensor,

$$\boldsymbol{\tau}^p = -nH\sum_{j=1}^{N-1}\boldsymbol{\sigma}'_j + (N-1)nk_BT\,\mathbf{1}\,, \tag{25}$$

can be found by solving Eq. (21). This implies that all the material functions in arbitrary homogeneous flows can also be found.

It is appropriate now to conclude the above dicussion by displaying one of the viscometric functions, namely, the viscosity obtained with the twofold normal approximation in steady simple shear flow. A comparison with the

Gaussian approximation, both in terms of property prediction and computational time, may then be made.

Steady shear flows are described by a $\boldsymbol{\kappa}$ tensor which has the following matrix representation in the laboratory-fixed co-ordinate system:

$$\boldsymbol{\kappa} = \dot{\gamma} \begin{pmatrix} 0 & 1 & 0 \\ 0 & 0 & 0 \\ 0 & 0 & 0 \end{pmatrix} . \tag{26}$$

Here $\dot{\gamma}$ is the constant shear rate. The three independent material functions used to characterise such flows are the viscosity, η_p, and the normal stress differences, Ψ_1 and Ψ_2. These functions are defined by the following relations [2]:

$$\tau_{xy}^p = -\dot{\gamma}\eta_p \,, \tag{27a}$$

$$\tau_{xx}^p - \tau_{yy}^p = -\dot{\gamma}^2\Psi_1 \,, \tag{27b}$$

$$\tau_{yy}^p - \tau_{zz}^p = -\dot{\gamma}^2\Psi_2 \,. \tag{27c}$$

In the Gaussian approximation [10], steady state viscometric functions are obtained by numerically integrating the system of differential Eqs. (13) for the independent components of the tensor, $\boldsymbol{\sigma}_{jk}$, with respect to time until steady state is reached. The CPU time required scales with the chain length as $N^{4.5}$. The twofold normal approximation sets the derivatives, $d\boldsymbol{\sigma}_j'/dt = 0$, in Eq. (21) and adopts a successive approximation procedure to solve the resultant set of nonlinear algebraic equations. The CPU time required to obtain the steady state solution with this scheme scales as N^x with $3.15 \leq x \leq 3.3$. This significant reduction makes it possible to examine chains with as many as 100 beads. The reasons for the CPU time reduction are discussed in some detail in reference [1].

Figure 1 compares the prediction of $\eta_p/\lambda_H n k_B T$ made by the Gaussian approximation with the prediction of the twofold normal approximation for parameter values $N = 45$ and $h^* = 0.15$ at low shear rates, and for $N = 15$ and $h^* = 0.15$ at high shear rates (see inset in Fig. 1). Results at high shear rates were not obtained for $N = 45$ with the Gaussian approximation due to the excessive requirement of computational time. As in all the figures below,

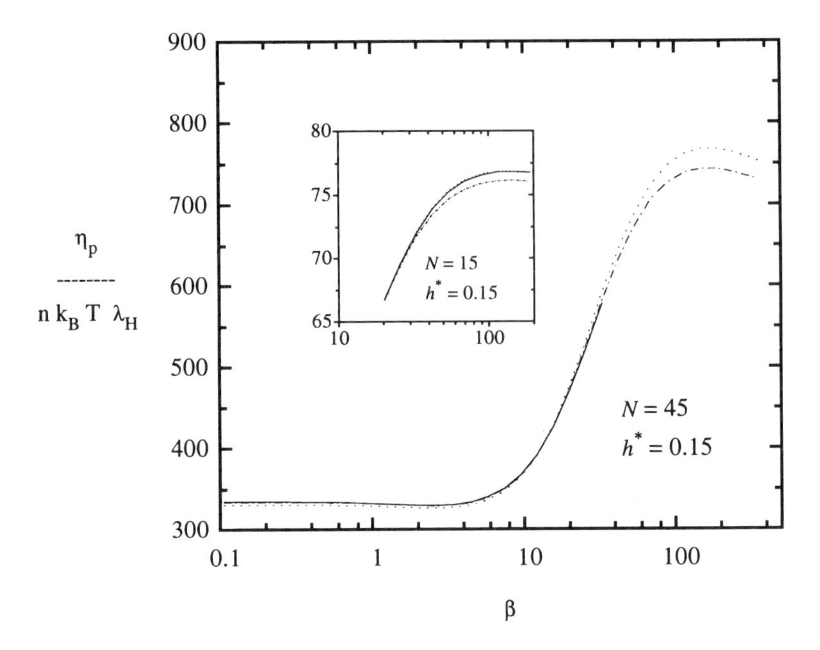

Fig. 1. Polymer contribution to the viscosity as a function of the reduced shear rate, β, for $N = 45$ at low shear rates and $N = 15$ at high shear rates (inset), reproduced from Prakash and Öttinger [1]. Continuous lines represent the Gaussian approximation while the chain-dashed and dotted lines represent the twofold normal Zimm and Rouse approximations, respectively.

the viscosity is plotted against the reduced shear rate, β, a dimensionless quantity which is defined as

$$\beta = \frac{[\eta]_0 \, \eta_s \, M \dot{\gamma}}{N_A \, k_B T} , \tag{28}$$

where M is the molecular weight of the polymer, N_A is Avagadros number, and $[\eta]_0$ is the intrinsic viscosity in the limit of zero shear rate. It can be shown that for dilute solutions, $\beta = \eta_{p,0} \, \dot{\gamma} / n k_B T$, where $\eta_{p,0}$ is the zero shear rate viscosity. For both the Gaussian approximation and the twofold normal approximation, $\eta_{p,0}$ can be evaluated explicitly [1, 7]. For the purpose of comparison, the value of $\eta_{p,0}$ given by the Gaussian approximation is used to calculate the reduced shear rate in Fig. 1. Note that the use of the Zimm orthogonal matrix leads to results indistinguishable from the Gaussian approximation at low to moderate shear rates. However, at high shear rates, the twofold normal

Zimm approximation predicts a slightly lower value of η_p (see inset). The Rouse orthogonal matrix causes a small error at low to moderate shear rates which becomes negligible at high shear rates. This might be understood by noting that the Zimm orthogonal matrix captures at least some aspects of hydrodynamic interaction. It therefore leads to more accurate results at low shear rates where hydrodynamic interaction is important. At high shear rates, the beads of the chain are far apart and hydrodynamic interaction is nearly switched off. As a result, the viscosity approaches its Rouse value. This does mean, of course, that the model has become poor, and one would have to consider chains with a larger number of beads in order to properly incorporate hydrodynmic interaction at high shear rates.

The Gaussian approximation has been implemented for values of N up to $N = 45$ with the help of an IBM RISC 6000/560 workstation [1]. The Gaussian approximation data in Fig. 1 took 679 hours of CPU time to generate, while the twofold normal approximation data was produced in about 17 minutes. The great reduction in computational time has enabled the examination of chains with $N \leq 100$ beads using the twofold normal approximation. The extrapolation of finite chain results to the infinite chain limit is discussed in the next section.

Universal Ratios and Viscometric Functions

The successful prediction of the various qualitative aspects of a dilute polymer solution's behaviour using bead-spring chain models, which account for the capacity of macromolecules to stretch and orient themselves under the influence of a flow field, justifies the expectation that these features of the macromolecule are the primary cause of the solution's viscoelasticity. However, as pointed out earlier, the predicted material functions are not parameter-free. Prakash and Öttinger [1] have recently shown numerically that in the limit of $N \to \infty$, the results of the twofold normal approximation are parameter-free and therefore universal. In this section, the universal results obtained in that paper are presented.

It is worth noting that the universal dependence of all material properties were obtained by extrapolating values accumulated for finite chains as a function of $N^{-0.5}$ to the $N \to \infty$ limit. This is based on results obtained in the generalised Zimm model [5] where it was shown analytically that leading corrections to the infinite chain limit are of order $N^{-0.5}$.

Table 1. Universal ratios in the limit of zero shear rate reproduced from Prakash and Öttinger [1]. Zimm, GA, and TNZ stand for the Zimm, the Gaussian approximation, and the twofold normal Zimm models, respectively. Numbers in parentheses indicate uncertainty in the last figure.

	$U_{\eta\lambda}$	$U_{\eta R}$	$U_{\Psi\eta}$	$U_{\Psi\Psi}$
Zimm	2.39	1.66425	0.413865	0.0
GA	1.835 (1)	1.213 (3)	0.560 (3)	−0.0226 (5)
TNZ	1.835 (1)	1.210 (2)	0.5615 (3)	−0.0232 (1)

Linear viscoelastic properties, such as the material functions for small amplitude shear flow and the viscosity and first normal stress difference in the limit of zero shear rate, can be obtained in twofold normal approximation by carrying out a perturbation expansion of the second moment in terms of the velocity gradient [1]. Table 1 reproduces the values obtained for the universal linear viscoelastic ratios, $U_{\eta\lambda}$, $U_{\eta R}$, $U_{\Psi\eta}$, and $U_{\Psi\Psi}$ which are defined by [15]:

$$U_{\eta\lambda} = \frac{\eta_{p,0}}{nk_BT\lambda_1};$$

$$U_{\eta R} = \lim_{n\to 0} \frac{\eta_{p,0}}{n\eta_s(4\pi R_g^3/3)};$$

$$U_{\Psi\eta} = \frac{nk_BT\Psi_{1,0}}{\eta_{p,0}^2};$$

$$U_{\Psi\Psi} = \frac{\Psi_{2,0}}{\Psi_{1,0}}.$$

The symbols, $\Psi_{1,0}$ and $\Psi_{2,0}$, represent the zero shear rate of the first and second normal stress differences, respectively, w hile λ_1 represents the longest relaxation time and R_g the root-mean-square radius of gyration at equilibrium. Details on the calculation of these ratios may be found in reference [1] along with a discussion of computational requirements. The values predicted by the Gaussian approximation and the exact Zimm model are also displayed in Table 1 for comparison.

Figures 2 and 3 display the results of extrapolating $\eta_p/\eta_{p,0}$ and Ψ_2/Ψ_1, respectively, as a function of $N^{-0.5}$ (accumulated for chains with $N \leq 100$) to the $N \to \infty$ limit at various values of the reduced shear rate, β. Beyond a reduced shear rate of about 25, the curves for the two values of h^* in Fig. 2 begin to diverge, indicating that the present data accumulated for chains with $N \leq 100$ is insufficient to carry out an accurate extrapolation at higher shear rates. The universal curve for $\Psi_1/\Psi_{1,0}$ is similar in shape to Fig. 2 for the reduced viscosity (see [1]).

In the same figures, the result of a renormalisation group calculation by Zylka and Öttinger [12] is also presented. The two methods appear to lead to significantly different results at moderate to high shear rates. The Gaussian and twofold normal approximations are "uncontrolled" approximations, and they are essentially non-pertubative theories. The renormalisation group method, on the other hand, is a procedure for refining a first-order perturbation expansion in the hydrodynamic interaction parameter [11].

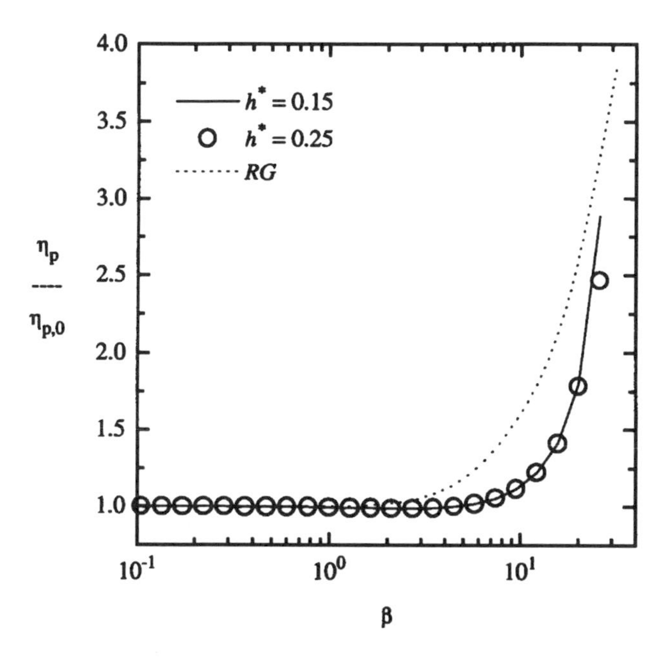

Fig. 2. Universal predictions of the reduced viscosity, $\eta_p/\eta_{p,0}$, as a function of the reduced shear rate, β, reproduced from Prakash and Öttinger [1]. Results obtained by extrapolating the twofold normal Zimm approximation data for $h^* = 0.15$ are represented by the continuous line while the symbol, (○), represents the results for $h^* = 0.25$. The dotted line is the result of a renormalisation group (*RG*) calculation obtained by Zylka and Öttinger [12].

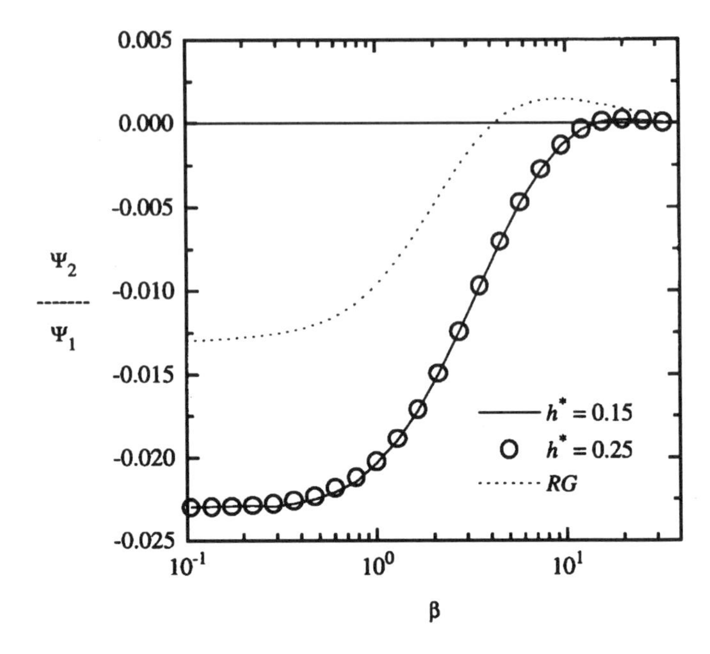

Fig. 3. Universal predictions of the ratio, Ψ_2/Ψ_1, as a function of the reduced shear rate, β, reproduced from Prakash and Öttinger [1]. Parameter values and symbols are the same as thoze found in the caption to Fig. 2.

Conclusions

This paper summarises the traditional treatments of hydrodynamic interaction with reference to their improved features relative to earlier treatments and their residual shortcomings. This is done both in the context of accuracy as compared to the exact results obtained with Brownian dynamics simulations, and in terms of computational efficiency. The recently introduced twofold normal approximation is discussed with regard to its accuracy and capacity to significantly reduce the computational intensity. The fact that this reduction in CPU time makes it possible to obtain universal predictions of viscometric functions in shear flow by extrapolating the finite chain length results to the limit $N \to \infty$ is highlighted.

Universal predictions, reproduced from Prakash and Öttinger [1], for the ratios, $U_{\eta\lambda}$, $U_{\eta R}$, $U_{\Psi\eta}$, and $U_{\Psi\Psi}$ defined in the linear viscoelastic limit are dispayed in Table 1. The universal dependence of $\eta_p/\eta_{p,0}$ and Ψ_2/Ψ_1 on the reduced shear rate, β, are displayed in Figs. 2 and 3.

Discussion

S. Ramaswamy How was the extrapolation to infinite size done? Did you do a systematic finite size scaling procedure?

J. R. Prakash In the Zimm theory and self-consistent averaging theory, it can be shown analytically that leading corrections to the $N \to \infty$ limit are of order $N^{-\frac{1}{2}}$ [5]. Expecting this behaviour to occur in the present case as well results were plotted as a funtion of $N^{-\frac{1}{2}}$ and extrapolated to the infinite chain limit using a rational function extrapolation algorithm [1].

T. C. B. McLeish Could you comment on the shear hardening behaviour of your results in the light of the usual shear thinning seen in experiments?

J. R. Prakash All theories that incorporate hydrodynamic interaction predict shear thickening if chains that are long enough are considered. For very short chains of up to four or five beads, they predict shear thinning. This can be understood by considering the fact that for such chains, the zero shear rate viscosity is higher than the free-draining Rouse value. At high shear rates, when hydrodynamic interaction is switched off because the beads are far apart, the predicted viscosity goes to its Rouse value and hence one sees shear thinning. However, this is an artefact of using short chains. For longer chains, since the free-draining Rouse value of viscosity is higher than the zero shear rate value, one predicts shear thickening. However, if effects such as finite extensibility and excluded volume are included, then one would predict shear thinning as observed in experiments.

T. C. B. McLeish So you think that current data on poly disperse solution viscosity is due to some quite subtle effect in theta solutions of destabilisation of theta conditions?

J. R. Prakash The present results are strictly for theta solvents and I believe there are experimental results which show shear thickening in theta solutions with very high molecular weight polymers.

E. J. Hinch Just a comment that modelling of shear flow is extremely delicate and it is quite easy to get curves to go up or down with little changes. One major deficiency with bead-spring models, which I use myself — I would never use in shear flow — is that the beads are points and there is no thickness for the beads. So if they line up in shear flow, they would just sit there in the flow plane. They don't feel the torque that would turn it around. I don't know if that is the right answer, but everything is too delicate in shear flow.

J. R. Prakash I would think that bead friction might account for the effect of torque.

E. J. Hinch You have a drag law using the velocity of the centre of the bead and you don't feel the thickness of the bead and the velocity difference across the bead. I think that is necessary to spin the thing around. When you spin something around in shear flows, it tends not to go so fully extended and the stress is lower and can be interpreted as shear thinning.

R. G. Larson With regard to the earlier comment about shear thickening, if you look very carefully at the results, you see a very weak amount of shear thinning at low Weisenberg numbers, and then at very high Weisenberg numbers of order 10 or more, you get the shear thickening which looked like an enormous effect the way it was plotted there. However, if you put in finite extensibility, and you then try to map the results to chains of realistic length, what you find is that, say for polystyrene, unless the molecular weight exceeds an order of five to 10 million, the shear thinning on account of finite extensibility will completely wash out the shear thickening and won't exist at all. For very high molecular weights, and Kishbaugh and McHugh [14] did this, you predict this very tiny shear thinning, a little weak maximum, and then shear thinning from finite extensibility. Data for the longest chains do show that kind of shape, a weak maximum in the first normal stress difference before the final thinning due to finite extensibility. So the effect is fairly washed out by finite extensibility for realistic chains of realistic length.

M. E. Cates You talked about finite extensibility a bit. Presumably, you can plug it into these calculations without stretching yourself?

J. R. Prakash No, you can't do it with the Gaussian approximation; but you can do it at the level of consistent averaging, which is what Kishbaugh and McHugh [14] have done, and those are the results that Dr. Larson was talking about.

S. Ramaswamy How was the renormalisation group done, was it near four dimensions, and was hydrodynamic interaction your only expansion parameter?

H. C. Öttinger It was an expansion near four dimensions, keeping only the lowest order terms. In the general renormalisation group theory, there was also excluded volume [12].

J. R. Prakash Not in the results presented here though.

H. C. Öttinger No, otherwise there would have been a little more shear thinning at low shear rates.

J. R. A. Pearson Will somebody help me out with my difficulty? Talking about finite extensibility, what happens if you just take jiggling motions, if you just looked at G', G''? On the basis of these models, is that a realistic way of testing your computations, or have I missed something crucial?

J. R. Prakash Well, G', G'' are very good with the Zimm theory, which doesn't have finite extensibility.

J. R. A. Pearson Yes, that is why I suggested that you do them. Do you get anything by comparing one method of computation with another? Looking at something where finite extensibility is not going to play an important part.

J. R. Prakash The present theory is excellent. It does as well as the Zimm theory. It is only in steady shear flow that one notices differences.

H. C. Öttinger A comment on that is that the curves, G' and G'', would look roughly the same if you look at them just as curves. But you can analyse the spectra, and where the Zimm model has the spectrum of N relaxation times with the Gaussian approximation, you have N^2 relaxation times. So we see some coupling between the modes which arise in the Gaussian approximation. So it is very different in that sense, but it doesn't change the curves.

J. R. A. Pearson So it is the hydrodynamic interaction that couples the modes, and the modes here are the normal modes?

H. C. Öttinger The fluctuations in the hydrodynamic interaction.

T. C. B. McLeish That is a very interesting point. In a certain sense, the Zimm theory gets the relaxation spectrum, $G(\omega)$, right sort of by accident. What you have shown is that taking the pre-averaging approximation is fundamentally misleading because the fluctuations are huge. We know that fluctuations in hydrodynamic interaction are huge and actually dominate the relaxation spectrum. Or is it getting it right for the right reasons?

H. C. Öttinger I think the comment would be that going to the spectrum is an ill-posed problem, so you shouldn't pay too much attention to the spectrum itself, but rather to the curves G' and G''. I don't like spectrology in general, and so this is an argument against it.

J. R. Prakash It is worth noting that even in the twofold normal approximation, you have $N-1$ relaxation times, just as in the Zimm theory, because you have introduced normal co-ordinates. However, in the Gaussian approximation, you have $(N-1)^2$ relaxation times. While they are strictly independent relaxation times in the Zimm theory, they are not independent of each other in the twofold normal approximation [1].

M. E. Cates Presumably, you could do a star polymer, a polymer that is strictly not a linear chain, within this general framework? There is no insuperable obstacle?

J. R. Prakash Yes, it could be done.

M. E. Cates I don't know if there are enough data on that to make it worthwhile, but it could be a reasonably stringent test. You are talking about, in many cases, fairly small differences between different models. It might be that looking at T-shaped polymers or three-armed star polymers might get you something. The chain architecture being linear is not actually essential to what you are doing, so it would be a small extra complication, though not a big one.

T. C. B. McLeish John Schrag has done experiments and some theory on different models, different objects, and you do get large changes in the spectrology.

M. E. Cates Yes, but what would be just as interesting is if you got some significant differences in these models.

References

1. Prakash, J. R. & Öttinger, H. C. (1997) Universal viscometric functions for dilute polymer solutions. *J. Non-Newtonian Fluid Mech.* **71**, 245–272
2. Bird, R. B., Curtiss, C. F., Armstrong, R. C. & Hassager, O. (1987) Dynamics of polymeric liquids. In *Kinetic Theory, Vol. 2*, 2nd edn. New York: John Wiley
3. Rouse, P. E. (1953) A theory of the linear viscoelastic properties of dilute polymer solutions of coiling polymers. *J. Chem. Phys.* **21**, 1272–1280
4. Zimm, B. H. (1956) Dynamics of polymer molecules in dilute solution: viscoelasticity, flow birefringence and dielectric loss. *J. Chem. Phys.* **24**, 269–281
5. Öttinger, H. C. (1987) Generalised Zimm model for dilute polymer solutions under theta conditions. *J. Chem. Phys.* **86**, 3731–3749
6. Öttinger, H. C. (1987) A model of dilute polymer solutions with hydrodynamic interaction and finite extensibility. I. Basic equations and series expansions. *J. Non-Newtonian Fluid Mech.* **26**, 207–246

7. Öttinger, H. C. (1989) Gaussian approximation for Rouse chains with hydrodynamic interaction. *J. Chem. Phys.* **90**, 463–473
8. Wedgewood, L. E. (1989) A Gaussian closure of the second moment equation for a Hookean dumbbell with hydrodynamic interaction. *J. Non-Newtonian Fluid Mech.* **31**, 127–142
9. Zylka, W. & Öttinger, H. C. (1989) A comparison between simulations and various approximations for Hookean dumbbells with hydrodynamic interaction. *J. Chem. Phys.* **90**, 474–480
10. Zylka, W. (1991) Gaussian approximation and Brownian dynamics simulations for Rouse chains with hydrodynamic interaction undergoing simple shear flow. *J. Chem. Phys.* **94**, 4628–4636
11. Öttinger, H. C. & Rabin, Y. (1989) Renormalisation group calculation of viscometric functions based on conventional polymer kinetic theory. *J. Non-Newtonian Fluid Mech.* **33**, 53–93
12. Zylka, W. & Öttinger, H. C. (1991) Calculation of various universal properties of dilute polymer solutions undergoing shear flow. *Macromolecules* **24**, 484–494
13. Magda, J. J., Larson, R. G. & Mackay, M. E. (1988) Deformation-dependent hydrodynamic interaction in flows of dilute polymer solutions. *J. Chem. Phys.* **89**, 2504–2513
14. Kishbaugh, A. J. & McHugh, A. J. (1990) A discussion of shear thickening bead-spring models. *J. Non-Newtonian Fluid Mech.* **34**, 181–206
15. Öttinger, H. C. (1996) *Stochastic Processes in Polymeric Fluids.* Berlin: Springer-Verlag

Dynamics of Complex Fluids, pp. 176–187
ed. M. J. Adams, R. A. Mashelkar, J. R. A. Pearson & A. R. Rennie
Imperial College Press–The Royal Society, 1998

Chapter 11

Polymeric Liquids at High Shear Rates

G. MARRUCCI* AND G. IANNIRUBERTO
Department of Chemical Engineering, University Federico II,
Piazzale Tecchio, Napoli, 80125, Italy
**E-mail: marrucci@unina.it*

The theory of entangled polymers at high shear rates is reconsidered to account for the fact that, contrary to predictions based on the classical tube theory, experiments indicate agreement with the Cox-Merz rule. Two factors are introduced. The first factor is the convective constraint renewal which is due to the relative motion between entangled chains. The second effect invokes the affinity of tube deformation and, as a consequence, a reduction of the tube diameter. Both effects can be accounted for (at least approximately) in a theory for steady flows, and predictions in shear seem to be in good qualitative agreement with the observations. A simple case of a transient which can also be treated is the jump in strain, where it turns out that the two effects essentially cancel out one another, thus reproducing the constant-tube-diameter assumption of the Doi-Edwards theory.

Introduction

The rheological behaviour of entangled polymeric liquids at high shear rates is still poorly understood. The Doi-Edwards theory [1] predicts that the system should become unstable very soon, that is, at shear rates where the nonlinear behaviour has just set in. However, direct observation of this effect in ordinary polymers is lacking. What is normally observed in polymers is the obedience of the Cox-Merz rule in the nonlinear region throughout an extended range of shear rates without any sign of instability. Instabilities, either constitutive in nature or more probably due to the loss of adhesion, only appear at even higher shear rates.

One theoretical ingredient of the constitutive response of entangled polymers which has so far received little attention is the convective contribution to entanglement renewal [2], a mechanism which is expected to play a significant role in fast flows. The relative motion between neighbouring chains due to flow works in the direction of renewing the topology over and above the thermally driven relative motion (reptation). In fact, in fast flows, reptation is effectively frozen and the whole topology renewal is due to convection. In a recent paper [3], it is shown that accounting for such a relaxation mechanism in the tube model significantly modifies the asymptotic behaviour of the shear stress curve at large shear rates. Instead of a decay of the shear stress towards zero with increasing shear rate, a constant non-zero value is approached. However, the Cox-Merz rule is not recovered. Indeed, the asymptotic constant value remains below the plateau modulus by a factor of ca. 2. Moreover, depending on the value of an unknown numerical coefficient, the flow curve may still show a maximum and hence an unstable branch. It appears that the convective constraint release (CCR) mechanism alone is insufficient to explain the observed nonlinear behaviour.

Another aspect of the basic theory which can somehow be considered an open problem is whether or not the tube diameter also maintains the equilibrium value at very large deformations. With respect to the classical assumption, that is, the tube diameter remains constant, different suggestions have already been made [4–7]. In favour of the classical one, however, monodisperse sample data of stress relaxation after a large jump in strain exist [1, 8]. After an initial relaxation occurring over a time of the order of the Rouse time, the stress remains essentially stationary up to the reptation time. The plateau value is in good agreement with the Doi-Edwards theory.

In this paper, we first show that a different assumption on the tube diameter (of the type previously suggested [4–7]), when coupled to the CCR mechanism, can still fully explain the jump-strain relaxation data for monodisperse polymers. We then work out the predictions for steady shear flows in order to test for an agreement with the Cox-Merz rule.

Tube Behaviour in Step Strain

Let us consider a jump in strain occurring at time $t = 0$ starting from equilibrium. We assume that at time $t = 0^+$, the whole system, including the topological constraints, is deformed affinely. Hence, at $t = 0^+$, the curvilinear tube length, L, and the tube diameter, a, are given by

$$\mathrm{L}_{\mathrm{affine}} = \mathrm{L}_{\mathrm{eq}}\langle|\mathbf{E}\cdot\mathbf{u}|\rangle_0 \;\; ; \;\; a_{\mathrm{affine}} = a_{\mathrm{eq}}\langle|\mathbf{E}\cdot\mathbf{u}|\rangle_0^{-1/2} , \tag{1}$$

where $\mathbf{E}$ is the deformation gradient, $\mathbf{u}$ is a unit vector and the average is made over the isotropic distribution of $\mathbf{u}$. The relationship for L in Eq. (1) is well-known [1]; it is based on the assumption that the chain is sufficiently long so that the average along its length can be replaced by the ensemble average. It is also well-known that $\mathrm{L}_{\mathrm{affine}}$ is always larger than L_{eq}. The relationship for a is a consequence of volume conservation for the tube of constraints ($\mathrm{L}a^2 = \mathrm{const}$). Since the tube lengthens, its diameter must decrease [4–7].

At $t = 0^+$, the linear density of monomers along the tube is not at equilibrium (with the existing tube). Indeed, if n is the total number of monomers in the chain, the equilibrium condition for the monomer density, n/L, is that n/L must be proportional to the tube diameter [9]. Since n is fixed, the equilibration of the monomer density is achieved if, and only if, a and L obey

$$a\mathrm{L} = \mathrm{const} = a_{\mathrm{eq}}\mathrm{L}_{\mathrm{eq}} . \tag{2}$$

It is apparent from Eq. (1) that a_{affine} and $\mathrm{L}_{\mathrm{affine}}$ do not fulfil Eq. (2). Hence, a retraction process of the chain inside its own tube must take place. During such a process, however, the tube diameter *does not stay constant* at the a_{affine} value. Indeed, the retraction motion of the chains generates a mean-field effect analogous to the CCR mechanism. For any given chain, the disappearance of the constraints exerted by segments of other chains which have retracted determines an effective increase in tube diameter.

In order to determine how much a will increase as a consequence of this retraction process, we assume that constraint disappearance can be treated as an effective *dilution* process. As is well-known (cf Eq. 7.68 in [1]), dilution changes the tube diameter according to the power law, $a \propto \rho^{-1/2}$, where ρ is the polymer density. The role of a decreasing density is here played by length contraction. Hence, self-consistently, we will write

$$\frac{a}{a_{\mathrm{affine}}} = \left(\frac{\mathrm{L}}{\mathrm{L}_{\mathrm{affine}}}\right)^{-1/2} , \tag{3}$$

where both a and L are unknown. At the end of the retraction process, however, a and L must also satisfy the equilibrium monomer density condition of Eq. (2). Hence, eliminating L from Eqs. (2) and (3), a is calculated as

$$a = \frac{a_{\text{affine}}^2 \mathrm{L}_{\text{affine}}}{a_{\text{eq}} \mathrm{L}_{\text{eq}}} . \tag{4}$$

Finally, by using Eq. (1), Eq. (4) reduces to

$$a = a_{\text{eq}} . \tag{5}$$

The result of Eq. (5) leaves the predictions for the relaxation following a step strain unaltered with respect to the Doi-Edwards theory. However, the different meaning of Eq. (5) in the two contexts should be emphasised. Here, the tube diameter does not remain constant under all conditions. We are proposing that the deformation process *per se* reduces the tube diameter according to volume preservation. On the other hand, the relative motion due to chain retraction works in the opposite direction. It so happens that the two effects exactly compensate one another in the jump experiment (given that the assumptions made above really hold true). Hence, after a Rouse time has elapsed, the tube diameter has recovered the equilibrium value. The rest of the relaxation process then occurs in the classical way.

Steady Flows

During a steady flow, there exists a slip velocity between chain and tube which progressively increases from zero at the midpoint of the chain up to a maximum value at the end of the chain (cf Eq. 7.232 in [1]). This relative velocity is due to the fact that the tube wants to deform affinely. Hence it would increase in length, whereas the tube, which is actually occupied by the chain, is in fact stationary in length. Thus, the chains continuously *retract* inside their own tubes. On average, this retraction velocity is given by

$$\mathrm{v} \approx \mathrm{L}\mathbf{k}{:}\langle \mathbf{uu} \rangle , \tag{6}$$

where $\mathbf{k}$ is the velocity gradient of the flow, and the average on segment orientation, $\mathbf{u}$, is now made over the existing anisotropic distribution.

So far, we have been unable to develop for steady flows arguments similar to those used in the previous section for the step strain. Rather, we have used Eq. (6) in a previous work [3], and shall use it here as well in order to introduce the CCR effect in the tube model in terms of the change in the dominant relaxation time, τ, under flow. Since Eq. (6) indicates a renewal frequency of the topology due to chain retraction proportional to $\mathbf{k}{:}\langle\mathbf{uu}\rangle$, and CCR works in parallel to reptation, we write

$$\frac{1}{\tau} = \frac{1}{\tau_o} + \beta\mathbf{k}{:}\langle\mathbf{uu}\rangle\,, \tag{7}$$

where τ_o is the reptation time and β is an unknown numerical factor. It is apparent that the thermal and convective frequencies appearing on the right-hand side of Eq. (7) will dominate in slow and fast flows, respectively. The convective term will therefore modify the rheological response in the nonlinear range.

The average in Eq. (7) can be calculated using an equation containing the same tensor, $\mathbf{Q}$, of the Doi-Edwards theory [1], namely,

$$\langle\mathbf{uu}\rangle = \frac{1}{\tau}\int_0^\infty dt\exp(-t/\tau)\,\mathbf{Q}(t)\,. \tag{8}$$

The main difference here is that τ is now given by Eq. (7). Hence the two equations are coupled and must be solved simultaneously to obtain τ. A secondary difference is that, for the sake of simplicity, we have taken a single relaxation time. The details of the calculations for a steady shear are reported in [3].

The effect of the reduction in tube diameter due to deformation is now considered. It enters the picture through the equation for the stress, $\mathbf{T}$, which becomes

$$\frac{\mathbf{T}}{\mathrm{G}} = \frac{1}{\tau}\int_0^\infty dt\exp(-t/\tau)\,\mathbf{Q}'(t)\,, \tag{9}$$

where G is a constant modulus ($\mathrm{G} = 3\nu k_B T \mathrm{L}_{\mathrm{eq}}/a_{\mathrm{eq}}$) and the tensor, $\mathbf{Q}'$, differs from $\mathbf{Q}$ by a scalar factor arising from tube shrinkage [4–7]:

$$\mathbf{Q}' = \left(\frac{a_{\mathrm{eq}}}{a_{\mathrm{affine}}}\right)^2\mathbf{Q} = \langle|\mathbf{E}\cdot\mathbf{u}|\rangle_o\mathbf{Q}\,. \tag{10}$$

The second equality in Eq. (10) is readily obtained from Eq. (1).

The constitutive equation given by Eq. (9), with τ determined through the set of Eqs. (7–8), provides a first crude estimate (applicable to steady flows only) of how the tube model is modified in the nonlinear range by the simultaneous (and opposite) effects of: (i) convective constraint release; and (ii) tube shrinkage due to large deformations.

Predictions for Steady Shear

We now derive the explicit predictions for the case of steady simple shear. To begin with, the dependence of the relaxation time, τ, on the shear rate, $\dot{\gamma}$, is obtained through Eqs. (7) and (8) which yield

$$\frac{1}{\tau} = \frac{1}{\tau_o} + \beta\dot{\gamma}\frac{1}{\tau}\int_0^\infty dt \exp(-t/\tau)\, \mathrm{F}_1(\dot{\gamma}t)\,, \tag{11}$$

where $\mathrm{F}_1(\ldots)$ is the xy-component of the tensor, $\mathbf{Q}$, given explicitly by Eq. (3.4a) in [10]. Equation (11) defines the dependence of τ on $\dot{\gamma}$ (albeit implicitly).

In order to obtain the stress from Eq. (9), we need to calculate the components of the tensor, $\mathbf{Q}'$, which differ from those of $\mathbf{Q}$ because of the scalar factor in Eq. (10). In steady shear, such a factor is a function of the deformation, $\lambda = \dot{\gamma}t$, through the following integral:

$$\langle|\mathbf{E}\cdot\mathbf{u}|\rangle_o = \frac{1}{4\pi}\int_0^{2\pi} d\phi \int_0^{\pi} d\vartheta \sin\vartheta\sqrt{1+\lambda\sin^2\vartheta\sin\phi(2\cos\phi+\lambda\sin\phi)}\,. \tag{12}$$

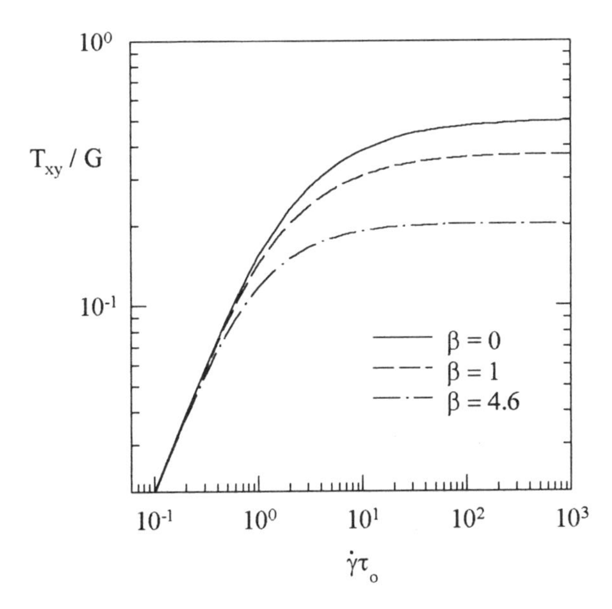

Fig. 1. The flow curve predicted by tube squeeze alone ($\beta = 0$). Several flow curves for $\beta \neq 0$ which also include the effect of the CCR mechanism. The value $\beta = 4.6$ gives complete agreement with the Cox-Merz rule.

Finally, the integration over time, as in Eq. (9), provides the stress as a function of $\dot{\gamma}$.

Figure 1 reports the results for the shear stress, T_{xy}. The curves in Fig. 1 refer to several values of the unknown constant, β, appearing in Eq. (11), with $\beta = 0$ corresponding to the case where the CCR mechanism is ignored. Figure 1 shows that the tensor, $\mathbf{Q}'$ (different from $\mathbf{Q}$), makes the shear stress a monotonically increasing function of the shear rate which approaches a constant value as $\dot{\gamma}$ goes to infinity. Although such a behaviour is qualitatively consistent with the Cox-Merz rule, the asymptotic value of the shear stress for $\beta = 0$ is too large for a full obedience of that rule, with the correct value for such a purpose being $T_{xy}/G = 0.2$. In order to fulfil the Cox-Merz rule completely, the CCR mechanism must be included with a β value as large as 4.6. It is perhaps important to emphasise that in order to find quantitative agreement, we need *both* CCR and tube contraction (the latter is accounted for by the tensor, $\mathbf{Q}'$, in place of $\mathbf{Q}$). Indeed, CCR alone, coupled to the classical assumption of a constant tube diameter, leads to asymptotic non-zero shear stress values which remain lower than 0.2 for all values of β [3].

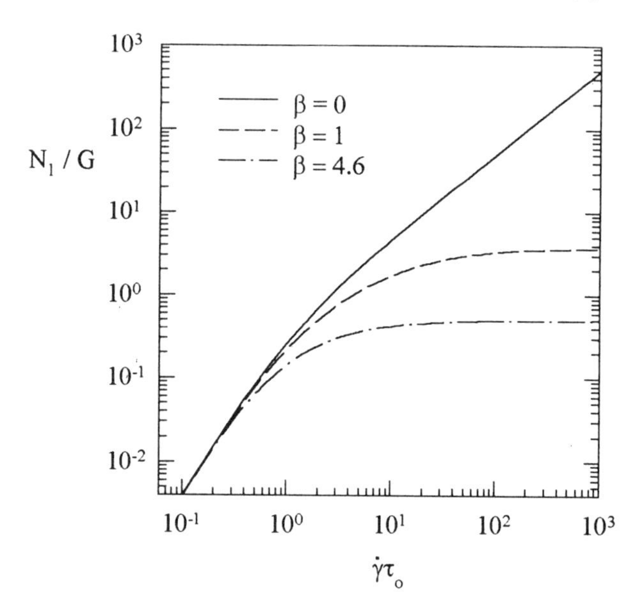

Fig. 2. The first normal stress difference for several β values. The $\beta = 0$ curve goes to infinity with increasing shear rate. For any other β, the first normal stress difference approaches a constant value.

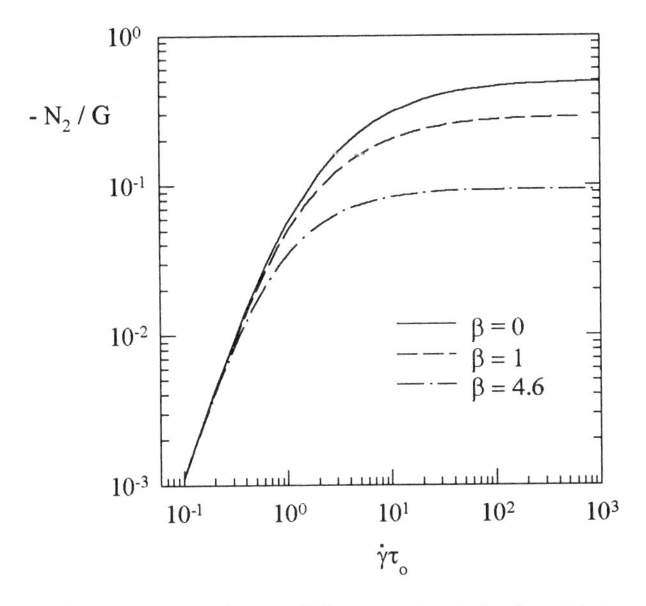

Fig. 3. Curves of the second normal stress difference. For all β values, N_2 becomes constant at high shear rates.

Figures 2 and 3 show the first and second normal stress differences, respectively. It is worth noting that the case $\beta = 0$ (no CCR mechanism) leads to a divergence of the first normal stress difference at high shear rates. Indeed, when reptation is effectively suppressed ($\dot{\gamma}\tau_o \gg 1$), and if CCR is also absent, there is no renewal mechanism of tubes. Therefore, when the tube shrinks in diameter with deformation, its length increases indefinitely and the stress diverges. However, only the normal component, T_{xx}, actually diverges because, simultaneously, the tubes align along the shear direction. It remains confirmed that, at high shear rates, the effect of CCR is particularly important since it provides the only renewal mechanism.

Conclusions and Future Work

We have shown that the complex nonlinear behaviour of entangled polymers can perhaps be explained if the classical tube model of Doi and Edwards is modified to incorporate *two* mechanisms which, in a way, act against one another. The two mechanisms are: (i) tube diameter reduction due to the affinity of the deformation; and (ii) tube renewal due to flow- or deformation-induced constraint release.

Separately, each of these two mechanisms had been considered previously [3–7]. However, the reduction in tube diameter, taken alone, predicts a damping function contradicted by stress relaxation data on monodisperse systems [1, 8]. On the other hand, the convective constraint release (CCR) mechanism has been considered only recently [2, 3] (except for the papers on stress relaxation by Viovy [11, 12]) and it was shown that, in conjunction with a constant tube diameter, agreement with the Cox-Merz rule was approached but not quite reached [3]. The results of this work show that if the two effects are taken *together*, the apparently conflicting evidence of a strong damping function in stress relaxation on the one hand, and of the Cox-Merz rule on the other, can perhaps be reconciled.

We have argued that in jump strain experiments, although tubes instantaneously contract in diameter, they enlarge again in the course of the first relaxation process (occurring in a Rouse time) because of constraint release induced by tube-length contraction. In particular, we have shown that with sensible assumptions, the diameter at the end of the contraction process goes back to the equilibrium value so that in this case, the classical Doi-Edwards damping function is recovered exactly. On the other hand, for a steady shear flow, the two effects do not cancel out one another. In fact, the calculations reported here and in a previous paper [3] show that *either* effect generates a non-zero asymptotic value of the shear stress at high shear rates (as opposed to a vanishing stress of the classical theory [1]) which is in qualitative agreement with the Cox-Merz rule (for a system with a single relaxation time). When both effects are taken together, the agreement with the Cox-Merz rule even becomes quantitative (for an appropriate choice of the adjustable parameter, β).

It remains to be said that in the two cases considered here (step strain and steady flows), the CCR mechanism has been dealt with in a somewhat different way. In the strain jump case, we have argued in terms of the tube diameter growing up again (after the reduction due to deformation). In steady flows, we have taken the approach of an increased renewal frequency of tube segments due to flow. As a theme for future work, we plan to develop a calculation for steady shear which is more consistent with the picture adopted for the jump in strain. In other words, an attempt will also be made to deal with tube enlargement due to constraint release in steady flow situations. In this way, we hope to get rid of the adjustable parameter, β.

Acknowledgements

This work was supported by the EEC under contracts Nos. CHRX-CT93-0200 and CHRX-CT94-0486.

Discussion

J. R. A. Pearson Would the following summary be correct? You have altered the precise assumptions of the traditional Doi-Edwards reptation model and shown that you predict qualitative agreement with both step-strain and continuous shear data, thus resolving the problem with the steady shear stress predictions of the traditional model. I then wish to emphasise the following. If you have an experiment, and build up a theory which happens to agree with it, it would be dangerous to conclude that the theory is proved because many different theories could do equally well.

G. Marrucci Although I agree with your statements in general terms, I wish to emphasise that the main message of the Doi-Edwards theory, that is, that polymer chains are embedded in the material and cannot move sideways, is maintained. However, the details of the basic theory, like a constant tube diameter in the nonlinear range, are open to question. Here, I have tried to combine the effect of two mechanisms operating on the tube diameter which, separately, have been around for a while.

U. S. Agarwal Is your treatment sufficiently general as to include other flows like, for example, extensional flows?

G. Marrucci So far the calculations can only be made for steady flows and for the very special transient which is the jump in strain. We do not have a complete constitutive equation yet.

E. J. Hinch I recall that Wagner considered the mechanism of tube diameter reduction. Is the one you are proposing the same?

G. Marrucci To be precise, I proposed the diameter reduction idea (together with De Cindio and Hermans) before Wagner did. In any event, the answer to your question is yes. The diameter reduction we are proposing here is quantitatively the same as in Wagner. Notice, however, that if the effect is taken alone, it leads to stress divergences whereas when it is coupled to CCR, it seems to behave properly.

T. C. B. McLeish When considering changes in tube diameter, problems may arise on time-strain separability. If the diameter and tube length change, the terminal relaxation time will change. Have you considered that aspect?

G. Marrucci I do not see that problem in the present context. Indeed, for the relaxation following a jump in strain, after a Rouse time has elapsed, the tube diameter goes back to the equilibrium value. Hence the terminal relaxation is not affected. On the other hand, in fast steady flows, reptation is suppressed altogether and the only relaxation mechanism is CCR.

R. G. Larson Does the shrinkage of the tube change the stress prediction immediately after a step strain?

G. Marrucci No, because at $t = 0^+$, the stress is determined by the affine deformation of the chain segments, and not by the tube diameter at all. Although the tube diameter here at $t = 0^+$ is smaller than in the classical theory, the linear monomer density of the chain along the tube is still *smaller* than the equilibrium value corresponding to the *existing* tube diameter. In such a situation (here, as in the classical theory), the stress is insensitive to the diameter value.

S. F. Edwards You may be interested in the original thinking on the constant tube diameter assumption. To simplify the picture imagine a polymer which is straight. The tube diameter is the distance the surrounding polymers must travel before they feel each other. Now, if you pull the polymer, the other chains will not know. Therefore, the tube lengthens because that is the polymer, whereas the diameter stays constant because that is the other chains. Maybe this argument can be easily demolished, I know, but it can at least be historically valuable. A different argument requires a new picture, as Professor Pearson has said. However, concerning his comment, I would like to say that if something is very complicated and one makes a model which gives complicated predictions, it is very likely that these predictions are wrong. Conversely, if one finds agreement, one cannot say that the model *happens* to agree with experiments.

References

1. Doi, M. & Edwards, S. F. (1986) *The Theory of Polymer Dynamics.* Oxford: Clarendon Press
2. Marrucci, G. (1996) Dynamics of entanglements. A nonlinear model consistent with the Cox-Merz rule. *J. Non-Newtonian Fluid Mech.* **62**, 279–289

3. Ianniruberto, G. & Marrucci, G. (1996) On compatibility of the Cox-Merz rule with the model of Doi and Edwards. *J. Non-Newtonian Fluid Mech.* **65**, 241–246
4. Marrucci, G. & de Cindio, B. (1980) The stress relaxation of molten PMMA at large deformations and its theoretical interpretation. *Rheol. Acta* **19**, 68–75
5. Marrucci, G. & Hermans, J. J. (1980) Nonlinear viscoelasticity of concentrated polymeric liquids. *Macromolecules* **13**, 380–387
6. Wagner, M. H. (1990) The nonlinear strain measure of polyisobutylene melt in general biaxial flow and its comparison to the Doi-Edwards model. *Rheol. Acta* **29**, 594–603
7. Wagner, M. H. & Schaeffer, J. (1992) Nonlinear strain measure for general biaxial extension of polymer melts. *Rheol. Acta* **36**, 1–26
8. Urakawa, O., Takahashi, M., Masuda, T. & Ebrahimi, N. G. (1995) Damping functions and chain relaxation in uniaxial and biaxial extensions: comparison with the Doi-Edwards theory. *Macromolecules* **28**, 7196–7201
9. Doi, M. & Edwards, S. F. (1978) Dynamics of concentrated polymer systems. Part 2. Molecular motion under flow. *J. Chem. Soc., Faraday Trans.* **74**, 1802–1817
10. Doi, M. & Edwards, S. F. (1979) Dynamics of concentrated polymer systems. Part 4. Rheological properties. *J. Chem. Soc., Faraday Trans.* **75**, 38–54
11. Viovy, J. L., Monnerie, L. & Tassin, J. F. (1983) Tube relaxation: a necessary concept in the dynamics of strained polymers. *J. Polym. Sci. Phys. Ed.* **21**, 2427–2444
12. Viovy, J. L. (1985) Tube relaxation: a quantitative molecular model for the viscoelastic plateau of entangled polymeric media. *J. Polym. Sci. Phys. Ed.* **23**, 2423–2442

Dynamics of Complex Fluids, pp. 188–192
ed. M. J. Adams, R. A. Mashelkar, J. R. A. Pearson & A. R. Rennie
Imperial College Press–The Royal Society, 1998

Chapter 12

Deuterium NMR Investigations of Liquid Crystals During Shear Flow

C. SCHMIDT*, S. MÜLLER AND H. SIEBERT

Institut für Makromolekulare Chemie, Universität Freiburg, Stefan-Meier-Str. 31, D-79104 Freiburg, Germany
**E-mail: schmidtc@ruf.uni-freiburg.de*

Deuterium NMR spectroscopy has been used to investigate the orientations of liquid crystals during shear flow. Examples are presented for different types of systems, including nematic side-chain liquid-crystalline polymers and hexagonal and lamellar lyomesophases of surfactant/water mixtures.

Introduction

To understand the rheological behaviour of anisotropic fluids, it is important to measure not only their macroscopic properties but also, on a microscopic scale, molecular orientation and phase structure during flow. In the past, mainly optical methods have been combined with rheological measurements. Here, we present a different approach, using deuterium NMR spectroscopy to provide the desired microscopic information.

Method

The quadrupole coupling of the deuterium nucleus is a convenient probe for the investigation of molecular orientations in solids and anisotropic liquids.

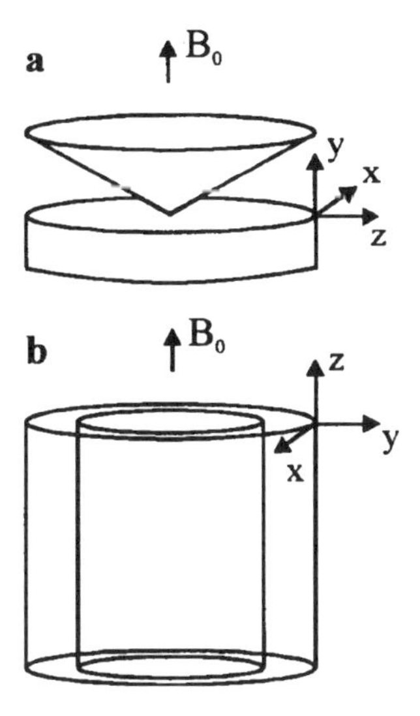

Fig. 1. Geometry of the cone-and-plate (a) and the Couette (b) shear cells used in the NMR experiments. The velocity, velocity gradient, and vorticity axes are labelled, x, y and z, respectively. For the cone-and-plate geometry, the plane spanned by the velocity and the velocity gradient (xy plane) is parallel to the magnetic field, B_0, whereas for the Couette geometry, the xy plane is perpendicular to the magnetic field.

Deuterium NMR has been used for many years to study the equilibrium properties of thermotropic and lyotropic liquid crystals. We demonstrate that NMR is also a useful tool for the investigation of the complex flow behaviour of liquid crystals [1–6]. To enable measurements of NMR spectra during shear, we constructed different shear cells that are integrated in conventional probeheads for a wide-bore superconducting magnet. Figure 1 shows the cone-and-plate and Couette geometries we used and the orientation of the cells with respect to the magnetic field.

Side-Chain Liquid-Crystalline Polymers

We studied several nematic polymers with mesogenic side-chains [2, 6]. According to the Ericksen-Leslie-Parodi (ELP) theory of nematic systems, two types can be distinguished: (i) the shear-aligning type with a stable director orientation at all shear rates; and (ii) the tumbling type which shows continuous director orientation unless tumbling is suppressed by a sufficiently strong magnetic field. The director orientation as a function of shear rate is displayed

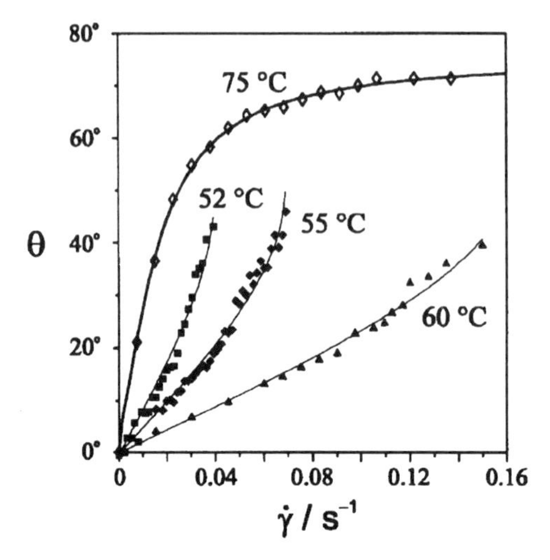

Fig. 2. Angle θ between nematic director and magnetic field as a function of shear rate, $\dot{\gamma}$, for a shear-aligning polysiloxane (open symbols) and a non-shear-aligning polymethacrylate in nematic solution (filled symbols), measured in the cone-and-plate cell.

in Fig. 2 for representatives of both types of nematics. The different flow behaviour can be identified qualitatively from the different curvatures. Fitting the data using the ELP theory yields two Leslie coefficients, α_2 and α_3, the relative sign of which determines the shear-aligning or tumbling character. The simultaneous measurements of orientations and shear viscosities allowed us to determine four out of the five independent viscosity parameters of the nematic polymers.

Lyomesophases of Surfactants

The investigation of the hexagonal phases of several non-ionic surfactant/D_2O mixtures in both the cone-and-plate and Couette geometries proves that shear induces an alignment of the hexagonal axis along the flow direction [3, 5]. At low shear rates, the hexagonal phase behaves like a solid: the alignment process depends on the shear strain and not on the applied shear rate. For a mono-domain re-orienting during shear, the director orientation can be calculated from the strain [5].

Our rheo-NMR experiments on lamellar (L_α) lyomesophases confirm the existence of previously described different states of orientation depending on

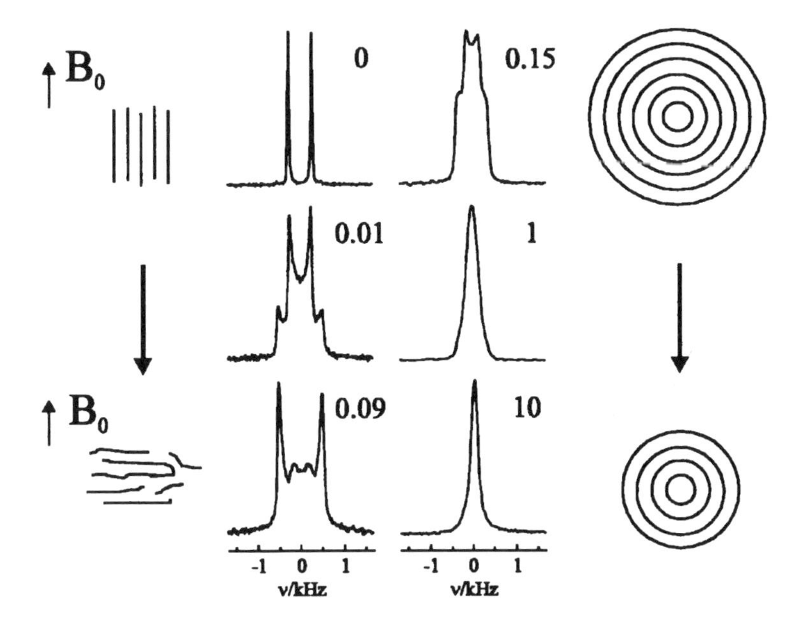

Fig. 3. Deuterium NMR spectra of a lamellar lyomesophase (35 wt% $C_{12}E_4$ in D_2O-enriched water) in the cone-and-plate cell at different shear rates, $\dot{\gamma}$, reflecting the change of the orientation state of the sample. The shear rates in s^{-1} are indicated to the right of each spectrum.

the applied shear rate [7]. At low shear rates, an alignment of the layers with their normal parallel to the shear gradient is observed, whereas at higher shear rates, a transition to multi-lamellar vesicles occurs. Figure 3 shows the change of the deuterium NMR lineshapes with increasing shear rate, evidence of the change of orientation and the formation of the vesicles [4], which become smaller and smaller as the rate of shear increases.

Conclusions

The examples presented here demonstrate that deuterium NMR spectroscopy can provide a wealth of information about sheared liquid crystals. A further analysis of the lineshapes will give more details of the orientational distribution of the director and the phase structure. Since the samples are probed on a local molecular scale, rheo-NMR is complementary to other techniques which yield information about mesoscopic structures, such as light scattering. Rheo-NMR can be applied to all materials, irrespective of their optical properties.

Acknowledgement

This work was supported by Deutsche Forschungsgemeinschaft.

References

1. Nakatani, A. I., Poliks, M. D. & Samulski, E. T. (1990) NMR investigations of chain deformations in sheared polymer fluids. *Macromolecules* **23**, 2686
2. Grabowski, D. A. & Schmidt, C. (1994) Simultaneous measurement of shear viscosity and director orientation of a side-chain liquid-crystalline polymer by rheo-NMR. *Macromolecules* **27**, 2632
3. Lukaschek, M., Grabowski, D. A. & Schmidt, C. (1995) Shear-induced alignment of a hexagonal lyotropic liquid crystal as studied by rheo-NMR. *Langmuir* **11**, 3590
4. Lukaschek, M., Müller, S., Hasenhindl, A., Grabowski, D. A. & Schmidt, C. (1996) Lamellar lyomesophases under shear as studied by deuterium nuclear magnetic resonance. *Coll. Polym. Sci.* **274**, 1
5. Müller, S., Fischer, P. & Schmidt, C. (1997) Solid-like director re-orientation in sheared hexagonal lyotropic liquid crystals as studied by nuclear magnetic resonance. *J. Physique II.* **7**, 421
6. Siebert, H., Grabowski, D. A. & Schmidt, C. (1997) Rheo-NMR study of a non-flow-aligning side-chain liquid crystal polymer in nematic solution. *Rheol. Acta* **36**, 618
7. Diat, O., Roux, D. & Nallet, F. (1993) Effect of shear on a lyotropic lamellar phase. *J. Physique II.* **3**, 1427

Dynamics of Complex Fluids, pp. 193–198
ed. M. J. Adams, R. A. Mashelkar, J. R. A. Pearson & A. R. Rennie
Imperial College Press–The Royal Society, 1998

Chapter 13

Formation of Polymer Brushes

J. WITTMER AND A. JOHNER*
Institut Charles Sadron (UPR CNRS 022),
6 rue Boussingault, 67083 Strasbourg Cedex, France
**E-mail: Johner@ics-crm.u-strasbg.Fr*

In systems such as block copolymer mesophases or physical gels formed by associating copolymers, the dynamical properties are often controlled by the extraction/association of a sticking group. We propose a description of the extraction/association process of a single sticker. The statistical physics of these associated systems is usually dominated by stretched brush-like regions. A sticker has to overcome a potential barrier both to penetrate the stretched structure or to escape a favourable region built by associated stickers. Our main result shows that these barriers are crossed by tension fluctuations and the corresponding processes are thus local with a friction independent of molecular weight. When the potential barriers are high, the very stretched equilibrium structures are not likely to develop with reasonable time scales. Stretched model systems may also be grown *in situ* from nuclei bearing initiating groups. These irreversibly bound structures are also briefly discussed.

Molten block copolymers self-assemble to form various structures depending mainly on their asymmetry [1, 2]. In a selective solvent, say, good for A and bad for B, the A-blocks are swollen by the solvent whereas the B-blocks assemble in almost solvent-free domains [3]. Soluble polymers decorated with insoluble stickers form a physical gel where the temporary crosslinks are built by aggregated stickers [4]. For such a material to flow, stickers have to be extracted from aggregates. There is usually a high energy barrier, E, to overcome during the extraction process. In the case where the stickers are small, insoluble B-blocks, $E \sim \gamma N_B^{2/3}$, and a high tension τ of order $a\gamma$ with a being the monomer size and γ the $B/solvent$, surface tension is needed for non-activated chain extraction. Thermally activated chain extraction (or sticker desorption) is thus an important issue. Long soluble blocks interact strongly in the vicinity of an insoluble domain and stretch to avoid one another [5]. Even in the case of associating polymers forming a physical gel, it is believed that the star-like regions surrounding small insoluble domains dominate the statistical physics of the network [6]. The extraction of a sticker belonging to a locally stretched chain is thus of rather general relevance.

In early work over the past decade, the random motion of the sticking group has been described as that of a point-like particle with the friction, $N\zeta$, being relevant to the overall motion of the chain. Here we argue that the internal modes of the chain are important and that the relevant friction is much lower. It corresponds to the first correlation length [7], ξ, linked to the aggregate parameters through $p\xi^{d-1} \sim S$, where p is the functionality of the aggregate and S its area [8] (for insoluble blocks $S \sim (pN_B)^{2/3}$). The physical argument is as follows: once the had group is off the aggregate, it diffuses almost freely over the first correlation length, ξ, and bounces back on the aggregate several times; at distances larger than ξ, the equilibrium tension, $\tau_e \sim k_b T/\xi$, drives the motion of the sticker and extraction is then achieved. This is to say that extraction is a local process independent of the overall chain length, N.

The idea that the relevant fiction should be local can be tested on the simpler case where a chain is cut off from the grafting surface in a polymer brush (with no stickers left). Due to the retraction of the chain, there is a drift at the chain end. The relevant scales are the height, h, of the brush and the relaxation time of the typical stretched configuration, τ_r. The following scaling form must thus hold for the motion of the cut end:

$$\langle z \rangle = h\mathrm{f}(t/\tau_r) \quad \text{with:} \quad \tau_r \sim \left(\frac{h}{\xi}\right)^2 \xi^d , \tag{1}$$

where d is the dimension of space, $d = 3$ corresponds to excluded volume statistics and Rouse-Zimm dynamics, and $d = 4$ corresponds to mean-field statistics and Rouse dynamics. Chain retraction, being a local process, $\mathrm{f}(x) \sim x^{1/2} (x \ll 1)$, and for $t \ll \tau_r$:

$$\langle z \rangle \sim \xi \left(\frac{t}{\xi^d}\right)^{1/2} . \tag{2}$$

The early motion of the chain is nonetheless dominated by anomalous diffusion: the chain end excites longer and longer modes whilst moving and the friction increases. This leads to the classical dispersion law [9]:

$$\langle (z - \langle z \rangle)^2 \rangle \sim t^{2/d} . \tag{3}$$

The motion is thus driven by the tension for $\langle z \rangle > \xi$ (or $t > \xi^d$) when the fluctuation around the average position, $\langle z \rangle$, is negligible. It is then unlikely that the chain end will hit the grafting surface again. These ideas are supported by Monte Carlo simulation and a Rouse analysis [10].

Similar ideas hold for thermally activated had group desorption from a flat surface. Once the head group is off the grafting surface, desorption is promoted by chain retraction and becomes irreversible for $z > \xi$. Being local again, the relevant time scale depends on ξ. Assuming that at the scale of the correlation length the only time scale is the co-operative relaxation time, ξ^d, the outwards flux obeys

$$J_{out} \sim \xi^{-d} \exp(-E)\sigma \sim \xi^{1-2d} \exp -E \tag{4}$$

and is independent of chain length. The grafting density, σ, is also linked to ξ via $\sigma \sim \xi^{1-d}$. The characteristic lifetime of a bound sticker, T_-, is deduced from rate Eq. (4) as

$$T_- \sim \xi^d \exp E . \tag{5}$$

This also applies to curved surfaces and the result is supported by Monte Carlo simulations [10].

The inward flux, J_{in}, is limited by the barrier of height μ opposed to sticker penetration by already grafted chains. Assuming that the incoming sticker also crosses the barrier by a local tension fluctuation, ξ is the only relevant scale. The inwards flux thus obeys

$$J_{in} \sim \xi^{1-d} \exp(-\mu) c_b , \tag{6}$$

with c_b being the bulk chain concentration. Note that the kinetic Eqs. (4) and (6) correspond to the isotherm

$$\mu_{eq} - E \sim \log(c_b/c^\star) - \frac{d}{d-1} \log(\sigma/\sigma^\star) , \tag{7}$$

where $\sigma^\star \sim R^{1-d}$ and $c^\star \sim R^{-d}$ are the overlap concentrations for grafted chains and free chains, respectively. The chemical potential increment, μ, for a grafted chain with respect to a free chain is $\mu \sim h/\xi$ to leading order. It also contains logarithmic corrections involving enhancement exponents.

Due to the activation barrier, the extraction/aggregation process is slow. The relaxation time of small fluctuations in the aggregation number (inside the peak of the size distribution) usually lies in the minute range for diblock copolymers [11, 12]. In contrast, large fluctuations which do not conserve the number of aggregates are found to relax extremely slowly [12]. Conversely, the existence of large equilibrium aggregates is questionable. One way to overcome this difficulty is to proceed with concentrated solutions where the excluded volume is screened and to swell the system in the solvent afterwards [13]. It is, however, unclear how the system relaxes topological constraints upon swelling. For some purposes, such as colloid coating (coated colloids are used in filled rubbers), chains can be irreversibly grafted. An effective way to achieve fairly high grafting densities is to grow the layer *in situ* monomer by monomer from a functionalised nucleus carrying polymerisation initiators. We consider a flat surface densely covered with initiators. There is some similarity between needle growth [14], governed by classical D. L. A [15] without branching, and polymer growth. Nonetheless, polymer chains can relax their configurations, and this sets an additional dynamical time scale. We consider growth rates low enough such that the chain configurations are completely relaxed and the grown structure is in internal thermodynamic equilibrium at any time. This provides an additional relationship between the bound monomer concentration and the density of end points. In a mean field approach marginally valid in 3d space [16], the free monomer density, u, the bound monomer density, ϕ, and the end point density, ρ, are determined by the following set of equations:

$$\begin{aligned} \rho &= -l\,\partial_z \phi^{1/\epsilon}\,; \\ \partial_t \phi &= k\,u\,\rho\,; \\ \partial_t \phi &= -\partial_t u + D\Delta u = D\Delta u\,; \end{aligned} \tag{8}$$

where l is a microscopic length of the order of the monomer size, $\epsilon = \frac{d-1/\nu}{d-1}$, an exponent linked to equilibrium properties [17, 5] (ν is the Flory exponent, $\epsilon \approx 2/3$), k is a kinetic constant, and D is the free monomer diffusion constant. While the first equation expresses the internal equilibrium of the layer at any time, the second equation describes the kinetics of the polymerisation reaction between chain ends and free monomers, and the third is the monomer conservation law and it embodies the *adiabatic* approximation (for a discussion of this point, see [18]). These equations are supplemented by following boundary conditions:

$$u(0,t) = 0, \quad \lim_{z \to \infty} D\partial_z u = a^d j_\infty \,. \tag{9}$$

The density of reacting chain ends is found to (formally) diverge at the wall and the flux of free monomers can be neglected there: $\partial_z u(0,t) = 0$. We set $1/k = l = 1$, thereby choosing the time and length units. We further set $D = 1$ by an appropriate rescaling of the fields u, ρ, and ϕ. The height of the structure, $H(t)$, is defined by the first moment, $H(t) = \int_0^\infty z\phi(z)\mathrm{d}z / \int_0^\infty \phi(z)\mathrm{d}z$. We now seek to derive a solution of the scaled form,

$$\phi = z^{-\alpha}\mathrm{f}(x) \quad \rho = z^{-\beta}\mathrm{h}(x) \quad u = z^\gamma \mathrm{g}(x)\,, \tag{10}$$

in the variable, $x = z/H(t)$. The aggregation process, Eqs.(8) and (9), imposes $\beta = 2$, $\gamma = 1$ and $\alpha = \epsilon$ where the scaling function, $\mathrm{f}(x)$, is assumed to be finite at zero and to vanish at infinity on physical grounds. The height of the brush, $H(t)$, increases as a power law with time: $H(t) \sim t^{\frac{1}{1-\epsilon}}$.

The scaling functions are then determined by solving Eqs. (8) numerically. In fact, the constitutive equation linking the monomer density, ϕ, and the end density, ρ, breaks down in the outermost correlation length. Our description, being coarse grained on the scale of the local correlation length, has to allow for the function, $\mathrm{f}(x)$, to jump to zero at the brush edge. The results of this coarse grained analytical mean field theory are nicely confirmed by Monte Carlo simulations [18]. The grown structure is densely grafted, polydisperse and highly stretched. It should be a good candidate for stabilisation purposes. There are early *in situ* grafting experiments [19] and more recent ones with a more detailed analysis of the obtained structure (mostly unpublished) [20]. The latter use thermally controlled radical precursors and the initiator formation is rate limiting. The chains are mainly grown one by one and the layers seem less densely grafted with very long, well stretched and rather monodisperse chains. There is hope that *in situ* growth will allow for well stretched layers with various grafting densities, mean chain lengths in plane structures and polydispersities controlled by the nature (anionic polymerisation has been reported recently) and density of the initiators (by temperature or irradiation eventually) and by the bulk monomer concentration.

References

1. Leibler, L. (1980) Theory of microphase separation in block copolymers. *Macromol.* **13**, 1602
2. Semenov, A. N. (1985) Contribution to the theory of microphase layering in copolymer melts. *JETP Lett.* **61**, 733
3. Marques, C., Joanny, J.-F. & Leibler, L. (1988) Adsorption of block copolymers in selective solvents. *Macromol.* **21**, 1051
4. Semenov, A. N., Joanny, J.-F. & Khokhlov, A. R. (1995) Associating polymers: equilibrium and linear viscoelasticity. *Macromol.* **28**, 1066
5. Alexander, S. (1977) Adsorption of chain molecules with a polar head: a scaling description. *J. Physique* **38**, 1983
6. Duplantier, B. (1989) Statistical mechanics of polymers of any topology. *J. Stat. Phys.* **54**, 581
7. de Gennes, P.-G. (1991) *Scaling Concepts in Polymer Physics.* Ithaca, N.Y: Cornell University Press
8. Daoud, M. & Cotton, J.-P. (1982) Star-shaped polymers: a model for the conformation and its concentration dependence. *J. Physique* **43**, 531
9. Doi, M. D. & Edwards, S. F. (1986) *The Theory of Polymer Dynamics.* Oxford: Oxford University Press
10. Wittmer, J., Johner, A., Joanny, J.-F & Binder, K. (1994) Chain desorption from a semi-dilute polymer brush: a Monte Carlo simulation. *J. Chem. Phys.* **101**, 4379
11. Tuzar, Z. (1992) Experiments on association of block copolymers in solution. *J. Macromolecular Science — Pure and Applied Chemistry* **A29** *(suppl.2)*, 173
12. Johner, A. & Joanny, J.-F. (1990) Block copolymer adsorption in a selective solvent: a kinetic study. *Macromol.* **23**, 5299
13. Auroy, P., Auvray, L. & Leger, L. (1991) Building of a grafted layer: 1. Role of the concentration of free polymers in the reaction bath. *Macromol.* **24**, 5158
14. Cates, M. E. (1986) Diffusion-limited aggregation without branching in the continuum approximation. *Phys. Rev. A* **36**, 5007
15. Witten, T. & Sanders, L. M. (1983) Diffusion-limited aggregation. *Phys. Rev. B* **27**, 5686
16. Krug, J., Kassner, K., Meakin, P. & Family, F. (1993) Laplacian needle growth. *Europhys. Lett.* **24**, 527
17. Milner, S., Witten, T. & Cates, M. E. (1988) Theory of the grafted polymer brush. *Macromol.* **21**, 2610
18. Wittmer, J. P., Cates, M. E., Johner, A. & Turner, M. S. (1996) Diffusive growth of a polymer layer by *in situ* polymerisation. *Europhys, Lett.* **33**, 397
19. Vidal, A. & Donnet, J.-B. (1985) Le greffage sur surface solides de molecules organiques et de polymères (French). *Bull. Soc. Chim. France* **6** 1088
20. Rribbe, A., Prucker, O. & Rühe, J. (1996) Imaging of polymer monolayers attached to silica surfaces by element specific transmission electron microscopy. *Polymer* **37**, 1087

Dynamics of Complex Fluids, pp. 199–212
ed. M. J. Adams, R. A. Mashelkar, J. R. A. Pearson & A. R. Rennie
Imperial College Press–The Royal Society, 1998

Chapter 14

Dynamics of Adsorbed Polymer Layers

J. L. HARDEN AND M. E. CATES*

Department of Physics and Astronomy, University of Edinburgh,
JCMB, King's Buildings,
Mayfield Road, Edinburgh EH9 3JZ, UK
**E-mail: mec@ph.ed.ac.uk*

We present the salient features of recent theoretical studies on the deformation and desorption of end-adsorbed polymer layers in strong flows of good solvent. Our work uses a scaling theory, based on the ansatz that all chains behave alike in flows, which addresses the non-uniform deformation of grafted chains in strong solvent flows. For the case of permeation flows, we use this approach to calculate the dependence of chain deformation on the solvent flow rate through the brush and the solvent pressure-flux relation of the brush system. For the case of shear flows, we calculate the deformation of grafted chains and the solvent flow profile within the brush in a mutually consistent fashion, and we discuss the effects of shear flow on the rate of chain desorption. Chain fractionation effects are briefly considered.

Introduction

The properties of adsorbed polymer layers have been the subject of extensive studies during the past two decades. Their equilibrium properties are now fairly well understood [1–3]. In recent years, studies have focused on the behaviour of adsorbed layers in non-equilibrium conditions [4, 5]. Many such studies have considered the steady-state response of adsorbed layers to applied perturbations, such as the hydrodynamic drag due to solvent flows in the neighbourhood of an adsorbed layer and the frictional forces due to the relative motion of polymer-coated surfaces in contact [6–19]. The response of adsorbed polymer layers to strong flows has important implications for the rheological behaviour of sterically stabilised colloidal dispersions, and for the lubrication properties of polymer-coated surfaces. Of particular interest is the limit of strong flows, in which there is significant deformation of the layers. In this paper, we describe some of our recent theoretical work on the deformation and desorption of adsorbed polymers in strong flow conditions. We will focus on the case of adsorbed layers of neutral polymers attached by one end to a surface at high grafting density.Such polymers may be attached either by a strongly adsorbing end group (sticker) or by adsorption of an insoluble block in a copolymer. In both cases, the unattached part forms a "polymer brush".

Strong flow conditions can give rise to novel and unexpected brush behaviour. For example, in the case of neutral polymer brushes subjected to shear flows, there is experimental evidence that brushes can swell in the direction normal to the grafting surface in response to sufficiently strong flows parallel to the grafting surface [5–9]. Unlike previous treatments [12–15], our theoretical approach calculates the deformation of grafted chains and the solvent flow characteristics within the layer in a mutually consistent fashion. Thus, it provides useful insights into the detailed deformation behaviour of densely grafted chains in strong flows and the nature of the coupling between flow and brush deformation.

In the next section, we introduce our model and apply it to brush deformation in uniform permeation flows of a good solvent perpendicular to the grafting surface. Following this, we extend the model to the case of brush deformation in simple shear flows parallel to the grafting surface and discuss the effects of shear flow on the rate of chain desorption. We then briefly outline the recent work of Aubouy and ourselves on fractionation effects, Finally, some brief concluding remarks are made.

Permeation Flows

Consider a brush consisting of monodisperse neutral chains of degree of polymerisation, N, and monomer size, a, grafted at areal density, $\sigma \sim \xi_0^{-2}$, to a flat porous medium in good solvent conditions. We consider the case of extending permeation flows, in which the brush is subjected to a flow, $\vec{V} = +V\hat{z}$, perpendicular to the grafting surface, as sketched in Fig. 1(a). For the sake of simplicity, we adopt the Alexander–deGennes ansatz that all the chains in

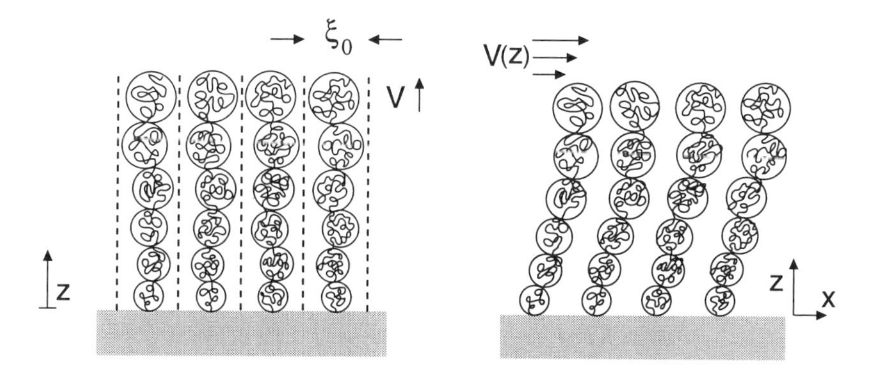

Fig. 1. A sketch of a deformed polymer brush in strong solvent flows: (a) permeation flow perpendicular to the grafting surface, and (b) shear flow $\vec{V}(z) = \dot{\gamma} z \hat{x}$ parallel to the grafting surface.

a brush behave alike [20, 21]. Each chain in the perturbed brush is a string of excluded volume blobs; all chain ends are at the outer edge of the brush. In the absence of flow, this scaling approach results in a brush of thickness, $H_0 = N\xi_0^{-2/3}a^{5/3}$, formed of close-packed blobs of size ξ_0.

Under flow, we represent the chains in the perturbed brush as strings of "Pincus blobs" [24]. Since the chain extension is due to the integrated hydrodynamic drag on each chain rather than a force applied to the chain ends, the blob size is a function, $\xi(z)$, of the distance, z, from the grafting plane which must be determined. This situation is analogous to the model of single chain extension in strong solvent flows studied in (see also [26]). [25] (see also [26]). In the strong extension regime, we generally have $\xi(z) < \xi_0$. This fact allows us to adapt several concepts introduced by Rabin and Alexander [12] in constructing our model free energy. The free energy per chain is the sum of two terms [17]: an elastic term, F_{el}, and an osmotic term, F_{int}.The elastic energy per chain is obtained by considering each chain as an elongated string of blobs, and it takes the form

$$\frac{F_{el}}{k_B T} \simeq \int_0^H \frac{dz}{\xi(z)}, \tag{1}$$

where H is the thickness of the deformed brush. The blob size, ξ, is related to the number of monomers in a blob, g, via $\xi \sim g^{3/5}a$. The local chain stretching can be obtained as a function of ξ by viewing a small section of chain containing $dn \ll N$ monomers as a linear string of Pincus blobs of length, $dz \simeq dn(\xi/a)^{-5/3}\xi$:

$$\frac{dz}{dn} \simeq \xi(z)^{-2/3} a^{5/3} . \tag{2}$$

Now consider a slab of thickness, $\xi(z)$, at distance, z, from the grafting plane. The local interaction between chains may be approximated by a virial expansion in the blob density [12, 13]. This leads to an interaction energy per chain of the form

$$F_{int} \simeq kT \int_0^H \frac{dz}{\xi(z)} \frac{\xi(z)^2}{\xi_0^2} = \frac{kT}{\xi_0^2} \int_0^H dz\, \xi(z) . \tag{3}$$

Our estimate of the free energy per chain is the sum of contributions from Eqs. (1) and (3). In these formulae, and those below, various order-unity numerical prefactors have been suppressed.

Following [17], we now define an effective local chain tension, $t(z)$, as follows. Consider a short section of chain of length Δz containing Δn monomers. Equations (1) to (3) give the free energy of this section of chain as

$$\Delta F \sim kT \left(\frac{1}{\xi(z)} + \frac{\xi(z)}{\xi_0^2} \right) \Delta z . \tag{4}$$

This expression may be written exclusively in terms of Δz and Δn by using the Pincus law for chain stretching, Eq. (2). Subsequent variation of ΔF, with respect to Δz at fixed Δn, yields the local effective chain tension:

$$t(z) = kT \left(\frac{1}{\xi(z)} - \frac{\xi(z)}{\xi_0^2} \right) . \tag{5}$$

This effective tension determines the local departure fromequilibrium of a representative chain under conditions whereall the chains are constrained to behave identically. It already contains osmotic terms and so should not be confused with the(purely elastic) tension in a stretched chain which arises in most mean field approaches to brushes[22, 23].

The local Pincus blob size, $\xi(z)$, is obtained by balancing the differential chain tension across a blob with the total effective hydrodynamic drag on the blob. For Zimm-Stokes blobs, this drag per blob scales as $f_h \sim \eta\xi(z)V(z)$, where η is the solvent viscosity. In mechanical equilibrium, the drag force, f_h, is balanced by the differential tension, $\Delta t = (\partial t/\partial z)|_n \xi(z)$, across a blob. Using Eq. (5), this force balance yields a differential equation for $\xi(z)$ in terms of the solvent velocity, $V(z)$. For a uniform permeation flow, $V(z) = V$, we obtain

$$\xi(z) = \frac{\xi_0}{2}\left\{\left(4 + \frac{\xi_0^2}{\xi_h^4}(H-z)^2\right)^{1/2} - \frac{\xi_0}{\xi_h^2}(H-z)\right\}, \tag{6}$$

where $\xi_h = (kT/\eta V)^{1/2}$ is a "hydrodynamic blob size" [25]. For low V, ξ is weakly dependent on z while for high V, $\xi(z) \sim 1/(H-z)$, as for an isolated chain in a uniform flow [25]. (Note that our expression for $\xi(z)$, Eq. (6), breaks down at $z \simeq H - \xi_{\max}$, where $\xi_{\max}$ is the size of the outermost blob. However, this does not affect any of the conclusions below.) Given $\xi(z)$, we may now determine the brush thickness, $H(N,V)$, by demanding that a stretched chain has N monomers:

$$(7) N \simeq \int_0^H \frac{dz}{\xi}\left(\frac{\xi}{a}\right)^{5/3}. \tag{7}$$

From the above equations, one may extract a characteristic velocity scale, $V_* = kT/(n_b \eta \xi_0^2)$, separating the weak and strong deformation regimes (here $n_b = N(a/\xi_0)^{5/3}$ is the number of blobs in the unperturbed brush). For $V \ll V_*$, we find $H \sim N$ while for $V \gg V_*$, $H \sim N^3 V^2$, as in the single chain case discussed in [25]. The rapid increase of H with V in the latter regime will in practice soon saturate at the maximum chain extension, $H_{\max} = Na$, which occurs at some maximum flow rate, $V_{\max}$ [27]. Our calculations are applicable for $V_* < V < V_{\max}$.

These results may be used to discuss the flux-pressure curve for the hydrodynamic flow of fluid through a polymer layer grafted to a porous support [17]. ($V \gg V_*$), one finds a nonlinear pressure-flux law, $V \sim V_* \left(\Delta P/P_*\right)^{1/3}$, where $P_* = kT/\xi_0^3$, and for compressional flows with $\vec{V} = -V\hat{z}$ (discussed in detail in [17]), $V \sim V_* \left(\Delta P/P_*\right)^{7/9}$ for $V \gg V_*$. Finally, we should mention that our scaling results for $H(N,V)$ and $\Delta P(N,V)$ may also be obtained using a global "monoblock" [25] balance of forces [17].

Shear Flows

We now consider the case of shear flows, in which the fluid above the brush obeys $\vec{V} = \dot{\gamma} z \hat{x}$ parallel to the grafting surface. Within the brush, this is modified to $\vec{V} = V(z)\hat{x}$ where $V(z)$ remains to be found. For strong flows, the grafted chains also stretch and tilt away from the normal direction. We picture this deformed brush as consisting of tilted chains of Pincus blobs [24], as sketched in Fig. 1(b). As before, the hydrodynamic Pincus blob size ξ (and now also chain tilt angle, θ) are functions of the distance from the grafting plane to be determined along with the solvent flow profile in the brush. The free energy per chain is again the sum of an elastic term, F_{el}, involving chain deformation and an osmotic term, F_{int}, involving interactions between blobs. As explained in [18], one can again derive an effective chain tension, $\vec{t}(s)$, whose components obey

$$\frac{t_x(s)}{k_B T} = \frac{\sin\theta(s)}{\xi(s)} + \frac{\xi(s)}{\xi_0^2}\tan\theta(s)\,, \tag{8}$$

$$\frac{t_z(s)}{k_B T} = \frac{\cos\theta(s)}{\xi(s)} - \frac{\xi(s)}{\xi_0^2}\left(\frac{2-\cos^2\theta(s)}{\cos^2\theta(s)}\right), \tag{9}$$

where s is an arclength co-ordinate and $\theta(s)$ is the local tilt angle measured from the $\hat{z}$ direction. Once again, this effective tension determines the local departure from equilibrium of a representative chain under conditions where all the chains are constrained to behave identically.

In mechanical equilibrium, the total hydrodynamic drag on a blob, $\xi(s)$, at $z(s)$ must be balanced by the differential tension across the blob. Assuming laminar flow within the brush, the total hydrodynamic drag on a blob is also in the $\hat{x}$ direction and scales as $\vec{f_h} \simeq \eta\xi(s)V(s)\hat{x}$. Mechanical force balance then implies $t_z = 0$ and $\Delta t_x = (\partial t_x(s)/\partial s)\xi(s) \simeq \eta V(s)\Delta s$. These requirements lead to

$$\xi(z) = \frac{\cos^{3/2}\theta(z)}{(2-\cos^2\theta(z))^{1/2}}\,\xi_0\,, \tag{10}$$

$$\frac{t_x(z)}{k_B T} = \frac{\sin\theta(z)}{\xi(z)} + \frac{\xi(z)}{\xi_0^2}\tan\theta(z) = \frac{\eta}{k_B T}\int_z^H dz'\frac{V(z')}{\cos\theta(z')}\,, \tag{11}$$

where H is the thickness of the deformed brush, and where we have changed the independent variable to z, the normal distance from the grafting surface.

The shear stress, σ_{xz}, in the brush is the sum of the contribution from the viscous solvent flow, $\sigma_{xz}^{(s)} = \eta dV/dz$, and that from the elastic deformation of polymer chains, $\sigma_{xz}^{(p)}$. The polymer contribution, $\sigma_{xz}^{(p)}$, is the product of the effective lateral chain tension, $t_x(z)$, and the number density of chains crossing the plane at height z: $\sigma_{xz}^{(p)} \simeq t_x(z)/\xi_0^2$. Requiring that the total shear stress is uniform throughout the grafted layer gives an additional relation between the chain conformations and the solvent velocity in the brush:

$$\frac{d^2V}{dz^2} = \frac{1}{\xi_0^2}\frac{V(z)}{\cos\theta(z)}\,. \tag{12}$$

Note that although Eq. (12) resembles a Brinkman equation (for flow through a rigid porous medium), its origin and interpretation are obviously quite different.

To solve these equations, we assume no-slip boundary conditions at the grafting surface and continuity of shear stress at the free surface of the brush. Furthermore, we assume that the effective tension, Eq. (11), vanishes at the free end of each chain. Thus, our boundary conditions are $V = 0$ at $z = 0$, $dV/dz = \dot{\gamma}$ and $\theta = 0$ at $z = H$.

For a given brush thickness H, this procedure in principle uniquely determines $V(z)$, $\xi(z)$ and $\theta(z)$ as a function of the solvent shear rate, $\dot{\gamma}$, outside the brush and the equilibrium brush parameters N and ξ_0. The brush thickness is again obtained by demanding that a chain of blobs stretched to height H has N monomers. Further progress now requires a numerical approach [18]. We present our results as a function of the dimensionless shear rate, $\dot{\gamma}\tau$, and the height of the unperturbed brush, $H_0 = n_b\xi_0$. Here, $\tau = \eta\xi_0^3/k_BT$ is of order the characteristic relaxation time of a Zimm blob of radius ξ_0. Figure 2 shows plots of the relative brush swelling $\delta h = (H - H_0)/H_0$ versus $\dot{\gamma}\tau$ for brushes initially with $n_b = 5$, 10 and 15 blobs. We find the onset of significant brush swelling at $\dot{\gamma}\tau \simeq 1$, followed by a saturation of swelling at large $\dot{\gamma}\tau$. Figure 3(a) shows the scaled solvent velocity profiles V/V_0 versus z/H for the case of $\dot{\gamma}\tau = 2$, where $V_0 = k_BT/(\eta\xi_0^2)$. Figure 3(b) shows the corresponding plots of the tilt angle and the (reduced) blob size.

Our results [18] show that, roughly speaking, stretched chains can be divided into two regions: (i) an interior region of essentially uniform stretching

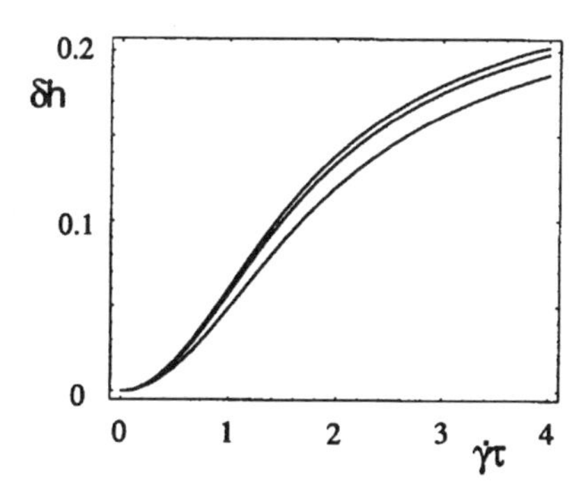

Fig. 2. Plots of the relative brush swelling, $\delta h = (H - H_0)/H_0$ versus $\dot{\gamma}\tau$, for brushes initially with $n_b = 5$ blobs (lowest curve) $n_b = 10$ blobs (middle curve), and $n_b = 15$ blobs (highest curve).

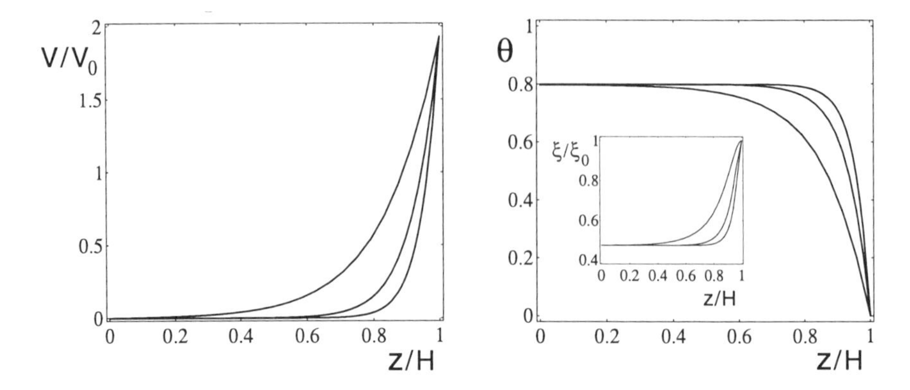

Fig. 3. (a) Plots of the scaled solvent velocity profiles, V/V_0, as a function of the reduced distance, z/H, from the grafting surface for brushes initially with $n_b = 5$ blobs (lowest curve), $n_b = 10$ blobs (middle curve) and $n_b = 15$ blobs (highest curve) sheared at $\dot{\gamma}\tau = 2$. (b) Plots of the tilt angle, θ, and scaled blob size, ξ/ξ_0 versus z/H, for $\dot{\gamma}\tau = 2$. The main plot shows θ for brushes with $n_b = 5, 10, 15$ (order as above) while the inset shows the analogous curves of ξ/ξ_0 in the reverse order.

and tilt in which V is very small; and (ii) a boundary region of thickness $\Delta z \simeq \xi_0$ of weakly stretched and tilted chains in which $\xi(z)$ and $\theta(z)$ vary rapidly. The thickness of this boundary layer corresponds roughly to the hydrodynamic penetration depth $l_p \equiv [V(H)/V'(H)] \simeq \xi_0$ calculated from the velocity profiles. Our results can be rationalised in terms of a "quasi-monoblock" picture in which uniformly tilted and stretched chains with constant blob size, ξ_{mb}, and tilt angle, θ_{mb}, terminate in an unperturbed *virtual* end-blob of size $\xi_{end} \simeq \xi_0$. If the drag force on this end blob, $f_{\parallel} \simeq \eta V(H)\xi_0$, obtained from the numerically computed velocity profile, $V(z)$, is used to compute the uniform stretching of the remainder of the chain, the resulting ξ_{mb} and θ_{mb} are in excellent accord with the numerically obtained ξ and θ in the interior region. We will make use of this quasi-monoblock concept below in our discussion of chain desorption and fractionation effects. We note that there is both a qualitative similarity and a quantitative difference between our predictions and those made earlier by Barrat [13] (see [18]).

Chain Desorption

We now discuss the qualitative features of the desorption of individual grafted chains in a brush subjected to shear flow. This issue has been considered in some detail by Aubouy and ourselves [19]. Here we give a somewhat simplified version by retaining the Alexander–deGennes ansatz. However, as emphasised in [19], the results are strongly sensitive to "fractionation effects", which we discuss very briefly in the next section. In general, one expects the chain desorption rate to be increased by putting grafted chains under external tension. Below, we consider this effect first for the case of a compact end-sticker, and then for the case of diblock copolymers grafted via an insoluble block. It turns out that the impact of flow on desorption is quite different in the two cases.

The desorption of one chain initially attached by a compact end-sticker to a surface supporting an equilibrium brush has been recently studied by Wittmer*et al.* [28]. Their results for Rouse chains in good solvent conditions indicate that the release of one chain can be interpreted in terms of a two-step process. First, the sticker must detach from the grafting surface and escape from a potential well of depth of order $k_B T$ by diffusing a distance of the order of ξ_0, the size of the first blob for equilibrium brushes. Once the end-sticker has moved a distance of order ξ_0 away from the grafting surface, the motion becomes deterministic and the chain is expelled from the brush by a net force due to chain tension and osmotic pressure effects. The crucial point is that the limiting step is the first one: the rate of desorption then takes the Kramers form [29],

$$\mathcal{R}_{\text{eq}} \cong \frac{1}{\tau_{\text{eq}}} \exp\left[-\Delta U / k_B T\right], \tag{13}$$

where τ_{eq} is the characteristic time for the unbound sticker to move one blob diameter from the surface and ΔU is the effective barrier height.

Equation (13) indicates that desorption is dominated by a local process independent of N, which is related to the dynamics of the first blob. One possible estimation for this time is $\tau_{\text{eq}} = \tau \simeq \eta \xi_0^3 / k_B T$, the characteristic Zimm relaxation time of the blob. The results of [28] for Rouse chains are in rough agreement with this choice. For Zimm chains, this choice for τ_{eq} in Eq. (13) leads to $\mathcal{R}_{\text{eq}} \sim \xi_0^{-3}$. The underlying physics of the equilibrium desorption process studied in [28] is that the important characteristic length is that of a section of chain storing $k_B T$ of stretching energy, that is, the Pincus blob size.

We may generalise this approach to sheared layers by using the appropriate Pincus blob size, ξ_w, at the grafting surface in place of ξ_0, leading to $\mathcal{R} \simeq \xi_w^{-3}$. The Pincus blob size at the grafting surface, ξ_w, is a decreasing function of the chain tension and hence of the shear rate. Thus, the desorption rate increases with $\dot{\gamma}$. Within the quasi-monoblock picture for sheared Alexander–deGennes brushes discussed in the previous section, the total drag force on a chain scales as $f_{\parallel} \sim \dot{\gamma}$, implying that $\xi_w \sim (\dot{\gamma}\tau)^{-1}$. Hence, the desorption rate for compact end-stickers scales as

$$\mathcal{R} \sim (\dot{\gamma}\tau)^3 . \tag{14}$$

The case of diblock copolymers (polymers A grafted to the surface by an insoluble block of polymer B) is quite different.We presume that the solvent is good for polymer A and poor for the anchor. In a simplified view, we assume that the B-chains form a dense, molten layer near the surface from which we may extract any of the anchoring blocks independently. Thus, desorption proceeds via the continuous extraction of the insoluble B-block from the matrix, which is penalised by a free energy cost due to partial exposure of the B-chain to a poor solvent. The extracted B-chain is pictured as a string of collapsed molten blobs of size $d \sim (k_B T/\Gamma)^{1/2}$, where Γ is the effective surface tension of the B-monomer/solvent interface. The energy of a string of such exposed blobs of length l then scales as

$$\mathcal{F}_{ch} \cong k_B T l/d . \tag{15}$$

This free energy cost is partially compensated by the shear-induced tension in the chain, leading to a reduced barrier height to desorption under shear. In the limit of high shear rates, this tension is proportional to the shear force on the last blob, $f_{\parallel} \sim \dot{\gamma}$, giving a desorption rate for a diblock-anchored chain which scales as

$$\mathcal{R} \sim \exp\left[k\dot{\gamma}\tau\right] , \tag{16}$$

where k is a shear-rate independent quantity.

The situation is in fact much more complicated when one goes beyond the Alexander–deGennes ansatz and accounts for chain fractionation effects [19]. However, the basic physics remains that, within a Kramers approach, the shear-induced enhancement of chain desorption for the case of a compact end-stickers is due to the influence of shear on a characteristic local relaxation rate; while for the case of diblock copolymers in selective solvents, the primary effect of shear is to lower the effective barrier height to desorption.

Fractionation Effects

So far, we have considered brush deformation and chain desorption processes by assuming that all chains in a brush behave alike (the Alexander–deGennes ansatz). For equilibrium brushes, however, self-consistent field calculations [22, 23] have shown that the free ends are in fact distributed throughout the layer, rather than concentrated at its extremity. One might also expect this to be the case for brushes subjected to solvent flows. The issue can be addressed, at least qualitatively, within a simple "dual chain" picture devised by Aubouy [19].

In this model, only two types of chains are present: a fraction, f, of the chains which are extended and tilted by the shear flow (the "dragged" chains), and the remaining fraction, $1-f$,which lies deeper in the adsorbed layer where the flow is screened(the "quiescent" chains). We assume, following the results in the third section, that the dragged chains form aquasi-monoblock-like layer, while the screened fraction form a quiescent Alexander–deGennes type brush amid the lower sections of the strongly deformed chains. Within this picture, the brush is subject to hydrodynamic drag forces which are assumed to affect only the outermost blobs of the fraction, f, of the dragged chains, each of which supports a force, $F_{\parallel} \cong \eta\dot{\gamma}/\left(f\sigma\right)$. Following [13], one can then write an effective free energy, $\mathcal{F}$, as the sum of osmotic and elastic contributions for each species (including cross terms) and a hydrodynamic potential coupled to the "dragged" chains alone [19]. This effective free energy is minimised with respect to the steady-state conformation of each type of chain and the fraction, f; there is an additional constraint that the quiescent fraction of chains is always screened from the shear flow.

The result of this procedure is that f falls rapidly with increasing $\dot{\gamma}\tau$, approaching $f_{min} \simeq 1/n_b$ for $\dot{\gamma}\tau \gg 1$. Within the dual chain model of Aubouy, the rapid decrease in f is actually, for large n_b, a discontinuous jump. Whatever the details of this, the model suggests that shear flow may pick out a few chains and pull them very hard. This has some interesting consequences for the deformation, and (especially) desorption behaviour, of a brush in shear flow which are explored more fully in [19].

Conclusions

We have presented a semi-phenomenological model for the behaviour of grafted polymer layers in strong normal and shear flows of good solvent. Using the Alexander–deGennes ansatz, in which all chains behave alike, this approach can address the non-uniform deformation of grafted chains in strong solvent flows. Moreover, in the case of shear flows, the solvent flow profile within the layer and the deformation of the layer are obtained in a mutually consistent fashion. Our work offers significant improvements over previous approaches to brush deformation [12–14]which ignore the details of solvent flow inside a brush. Instead, they model the solvent-brush frictional forces by *ad hoc* forces applied to the free surface. Nonetheless, one conclusion of our work is that these approaches (if interpreted carefully) are reasonably well-founded. We have also found that the dependence of the rate of chain desorption on shear rate may be predicted for both functionalised end-group "stickers" and block copolymer grafting mechanisms. In particular, the latter results are sensitive to the use of the Alexander–deGennes ansatz. This sensitivity is revealed by studying the dual-chain model of Aubouy [19].

Our calculations of brush bevaviour in strong flows of course have certain inherent limitations. First, our neglect of order unity prefactors means we cannot provide precise numerical predictions. Second, a relaxation of the Alexander–deGennes ansatz ultimately requires a fully self-consistent treatment that allows for an arbitrary dispersion in the chain end positions. Third, our description of hydrodynamic forces might also be improved, perhaps along the lines of the multiple scattering theory of [30] for adsorbed homopolymer layers. This, however, would be a formidable numerical task. Finally, the present approach is limited by the finite extensibility of the grafted polymer chains. We have shown that for sufficiently high flow rates, the chains approach full extension and the blob concepts we employed break down. With these provisos, our approach may be readily extended to address several other problems. Work in progress is considering the details of chain deformation in grafted polyelectrolyte layers [31] and in more general adsorbed polymer layers [32] in strong flows. The latter complements recent studies in the linear regime [30, 33].

Discussion

T. McLeish What is the effect of the swelling of brushes under shear on the collision between two coated colloidal particles?

M. Cates To answer that properly, we would need to do the calculations explicitly with two layers present. These will interact via both direct and hydrodynamic forces in a non-additive way. We have not done this for brushes, though we do have some results for adsorbed homopolymer layers; see [34].

V. Kumaran Under strong shear flows, the chains are nearly horizontal and can be extended to several times their usual length. Therefore, a small uncertainty in the angle, θ, could lead to a big change in the brush height. How robust are the predictions, and how tilted are the chains predicted to be compared with the earlier calculations of Barrat?

M. Cates The degree of tilt is similar to that found by Barrat, even though he did not account for the large size of the terminal blob. We therefore obtain a stronger effect for any given shear rate. It is true that a small error in the angle could give a large change in height. However, a check of robustness is provided by the dual chain model of [19]. That also shows swelling (by a rather large amount, in fact) which suggests that the effect is probably real. The experimental consensus seems to be in favour of swelling, though some of the evidence is rather indirect.

B. Costello It is not easy to measure adsorbed layer thickness in shear, but the hydrodynamic layer thickness could be measured, even at very high shear rates, for example, in a capillary. Does your theory give hydrodynamic thickness, and what would you expect to observe?

M. Cates Since the flow profile is found, the hydrodynamic thickness can indeed be calculated from our results. It will be less than what we call the "brush height" (H), but not by very much (about half the size of the outermost blob). For brushes with many blobs initially, the fractional swelling of the hydrodynamic thickness follows that of H itself. The position is more complex for adsorbed homopolymers; see [34].

Acknowledgements

This work was supported in part by the EPSRC and the DTI Colloid Technology Programme. We thank Miguel Aubouy for valuable insights.

References

1. Milner, S. T. (1991) *Science* **251**, 905
2. Halperin, A., Tirrell, M. & Lodge, T. P. (1992) *Adv. Polym. Sci.* **100**, 33
3. Fleer, G. J., Cohen Stuart, M. A., Scheutjens, J. M. H. M., Cosgove, T. & Vincent, B. (1993) *Polymers at Interfaces* London: Chapman & Hall
4. Cohen Stuart, M. A. & Fleer, G. J. (1996) *Annu. Rev. Mater. Sci.* **26**, 463
5. Klein, J. (1996) *Annu. Rev. Mater. Sci.* **26**, 581
6. Klein, J., Perahia, D. & Warburg, S. (1991) *Nature* **352**, 143
7. Klein, J., Kamiyama, Y., Yoshizawa, H., Israelachvili, J. N., Fredrickson, G. H., Pincus, P. & Fetters, L. J. (1993) *Macromolecules* **26**, 5552
8. Kleinm, J. (1994) *Colloids and Surfaces* *A***86**, 63

9. Klein, J., Kumacheva, E., Perahia, D., Mahalu, D. & Warburg, S. (1994) *Faraday Discussions* **98**, 173
10. Fredrickson, G. H. & Pincus, P. (1991) *Langmuir* **7**, 786
11. Milner, S. T. (1991) *Macromolecules* **24**, 3704
12. Rabin, Y. & Alexander, S. (1990) *Europhys. Lett.* **13**, 49
13. Barrat, J.-L. (1992) *Macromolecules* **25**, 832
14. Kumaran, V. (1993) *Macromolecules* **26**, 2464
15. Birshtein, T. M. & Zhulina, E. B. ((1992) *Makromol. Chem., Theory Simul.* **1**, 193
16. Williams, D. R. M. (1993), *Macromolecules* **26**, 5808
17. Harden, J. L. & Cates, M. E. (1995) *J. Phys. II France* **5**, 1093; **5**, 1757
18. Harden, J. L. & Cates, M. E. (1996) *Physical Rev. E* **53**, 3782
19. Aubouy, M., Harden, J. L. & Cates, M. E. (1996) *J. Phys. II France* **6**, 969
20. Alexander, S. (1977) *J. Phys. Paris* **38**, 983
21. de Gennes, P. G. (1976) *J. Phys. Paris* **37**, 1443; *Macromolecules* **13**, (1980) 1069
22. Milner, S. T., Witten, T. A. & Cates, M. E. (1988) *Europhys. Lett.* **5**, 413; *Macromolecules* **21**, (1988) 2610
23. Skvortsov, A. M., Gorbunov, A. A., Pavlushkov, I. V., Zhulina, E. B., Borisov, O. V. & Priamitsyn, V. A. (1988) *Polym. Sci. USSR* **30**, 1706; Zhulina, E. B., Borisov, O. V. & Pryamitsyn, V. A. (1990) *J. Coll. Int. Sci.* **137**, 495
24. Pincus, P. (1976) *Macromolecules* **9**, 386
25. Brochard-Wyart, F. (1993) *Europhys. Lett.* **23**, 105; Brochard-Wyart, F., Hervet, H. & Pincus, P. (1994) *Europhys. Lett.* **26**, 511
26. Ajdari, A., Brochard-Wyart, F., de Gennes, P. G., Leibler, L., Viovy, J.-L. & Rubinstein, M. (1994) *Physica A* **204**, 17; Rubinstein, M., Ajdari, A., Leibler, L., Brochard-Wyart, F. & de Gennes, P. G. (1993) *C. R. Acad. Sci. Paris II* **316**, 317
27. Scaling arguments given in Ref. [17] indicate $V_{\max} \simeq [3/2 + (\xi_0/a)^{1/3}]\, V_*$
28. Wittmer, J., Johner, A., Joanny, J.-F. & Binder, K. (1994) *J. Chem. Phys.* **101**, 4379
29. Kramers, H. A. (1940) *Physica* **7**, 284
30. Wu, D. T. & Cates, M. E. (1993) *Phys. Rev. Lett.* **71**, 4142
31. Harden, J. L., Borisov, O. V. & Cates, M. E. (1997) *Macromolecules*, **30**, 1179
32. Harden, J. L., Aubouy, M. & Cates, M. E., unpublished
33. Sens, P., Marques, C. M. & Joanny, J.-F. (1994) *Macromolecules* **27**, 3812
34. Wu, D. T. & Cates, M. E. (1996) *Macromolecules* **29**, 4417

Dynamics of Complex Fluids, pp. 213–229
ed. M. J. Adams, R. A. Mashelkar, J. R. A. Pearson & A. R. Rennie
Imperial College Press–The Royal Society, 1998

Chapter 15

The Molecular Dynamics Simulation of Boundary-Layer Lubrication: Atomistic and Coarse-Grained Simulations

D. J. TILDESLEY
Department of Chemistry,
The University of Southampton, Southampton S017 1BJ
E-mail: d.j.tildesley@soton.ac.uk

This paper describes the use of non-equilibriun molecular dynamics simulations to study the friction of surfaces coated with amphiphiles. The first model contains a full atomistic representation of the surfactant, DODAB (dimethyl distearyl ammonium bromide). Two flat surfaces covered with DODAB are brought into contact and the surfaces sheared at a constant speed and at constant surface separation. The friction coefficient is calculated as a ratio of the tangential force in the direction of the shear to the normal force. Coarse-grained simulation, in which the amphiphile is represented by a grafted bead-spring model, are also performed in the presence of solvent. The viscosity profile between the surfaces is calculated as a function of the shearing rate and chain stiffness.

Atomistic Simulations of DODAB Bilayers

Considerable interest in the rheology of molecularly thin films has been stimulated by experiments conducted to determine frictional forces [1–9]. These experiments reveal that the rheology of thin films of organic lubricants can be dramatically different from that of the corresponding bulk liquids.

A detailed understanding of the properties of surfactant films in confined geometries is fundamentally important to problems such as adhesion, capillarity, contact formation, friction, lubrication, wear and modifications of surfaces [10–21].

We report simulation studies of the softening of fabric by the addition of cationic surfactants. A number of commercial fabric softeners consist of dispersions of dialkyl quaternary ammonium salts, such as dioctadecyldimethylammonium chloride (DODAB) (commercially known as Arquad 2HT or TA100). It has been suggested that the perceived softness is brought about by the lubrication of the fibre surface by the surfactant so that the yarns of the fabric slide over one another more easily [22–24]. A typical liquid rinse conditioner consists of a concentrated dispersion of gel phase (L_β) particles in water. These positively charged liposomal particles are attracted to the negatively charged cotton surface. Recent studies by Jones [25] have helped to elucidate the mechanism of surface deposition. As the fabric is removed from solution, the water dries from the surface resulting in a thin layer around each fibre. As drying continues, the air-water interface is ruptured by the deposited particle. As a consequence of the Marangoni effect [26], surfactant molecules spread rapidly from the air-water-particle three-phase contact line across the air-water interface as the interface moves past the particles. The surfactant molecules form a well-ordered Langmuir film at the air-water interface. As the remaining water supporting this monolayer evaporates, the surfactant molecules are transferred as a well-ordered film to the fabric surface.

Thus, some basic understanding of fabric conditioning can be gained by studying the behaviour of one well-ordered monolayer of surfactant adsorbed on a surface as it moves across a second adsorbed surfactant layer. Experiments of this kind have been performed on TA100 by Chugg and Chaudhri [27, 28]. They use the technique of sliding a spherical stylus on film-coated flat substrates. Surfactant films of various thicknesses are created by Langmuir Blodgett deposition or retraction from organic solvents. These experimental studies have measured the effect of contact pressure, the number of friction cycles, the substrate temperature and the sliding velocity on the friction coefficient of thin DODAB films.

Recently, we have used the molecular dynamics simulation technique to model the friction coefficient between two layers of DODAB [29]. An atomistic model of the bilayer is constructed and the distance between the two layers is adjusted to produce the required contact pressure. The top layer is moved at a fixed shearing velocity with respect to the lower surface and the heat produced is removed using a thermostat. The inter-layer force in the direction of the shear is measured and used to calculate the friction coefficient. Glosli and McClelland [30] have used this technique to study friction between hexane layers anchored rigidly to a flat surface and Harrison *et al.* [31, 32] have simulated the friction between two hydrogen-terminated (111) surfaces of diamond using the same approach.

Our simulation code has been written for a distributed memory parallel machine so that we can study low shearing velocities and large system sizes. The details of the parallelisation of the molecular dynamics code and its implementation on a number of shared memory and distributed memory parallel machines can be found in reference [33].

Our model of the surfactant uses the anisotropic, united-atom model of Toxvaerd [34] for the hydrocarbon backbone. The long-range interactions between charges on head groups of the amphiphiles are calculated using the Ewald-like method of Hautman and Klein [35]. The amphiphiles are not tethered to the underlying surface and are free to adopt any orientational, conformational or translational structure with respect to the substrate.

The model of the dioctadecyldimethylammonium chloride is described in detail elsewhere [36]. Briefly, the bond-lengths are fixed at their equilibrium values by using constraint dynamics implemented with the SHAKE algorithm [37]. The three-atom valence angle distortions are controlled using the normal harmonic potential. The three torsional potentials, CCCC, CCNC and CCCN, are described by potentials of the form used by Ryckaert and Bellemans for butane [38].

In [29], we have reported simulations of $C_{18}C_{18}$ bilayers at constant surface separation. We have studied the effect of changing the sliding velocity on friction coefficient. A number of simulations were performed at the same normal pressure but at different amphiphile head group areas to study the effect of surface density on the friction. Finally, we have increased the contact pressure between the layers at fixed temperature and sliding velocity to examine the effect of compression on the friction.

At a head group area of 50 Å^2 molecule^{-1}, the layers form a structure where the amphiphiles are tilted at an angle of 45° to the surface normal in which one of the chains in the amphiphile contains a $g^+g^+tg^+g^+$ conformational defect close to the head group. This allows it to bend around and follow the direction of the second all-trans chain.

In the simulations, the top layer is sheared with respect to the bottom layer by moving the atoms on the top surface at a distance $\Delta x = v_x \delta t$, where δt is the time-step of 2.5×10^{-15} s and v_x is the sliding velocity. We have studied shearing velocities of between one and 100 ms^{-1}. In the case of the slowest shearing velocity, we have also displaced the two layers simultaneously at $(1/2)\Delta x$ in opposite directions as a check on the invariance. The layer is initially sheared in the direction orthogonal to the molecular tilt. At the lowest shearing velocity, a run of 10^6 time-steps (2.5 ns) is equivalent to displacing the top surface 25 Å, or approximately four lattice spacings. In simulating at higher shearing velocities, we have chosen to keep the total lateral movement of the top layer fixed at 25 Å and to reduce the length of the simulation in proportion. In a few cases, we performed much longer runs at higher shearing velocities to check on the time-dependence of the friction.

This range of sliding velocities is appropriate to real problems in friction. TA100 and similar surfactants are used to lubricate fibres during the spinning process at speeds of up to 20 ms^{-1}. Equally, speeds of between 0.1 and 1 ms^{-1} are appropriate for conditioning fabric and hair. Most of the experimental work in measuring friction coefficient of adsorbed layers is performed at much lower velocities [28, 1], in the range of 0.01 to 10 mm s^{-1}. During the course of the simulation, the heat is removed from the system using a uniform thermostat and all the runs are performed close to room temperature. In determining the instantaneous temperature, the drift velocity, v_x^{drift}, in the shearing direction is calculated separately for the two layers. The kinetic energy of the system is determined with the calculated drift velocity adjustment applied to the molecules in each layer.

The frictional force, F_x, is defined as the force required to initiate or maintain relative motion between two surfaces and is given by Amonton's law, $F_x = -\mu F_z$, where F_z is the normal force applied by one body on another and μ is the coefficient of friction. The in-plane force required to start the motion is the static frictional force while that required to sustain the motion at a constant velocity is the dynamic frictional force. The static force is usually greater than the dynamic force. Coulomb's law suggests that the frictional force is proportional to the normal force (that is, μ is independent of F_z). The calculation of the friction coefficient in the simulation is straightforward. During the course of the simulation, we separate the force into an intra-layer and inter-layer component. During the course of these simulations, molecules always remain associated with the same layer and it is easy to identify the force which acts across the boundary layer. We average the normal force, F_z, and the tangential force in the direction of the shear, F_x, over all molecular configurations during the shearing phase of the simulation.

For the system at 50 Å^2 and 1 ms^{-1}, the estimated friction coefficient is 0.15 compared to the extrapolated experimental value of 0.11.

We observe that the friction coefficient (a) decreases with increasing normal force; (b) increases with decreasing amphiphile density; and (c) increases with increasing shearing velocity.

During the shearing, the electrostatic interaction between the charged head groups and counter-ions attached to the surface means that there is very little slip in the amphiphile surface region. The ordered hexagonal structure of the counter-ions induces a solid head group structure but even at high density, the radial distribution of the methyl tail groups in the plane of the surface is liquid-like. At all the shearing velocities studied in this paper, the amphiphiles align with the flow direction.

This work has been extended to study the effect of the structure of the amphiphile on the inter-layer friction by simulating dialkyl ammonium surfactants with different chain lengths at the same normal pressure [39]. In particular, we have simulated $C_{18}C_{18}$, $C_{10}C_{18}$ and $C_{10}C_{10}$ bilayers at 298.15 K and fixed wall separations and a shearing velocity of 1 ms^{-1}.

Figure 1(a) shows a snapshot of the $C_{10}C_{18}$ bilayer at a head group area of 50 Å^2 after shearing at 1 ms^{-1} for 50,0000 time-steps. The white spheres are the nitrogen atoms of the head group, the black lines represent the bonds of the C_{10} chain and the last eight atoms in the C_{18} chain are joined by lighter grey lines. Although there is a significant interaction between the C_{18} chains, it is clear that there is no direct steric interactions between the C_{10} chains of the two layers. We note that the all-trans length of a C_{10} chain is approximately 11 Å and the combined length of two C_{10} chains and the head groups is not sufficient for them to touch at a wall separation of 34 Å. In the interface region, the C_{18} chains are inter-digitated at the required normal pressure and the chain tails exhibit a "liquid-like" structure. As a result of this overlap, the friction coefficient of the $C_{10}C_{18}$ system is significantly higher than that of the dialkyldimethylammonium cationic surfactants with chains of equal lengths at the same normal pressure, coverage and shearing velocity.

For the system at a head group area of 77 Å^2 per molecule, a wall separation distance of 21.05 Å is required to obtain a normal pressure of 210 MPa. This separation is also small enough to allow both the C_{10} and C_{18} chains to be in contact (Fig. 1(b)). The more compact interactions between both sets of chains lead to a sharper interface and a smaller overlap. The friction coefficient

(a)

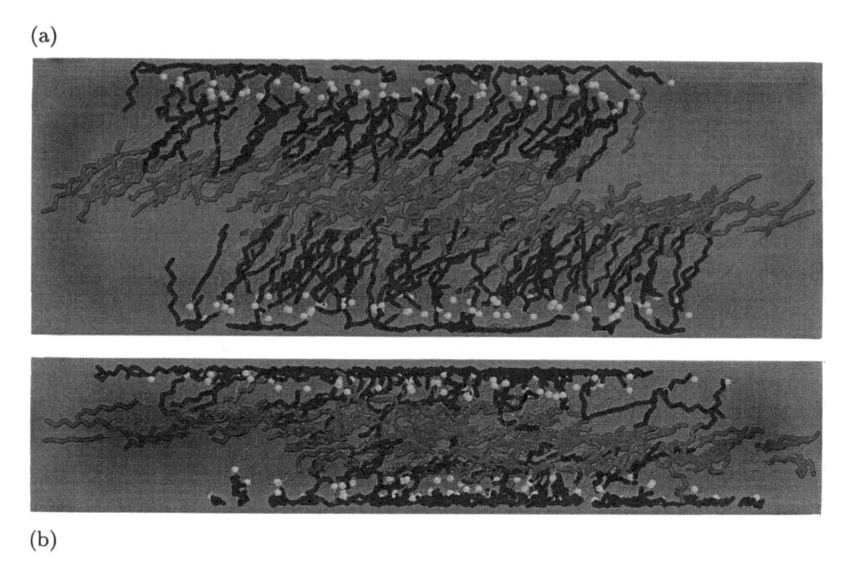

(b)

Fig. 1. Snapshots of the $C_{10}C_{18}$ bilayer at a normal pressure of 210 ± 10 MPa and 298.15 K after shearing at 1 ms^{-1} for 500,000 time-steps. The white spheres are the nitroge atoms of the head group and the dark lines represent bonds of the C_{10} chain. The last eight atoms in the C_{18} chain are joined by the grey lines: (a) the top snapshot shows the system at a coverage of 50 $Å^2$ per molecule; (b) the bottom snapshot shows the systems at 77 $Å^2$ per molecule.

decreases with decreasing head group area for the $C_{10}C_{18}$ bilayer. This is in contrast to the $C_{18}C_{18}$ and $C_{10}C_{10}$ systems where the friction coefficient decreases with increasing head group area.

In a bilayer of dialkyldimethylammonium cationic surfactants with equal chain length, the chain tails form a well-defined interface at a head group area of 50 $Å^2$. When the head group area is increased at a constant normal pressure, there is a larger overlap between the layers which leads to a higher friction coefficient. For the mixed chain length system, the decrease in the surface density requires a smaller wall separation at the same normal pressure. This produces a more compact structure at the interface, thus leading to a lowering of the friction coefficient.

Coarse-Grained Models of Hydrodynamic Lubrication

The work on atomic-scale modelling of friction restricts us to relatively small systems and short times. We have also used a mesoscale model comprising of a spring-bead representation of the amphilphile. This reduces the computational burden significantly and provides a route to extend the simulations beyond the nanosecond barrier. It also allows us to introduce a simple model of the solvent.

The model used in our coarse-grained studies is composed of solvent molecules, amphiphilic chains and two atomic boundary walls. The chains can be grafted to the wall or are free to move about in the solvent. The interaction among all the particles is described by a shifted and truncated Lennard-Jones potential known as the repulsive Weeks-Chandler-Andersen potential (WCA) [45]. It is given by

$$U_{WCA} = 4\epsilon\left[\left(\frac{\sigma}{r}\right)^{12} - \left(\frac{\sigma}{r}\right)^{6}\right] + \epsilon \qquad r \leq 2^{1/6}, \tag{1}$$

$$U_{WCA} = 0 \qquad r > 2^{1/6}, \tag{2}$$

where σ and ϵ are the collision diameter and the well-depth of the potential, respectively. This potential is truncated and shifted at the minimum of the full Lennard-Jones potential and includes only repulsive contributions. Additional attractive potentials are required for bead-bead interaction within the chains and to retain certain particles in the crystal lattice forming the walls.

The intra-molecular bead interactions are modelled using the bead-spring potential [46] where beads are connected by an anharmonic spring [47–48] of the form

$$U_c = -\frac{1}{2}k_c R_o^2 \ln\left[1 - \left(\frac{r}{R_o}\right)^2\right] \qquad r \leq R_o, \tag{3}$$

$$U_c = \infty \qquad r > R_o. \tag{4}$$

The values for k_c and R_o are $30\epsilon/\sigma^2$ and 1.5σ, respectively, and are taken from [48].

The stiffness of the chains is controlled by including one additional harmonic potential acting between beads i and $i+2$:

$$U_{bd,ik} = \frac{1}{2}k_{ik}(\mid \vec{r}_{ik} - \vec{r}_{eq.,ik} \mid)^2 \qquad k = i+2 \tag{5}$$

with $\vec{r}_{eq.,ik} = 1.9216\sigma$, which is twice the distance of the location of the minimum potential given by Eqs. (1) and (2). Two sets of simulations were performed using ($k_{ik} = 0$) (LJ+C: Lennard-Jones plus chains) and ($k_{ik} = 30\epsilon/\sigma^2$)

(LJ+SC: Lennard-Jones plus stiff chains). Solvent molecules composed of repulsive WCA spheres are included in the region between the chains.

In our initial study of this model [41], we performed non-equilibrium molecular dynamics simulations to measure the viscosity profiles of fluids between walls coated with amphiphiles. The physical properties of such a complex system are defined by a large number of parameters, that is, wall separation, chain length, chain stiffness, surface coverage and shearing velocity. We have focused on a small set of parameters which is feasible to simulate and which is sufficient to gain a basic understanding of the rheology of this model system. We have chosen a low surface coverage to examine the interplay between solvent and amphiphiles at a range of reduced shear rates which vary up to 0.101. The lowest reduced shear rate considered in this study is 0.005. Below this rate, it is not possible to calculate reliable stress and viscosity data because the response to shear is lost in the noise.

The most interesting result from this study is that even at this low coverage, the chains behave as an extension of the wall. This is reflected in the steep velocity profiles between the ends of the chains and in the remarkable changes to the mean square radius of gyration of the adsorbed polymer with increasing shear. The component of the mean squared radius of gyration perpendicular to the wall, $S_{\perp}$, drops significantly as we change from a shear rate of 0.0 to 0.005 and does not change with increasing shear rates thereafter. On the other hand, $S_{\|}$ increases continuously with increasing rate indicating that the tilt of the amphiphilic molecules, due to the shear induced flow, is accompanied by an elongation of the chains. Increasing the stiffness of the chains produces a significant change in the polymer geometry and only a small change in the rheology of the flow.

At low shear rate, the stress, $-P_{xz}$, is constant across the slab and can be computed directly by measuring the force across the wall in the direction, x, of the shear. Only at higher shear rates does the stress, influenced by the temperature profile, change across the pore. As the shear field increases, the temperature profiles become increasingly non-parabolic, indicating that the fluid and layer of amphiphiles have a different thermal conductivity. This effect is more pronounced for the stiffer chains.

In a second study using the coarse-grained model [49], we varied the surface coverage by randomly detaching amphiphiles from the wall. This is a process which is encountered, for instance, in the surface force apparatus when studying frictional forces between solid surfaces coated with polymer brushes [42]. We were interested in how the shearing and detachment of surfactant molecules change the structural characteristics and rheological properties of the system. Throughout the simulations, we applied a reduced shear rate of 0.0101, assuming that in the case of block polymers each unit is of size 100 Å with an average mass of 50,000 g/mole corresponding to a shear rate of the order of 10^{-3} ns^{-1}. Typical shear rates observed in experiments are in the range of 10^{-7} to 10^{-4} ns^{-1} in the surface force apparatus [43, 27]. Similar rates are proposed for spontaneous shape changes in bilayers leading to a significant increase in curvature [44].

As before, each wall is covered by a layer of amphiphilic molecules which are 20 units long and bound chemically to the surface. Simulations were performed at amphiphlic coverages ranging from 1/3 to zero, where 1/3 corresponds to a system with only grafted chains attached to every third wall atom. Coverages lower than 1/3 were obtained by randomly detaching chains from the wall.

As before, the grafted chains behave as an extension of the wall at coverages greater than zero, that is, for all coverages at which some chains are attached. For a pure amphiphile-solvent mixture (zero coverage), a slip evolves at the boundary between the fluid and the wall which is more pronounced for the stiffer surfactants. For the LJ+C and LJ+SC models, the friction coefficients are constant across the pore at all coverages studied.

Surprisingly, the frictional forces for the intermediate coverages are higher than those found for coverages of 0.33 and zero. At the intermediate coverages, there are both grafted and free chains in the pore region. The entanglement between these two types of chain, as indicated by the overlap of the corresponding density profiles, increases the frictional force. This effect is more pronounced for the LJ+SC model where the chains are less flexible and hence the amphiphiles cannot easily adapt to changes in the fluid flow.

Acknowledgements

I wish to acknowledge the collaboration with Günther Peters, Jose Alejandre and Colin Kong in performing the work described in this abstract. Their names are associated with the original publications referenced throughout. We wish to acknowledge a grant from the materials chemistry MPP consortium for a generous allocation of computer time on the CRAY T3D. YCK thanks the EPSRC for a postdoctoral fellowship (GR/K37734). DJT acknowledges a grant for computing equipment (GR/J74459) and and JA thank Unilever for financial support under the CREF Physical Sciences programme.

Discussion

J. R. A. Pearson Does the constraint on surface density used in the simulations apply in experiment and how does the temperature come into the simulations?

D. J. Tildesley I suspect that in the experiments of Chugg and Chaudhri [27, 28], the head group area per molecule of the adsorbed DODAB surfactant was fixed and measurable. In these experiments, it is possible to change the density of chains on the surface by changing the size of the head group, say, from an ammonium ion to a piperidine ring at close packing. This is equivalent, in the simulation, to changing the area per counter-ion in the surface from 55 Å^2 to approximately 65 Å^2. The temperature enters the simulation by setting the velocity of the particle. The velocity components are chosen from a Maxwell-Boltzmann distribution such that

$$\frac{3N - N_c}{2} k_{\mathrm{B}} T = \sum_{i=1}^{N} \frac{1}{2} m_i \mathbf{v}_i^2 , \tag{6}$$

where N is the number of atoms and N_c the number of constraints. The only slightly complicating factor in these simulations is that the streaming or mean velocity in the direction of the flow, which changes across the pore, has to be removed from the appropriate component of the atomic velocity before computing the temperature.

M. J. Adams A frictional force is usefully considered as the work per unit sliding distance. Would you describe in more detail the energy dissipation processes that you have observed in both the atomistic and coarse-grained models? The distinction appears to be that for the former, a solid-like model of chains against each other is applicable while for the latter, a hydrodynamically lubricated contact is developed.

D. J. Tildesley You are quite correct. The atomistic simulations involve two sets of hydrocarbon chains in direct contact. This is solid-like. However, the chains have considerable conformational freedom and the simulations demonstrate that the interface becomes slightly sharper as the normal pressure increases. This lowers the friction coefficient. The principal mechanism for dissipation of energy is the removal of heat. At much lower temperatures than those studied in this work, the chains move past one another in a highly ordered fashion. The picture that this evokes is that of two flexible, overlapping plates passing one another. The plates bend as they pass and then snap back into position releasing energy. The thermostat controlling the temperature in the simulation receives a regular pulse of energy at times corresponding to a movement through half the lattice spacing [30]. We did not observe any regular structure in the frictional force as a function of time in our simulations, although the $F_x(t)$ was subjected to a time-series analysis. At room temperatures, the methyl tails of the chain show a liquid-like structure and the removal of the heat energy is continuous. The coarse-grained simulations contain solvent between the chains and model hydrodynamic lubrication. Again, the heat generated in the shear is controlled by a thermostat but this is handled more realistically than in the atomistic simulations. Only the atoms in the wall are thermostated so that it is possible for a temperature gradient to build up across the box. At the lowest reduced shear rates studied, there is a slight maximum in the temperature profile but this is enhanced by stiffening the grafted chains.

M. E. Cates Your work addresses the issue of *microscopic* friction between perfectly flat coated surfaces. In practice, such surfaces are rare. Does anyone know how to "scale-up" microscopic friction data to predict the *macroscopic* friction between rough surfaces (which is probably dominated by the sharpest bump)? Under some conditions, you predict very high temperature gradients within the lubrication layer. Do your parameters for the surface atoms correspond to good or poor conductors of heat?

D. J. Tildesley I do not know how to scale up the friction to include the roughness of the surface and I think the chances of making such a connection are small. In one idealised model, we could imagine that the lubrication between irregular cotton fibres might involve a low direct surface area of contact, say, less than 0.1% of the total area of the fibre, and that the contact might be of the form shown in the sketch below.

In this instance, there could be regions of less than 0.1 μm which were touching directly. Given a model of amphiphile adsorption involving deposition from the drying air/liquid interface to form regular compact minelayers (as suggested by C. Jones, Unilever Research Port Sunlight), we might envisage well-ordered bilayers of amphiphiles in contact over distances of approximately 200 lattice spacing surrounded by significant regions of no contact. Assuming that these regions of contact were atomically flat, then the simulated friction coefficients would determine the macroscopic friction. Following discussions with Brian Briscoe (Chemical Engineering, Imperial College), it appears that this idealised model is unlikely, and that atomic force microscope measurements of this type of surface show clear steps and defects on distance scales of less than 0.1 μm. If this is the case, the prediction of the macroscopic friction is a difficult problem.

To answer the second part of your question, in coarse-grained simulations the walls are formed from atoms interacting via the repulsive WCA potential. These atoms are anchored to their lattice sites by springs with $k_w = 72\epsilon/(2^{1/3}\sigma^2)$, which is derived from the second derivative of the full Lennard-Jones potential evaluated at the minimum. The temperature of the wall atoms is controlled using a Nosé-Hoover thermostat. I believe that the choice of a low spring constant in the solid, with vibrational motions that are well-matched to those of the solvent atoms close to the wall, means that we have chosen a solid with a high thermal conductivity. If we take a reduced shear rate of $\dot{\gamma}^* = \dot{\gamma}\sqrt{m\sigma^2/\epsilon} = 0.040$ and simulate a pure WCA fluid against the wall, there is an observable but small increase in temperature in the middle of the pore. The temperature gradient goes up at a fixed shear rate when we introduce the grafted chains and it is more pronounced the stiffer the chain becomes. I suspect that the thermal coupling between the solvent and the wall is reduced by the chains, and it is the poor thermal conductivity of the stiffer chains that causes the sharp temperature rise at high shear rates. The characterisation of the thermal conductivity in the chain region and close to the wall is an interesting problem that deserves some more attention.

M. Lal You indicated that the presence of a flow field renders necessary the modification of the Kirkwood-Irving expression for the various components of the pressure tensor. Would the modification in question have some implications for the surface tension that is, would the modification result in shear rate dependent surface tension?

D. J. Tildesley I wish to correct a misconception that I may have caused during the lecture. I am not aware that the Irving-Kirkwood expression for the pressure tensor is modified by the presence of a flow field. Of course, this does not mean that the tangential pressure profiles are independent of the shear rate at fixed normal pressure, and the surface tension could certainly change with shear rate.

Todd *et al.* has recently pointed out that the Irving-Kirkwood expression for the pressure needs to be used with care in inhomogeneous fluids (B. D. Todd, D. J. Evans and P. J. Daivis (1995) *Phys. Rev.* **E52**, 1627). At time t,

$$\mathbf{P}(\mathbf{r},t) = \frac{1}{V}\left[\sum_i m_i(t)[\mathbf{v}_i(t) - \mathbf{u}(\mathbf{r}_i,t)][\mathbf{v}_i(t) - \mathbf{u}(\mathbf{r}_i,t)] + \sum_{i>j} \mathbf{r}_{ij}(t)\, \mathcal{O}_{ij}\, \mathbf{f}_{ij}(t)|_{\mathbf{r}_i(t)=\mathbf{r}}\right],$$

where V is the volume of the system, $\mathbf{v}_i$ is the particle velocity, $\mathbf{u}$ is the streaming velocity, $\mathbf{f}_{ij}$ is the force between atoms i and j, and $\mathcal{O}_{ij}$ is the operator

$$\mathcal{O}_{ij} = 1 - \frac{1}{2!}\mathbf{r}_{ij}\cdot\nabla_{\mathbf{r}} + \cdots + \frac{1}{n!}\left[-\mathbf{r}_{ij}\cdot\nabla_{\mathbf{r}}\right]^{n-1} + \cdots .$$

For uniform fluids, $\mathcal{O}_{ij} = 1$ is exact. However, for calculations close to a solid wall where the density is changing rapidly, it is important to use the full operator, $\mathcal{O}_{ij}$, and Todd *et al.* have described a technique called the method of planes that includes all the terms in $\mathcal{O}_{ij}$.

S. Ramaswamy If the system is in a "broken-symmetry" state (as Mike Cates suggested), the characteristic time scale should be determined by the lateral size of the system or the size of the oriented domains, whichever is smaller. Therefore, the η versus $\dot{\gamma}$ should start to show shear-thinning at a value of $\dot{\gamma}$ which depends on this length scale.

D. J. Tildesley I find it difficult to add much to this question. I would comment that the amphiphiles are highly ordered in the plane of the surface and the central simulation box is reproduced periodically in this direction. The simulations are too small for the amphiphiles to exhibit grain boundaries between orientated domains. I can reiterate what we observe in the simulations. In the direct interaction of the two amphiphile layers, the friction coefficient appears to decrease slowly with decreasing shearing velocity for velocities of between 100 and 1 ms^{-1} (see Fig. 3 of [29]). In the hydrodynamic lubrication of the coarse-grained simulations, the viscosity profiles across the pore are parabolic. The minimum viscosity in the centre of the pore is almost independent of shear rate, except in the case of the stiffest chains where the viscosity decreases slightly with increasing shear rate (see Fig. 8 of [49]).

B. Costello Professor Tildesley asked why the surfaces forces apparatus had not been used at the shear rates envisaged in the simulation and suggested that this might be due to the problem of dissipation of heat. This may well be true, but it is also the case that the SFA is of a finite size and therefore easier to use in oscillatory rather than steady-state mode. The advantage of this is that in phase and out of phase, components are obtained and these can be used to calculate the energy stored and dissipated through friction. Would it be possible to perform the simulations in oscillatory mode?

D. J. Tildesley The length of the simulations is controlled by the molecular dynamics time-step of 2.5 fs. In a run of 10^6 steps, the total real time simulated is 2.5 ns and the total lateral displacement at 1 ms^{-1} is approximately 25 Å or four lattice spacings. In a typical SFA, the lateral movement of the tip is 7 μm so that the whole simulation lies well within one experimental cycle. It would be possible to perform the simulations in oscillatory mode but only at a very high frequency. If the experimental frequency is 15 Hz, then the experimental shearing velocity is approximately 2×10^{-4} ms^{-1}. In the simulations, it would be difficult to use shearing velocities of less than 0.1 ms^{-1} as the component of the force in the shearing direction is indistinguishable from the noise at reasonable normal pressures. The comparison between simulation and the SFA experiments at very low shearing speeds is probably best made by switching to static Monte Carlo simulation techniques with applied tangential and normal stresses (see, for example, M. Lupowsky and F. van Swol (1991) *J. Chem. Phys.*, **95**, 1995 and M. Schoen, S. Hess and D. J. Diestler (1995) *Phys. Rev.*, **E52**, 2587).

Ronald Larson Have you analysed the dependence of the fluctuations in pressure on the system size?

D. J. Tildesley In the first set of simulations, we were unable to perform system size checks on the pressure because of the cost of the runs. Running on 32 nodes of the CRAY T3D at Edinburgh, 10^6 timesteps for the atomistic model require 132 hours of processing (4224 processor hours). Doubling the system size would not have been feasible. In the coarse-grained simulations, we were able to check the effects of system size by increasing the surface area without changing the separation of the surfaces. As the number of grafted amphiphiles changes from 68 to 136 at a fixed coverage of 1/3 and at a fixed reduced density of 0.825, the temperature, pressure and viscosity profiles remain unchanged.

References

1. Yoshizawa, H., Chen, Y.-L. & Israelachvili, J. (1993) Recent advances in molecular level understanding of adhesion, friction and lubrication. *J. Phys. Chem.* **97**, 4128
2. Yoshizawa, H. & Israelachvili, J. (1993) Fundamental mechanisms of interfacial friction. 2. Stick-slip friction of spherical and chain molecules. *J. Phys. Chem.* **97**, 11300
3. Thompson, P. A. & Robbins, M. O. (1990) Origin of stick-slip motion in boundary lubrication. *Science* **250**, 792
4. Schmitt, V., Lequeux, F., Pousse, A. & Roux, D. (1994) Flow behaviour and shear-induced transition near an isotropic-nematic transition in equilibrium polymers. *Langmuir* **10**, 955
5. Gee, M. L., McGuiggan, P. M., Israelachvili, J. & Homola, A. M. (1990) Liquid to solid-like transitions of molecularly thin films under shear. *J. Chem. Phys.* **93**, 1895
6. Horn, R. G., Hirz, S. J., Hadziioannou, G., Frank, C. W. & Catala, J. M. (1989) A re-evaluation of forces measured across thin polymer films. Non-equilibrium and pinning effects. *J. Chem. Phys.* **90**, 6767
7. Van Alsten, J. & Granick, S. (1990) Tribology studied using atomically smooth surfaces. *Tribology Transactions* **33**, 443
8. Klein, J., Kumacheva, E., Mahalu, D. Perahia, D. & Fetters, L. J. (1994) Reduction of frictional forces between solid surfaces bearing polymer brushes. *Nature* **370**, 634
9. Hu, H.-W. & Granick, S. (1992) Viscoelastic dynamics of confined polymer melts. *Science* **258**, 1339
10. Horn, R. G., Israelachivili, J. N. & Pribac, F. (1987) Measurement of the deformation and adhesion of solids in contact. *J. Coll. Interf. Sci.* **115**, 480
11. Landman, U., Luedtke, W. D., Burnham, N. A. & Colton, R. J. (1990) Atomistic mechanisms and dynamics of adhesion, nanoindentation, and fracture. *Science* **248**, 454
12. Landman, U., Luedtke, W. D. & Ringer, E. M. (1992) Atomistic mechanisms of adhesive contact formation and interfacial processes. *Wear* **153**, 3
13. Israelachivili, J. N. (1992) Intermolecular and Surface Forces. London: Academic Press
14. Adamson, A. W. (1989) Physical Chemistry of Surfaces. New York: Wiley
15. Smith, J. R., Bozzolo, G., Banerjea, A. & Ferrante, L. (1989) Avalanche in adhesion. *Phys. Rev. Lett.* **63**, 1269

16. Ribarsky, M. W. & Landman, U. (1988) Dynamical simulations of stress, strain, and finite deformations. *Phys. Rev. B* **38**, 9522
17. Rabinowicz, E. (1965) *Friction and Wear of Materials.* New York: Wiley
18. Mate, C. W., McClelland, G. M., Erlandsson, R. & Chiang, S. (1987) Atomic-scale friction of a tungsten tip on a graphite surface. *Phys. Rev. Lett.* **59**, 1942
19. Han, K. K., Cushman, J. H. & Diestler, D. J. (1993) Grand canonical Monte Carlo simulations of a Stockmayer fluid in a slit micropore. *Mol. Phys.* **79**, 537
20. Wang, Y., Hill, K. & Harris, J. G. (1994) Confined thin films of a linear and branched octane: a comparison of the structure and solvation forces using molecular dynamics simulations. *J. Chem. Phys.* **100**, 3276
21. Cohen, S. R., Neubauer, G. & McClelland, G. M. (1990) Nanomechanics of Au-Ir contact using a bi-directional atomic force microscope. *J. Vac. Sci & Technol. A* **8**, 3449
22. Gralen, N. & Olofsson, B. (1947) Measurement of friction between single fibres. *Textile Res. J.* **17**, 488
23. Röder, H. L. (1955) Measurement on the influence of finishing agents on the friction of fibres. *J. Text. Inst.* **44**, 85
24. Sebatian, S. A. R. D., Bailey, B. J., Briscoe, B. J. & Tabor, D. (1986) Effect of a softening agent on yarn pull-out force of a plain weave fabric. *Text Res. J.* **56**, 604
25. Jones, C. C. (1993) Unilever Port Sunlight Research Laboratory, private communication
26. Clint, J. H. (1992) *Surfactant Aggregation.* London: Blackie
27. Chugg, K. J. (1990) *The Mechanisms of Fabric Softening.* PhD thesis, St Edmunds College, Cambridge, United Kingdom
28. Chugg, K. J. & Chaudhri, M. (1993) Boundary lubrication and shear properties of thin solid films of dioctadecyl dimethyl ammonium-chloride (TA100). *J. Phys. D. Appl. Phys.* **26**, 1993
29. Kong, Y. C., Tildesley, D. J. & Alejandre, J. A. (1997) The molecular dynamics simulation of boundary layer lubrication. *Mol. Phys.*, **92**, 7
30. Glosli, J. N. & McClelland, G. M. (1993) Molecular dynamics study of sliding friction of ordered organic monolayers. *Phys. Rev. Lett.* **70**, 1960
31. Harrison, J. A., White, C. T., Colton, R. J. & Brenner, D. (1992) Molecular dynamics simulations of atomic-scale friction of diamond surfaces. *Phys. Rev. B* **46**, 9700
32. Harrison, J. A., White, C. T., Colton, R. J. & Brenner, D. (1993) Effects of chemically bound, flexible hydrocarbon species on the frictional properties of diamond surfaces. *J. Phys. Chem.* **97**, 6573
33. Surridge, M., Tildesley, D. J., Kong, Y. C. & Adolf, D. B. (1996) A parallel molecular dynamics simulation code for dialkyl cationic surfactants. *Parallel Computing* **22**, 1053
34. Toxvaerd, S. (1990) Molecular dynamics calculation of the equation of state of alkanes. *J. Chem. Phys.* **93**, 4290

35. Hautman, J. & Klein, M. L. (1992) An Ewald summation method for planar surfaces and interfaces. *Mol. Phys.* **75**, 379
36. Adolf, D. B., Tildesley, D. J., Pinches, M. R. S., Kingdon, J. B., Madden, T. & Clark, A. (1995) Molecular dynamics simulations of dioctadecyldimethylammonium chloride monolayers. *Langmuir* **11**, 237
37. Ryckaert, J. P., Ciccotti, G. & Berendsen, H. C. (1977) Numerical integration of the Cartesian equations of motion of a systems with constraints: molecular dynamics of *n*-alkanes. *J. Comp. Phys.* **23**, 327
38. Ryckaert, J. P. & Bellemans, A. (1975) Molecular dynamics of liquid *n*-butane near its boiling point. *Chem. Phys. Lett.* **30**, 123
39. Kong, Y. C. & Tildesley, D. J. (1998) The effect of molecular geometry on boundary layer lubrication. *Mol. Sim.*, submitted for publication
40. Jorgensen, W. L. & Gao, J. (1986) Monte Carlo simulations of the hydration of ammonium and carboxylate ions. *J. Phys. Chem.* **90**, 2174
41. Peters, G. H. & Tildesley, D. J. (1995) Computer simulation of the rheology of grafted chains under shear. *Phys. Rev. E* **52**, 1882
42. Klein, J., Kumacheva, E., Mahalu, D., Perahia, D. & Fetters, L. J. (1994) Reduction of frictional forces between solid surfaces bearing polymer brushes. *Nature* **370**, 634
43. Gee, M. L., McGuiggan, P. M., Israelachvili, J. & Homola, A. M. (1990) Liquid to solid-like transitions of molecularly thin films under shear. *J. Chem. Phys.* **93**, 1895
44. Evans, E. & Yeung, A. (1994) Hidden dynamics in rapid changes of bilayer shape. *Chem. Phys. Lip.* **73**, 39
45. Weeks, J. D., Chandler, D. & Andersen, H. C. (1971) Role of repulsive forces in determining the equilibrium structure of simple fluids. *J. Chem. Phys.* **54**, 5237
46. Grest, G. S. & Kremer, K. (1986) Molecular dynamics simulation for polymers in the presence of a heat bath. *Phys. Rev. A* **33**, 3628
47. Kremer, K. & Grest, G. S. (1990) Dynamics of entangled linear polymer melts: a molecular dynamics simulation. *J. Chem. Phys.* **92**, 5057
48. Kremer, K. (1993) In *Computer Simulation in Chemical Physics*, ed. M. P. Allen & D. J. Tildesley, London: Kluwer Academic Publishers
49. Peters, G. & Tildesley, D. J. (1996) Computer simulation of the rheology of grafted chains under shear. 2. Depletion of chains at the wall. *Phys. Rev. E* **54**, 5493

Dynamics of Complex Fluids, pp. 230–242
ed. M. J. Adams, R. A. Mashelkar, J. R. A. Pearson & A. R. Rennie
Imperial College Press–The Royal Society, 1998

Chapter 16

Direct Simulation of Surfactant Solution Phase Behaviour

R. G. LARSON
AT&T Bell Laboratories,
Murray Hill, New Jersey 07974

The three-dimensional structures of all major types of phases formed by surfactants in water, including the bi-continuous Ia3d "gyroid" phase, are simulated by a direct Monte Carlo lattice method. The structure of the gyroid inferred from x-ray scattering is confirmed and its epitaxial transformation to hexagonal or lamellar phases is simulated by varying the composition. As the ratio of lengths of the hydrophilic head to the hydrophobic tail is varied, a "universal" progression in the composition-temperature phase diagram is computed, which closely matches that of the experimental phase diagrams of nonionic, anionic, cationic and zwitterionic single-tailed surfactants.

Introduction

Widely disparate surfactant molecules can show similar phase behaviour when mixed with water. For example, Fig. 1 depicts the similar temperature-concentration phase diagrams of four very different amphiphiles: a non-ionic, an anionic, a cationic and a zwitterionic surfactant. Many of the same phases recur in all of these phase diagrams, including lamellar (L_α), hexagonal cylindrical (H) and, in all cases but one, bi-continuous gyroid (G) with cubic Ia3d symmetry [1–4]. The progression of phases with increasing surfactant concentration is also similar in each case, that is, H $\rightarrow$ G $\rightarrow$ L_α. The "T"

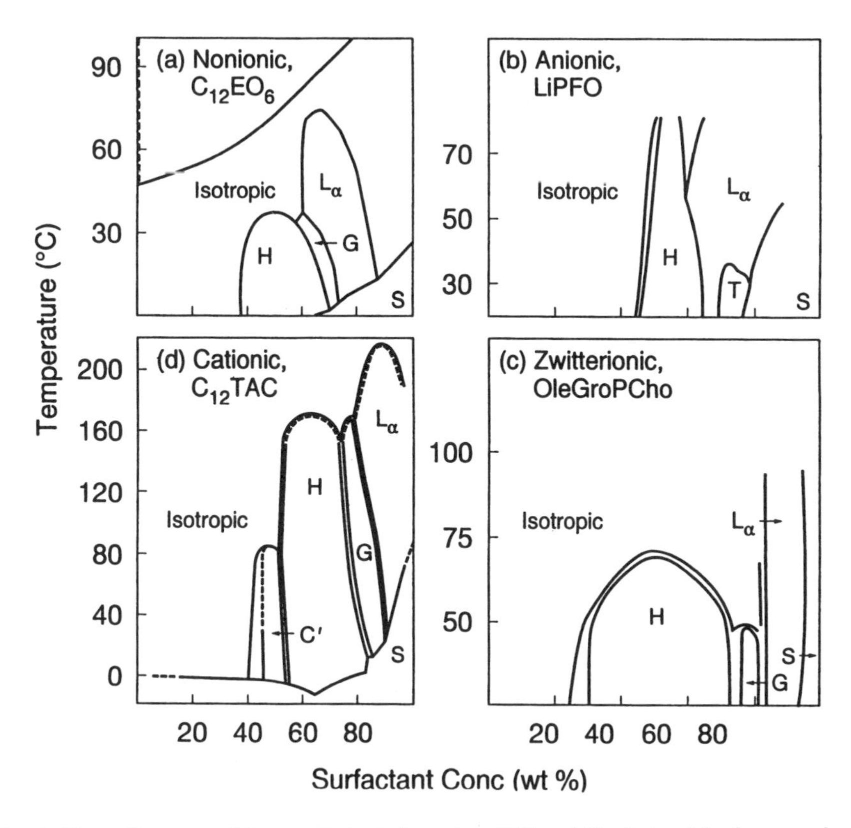

Fig. 1. Phase diagrams of four surfactants in water. "H" and "L_α" stand for hexagonal and lamellar phases. "G" and "T" are gyroid cubic and tetragonal mesh intermediate phases. "C′" is a cubic packing of spherical micelles. "S" refers to crystalline solid. (a) Hexa-ethyleneglycol mono *n*-dodecyl ether, $C_{12}EO_6$ (from Clerc *et al.* [4]); (b) lithium perfluo-rooctanoate, LiPFO (from Kekicheff & Tiddy [3]); (c) 1-oleoyl-*sn*-glycero-3-phosphocholine, OleGroPCho (from Arvidson *et al.* [2]); and (d) dodecyltrimethylammonium chloride $C_{12}TAC$ (from Balmbra *et al.* [1]).

phase, which occurs in Fig. 1(b) instead of the G phase, is a "mesh" phase, namely, lamellae perforated by tetragonally ordered holes. In other surfactant solutions, a mesh phase with rhombohedral (R) symmetry is formed. Note also that in Fig. 1, the order-disorder transition on heating or cooling always occurs at the highest temperature for the lamellar phase, the next highest temperature for the hexagonal phase, and at lower temperatures for the other phases. Finally, the order-order transitions are largely *lyotropic* in each case —

that is, they are driven mainly by changes in concentration, not temperature. Phase behaviour similar to that depicted in Fig. 1 is also observed in aqueous solutions of other single-tailed surfactants [5–7]. These similarities suggest that the phase transitions are driven mainly by "universal" features of surfactant mixtures, namely, volume-filling constraints, entropies and energies of mixing, and entropies of surfactant chain conformation.

There are also some differences among the phase diagrams in Fig. 1 particularly the type of intermediate phase (T versus G) that appears whether or not an ordered cubic micellar phase (C′) appears at low surfactant concentrations, as it does in Fig. 1(d), and in the ranges of composition over which the different phases appear. It is not entirely clear to what extent these differences among the phase diagrams reflect differences in the degree of influence of the "universal" features described above, or of the expression of other molecule-specific effects, such as ionic interactions, hydration layers and specific packing constraints.

To address these issues and examine in detail the structure of the more exotic G and T phases, computer simulations are described in this report which predict the self-assembled phases of a "minimal" molecular model that is bare except for the "universal" features of surfactant solutions described above. The simulated surfactant molecule, designated as H_i T_j, is a sequence of i "head" units connected to j "tail" units. Each unit occupies one site on the simple cubic lattice with periodic boundary conditions. A unit of the amphiphilic chain can be bonded to the nearest or diagonally nearest neighbour so that the co-ordination number is 26. The "water" molecules occupy single sites and interact with other neighbouring or diagonally neighbouring sites in the same way as the "head" units in the amphiphile chain. Thus, there is a single dimensionless interaction energy parameter, w, which is the interaction energy per head/tail contact divided by k_BT. For the sake of convenience, we shall refer to $1/w$ as the "temperature" of the system. The surfactant molecules can move in snake-like and kink-like motions, as discussed elsewhere [8, 9]. Moves are accepted or rejected using the usual Metropolis scheme. Equilibrium ordered or disordered phases are obtained by starting at infinite temperature ($w = 0$) and "cooling" the system by increasing w in small increments until the desired value of w is reached or an ordered phase is formed spontaneously.

Figure 2 shows the phase diagrams determined by simulation for the surfactants H_4T_7, H_4T_6, H_4T_4 and H_6T_4 in water. All runs were carried out on $30 \times 30 \times 30$ boxes except for simulations producing the gyroid (G) and rhombohedral-like (R) phases. The phases were identified by direct

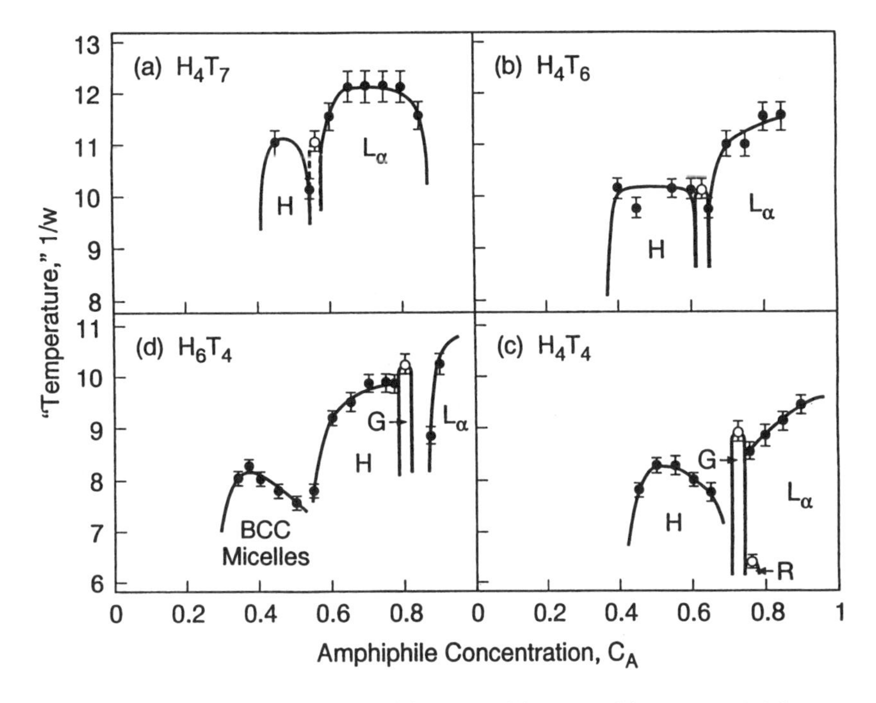

Fig. 2. Simulation phase diagrams of (a) H_4T_7, (b) H_4T_6, (c) H_4T_4 and (d) H_6T_4 in "water". "G" and "R" stand for cubic gyroid and rhombohedral-like mesh intermediate phases. "BCC micelles" is a body-centred cubic packing of spherical micelles. The closed circles are phase boundaries determined on $30 \times 30 \times 30$ lattices. The open circles were determined on lattices of other sizes, namely, $29 \times 29 \times 29$ for the gyroid phase for H_4T_7; $26 \times 26 \times 26$ for the gyroid phase for H_4T_6; $24 \times 24 \times 24$ for the gyroid phases for H_4 and H_6T_4; and $34 \times 34 \times 34$ for the mesh phase for H_4T_4. The gyroid phase for H_4T_7 is marked by dashed lines because it does not appear in every run, and may be metastable.

inspection. Images of simulated lamellar, cylindrical and BCC micellar phases are reported elsewhere [9, 10]. For H_4T_4 and other symmetric surfactants such as H_6T_6, the gyroid phase appears at amphiphile volume fractions (C_A) in a narrow range close to $C_A = 0.73 \pm 0.01$. It appears only on boxes that closely match (within one lattice unit) the gyroid's preferred *d*-spacing. (As shown elsewhere [10], lamellar and hexagonal phases can form within almost any simulation box that is larger than the minimum required to form a single unit cell.) The unit cell dimensions of the G phase correspond to a *d* spacing that is about 2.45 times the repeat spacing of the lamellar phase which forms at concentrations slightly higher than that for the gyroid phase, or about 2.12

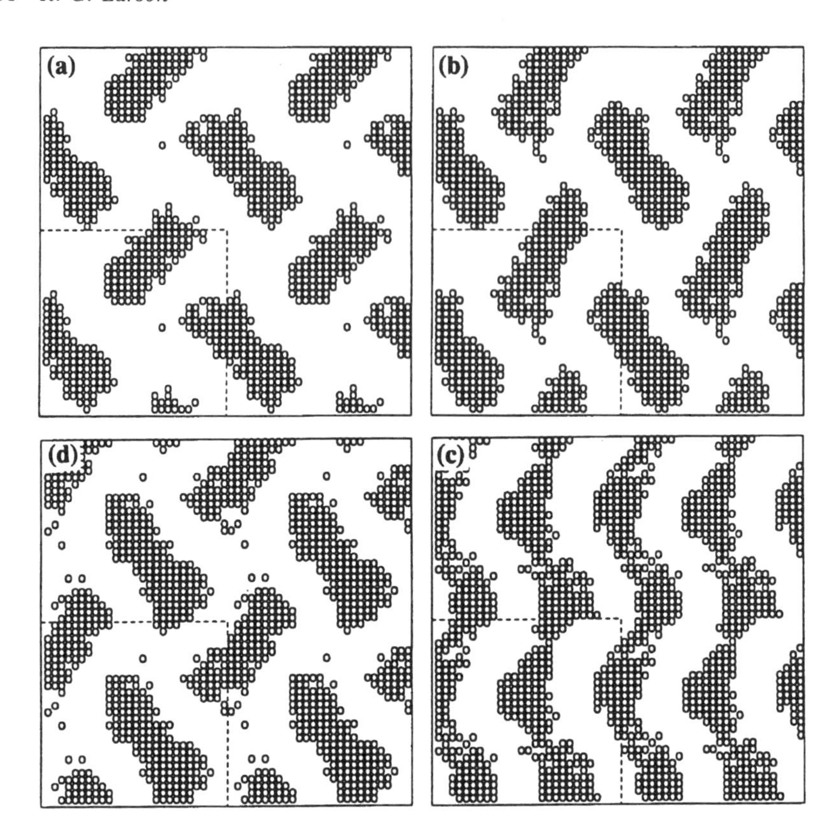

Fig. 3. Slices (a) 1, (b) 3, (c) 5 and (d) 7 of a gyroid phase formed by 73% H_6T_6 in water on a $27 \times 27 \times 27$ lattice. The circles are the tail groups; water and head units are omitted.

times that of the neighbouring hexagonal phase. These epitaxial relationships agree with those inferred from x-ray scattering data for experimental surfactant phases [4, 7, 11, 12].

The structure of the gyroid phase for 73% H_6T_6 on the $27 \times 27 \times 27$ lattice is depicted in Fig. 3, which shows parallel slices through layers 1, 3, 5 and 7 of the $26 \times 26 \times 26$ simulation box. The boundaries of the box are indicated by the dashed lines and those of the remaining images in Fig. 3 are formed by periodic replication. In slice 1 (Fig. 3(a)), the cross-sections of the tail-containing strut networks of the gyroid phase are oblong domains that form a herring-bone pattern. An experimental pattern similar to this can be found in an electron density map for the gyroid phase of dodecyl-trimethyl ammonium chloride (DTAC) [7]. In slice 5 (Fig. 3(c)), each domain *connects* vertically with its two image domains in the two neighbouring unit cells to form two separate zig-zag stripes per unit cell. These stripes break apart in slice 7 (Fig. 3(d)), leaving one domain that has rotated 90° and the other −90°, and each has shifted horizontally by half a unit cell with respect to slice 1. Further slices through the unit cell show further rotations of the oblong domains with totals of 270° and −270° respectively. Three further reconnections, one in the vertical and two in the horizontal directions, along with horizontal and vertical shifts so that the pattern in Fig. 3(a) is finally recovered at the last slice of the unit cell are shown. Tracing the domains from slice to slice, one finds *two interpenetrating, but non-intersecting*, networks of tail-containing units conforming to the structure proposed by Luzzati and co-workers [13, 14] on the basis of x-ray scattering patterns. A projection along the 111 direction of the unit cell of the simulated gyroid phase for 73% H_6T_6 appears in Fig. 4(a). It shows the "wagon wheel" morphology first seen in transmission electron micrograph images of this phase in block copolymers [15] (see Fig. 4(b)).

There are remarkable similarities between the computed phase diagrams in Fig. 2 and the experimental ones in Fig. 1. The experimentally observed phases: lamellar, gyroid, hexagonal, mesh and cubic micellar all occur in the simulations, and the ranges of compositions at which they appear correspond well with the experimental ranges. Further discussion and images of the simulated "mesh" tetrahedral (T) and rhombohedral-like (R) phases can be found in [9, 10]. The box size $24 \times 24 \times 24$ used to obtain the gyroid phases of H_4T_4 and H_6T_4 in Fig. 2 is significantly smaller than that used to obtain the neighbouring cylinder and lamellar phases, which is $30 \times 30 \times 30$. The relative smallness of the $24 \times 24 \times 24$ box, which contains only a single Ia3d unit cell, probably enhances the stability of the gyroid phase artificially and raises its ordering transition temperature. For H_4 T_6 and H_4 T_7, the box sizes used to obtain the gyroid phases are larger, being $26 \times 26 \times 26$ and $29 \times 29 \times 29$, respectively, and the transitions to the gyroid phase occur at lower temperatures than those to the hexagonal phase (see Figs. 2(a) and (b)).

Figure 2 also shows that each liquid crystalline region shifts towards a higher range of amphiphile concentrations when the ratio of the head to the tail length of the surfactant is increased from 4:7 to 6:4 in the clockwise progression from (a) to (d). Thus, the range of concentrations over which the hexagonal phase forms is shifted from 0.45–0.55 for H_4T_7 to 0.55–0.75 for H_6T_4, and the BCC phase appears for H_6T_4 in the concentration range, 0.38–0.50. This

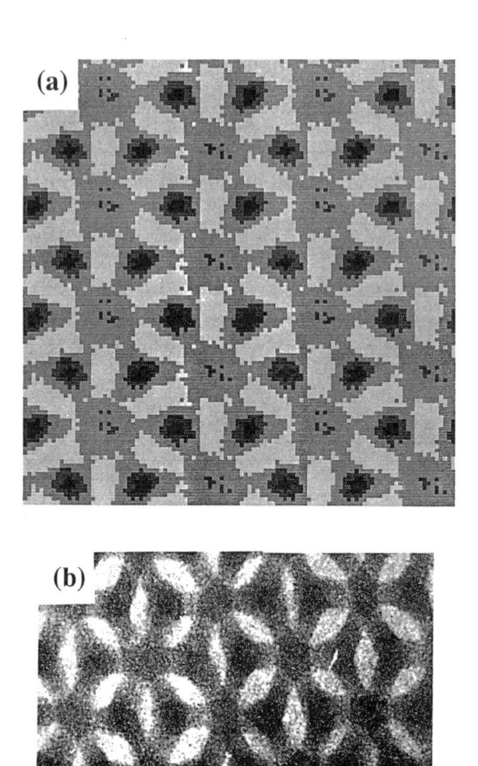

Fig. 4. (a) "Wagon wheel" image of gyroid phase formed by 73% H_6T_6 in water on a $27 \times 27 \times 27$ lattice which was extended periodically. The image was formed by projection along the 111 axis of the unit cube using a high contrast between the tail units against the head and water units. (b) Transmission electron micrograph obtained by Thomas *et al.* [15] of a cubic intermediate phase formed by a polystyrene-polyisoprene block copolymer. The phase was originally thought to be a "double diamond" phase, which is similar to the gyroid.

progression, with an increasing ratio of head-to-tail length, is similar to that observed by Mitchell *et al.* [16] on the phase behaviour of polyoxyethylene surfactants in water. Note that for H_4T_7, the range of compositions over which the hexagonal phase appears is rather narrow, indicating that the stability of this phase is reduced for H_4 T_7. The composition range for gyroid phase for this surfactant is marked with dashed lines (Fig. 6(a)) because it might be metastable. For one run on the 29 × 29 × 29 lattice, the gyroid phase is formed. However, in a repeat run for the same composition ($C_A = 0.57$) and box size, the lamellar phase is formed instead. For asymmetric surfactants with even longer tail groups, both the gyroid and the hexagonal phases disappear altogether (see below).

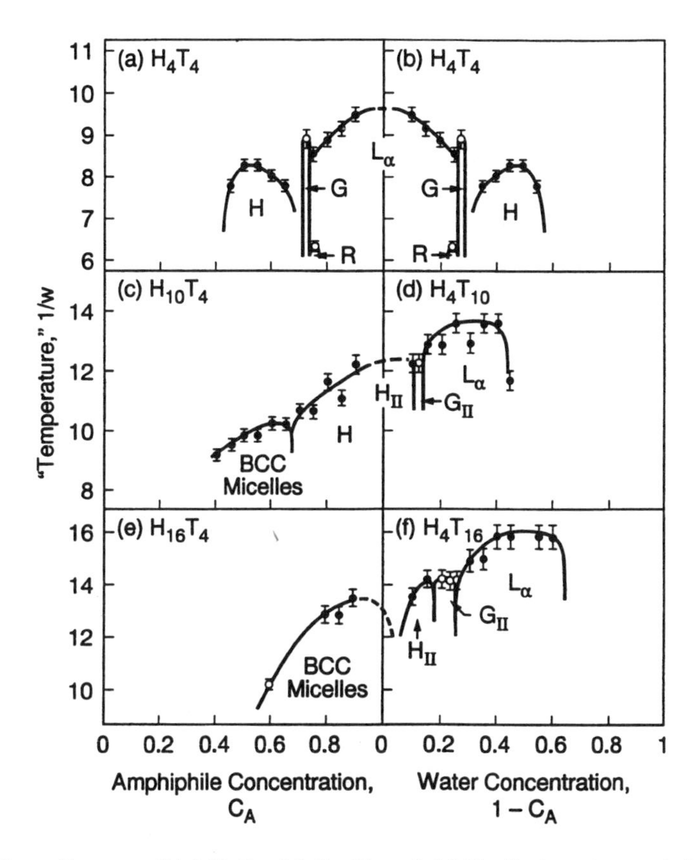

Fig. 5. Phase diagrams of (a) H_4T_4, (c) $H_{10}T_4$ and (e) $H_{16}T_4$ in water, and those of the "complementary" surfactants: (b) H_4T_4, (d) H_4T_{10} and (f) H_4T_{16}, plotted against water concentration, $1-C_A$. The symbols H, G, L_α, BCC and R mean the same as in Figs. 1 and 2. H_{II} and G_{II} are inverse hexagonal and inverse gyroid phases with heads and water inside the channels or struts. The closed circles are phase boundaries determined on 30 × 30 × 30 lattices, while the open circles were determined on lattices of others sizes, namely 24×24×24 for the gyroid phase and 34 × 34 × 34 for the mesh phase of H_4T_4; 28 × 28 × 28 for the inverse gyroid phase of H_4T_{10}; 33 × 33 × 33 for the BCC phase ($C_A = 0.60$) of $H_{16}T_4$; and 34 × 34 × 34 for the inverse gyroid and the lamellar phase ($C_A = 0.75$) of H_4T_{16}.

Figure 5 shows the phase diagrams of surfactants with more extreme ratios of head-to-tail lengths than those in Fig. 2, namely, those of H_{10} T_4, H_4 T_{10}, H_{16} T_4 and H_4 T_{16}, and compares them to that of H_4T_4. In Figs. 5(b), (d) and (f), the diagrams for H_4T_4, H_4 T_{10} and H_4 T_{16} are plotted against water concentration ($1-$ C_A) and presented as extensions of the diagrams, respectively, for the "complementary surfactants" with inverse ratios of head-to-tail lengths, H_4T_4, H_{10} T_4 and H_{16} T_4. The extended diagram for the symmetric surfactant, H_4T_4 (Figs. 5(a) to (b)), is qualitatively similar to a hypothetical phase diagram proposed by Seddon [17]. As one increases the asymmetry of the surfactant, there is a rightward shift for each extended diagram, a gain of a BCC phase on the left-hand side, and a loss of the hexagonal phase on right-hand side of the diagram. For the most asymmetric surfactant/complement, H_{16} T_4/H_4 T_{16}, only sphere packings are formed by H_{16} T_4 while for H_4 T_{16}, the normal hexagonal and gyroid phases are missing. However, *inverse* hexagonal (H_{II}) and *inverse* gyroid (G_{II}) phases on the surfactant-rich side of the lamellar phase exist. Note also that for the surfactant with the longest tail group, H_4 T_{16}, the size of the region of the gyroid phase is enlarged so that it spans a compositional range about 0.055 ± 0.015 wide compared to a width of only 0.025 ± 0.01 for more symmetric surfactants. Consistent with this trend in the simulations, long-tailed monoglycerides [17–20] and two-tailed lipids, such as phosphatidyl choline or di-dodecyl alkyl-β-D-glucopyranosyl-rac-glycerol, have large regions of inverse hexagonal and inverse gyroid and other bi-continuous inverse phases [21, 22].

These changes produced by the lengthening of the surfactant head group relative to the tail can be rationalised by considering of the *preferred curvature* of the surface defined by the average locus of the junctions between head and tail groups. The bulkier group, head or tail prefers to reside on the convex side of this curved surface. The agreement between Figs. 1 and 2 suggests that the phase behaviour of surfactants is dominated by the ratio of "effective" head to tail sizes as well as other phenomena that are captured by the lattice model, including volume-filling constraints, the configurational entropy of the surfactant chain, and the repulsive interactions between hydrophilic and hydrophobic moieties. Apparently, the physico-chemical details of various surfactants can to a significant degree be collapsed into the "minimal" model presented here. This basic approach of indexing surfactants with their ratios of effective head to tail sizes has been promoted by Israelachvili *et al.* [23] and appears to be remarkably fruitful.

In summary, a primitive lattice model for surfactant water mixtures is amazingly successful at capturing the general trends in the phase behaviour of such mixtures, including the presence of exotic phases such as the gyroid or ordered mesh phases. The success of the model makes it a useful standard against which to compare the phase behaviour of experimental surfactant solutions, and deviations from the model's predictions might then suggest the influence of molecule-specific factors, such as ionic interactions and hydration layers which are not included in the model. The similarity of the lattice model described here to those used to predict the self-assembly of heterostructural polymers, such as block copolymers and proteins [24–26], suggests that such simulation methods hold great promise for the prediction of even more complex self-assembly involved, for example, in lipid membranes [27], bio-mimetic nanocomposites [28] and protein chain folding [26].

Discussion

JRA Pearson How does temperature come into the simulations predicting the phase diagrams?

R. G. Larson There is a dimensionless interaction energy between hydrophilic and hydrophobic units which can be thought of as an inverse temperature.

V. M. Naik How far away is the simulation model from accounting for the hydrogen-bonding associations of the head groups of non-ionics with water (as a function of temperature) and ion condensation on the head groups of ionic surfactants as a function of concentration?

R. G. Larson At this point, the model accounts for such effects only to the extent that they can be lumped into an "effective head size" which includes tightly bound hydration layers. The success of the model in predicting the phase diagrams of non-ionic and ionic surfactants suggests that this "coarse-grained" approach has some validity. More realistic treatment of these effects will probably require off-lattice simulations.

C. J. Lawrence Is it possible, in the context of static simulations, to determine a relaxation time associated with the formation of structure, and if so, do different structures have different characteristic time scales?

R. G. Larson Monte Carlo methods are essential sampling methods, and one must be careful about imputing physical realism to "time" in such methods.

There are also long-range hydrodynamic interactions that are not accounted for in MC methods. Nevertheless, the relative rates at which various transitions occur in such simulations might be qualitatively related to those of real systems.

M. E. Cates To what extent might the "holey" lamellar phases be metastable states caused by the system's inability to phase separate in your simulations?

R. G. Larson I can't rule out the possibility that there are large bi-phasic regions in the phase diagrams that can't be resolved with the small boxes used in the simulations. Nevertheless, I believe that the bi-phasic regions are in fact small. This belief is based on the narrowness of the bi-phasic regions observed in the experimental diagrams, such as Fig. 1.

M. E. Cates Does anyone know why the miscibility gaps are so narrow? (This applies even for non-ionics. For coulombic systems, local charge neutrality combined with counterion entropy will force the gaps to be small, though not in the simulation.)

R. G. Larson I don't really know why the gaps are so small (although I suspect that the smallness of the "water" and "oil" solvent units plays a role by making segregation of such units entropically expensive. Ternary phase diagrams involving relatively long oligomeric hydrocarbon "oil" molecules tend to have large multi-phase regions).

D. J. Tildesley The hydration of the head-group is often important in determining the phase behaviour of lipid systems. Can you include a temperature-dependent head-group interaction in your model, and would this produce the necessary physics to describe the lipid phase diagrams?

R. G. Larson One could make head-group interactions with water different from water-water and head-water interactions and make these temperatures dependent. This might introduce the thermotropic order-order transitions seen in such systems, but I suspect that the large size of the gyroid regions for such surfactants might have something to do with the stiffness of the lipid tail groups. This stiffness could be introduced into the model and it would be interesting to see what effect it would have.

M. Lal The range of spatial correlations in the structure emerging in your simulations can exceed the size of the simulation cell, which would lead to spurious findings. What measures were taken to ensure that artifacts due to the long-range nature of the correlations did not exist?

R. G. Larson Simulations were in some cases repeated with different box sizes. The lamellar and hexagonal phases are insensitive to box size, but the cubic phases such as the gyroid only form on boxes commensurate with the d spacing of the phase. The phase transition temperatures are somewhat sensitive to box size.

M. Lal The triangular phase diagram shown suggests the possibility of the occurrence of co-existing phases. Would your simulation approach cope with the incidence of co-existing phases?

R. G. Larson The simulation of co-existing phases would require box sizes much larger than currently feasible. The existence of co-existing phases might be inferred, however, from simulations with an open, or partially open (that is, grand canonical) ensemble.

M. Lal How can the effects of flow be introduced in the simulations?

R. G. Larson This would be very hard to do on a lattice. It would be better to switch to off-lattice simulations if one wants to consider flow.

References

1. Balmbra, R. R., Clunie, J. R. & Goodman, J. F. (1969) *Nature* **222**, 1159
2. Arvidson, G., Brentel, I., Khan, A., Lindblom, G. & Fontell, K. (1985) *Eur. J. Biochem.* **152**, 753
3. Kekicheff, P. & Tiddy, G. J. T. (1989) *J. Phys. Chem.* **93**, 2520
4. Clerc, M., Levelut, A. M. & Sadoc, J. F. (1990) *Coll. Phys. C7* **51**, 97
5. Eriksson, P. O., Lindblom, G. & Arvidson, G. (1987) *J. Phys. Chem.* **91**, 846
6. Blackburn, J. C. & Kilpatrick, P. K. (1992) *J. Coll. Inter. Sci.* **149**, 450
7. Mariani, P., Amaral, L. Q., Saturni, L. & Delacroix, H. (1994) *J. Phys. II France* **4**, 1393
8. Larson, R. G. (1988) *J. Chem. Phys.* **89**, 1642
9. Larson, R. G. (1996) *J. Phys. II (Paris)* **6**, 1
10. Larson, R. G. (1994) *Chem. Eng. Sci.* **49**, 2833
11. Rancon, Y. & Charvolin, J. (1988) *J. Phys. Chem.* **92**, 2646
12. Clerc, M., Levelut, A. M. & Sadoc, J. F. (1991) *J. Phys. II France* **1**, 1263
13. Luzzati, V. & Spegt, P. A. (1967) *Nature* **215**, 701
14. Luzzati, V., Tardieu, A., Gulik-Krzywicki, T., Rivas, E. & Reiss-Husson, R. (1968) *Nature* **220**, 485
15. Thomas, E. L., Anderson, D. M., Henkee, C. S. & Hoffman, D. (1988) *Nature* **334**, 598
16. Mitchell, D. J., Tiddy, G. J. T., Waring, L., Bostock, T. & McDonald, M. P. (1983) *J. Chem. Soc. Far. Trans. I* **79**, 975

17. Seddon, J. M. (1990) *Biochim. Biophys.* **1031**, 1
18. Lutton, E. S. (1965) *J. Amer. Oil Chem. Soc.* **42**, 1068
19. Larsson, K. (1989) *J. Phys. Chem.* **93**, 7304
20. Fontell, K. (1990) *Coll. Polym. Sci.* **268**, 264
21. Turner, D. C., Wang, Z.-G., Gruner, S. M., Mannock, D. A. & McElhaney, R. N. (1992) *J. Phys. II France* **2**, 2039
22. Ström, P. & Anderson, D. M. (1992) *Langmuir* **8**, 691
23. Israelachvili, J., Mitchell, D. J. & Ninham, B. W. (1976) *J. Chem. Soc. Faraday Trans. I* **72**, 1525
24. Fried, H. & Binder, K. (1991) *J. Chem. Phys.* **94**, 8349
25. Pande, V. S., Grosberg, A. Y. & Tanaka, T. (1995) *Macromolecules* **28**, 2218
26. Chan, S. H. & Dill, K. A. (February, 1993) *Physics Today* 24
27. Ringsdorf, H. (1988) *Angew. Chem. Int. Ed. Engl.* **27**, 113
28. Firouzi, A., Kumar, D., Bull, L. M., Besier, T., Sieger, P., Huo, Q., Walker, S. A., Zasadzinski, J. A., Glinka, C., Nicol, J., Margolese, D., Stucky, G. D. & Chmelka, G. F. (1995) *Science* **267**, 1138

Part Three
Particulate Dispersions

Introduction		245
Chapter 17.	The Statistical Mechanics of Slow, Steady Sedimentation *S. Ramaswamy*	248
Chapter 18.	Effect of Collisional Interactions on the Properties of Particle Suspensions *V. Kumaran*	263
Chapter 19.	Migrational Instabilities in Particle Suspensions *J. D. Goddard*	280
Chapter 20.	Review of Chaotic Behaviour of Suspensions of Slender Rods in Simple Shear Flow *K. Satheesh Kumar, S. Savithri and T. R. Ramamohan*	286
Chapter 21.	Shear Thickening in Model Concentrated Colloids: The Importance of Particle Surfaces and Order Change Through Thickening *J. R. Melrose, J. H. van Vliet and R. C. Ball*	301
Chapter 22.	Micro-Structure of Colloidal Dispersions Under Flow *S. M. Clarke and A. R. Rennie*	315
Chapter 23.	Measurement of the Flow Alignment of Clay Dispersions by Neutron Diffraction *A. B. D. Brown, S. M. Clarke and A. R. Rennie*	330
Chapter 24.	The Rheology and Micro-Structure of Equine Blood *R. M. de Roeck and M. R. Mackley*	338

Introduction

This section contains articles on particulate systems. As complex fluids, they are frequently found in the form of colloidal dispersions, although a variety of soft solids and liquids contain objects of a similar size. The practical applications of this class of fluid range from foods, paints, pigments and soil mechanics to numerous other industrial and everyday materials. The extensions of these topics overlap with the contributions found in the next section on the viscoplasticity of soils and other materials.

The work described reflects largely the new interest of physicists and applied mathematicians in the study of colloidal materials which have, until recently, remained the domain of physical chemists. Approaches range from the application of statistical mechanics and analytical theories to computer simulation and the development of novel experimental methods. Several themes are apparent in a review of the recent progress in this area. The advance of theory is often closely coupled to experiment. In most cases, it has been necessary to consider systems that are idealised in one or more respects, for example, both calculations and experiments are primarily concerned with monodisperse particles with very simple interaction potential. The effects of polydispersity and mixtures can be included when some of the principles are clearly established in simple systems.

This section begins with an analysis of sedimentation by Ramaswamy. Charge stabilised dispersions are shown to cluster while sedimenting or moving in an idealised fluidised bed. Scaling phenomena can be identified in this model. The work by Kumaran highlights the different velocity distributions that may be found in colloidal dispersions. Binary mixtures may have a velocity distribution far from the Maxwell-Boltzmann form if collisions are inelastic. The subject of fluctuations and instabilities in flow are treated by Goddard. The shear-induced migration of particles (considered in the limit of zero Reynolds number) is shown to give rise to instabilities and the possible bifurcations of flow. The model suggests a mechanism for shear-banding in particulate dispersions.

An analysis of the shear flow of rod-like particles by Ramamohan *et al.* provides an account of a system that evolves chaotically. The combination of electric or magnetic fields with applied stress allows certain modes and patterns to be stabilised. The paper highlights how structures on a variety of length scales emerge in particulate dispersions under flow.

The advances in computer simulation of the rheology and structure of particulate dispersions are described by Melrose *et al.* This work emphasises the need to find computationally efficient techniques to calculate inter-particle interactions if systems that are large enough to display realistic behaviour or adequate statistical accuracy are to be computed. The use of moving mesh algorithms has been highly successful and has helped to identify mechanisms that will give rise to shear thickening. The importance of colloidal forces and inter-particle hydrodynamic interactions have been highlighted.

Analytical theories and computer models on the flow and hydrodynamic interactions between particles are complemented by experiments which test ideas and provoke new calculations. In many cases, it is difficult to reproduce in practice the idealised systems which are the subject of theoretical work. The last three papers describe experimental studies on the flow of complex fluids. The review by Clarke and Rennie describes scattering studies on colloidal dispersions which have been carefully chosen as model systems with monodisperse, spherical particles. The structural changes that occur during shear thinning observed in scattering experiments are compared with the simulations of Melrose and Ball.

The final two short papers describe experiments on complex, colloidal fluids using novel experimental techniques. Brown *et al.* describe the use of neutron diffraction to obtain data on particle orientation under conditions of flow. The unusual feature of these experiments on monocrystalline particles (clays) is that a complete distribution function can be obtained, rather than particular averages obtainable from measurements such as flow birefringence. They can also provide information on concentrated, opaque samples. De Roek and Mackley describe the rheology of equine blood and indicate that observations of the micro-structure in a rheo-optical cell are necessary to an understanding of viscoelasticity. The aggregation of blood cells is an important feature of the behaviour of this material.

This section gives an indication of the rich variety of phenomena that can be observed in the flow behaviour of particulate dispersions. The conclusion drawn from both experimental and theoretical work is that a consideration of microscopic (colloidal) forces is very important. The conservative potentials that govern equilibrium structure and the frictional or hydrodynamic interactions play a large role in the flow behaviour of these fluids. The delicate balance of these interactions will require careful attention to be paid to the detail of tests of theoretical or computer models. The development of efficient computation algorithms remains important in investigation of complex fluids. Experimentalists are also challenged to provide reliable data on systems that can be used to test the basic concepts of the models.

Dynamics of Complex Fluids, pp. 248–262
ed. M. J. Adams, R. A. Mashelkar, J. R. A. Pearson & A. R. Rennie
Imperial College Press–The Royal Society, 1998

Chapter 17

The Statistical Mechanics of Slow, Steady Sedimentation

SRIRAM RAMASWAMY
Department of Physics, Indian Institute of Science,
Bangalore 560 012 India
E-mail: sriram@physics.iisc.ernet.in

The steadily sedimenting state of hard-sphere as well as charge-stabilised suspensions is attracting a great deal of attention from condensed-matter physicists interested in the statistical mechanics of systems far from equilibrium. I present a summary of the theoretical work done with R. Lahiri on the effect of sedimentation on crystalline suspensions of charged colloids, and with A. J. Levine and R. Bruinsma on velocity fluctuations in disordered fluidised beds. Dramatic scaling phenomena and non-equilibrium phase transitions are uncovered.

Introduction

Sedimentation in General

The settling of particles under gravity in a viscous fluid is one of the most familiar non-equilibrium phenomena in nature [1]. In this lecture, I shall be concerned only with the *steady-state* properties of this process. That is, I shall pretend the particles are settling forever in a container without a bottom, and with the concentration of particles being held constant and uniform on the average. This is easily achieved in the "fluidised bed" geometry, in which the particles are *on average* at rest in the laboratory frame while the suspending fluid flows upward past them in an imposed counterflow. Of course, fluidised beds come in many shapes and sizes, and hence let me be specific about the kind that I am talking about. I will be discussing the settling or fluidisation of particles (generally synthetic, highly monodisperse spheres with specific gravity ranging from 1.05 for polystyrene to two for silica) whose radius, a, ranges somewhat from less than a μm to a few μm, and settling at speeds, v_{sed}, of up to a few μm per second in a liquid such as water (viscosity $\eta \simeq 1$ cp, density $\rho \simeq 1$ g cm^{-3}). The Reynolds number for such a situation is $\text{Re} \equiv 2v_{\text{sed}}a\rho/\eta \simeq 10^{-6}$ to 10^{-4} so that inertia and acceleration are utterly negligible. All motions are dominated by the friction of the fluid. The importance of gravity relative to thermal diffusion is measured by the Péclet number, $\text{Pe} \equiv v_{\text{sed}}a/6D$, where D is the self-diffusivity of the particle. For the systems we are interested in, $\text{Pe} > 10^4$, settling is exceedingly important.

For a single slowly settling non-Brownian particle, the problem is one which we have all seen in textbooks. Stokes solved it about a century and a half ago [2]. He showed that $v_{\text{sed}} = v_0 \equiv \frac{2}{9}a^2\Delta\rho g/\eta$, where $\Delta\rho$ is the mass density difference between the particle and the fluid, and g is the acceleration due to gravity. The problem of many-particle sedimentation is complicated for two reasons. First, the motion of a given particle sends diffusive shear into the surrounding fluid which moves the other particles — the hydrodynamic interaction. Second, the particles of a suspension, even while at rest, can interact directly via electrostatic forces and simply by excluded volume. The hydrodynamic interaction makes the motion of individual particles hard to predict even in the absence of Brownian motion. In fact, their motion acquires a random component, a phenomenon known as hydrodynamic dispersion. Electrostatic repulsion (as well as excluded volume), in suitable ranges of particle volume fraction and ionic strength of the solvent, can cause the particles of the suspension to order into a periodic lattice. I shall use the term *crystalline* for such ordered suspensions. Those with a disordered micro-structure with freely moving particles will be termed *fluid suspensions.* They are often called *disordered suspensions*, but the latter term is confusing because it could equally be applied to amorphous but rigid aggregates. Sedimentation has distinct, dramatic effects on both crystalline and fluid suspensions.

A major part of my motivation in presenting this work here is to argue strongly for the use of stochastic models, both lattice-gas models and generalised Langevin partial differential equations, in the analysis of the dynamics of suspensions, even when the particles are essentially non-Brownian. The long experience of condensed-matter theorists and practitioners of the statistical mechanics of time-dependent phenomena, and the many techniques developed in those areas over the years, have proved extremely valuable to us in our pursuit of solutions to the problems discussed here.

Crystalline Suspensions

Charge-stabilised suspensions have a phase diagram which is well understood in the equilibrium, that is, *neutrally buoyant* case. The control parameters are particle volume fraction, ϕ, and ionic impurity strength, n_i. The latter acts as a kind of temperature: a large n_i means a short Debye screening length, ξ_D, and hence a weaker interaction. Temperature itself is not a particularly convenient parameter to vary. This is because the temperature dependence of the dielectric constant of water makes the combination [interaction energy/temperature] almost independent of temperature. At low enough n_i and large enough ϕ, the suspensions we are interested in form a crystalline lattice, generally called *colloidal crystals*, with a lattice spacing of the order of a micron for the cases of interest.

Now suppose the crystal is made up of particles heavier than water. A fluidised bed experiment on such an object will allow us to study a lattice being *driven through a dissipative medium* by a constant external force. This important non-equilibrium steady-state arises, for example, not only in the steady sedimentation [1, 3] of a colloidal crystal [4], but also in the motion of a depinned flux lattice in a current-carrying superconductor as well. Lahiri and I asked a very simple question: what is the response of such a driven crystal to small, long wavelength disturbances? We found [5], remarkably, that the dominant linear response at long wavelengths is proportional to the *driving speed* of the lattice, and not to its elastic constants. Even more strikingly, we show that this response leads at high Pe to a non-equilibrium phase transition to a clumped and buckled state.

Fluid Suspensions

I shall take this opportunity to discuss our work on disordered or fluid suspensions, although time did not permit me to present it at the symposium. Hard-sphere suspensions, in which the only interactions are hydrodynamic and excluded volume, have been a subject of enduring interest to fluid dynamicists.

A good approximation to these can be achieved in the laboratory by taking charge-stabilised suspensions with large n_i so that $\xi_D \ll a$. Such suspensions remain fluid up to volume fractions of about 0.5. The sedimentation of these disordered suspensions poses a major puzzle. It is observed [6–8] that the relative fluctuation, σ_v, of the particle velocity, that is, standard deviation scaled by the mean, grows with volume fraction, ϕ, and is of order unity for $\phi > 0.1$. These suspensions are essentially non-Brownian. Hence thermal noise is not the origin of these fluctuations. Moreover, seemingly straightforward continuum treatments of the dynamics of concentration and velocity fields [9] for a fluidised bed with linear dimension, L, which *assume* spatially uncorrelated concentration fields, predict that $\sigma_v \sim L^{1/2}$. Experiments see no such dependence. There are two puzzles here: why are there large velocity fluctuations? And why is there no size dependence? Levine (Exxon), Bruinsma (UCLA) and I developed a coarse-grained formulation of the problem, and showed that the result of [9] arises because of the neglect of the advection of concentration fluctuations by the velocity fluctuations they themselves produce. The problem turns out to be difficult to solve in space dimension $d < 6$. We offered various directions to a solution, including an expansion in $\epsilon = 6 - d$ and an integral equation approach [10]. The preliminary results from that work are presented here.

Instability of Crystalline Fluidised Beds

Background

The work of Crowley, though we learnt of it long after we had developed our model, provides a useful starting point for our approach. In [11], carefully prepared periodic arrays of steel balls were released gently into turpentine oil and their motion was monitored by strobe photographs. The corresponding theory was worked out using low-Reynolds-number hydrodynamics [12] to determine the hydrodynamic interactions between the spheres. Experiment and theory (linear stability analysis) agreed: a regular horizontal array of sedimenting spheres was *unstable* to clumping and buckling. The mechanism at work here is as follows: start by having the spheres arranged as · · · · · · · · · · in a perfect periodic array along the x-axis and settling along the z-axis. Suppose an accidental event compresses the array locally by moving two adjacent spheres closer together. This pair will sink faster than the rest, as you can see either by intuition or by solving the Stokes equations. A few moments later, the array will look like ·····.·:····· as the perturbed spheres get ahead of their neighbours. The lines joining the perturbed spheres to their outer neighbours are then *tilted*. On the grounds of symmetry, these tilted regions cannot be expected to settle down in a straight manner, but must have a *lateral* component to their drift. Symmetry alone cannot say whether they will drift together or apart, but Crowley's calculation says that they drift *towards* each other. Physically, I think this can be rationalised by noting that a tendency to drift *along* rather than normal to the line joining their centres produces less viscous drag. In any case, we see that the compressed region is compressed further, which is the aforementioned linear instability.

So is a sedimenting colloidal crystal always unstable? We can't answer this question without including the effects of elastic forces and Brownian motion. These are present in a real suspension and would oppose compression and buckling, and they are absent from [11] (and obviously unimportant for an array of steel balls). The *linear* analysis of [11] is therefore in the limit of *infinite* Pe. By Pe, we mean the ratio, $\Delta\rho a g/G$, where G is a typical elastic modulus of the suspension. What happens at finite Pe, and what is the role of nonlinearities?

The Model

We could try to answer the question just posed by a direct Brownian-Stokesian simulation, or perhaps by a first-principles analytical calculation. However, the former would not provide the understanding that we are after. Some progress in the latter direction has been made in the form of a set of nonlinear stochastic partial differential equations for the elastic displacement field of the moving lattice [5]. However, we still lack a completely satisfactory way of handling the linearly unstable nature of the problem. Most of our answers were drawn from a completely different approach, which I will outline next.

A succinct description of the dynamics implied by the description in the previous sub-section is that a downtilt, \, favours a drift of material to the right and an uptilt, /, does the exact opposite. An excess concentration tends to sink and a deficit in the concentration, to float up. This picture inspired us to think of a highly simplified discrete dynamical model which, we believe, captures the essential physics of the problem and includes, in a natural way, the elasticity of the lattice, Brownian motion and nonlinear effects. For the sake of simplicity, let us restrict ourselves to a one-dimensional system with sites labelled by an integer, i, and describe the state of the lattice of spheres in terms of an array of two types of two-state variables: $\rho_i = \pm$, which tells us if the region around site i is compressed (+) or dilated (−) relative to the mean, and $\theta_i = \pm$, which tells us if the local tilt is up (+, which we will call '/') or down (−, which we will denote '\'). Let us put the two types of variables on the odd and even sub-lattices. A valid configuration would then look like $\rho_1\theta_1\rho_2\theta_2\rho_3\theta_3\rho_4\theta_4 = +\backslash + / - \backslash + \backslash - / - /$. An undistorted lattice is then a statistically homogeneous admixture of + and − for both the variables (a "paramagnet"). Crowley's instability is a phase separation of pluses from minuses and ups from downs.

Let us try to construct a set of update rules for the evolution of $\{\rho_i, \theta_i\}$, by careful imitation of the dynamics described in the previous sub-section. We will opt for stochastic dynamics since we wish to include the effects of Brownian motion. By construction, the dynamics must consist of *exchanges*, $+- \rightleftarrows -+$, or $/\backslash \rightleftarrows \backslash/$. In the absence of the driving force of sedimentation, these exchanges are unbiased. When sedimentation is taken into account, the rates (denoted $W[\,]$) are

$$W[+\backslash- \rightleftarrows -\backslash+] = e^{D\pm a}\,, \tag{1}$$

$$W[-/+ \rightleftarrows +/-] = e^{D\pm a}\,, \tag{2}$$

$$W[\backslash - / \rightleftarrows / - \backslash] = e^{D_1\pm a_1}\,, \tag{3}$$

$$W[/ + \backslash \rightleftarrows \backslash + /] = e^{D_2\pm a_2}\,. \tag{4}$$

The unbounded growth that goes with the linear instability has been automatically limited by our choice of variables. Elastic forces (compression as well as shear) are implicitly contained in the dynamics: a large macroscopic strain (an array of pluses followed by an array of minuses) will dissolve by inter-diffusion. In Eqs. (1)–(4), the upper (lower) sign applies to the upper (lower) arrow. The rates in Eqs. (1) and (2) are equal because the system is left-right symmetric. The remaining two rates are unequal because there is no up-down symmetry in the problem. The degree to which a local excess in the concentration increases the local settling speed need not be equal to that which a local rarefaction reduces it. The parameters, a, a_1 and a_2, as well as the difference between D_1 and D_2, contain the effects of sedimentation. To model a neutrally buoyant array, these would have to be zero. D, D_1 and D_2 are in general positive. This is because they are to be thought of physically as ratios of elastic constants to friction coefficients. However, a, a_1 and a_2 can, in principle, have either sign. If one stares at the update rules a bit, it becomes clear that to model the "unstable" dynamics of [11], they must all be positive.

Results

We [5] studied the model which keeps all the other parameters fixed and varies D. What we found was quite remarkable: for large D (that is, for large values of the elastic stiffness), the system consisted of a random, freely inter-mixing array of $+$ and $-$ and of $\backslash$ and $/$, that is, a lattice which was undistorted on the average. As D was lowered, a *non-equilibrium phase transition* took place. The system phase is also separated into macroscopic domains of pluses and minuses, and uptilts and downtilts. We applied various tests, all of which confirmed that the observed phase separation was a true phase transition and not merely a crossover. In a real, laboratory fluidised bed, the phase transition should be observable in charge-stabilised crystalline suspensions in the fluidised bed geometry. Start with a system that has a large Debye screening length, ξ_D, and keeping other parameters fixed, varies the screening length by adding ionic impurities. When n_i gets high enough, but well below the value needed to *melt* the crystal at equilibrium, the system should clump and fracture into smaller crystallites separated by regions of strong upflow. There appears to be some evidence of this remarkable non-equilibrium phase transition in experiments [13].

Neither the three-dimensional nature of the experimental systems nor the long-range hydrodynamic interactions should make the predictions of our one-dimensional, local model any less relevant: (i) the fact that the transition is seen in a low dimensional system makes it more likely that it will be seen in a high dimensional one; (ii) the hydrodynamic interaction can be made effectively short-ranged by working with a container whose y dimension is much smaller than the x and z dimensions. The *local* effects of the hydrodynamic interaction will produce the tilt-dependence and concentration-dependence underlying the instability. However, long-range effects will be screened by the walls. In any case, the shear elasticity of the lattice should itself screen the hydrodynamic interaction.

"Fluid" Fluidised Beds: Giant Velocity Fluctuations

Background

As I said in the introduction, experiments on fluidised beds of non-Brownian hard-sphere suspensions [3, 6, 7, 8], whose *equilibrium*, that is, neutrally buoyant state, is that of a fluid with negligible structural correlations, show large *non-thermal* fluctuations in the particle motion with the character of random stirring on a scale $\sim 50a$. The fluctuations, δv, in the settling speed are of the same order as the mean, $v_{\rm sed}$, for volume fractions, $\phi > 0.1$. Caflisch and Luke (CL) [9] showed that for a sedimenting system of linear dimension L, the assumption of purely *random* local concentration fluctuations led to long-range velocity fluctuations with $\delta v \sim L^{1/2}$. Experiments, however, find *no* dependence of δv on L [7, 8]. We will not try to explain the microscopic origin of the random motion, although I am fairly sure it lies in the instability of the arrays discussed in the previous section. However, this will provide the beginnings of a theory for the large-distance and long-time properties of correlation functions for these systems.

A New Approach: Coarse-Graining

The most important contribution that our work makes is its starting point, namely, a *stochastic* effective model. The virtues of this model are many: (i) it provides a basis for analytical calculations for this formidable many-body problem; (ii) it places the problem of sedimentation squarely in the domain of scaling phenomena in spatially extended driven systems; and (iii) it allows us to draw on many decades of experience in dynamical critical phenomena and on the great range of techniques developed therein. We are, of course, trying to model *non-Brownian suspensions* (or real, laboratory suspensions of large particles in which thermal motions are expected to play a negligible role). However, it is clear that despite the deterministic nature of the underlying dynamics, there is a substantial amount of random motion induced by the hydrodynamic interaction. We are thus dealing with a spatio-temporally chaotic system with many degrees of freedom. In such a situation, except perhaps in the singular limiting case of a point-particle suspension [14], it is rash to try to obtain analytical expressions for the statistical properties of the system starting from the exact microscopic dynamical equations. A coarse-grained approach [15] is called for, in which concentration fluctuations with wavelength shorter than a length, ℓ, which can be taken as the stirring scale seen in experiments, are eliminated to yield an effective equation of motion for the long wavelength modes of physical interest. The degrees of freedom thus eliminated contain velocity fluctuations *beyond* the point-force approximation and hydrodynamic reflections involving groups of particle with separation less than ℓ. Since the dynamics is chaotic, that is, there is hydrodynamic dispersion [16], the *coarse-grained* equations for the dynamics of the concentration field must: (i) be stochastic with short-time dynamics (that is, for time scales slightly larger than $\tau_\ell \equiv \ell/v_{\rm sed}$) governed by a number-conserving "bare" noise with no correlations on scales larger than ℓ; and (ii) have a "bare" diffusivity $D_o \sim v_{\rm sed}\ell$ on dimensional ground. In all other respects, as far as long wavelength properties are concerned, we are dealing in effect with a point-particle suspension.

Since this approach is unfamiliar to most fluid dynamicists and, indeed, has proved a source of confusion and unease at earlier presentations of this idea, let me elaborate on it. We know, empirically, that the dynamics is chaotic. Anything short of a *complete* specification of the initial conditions will lead, therefore, to indeterminacy in the dynamics. Indeed, it is really for the same reason that we use stochastic models in dynamical descriptions [17] of *equilibrium* statistical mechanics. Beginning with an isolated system assumed to be distributed micro-canonically, that is, with its phase-space co-ordinates obeying the *deterministic, reversible* equations of motion of classical mechanics and distributed uniformly on the energy surface because of dynamical chaos, one argues [18] that the dynamics of a subset of those variables (the "slow variables") must be *stochastic and dissipative.* While we do not carry out a rigorous derivation of the stochastic equations of motion for our case, the general arguments still hold. Now, what sort of stochasticity is appropriate here? Our elimination of degrees of freedom is over a limited range of length scales, ℓ (and hence time scales, τ_ℓ). The resulting noise source must perforce consist of random particle currents with no spatial correlations over scales larger than ℓ and no temporal correlations over scales larger than τ_ℓ. Note that we have only made an innocuous assumption about the short-time, local properties of the noise. We make no assumptions about the nature of the large-scale, long-time behaviour. This is what we aim to *calculate* from our coarse-grained equations of motion.

A Fluctuating Hydrodynamic Model

Once the above arguments are accepted, the following coupled equations for fluctuations, $\mathbf{v} = (\mathbf{v}_\perp, v_z)$, and c in the velocity and concentration fields, respectively, at zero Reynolds number for a suspension sedimenting steadily along $\hat{\mathbf{z}}$, it follows inevitably that

$$\frac{\partial c}{\partial t} + \mathbf{v} \cdot \nabla c = [D_\perp \nabla_\perp^2 + D_z \nabla_z^2] \, c + \nabla \cdot \mathbf{f}(\mathbf{r}, t) \,, \tag{5}$$

$$[\eta_\perp \nabla_\perp^2 + \eta_z \nabla_z^2] \, v_i(\mathbf{r}, t) = m_R g P_{iz} c(\mathbf{r}, t) \,. \tag{6}$$

Equation (5) is the randomly forced scalar advection-diffusion equation, The velocity fluctuation is itself a source for concentration fluctuations via the Stokes Eq. (6), which expresses the balance between the driving by gravity and dissipation by viscosity. Here $m_R g$ is the buoyancy-reduced weight of a particle. The pressure field has also been eliminated by imposing incompressibility via the transverse projector, $P_{ij} = \delta_{ij} - \nabla_i \nabla_j (\nabla^2)^{-1}$. Coarse-graining has given us "bare" anisotropic collective diffusivities, $D_z, D_\perp$, effective viscosities, $\eta_\perp, \eta_z$, and an additive Gaussian noise term, the divergence of the random current mentioned above with zero mean, and

$$\langle f_i(\mathbf{r}, t) f_j(\mathbf{r}', t') \rangle = N_\perp \delta_{ij}^\perp + N_z \delta_{ij}^z \delta(\mathbf{r} - \mathbf{r}') \delta(t - t') \,, \tag{7}$$

where δ_{ij}^z and $\delta_{ij}^\perp$ are, respectively, the projectors along and normal to the z-axis.

Some technical remarks are required here. (i) Although the noise, $\mathbf{f}$, will have correlations on scales smaller than ℓ, it has been taken as δ-correlated, as is appropriate for a description on scales much larger than ℓ. (ii) Since the system has only *uniaxial* symmetry, and it is not at thermal equilibrium, we have no grounds *a priori* for assuming either isotropy of, or any relation at all, among the noise strengths, diffusivities and viscosities. Their numerical values can be determined, in principle, only by a *numerical* projection of the microscopic equations of Stokesian dynamics onto the coarse-grained concentration field. (iii) In Eqs. (5) and (6), other possible nonlinearities, such as those arising from the concentration-dependence of mobilities and viscosities, as well as multiplicative noise terms of the form, $\nabla \cdot (c\mathbf{h})$, where $\mathbf{h}$ is a spatio-temporally white vector noise, can readily be shown, by power-counting, to be sub-dominant at small wavenumber relative to the advection term, $\nabla . \mathbf{v}c$. This is also true of advection by *thermal* fluctuations in the fluid velocity field [20].

Linearised Theory, and What is Wrong with It

The first, and easiest thing to see at this point is that the result of [9] can be recovered if we ignore the nonlinear term in Eq. (5) and the Fourier-transform in Eqs. (5) and (6), and use Eq. (7) to solve for the velocity variance. From Eq. (5), we see that

$$S(q) \equiv \langle |c(q)|^2 \rangle = \frac{N_\perp q_\perp^2 + N_z q_z^2}{D_\perp q_\perp^2 + D_z q_z^2}, \tag{8}$$

that is, anisotropy apart, concentration fluctuations are effectively structureless. Equation (6) tells us that

$$\langle |v_i(q)|^2 \rangle = \frac{(m_R g)^2 |P_{iz}(q)|^2}{[\eta_\perp q_\perp^2 + \eta_z q_z^2]^2} \langle |c(q)|^2 \rangle . \tag{9}$$

Thus, $\langle |v(q)|^2 \rangle \sim S(q)/q^4$. This leads, in d space dimensions, to a variance of the real-space velocity field, $\langle v^2 \rangle \sim \int d^d q S(q)/q^4$. Since from Eq. (8) the $S(q)$ is flat, we see that for a system with the smallest linear dimension, L, for any $d < 4$:

$$\langle v^2 \rangle \propto L^{4-d} . \tag{10}$$

For $d = 3$, this is just the result of [9]. In other words, the treatment is equivalent to ignoring the large-scale advection of the concentration fluctuations by the velocity fluctuations that they themselves produced! Of course, the *local* effect of the hydrodynamic interaction is implicitly included. Without it, there would have been no hydrodynamic dispersion and hence neither a diffusivity nor a noise in the coarse-grained equations.

Including Nonlinearities

We must therefore learn how the advective nonlinearity changes the correlation function and propagator of the concentration field from its linearised form in Eq. (8). As argued by Koch and Shaqfeh [20], if multi-particle correlations caused $S(q)$ to vanish even slightly faster than q for small q, the variance in Eq. (10) would be finite in $d = 3$. In our framework, once nonlinearities are included, we can still write $S(q)$ in the form of Eq. (8), provided $N_\perp$, N_z, $D_\perp$ and D_z are replaced by "renormalised" quantities which depend on wavenumber and are analogous to eddy diffusivities in turbulence. (The coefficient of the advective nonlinearity is obliged to remain at unity, for reasons of Galilean invariance [10] which I cannot discuss here.) Calculating the wavenumber dependence proves to be difficult. This is because a perturbation expansion finds contributions that diverge at the small wavenumber end for any dimension

$d < 6$. We can do one of two things: calculate just below $d = 6$ and examine the trend as you approach $d = 3$, which seems to be towards a reduced size-dependence [10], or solve a system of simultaneous nonlinear integral equations for the renormalised noise strengths and diffusivities, which is work in progress [10]. I will not discuss both approaches here. Instead, note that the equations have an interesting rescaling property. By ignoring anisotropy, rescaling $r \to br$, $t \to b^z t$, $c \to b^{\chi_c} c$, $v \to b^{\chi_v} v$, normalising to preserve the form of the first term in Eq. (5), and keeping the coefficient of the advective nonlinearity fixed, yields $\chi_c + z + 1 = \chi_v + z - 1 = 0$. If we *assume*, in addition, that there is no singular renormalisation (except possibly logarithms) of the noise spectrum, we get for $d < 6$, $z = d/3$, and $S(q) \sim q^{2-d/3}$, which in three dimensions is the result argued for by [20].

An Analogy with Turbulence

A tempting argument for this last result can be made, by analogy with those used in the theory of turbulence [21], but assuming a scale-independent *velocity* rather than dissipation. That is, since this is a forced flow of fluid past particles, there is a natural velocity scale, v_{sed}. If we then guess that *all* physical quantities in the Pe $\to \infty$, Re $\to 0$ limit are built from v_{sed} and wavenumber q, we are forced to conclude that the "renormalised diffusivity" $D(q) \sim v_{\text{sed}}/q$, the velocity variance $\langle v^2 \rangle \sim v_{\text{sed}}^2$, and the mean square displacement of particles $\sim v_{\text{sed}}^2 t^2$ up to logarithmic factors and volume-fraction dependent coefficients. It remains to be seen whether the more detailed calculations now in progress bear out this attractive physical picture.

Summary

In this lecture, I have presented a fairly detailed summary of two striking, incompletely understood non-equilibrium phenomena that arise in steady sedimentation in the limit of zero Reynolds number, where the physics is dominated by the competition between settling and diffusion. Specifically, I have shown [5] that a fluidised crystalline suspension should undergo a non-equilibrium phase transition to a clumped state. I have also offered the first results from a new stochastic approach [10] to the old problem of size-dependence of the velocity dispersion in hard-sphere fluidised beds. Apart from the specific problems discussed, if I have left the reader with some sense of the extraordinary range of statistical-mechanical phenomena that can be found in these systems, my task will have been accomplished.

Acknowledgements

I am very grateful to my collaborators Rangan Lahiri, Alex Levine and Robijn Bruinsma for their permission to present our joint unpublished work.

Discussion

M. E. Cates Did you include elasticity in the model?

S. Ramaswamy What these equations have is a height field with elasticity and a concentration field.

M. E. Cates Do they really describe a sedimenting colloidal crystal?

S. Ramaswamy The system for which these equations would be most appropriate is a one-dimensional interface with a height field *and* a concentration field living in the interface. It is a highly simplified description of a sedimenting crystal which would actually require a concentration field and a three-dimensional displacement field.

V. Kumaran I don't get the relation between your model and the system. If all pluses move to one side, then it's still a region of high concentration and it's going to go higher and higher.

S. Ramaswamy It can't go higher and higher. The reason I resorted to a discrete model is that in a real system, the concentrations can't get arbitrarily big. There are two ways to do this in a model. One is to introduce a chemical potential which depends sufficiently and nonlinearly on the concentration so as to give saturation. The other is to use Ising variables where the concentration fluctuation and tilt can never lie outside ± 1. The relation between the discrete variables and the continuum model (actually the first x-derivative of the equations I presented) is the same as that between the Ising and Ginzburg-Landau models for an equilibrium phase transition.

D. Tildesley Could you see your first-order phase transition in experiment?

S. Ramaswamy It's second-order. Yes, you should see it, preferably by changing the Peclet number without changing anything else. If you simply change the counterflow velocity, the volume fraction will change. You could do it that way as long as you keep track of it.

M. E. Cates You could change the viscosity.

S. Ramaswamy You could, I suppose.

M. E. Cates You could do it by varying the temperature.

S. Ramaswamy But then many things would change.

M. E. Cates There must be some dimensionless group.

S. Ramaswamy That's the Peclet number.

M. E. Cates You could vary gravity.

S. Ramaswamy By decree ...

M. E. Cates No, with a centrifuge.

S. Ramaswamy How would you do your light-scattering then?

J. F. Brady It would seem to me that there is a correlation between the tilt and the concentration.

S. Ramaswamy There is, because of the nature of the update rules. Regions with a large gradient in the concentration have an appreciable tilt and so forth.

References

1. Blanc, R. & Guyon, E. (1991) Sedimentation physics. *La Recherche* **22**, 866
2. Stokes, G. G. (1845) On the theory of the internal friction of fluids in motion and of the equilibrium and motion of elastic solids. *Proc. Camb. Phil. Soc.* **8**, 287; On the effect of internal friction on the motion of pendulums. *ibid.* **9**, (1851) 8
3. Rutgers, M. A., Xue, J. Z., Herbolzheimer, E., Russell, W. B. & Chaikin, P. M. (1995) Crystalline fluidised beds. *Phys. Rev. E.* **51**, 4674; Rutgers M. A. (1995) *Experiments on Hard, Soft and Hydrodynamically Interacting Spheres.* PhD Thesis. Princeton University
4. Sood, A. K. (1991) Structural order in colloids. In *Solid State Physics Vol. 45*, ed. H. Ehrenreich & D. Turnbull, pp. 1. New York: Academic Press
5. Lahiri, R. & Ramaswamy, S. (1997) Are moving crystals unstable? *Phys. Rev. Lett.* **79**, 1150; Lahiri, R. (1996) PhD Thesis. Indian Institute of Science
6. Tory, E. M., Kamel, M. T. & Chan, Man Fong (1992) Sedimentation is container size dependent. *Powder Technology* **73**, 219
7. Nicolai, H. & Guazzelli, E. (1995) Effect of the vessel size on the hydrodynamic diffusion of sedimenting spheres. *Phys. Fluids* **7**, 3
8. Xue, J. Z., *et al.* (1992) Diffusion, dispersion and settling of hard spheres. *Phys. Rev. Lett.* **69**, 1715
9. Caflisch, R. E. & Luke, J. H. C. (1985) Variance in the sedimentation speed of a suspension. *Phys. Fluids* **28**, 759

10. Levine, A. J., Ramaswamy, S., Frey, E. & Bruinsma, R. E. (1998) Screened and unscreened phases in sedimenting suspensions, cond-mat9801164. *Phys. Rev. Lett.*, submitted for publication
11. Crowley, J. M. (1971) Viscosity-induced instability of a one-dimensional lattice of falling spheres. *J. Fluid Mech.* **45**, 151; Clumping instability of a falling horizontal lattice. *Phys. Fluids* **19** (1976) 1296
12. Happel, J. & Brenner, H. (1965) *Low Reynolds Number Hydrodynamics.* Englewood Cliffs, N.J.: Prentice-Hall
13. Poon, W. C.-K., Pirie, A. & Pusey, P. N. (1995) Gelation in colloid-polymer mixtures. *Faraday Discuss.* **101**, 65; Poon, W. C.-K. & Meeker, S. P. (1996) unpublished; Chaikin P. M., personal communication
14. Koch, D. (1994) Hydrodynamic diffusion in a suspension of sedimenting point particles with periodic boundary conditions. *Phys. Fluids* **6**, 2894
15. Hayot, F., Jayaprakash, C. & Pandit, R. (1993) Universal properties of the two-dimensional Kuramoto-Sivashinsky equation. *Phys. Rev. Lett.* **71**, 12
16. Bossis, G. & Brady, J. F. (1990) Diffusion and rheology in concentrated suspensions by Stokesian dynamics. In *Hydrodynamics of Dispersed Media*, ed. J. P. Hulin, A. M. Cazabat & E. Guyon. Elsevier
17. Ma, S.-K. (1976) *Modern Theory of Critical Phenomena.* Benjamin, Reading MA
18. Mori, H. & Fujisaka, H. (1973) Nonlinear dynamics of fluctuations. *Prog. Theor. Phys.* **49**, 764
19. Ramaswamy, S. (1997) unpublished
20. Koch, D. L. & Shaqfeh, E. S. G. (1991) Screening in sedimenting suspensions. *J. Fluid Mech.* **224**, 275
21. Tennekes, H. & Lumley, J. L. (1972) *A First Course in Turbulence.* Cambridge: MIT Press

Dynamics of Complex Fluids, pp. 263–279
ed. M. J. Adams, R. A. Mashelkar, J. R. A. Pearson & A. R. Rennie
Imperial College Press–The Royal Society, 1998

Chapter 18

Effect of Collisional Interactions on the Properties of Particle Suspensions

V. KUMARAN
Department of Chemical Engineering,
Indian Institute of Science,
Bangalore, 560 012, India
E-mail: kumaran@chemeng.iisc.ernet.in

Velocity distribution functions are determined for a bidisperse sedimenting suspension of particles in a gas and for a sheared suspension of inelastic particles. The distribution functions are determined in two limits. In the kinetic limit, the dissipation of energy due to inelasticity during a collision or viscous drag between successive collisions is small compared to the energy of the particles. In this limit, the distribution function is close to a Maxwell-Boltzmann distribution and the velocity moments are determined using a perturbation expansion about this distribution. In the dissipative limit, the energy dissipation due to inelasticity during a collision or viscous drag between successive collisions is of the same magnitude as the energy of particle velocity fluctuations. In this limit, the distribution function is very different from the Maxwell-Boltzmann distribution and the analytical technique used is specific to the system under consideration.

Introduction

Suspensions of particles in a gas are encountered in many applications, such as solids handling and transport, fluidised beds in chemical unit operations, and in natural systems such as rock slides and snow avalanches. The flow of these suspensions can be broadly classified into two types:

1. In *slow flows*, the distance between the particles is typically small compared to the particle size. In these flows, there is extended contact between the particles, and momentum and energy transport occur due to tangential and normal frictional forces. The examples include flows in bunkers and hoppers in solids handling systems.
2. In *rapid flows*, the particles are widely spaced and the inter-particle distance is usually larger than the particle size. The particles are in vigorous motion, and momentum and energy transfer takes place due to instantaneous particle-particle and particle-wall collisions. These examples include fluidised beds, pneumatic transport and the vigorous motion of a thin layer of particles in rock slides and snow avalanches.

The models currently used for slow flows are continuum models and the constitutive relations are adapted from the yield stress equations used in soil mechanics. At present, no microscopic description is available for slow flows. Microscopic models for rapid flows have been derived by drawing an analogy between the vigorous motion of the particles in a suspension and the fluctuating motion of molecules in a gas which is not at equilibrium. These microscopic descriptions for rapid flows are the subject of the present article.

Microscopic Description of Rapid Flows

The fundamental quantity of interest in a microscopic description of a system of particles is the "distribution function". It provides the density of particles in phase space because mechanical properties such as pressure, shear stress and the energy dissipation rate can be derived from the distribution function. Formally, the single particle distribution function, $f(\mathbf{x}, \mathbf{u}, t)$, is defined such that $nf(\mathbf{x}, \mathbf{u}, t) d\mathbf{x}\, d\mathbf{u}$ is the number of particles whose centres are in the differential volume $d\mathbf{x}$ about the position $\mathbf{x}$, and whose velocities are in the differential volume $d\mathbf{u}$ about $\mathbf{u}$ at time t. Here, $\mathbf{x}$ and $\mathbf{u}$ are the position and velocity, respectively, and n is the number density of the particles. A conservation equation for the single particle distribution function can be written as

$$\frac{\partial(nf)}{\partial t} + \frac{\partial(u_\alpha nf)}{\partial x_\alpha} + \frac{\partial(a_\alpha nf)}{\partial u_\alpha} = \frac{\partial_c(nf)}{\partial t}, \tag{1}$$

where $\mathbf{a}$ is the particle acceleration, Greek subscripts are used to denote the components of a vector, and repeated subscripts represent a dot product. In Eq. (1), the terms on the left-hand side are the time rate of change of the distribution function, the change in the distribution due to convective transport in real space, and the change due to convective transport in velocity space, respectively. The term on the right-hand gives the change in the distribution function due to particle collisions. This can be written as [1]

$$\frac{\partial_c(nf)}{\partial t} = n^2 \int d\mathbf{k} \int d\mathbf{u}^* [f^{(2)}(\mathbf{x}, \mathbf{u}'; \mathbf{x} + r\mathbf{k}, \mathbf{u}^{*\prime}; t) - f^{(2)}(\mathbf{x}, \mathbf{u}; \mathbf{x} + r\mathbf{k}, \mathbf{u}^*; t)] \times (4\pi r^2 \mathbf{w}.\mathbf{k}) \,, \tag{2}$$

where r is the particle radius, $\mathbf{k}$ is the unit vector in the direction of the line joining the centres of the particles at the point of collision, $\mathbf{u}$ and $\mathbf{u}^*$ are the velocities of the particles before collisions, $\mathbf{u}'$ and $\mathbf{u}^{*\prime}$ are the velocities of the particles after collision, and $\mathbf{w} = \mathbf{u} - \mathbf{u}^*$ is the difference in the velocities of the colliding particles.

Equation (1) for the single particle distribution, $f(\mathbf{x}, \mathbf{u}, t)$, can only be solved if the pair distribution function, $f^{(2)}(\mathbf{x}, \mathbf{u}; \mathbf{x}+r\mathbf{k}, \mathbf{u}^*, t)$, is known. However, a conservation equation for the pair distribution function contains three particle distribution functions. In general, a conservation equation for an n particle distribution function contains an $(n + 1)$ particle distribution function, and one obtains an infinite hierarchy of equations known as the BBKGY hierarchy [1]. General methods for solving this hierarchy of equations are not available. In the kinetic theory of gases, the positions and velocities of the colliding particles are considered to be uncorrelated (the assumption of "molecular chaos") and the two particle distribution functions are the product of the single particle distribution function. This assumption is valid when the mean free path, which is the distance a molecule travels between successive collisions, is large compared to the size of the molecule. Hence there are no repeated collisions between the same pair of particles. With this assumption, a closed equation for the single particle distribution function, called the "Boltzmann equation", is obtained [2]:

$$\frac{\partial(nf)}{\partial t} + \frac{\partial(u_\alpha nf)}{\partial x_\alpha} + \frac{\partial(a_\alpha nf)}{\partial u_\alpha} = n^2 \int d\mathbf{k} \int d\mathbf{u}^* [f(\mathbf{x}, \mathbf{u}') f(\mathbf{x} + r\mathbf{k}, \mathbf{u}^{*\prime}, t) - f(\mathbf{x}, \mathbf{u}) f(\mathbf{x} + r\mathbf{k}, \mathbf{u}^*, t)] (4\pi r^2 \mathbf{w}.\mathbf{k}) \,. \tag{3}$$

This equation is a non-linear integro-differential equation, and is difficult to solve in general. However, it can be shown [2] that for a system at steady state, in the absence of external forces, the distribution function is a

Maxwell-Boltzmann (MB) distribution. For dense systems where the particle positions are correlated before a collision, such a simplification is not possible and some sophisticated mathematical techniques, called cluster expansions, have been developed for a dense system with hard-sphere molecules. There are no equally successful methods for a dense system of particles. In the present study, we will deal exclusively with dilute suspensions and the Boltzmann equation is the starting point of the description.

While drawing an analogy between gases and dilute suspensions, it should be noted that there is an important difference: the energy of the molecules in a gas at equilibrium is conserved but the motion of the particles in a suspension can only be sustained if there is a continuous source of energy. This is because dissipation exists due to inelastic collisions or the drag force of the gas. Based on this distinction, there are two limiting cases for the dynamics of a dilute suspension:

1. The dynamics will resemble that of the molecules of a gas if the dissipation of energy during a binary collision (due to inelasticity), or between successive collisions (due to viscous drag), is small compared to the average energy of the particle velocity fluctuations. This limit is known as the *kinetic* limit.
2. In the complementary limit, called the *dissipative* limit, the change in energy during a collision or between successive collisions is of the same magnitude as the energy of fluctuations. The properties are very different from that of a gas at equilibrium.

The dynamics of suspensions in the kinetic limit are obtained by assuming that the distribution function is a small perturbation about the MB distribution for a gas at equilibrium. The deviations from the MB distribution are evaluated from the velocity moments of the Boltzmann equation. These methods are fairly standard and are therefore not discussed in detail here. However, the extension of the Boltzmann H-Theorem to dissipative systems and its consequences and the calculation of the distribution function for suspensions in the dissipative limit are examined in the next section. Two systems, a bidisperse suspension of particles settling in a gas and a sheared suspension of inelastic particles, are considered. The analysis is restricted to spatially homogeneous suspensions at steady state. Hence the distribution function is independent of time and the spatial co-ordinates.

Distribution Function in the Kinetic Limit

In the kinetic limit, the collisional transport in phase space (the two terms on the right-hand side of Eq. (3)), which represents the rate of transport of particles into and out of a differential volume in velocity space, is large compared to the terms on the left-hand side. In addition, the particles are nearly elastic. In this case, a perturbation expansion can be used where the system is considered to be a collection of elastic particles in the leading approximation, and the effects of inelasticity, drag and body forces are included in higher order corrections in a systematic fashion. The leading order distribution function is a Maxwell-Boltzmann distribution [2]:

$$F = \frac{1}{(2\pi T)^{3/2}} \exp\left(\frac{-mc_\alpha^2}{2T}\right) \tag{4}$$

where m is the mass of the particle, the fluctuating velocity $c_\alpha = u_\alpha - U_{m\alpha}$ is the difference between the particle velocity and the mean velocity, and T is the "temperature". Unlike the case of molecular gases, the temperature is not specified *a priori*, but is determined by a balance between the source and dissipation of energy.

Bidisperse Particle-Gas Suspension

The system consists of a suspension of particles with masses m_1 and m_2, radii r_1 and r_2, and terminal velocities $\mathbf{U}_1$ and $\mathbf{U}_2$ settling in a gas. The drag force on the particles is considered to be a linear function of the particle velocity and the acceleration is

$$a_{i\alpha} = -(\mu_i/m_i)(u_{i\alpha} - U_{i\alpha}), \tag{5}$$

where the drag coefficient, μ_i, is $(6\pi\eta r_i)$ in the Stokes regime. The inertia of the gas is neglected compared to that of the particle. Hydrodynamic interactions are also neglected so that the dominant effects comprise the inertia of the particles and viscous drag due to the gas. There are two important time scales: the viscous relaxation time, $\tau_v = (m_1/\mu_1)$, which is the time taken by a particle to relax to its terminal velocity after a collision, and the collision time, $\tau_c = (1/(n_1 d_{12}^2 c))$, is the time that elapsed between successive collisions. Here, $d_{ij} = r_i + r_j$ and c is the magnitude of the fluctuating velocity. In the kinetic limit, the collision time is small compared to the viscous relaxation time.

As noted earlier, the distribution function is a Maxwell-Boltzmann distribution in the leading approximation. The mean velocities and "temperatures" for the two species are also equal. The first correction to the distribution function can be obtained using an asymptotic analysis where the distribution function is expressed as

$$f_i(\mathbf{c}_i) = F_i(\mathbf{c}_i)[1 + \delta\Phi_i(\mathbf{c}_i)]\,. \tag{6}$$

The small parameter, δ, will be specified a little later. When this is inserted into the Boltzmann equation, a linear equation for the perturbation, ϕ_i, is obtained:

$$\frac{\partial(a_{i\alpha}F_i)}{\partial c_{i\alpha}} = \delta\sum_{j=1}^{2} n_j \int d\mathbf{k}\int d\mathbf{c}_j : [F(\mathbf{c}_i)F(\mathbf{c}_j)[\Phi(\mathbf{c}_i') + \Phi(\mathbf{c}_j') - \Phi(\mathbf{c}_i) - \Phi(\mathbf{c}_j)](\pi d_{ij}^2 \mathbf{w}.\mathbf{k})\,, \tag{7}$$

where $\mathbf{c}_i$ and $\mathbf{c}_j$ are the velocities of the colliding particles before the collision and $\mathbf{c}_i'$ and $\mathbf{c}_j'$ are the velocities after the collision. Note that the velocity coordinate has been transformed from the particle velocity, $\mathbf{u}_i$, to the fluctuating velocity, $\mathbf{c}_i$. This transformation is trivial because the mean velocities of the two species are equal in the leading approximation.

Equation (7) does not provide the magnitude for δ. However, it can be obtained from the equivalent of the Boltzmann H-Theorem [2] for this system. The function, H, is defined as

$$H = \sum_i \int d\mathbf{x} \int d\mathbf{c}_i f_i \log(f_i)\,. \tag{8}$$

The time derivative of H is

$$\begin{aligned} \frac{dH}{dt} &= \sum_i \int d\mathbf{c}_i \left[\int d\mathbf{x}(1 + \log(f_i))\frac{\partial f_i}{\partial t}\right] \\ &= \left[\sum_i \int d\mathbf{x}\int d\mathbf{c}_i \left((1+\log(f_i))\frac{\partial_c f_i}{\partial t} - f_i\frac{\partial a_{i\alpha}}{\partial c_\alpha} - \underline{\frac{\partial}{\partial c_\alpha}(\mathbf{a}_i f_i \log f_i)}\right)\right]. \end{aligned} \tag{9}$$

The underlined term in the above expression can be reduced to a surface integral in velocity space which is zero, and therefore

$$\frac{dH}{dt} = \sum_i \int d\mathbf{x} \int d\mathbf{u}_i \left((1 + \log(f_i)) \frac{\partial_c f_i}{\partial t} - f_i \left(\frac{\partial a_{i\alpha}}{\partial c_\alpha} \right) \right) . \tag{10}$$

When the viscous relaxation time is large compared to the time that elapsed between collisions, the asymptotic expansion in Eq. (6) can be used for the distribution function. The leading order equation for the rate of change of H is

$$\left. \frac{dH}{dt} \right|_0 = \sum_{ij} \int d\mathbf{x} \int d\mathbf{k} \int d\mathbf{u}_i \int d\mathbf{u}_j \left[(F'_i F'_j - F_i F_j) \log \left(\frac{F_i F_j}{F'_i F'_j} \right) (\mathbf{w}.\mathbf{k}) \right] . \tag{11}$$

It can be shown [2] from the above equation that the leading order distribution function is a Maxwell-Boltzmann distribution, Eq. (7), from the Boltzmann H-Theorem. The first correction to $(dH/dt)_1$ is

$$\left. \frac{dH}{dt} \right|_1 = -\delta^2 \sum_{ij} \int d\mathbf{x} \left[\int d\mathbf{c}_i \int d\mathbf{c}_j F_i(\mathbf{c}_i) F_j(\mathbf{c}_j) [(\Phi_i(\mathbf{c}'_i) + \Phi_j(\mathbf{c}'_j) - \Phi_i(\mathbf{c}_i) - \Phi_j(\mathbf{c}_j))^2 (\mathbf{w}.\mathbf{k})] - \int d\mathbf{c}_i F_i \left(\frac{\partial a_{i\alpha}}{\partial c_{i\alpha}} \right) \right] . \tag{12}$$

At steady state, the first correction to (dH/dt) is also zero. In the above equation, the first term on the right-hand side is proportional to $\delta^2 \tau_c^{-1}$ while the second term is proportional to τ_v^{-1}. Thus, it can be inferred that $\delta = (\tau_c/\tau_v)^{1/2}$. In addition, a comparison of Eqs. (7) and (12) shows that

$$\frac{\partial a_{ia}}{\partial c_\alpha} \sim \frac{\delta a_{ia}}{\tau_v} . \tag{13}$$

This provides the estimate, $c \sim \delta U_m$, for the fluctuating velocity and $T \sim \delta^2 m_i U_m$ for the temperature. (Here, it has been assumed that the mean velocity of the suspension and the terminal velocities of the two species are of the same magnitude.)

The first correction to the distribution function is obtained by solving Eq. (7) using the Enskog expansion [2] for the present case. The following functional form is assumed for Φ_i:

$$\Phi_i(\mathbf{c}_i) = A_i(\mathbf{C}_i)\mathbf{C}_i.\mathbf{U}_m \tag{14}$$

where $\mathbf{C}_i = (m_i^{1/2}\mathbf{c}_i/T^{1/2})$ is $O(1)$. It is not possible to obtain an explicit solution for $A(\mathbf{C}_i)$. However, this can be expanded in an appropriate orthogonal function space and the series solution can be obtained. However, the magnitude of the difference in the mean velocities of the two species can be obtained without explicitly solving the equation. It can be easily seen that the difference between the mean velocity of species, i, and the mean velocity of the suspension, $\int d\mathbf{c}_i F_i \delta\Phi_i$, is $O(\delta^2 U_m)$ since $\mathbf{c}_i \sim \delta U_m$.

To determine the exact values, it is necessary to use the moment expansion method [3] where the Boltzmann equation is multiplied by different moments of the velocity distribution to obtain conservation equations for the velocity moments. However, the present analysis provides a clearer insight into the effect of velocity-dependent forces on the dynamics of the system in the kinetic limit.

Sheared Suspension of Inelastic Particles

The shear flow of a suspension of slightly inelastic particles in the kinetic limit has been studied in detail. The analysis is very similar to that for a gas of hard-sphere molecules in shear flow [2]. Hence the details of the analysis are not discussed here. The major differences are that the collisions between particles are inelastic and the temperature of the suspension is determined by a balance between the input of energy and dissipation due to shear flow and inelastic collisions, respectively. The Enskog expansion is used to determine the deviation of the distribution function from the MB distribution in Eq. (4):

$$f = F(1 + \delta\Phi) \tag{15}$$

where Φ, the deviation from the Maxwell-Boltzmann distribution, has the following form in a shear flow:

$$\Phi = A(\mathbf{C})C_\alpha \partial_\alpha T + B(\mathbf{C})C_\alpha C_\beta (\partial_\alpha U_\beta) \tag{16}$$

where $\mathbf{C} = (m\mathbf{c}/T)^{1/2}$, and $\partial_\alpha U_\beta$ and $\partial_\alpha T$ are the gradients in the mean strain rate and temperature, respectively. Using the above expansion, constitutive equations can be derived for the density, momentum and "temperature" of the suspension. The conservation equation for the granular temperature contains a source term and dissipation term due to the shear work and inelastic collisions, respectively. A balance between these gives the granular temperature at steady state. To extend the analysis to the dense limit, attempts have been made to use a pair distribution function which is not just a single particle distribution function, but also includes the effect of excluded volume. The distribution function most commonly used is the Carnahan-Starling approximation for a system of dense gases.

Distribution Functions in the Dissipative Limit

In contrast to the kinetic limit, no standard methods exist to determine the distribution function in the dissipative limit. The method used has to be designed for the system under consideration. This is illustrated in the examples that follow.

Bidisperse Particle-Gas Suspension

In the dissipative limit, the number density of the particles is sufficiently small. Thus, the viscous relaxation time, $\tau_{vi} = (m_i/\mu_i)$, is small compared to the time that elapsed between successive collisions, $\tau_{cij} = 1/(n_i d_{ij}^2 (U_1 - U_2))$. In this limit, a perturbation expansion in the small parameter, $\epsilon = (\tau_{v1}/\tau_{c12})$, is used to calculate the distribution function. In the leading approximation, the effect of collisions is neglected and the particles are considered to settle at their terminal velocities. In this case, the distribution functions are delta functions at the terminal velocities of the two species.

The distribution function that includes the effect of collisions between particles settling at their terminal velocities can be determined using a flux balance in velocity space. The balance equation for the distribution function is

$$\frac{\partial a_{i\alpha} f}{\partial u_{i\alpha}} = N_i^{in}(\mathbf{u}_i) - N_i^{out}(\mathbf{u}_i)\,, \tag{17}$$

where N_i^{in} and N_i^{out} are the flux of particles entering and leaning a differential volume due to collisions. In the collisional limit, the number of particles with velocities $O(U_1 - U_2)$ that are different from their terminal velocities is small. Therefore, in the calculation of the leading order estimate of N_i^{in} and N_i^{out}, it is assumed that the colliding particles are moving at their terminal velocities. The collisional fluxes are determined by relating the angle made by the line joining the centres of the particles at the point of collisions to the change in the velocity. The details of the calculation are found in Kumaran and Koch [7]. The collisional fluxes are inserted into Eq. (17) to determine the distribution function

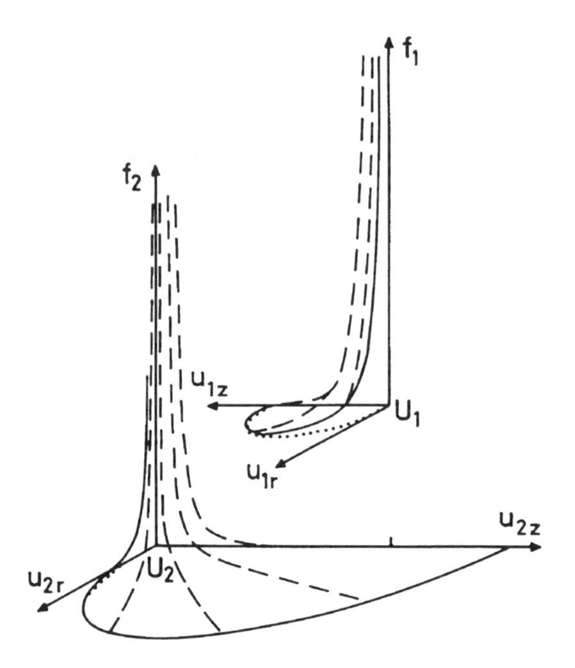

Fig. 1. Schematic of the shape of the distribution function in a bidisperse suspension. The zero levels of the distribution function of the two species have been separated for clarity. The dotted line represents the projection of the surface onto the (u_x, u_y) plane and the solid line shows the distribution function on this surface.

$$f_i = (\epsilon\gamma_{ik}/\pi)(\cos(\chi_i))^{1-\epsilon\gamma_{ik}}(2M_i)^{-\epsilon\gamma_{ik}}v_i^{(\epsilon\gamma_{ik}-3)}, \tag{18}$$

where $\epsilon = (\tau_{v1}/\tau_{c12})$ and $\gamma_{ik} = (\tau_{vi}/\tau_{v1})(\tau_{c12}/\tau_{cik})$.

The distribution function, Eq. (18), shown in Fig. 1, is very different from the MB distribution. Some of its salient features are described here. The distribution is non-zero only in finite regions of the velocity space and has a divergence at the terminal velocities of the two species. The first correction to the velocity moments can be determined using the distribution function, Eq. (18). The analysis shows that the difference between the particle velocity and the mean velocity, $\langle v_z \rangle$, is $O(\epsilon)$ smaller than the terminal velocity in the limit $\epsilon \ll 1$. Meanwhile, the mean square velocities are $O(\epsilon)$ smaller than the square of the terminal velocity. The distribution function is highly anisotropic. The ratio of the mean square velocity in the vertical and horizontal directions four in the limit $\epsilon \to 0$. In addition, the distribution function is highly skewed. Hence the ratio, $(\langle v_z^3\rangle/\langle v_z^2\rangle^{3/2}$, diverges proportional to $\epsilon^{-1/2}$ in the limit $\epsilon \rightarrow 0$.

Sheared Suspension of Inelastic Particles

For the sake of simplicity, the system considered here is a two-dimensional suspension of inelastic disks. Nevertheless, the analysis can easily be extended to a three-dimensional suspension of spherical particles. The disks are of radius r, number density n and coefficient of elasticity e in a channel with width L. The channel is bounded by walls at $y = (L/2)$ and $y = -(L/2)$ moving with velocities $+U_w$ and $-U_w$ in the x direction, respectively. Here, the co-ordinate y is perpendicular to the walls of the channel and the x co-ordinate is along the flow direction. The particle-particle collisions are described by the standard laws for collisions between smooth elastic disks. The change in the particle velocity due to a wall collision is given by

$$u'_x - u_x = (1 - e_t)(\pm U_w - u_x)_y \; u'_y - u_y = -(1 - e_n)u_y \,, \tag{19}$$

where (u_x, u_y) and (u'_x, u'_y) are the particle velocity before and after the wall collision, respectively, and e_t and e_n are the tangential and normal coefficients of restitution which are less than one. In the equation for $u'_x - u_x$, the positive sign for U_w is used for a collision with the wall at $y = +(L/2)$ and the negative sign for the wall at $y = -(L/2)$.

In the dissipative limit, $(nrL) \ll 1$, particle-wall collisions are more frequent than particle-particle collisions. In the absence of inter-particle collisions, a particle with a non-zero velocity in the y direction collides repeatedly with the walls. Its velocity after i collisions evolves as

$$u_x + (-1)^i(1 + (-1)^{(i-1)}e_t^i)U = e_t^i u_x^{(0)}, \quad u_y = (-1)^i e_n^i u_y^{(0)} \,, \tag{20}$$

where $U = (1 - e_t)U_w/(1 + e_t)$, $u_x^{(0)}$ and $u_y^{(0)}$ are the particle velocities before the first collision, and the first collision is assumed to take place with the wall at $y = +(L/2)$. It can be seen that in the limit of large i, the particle velocity converges towards $(\pm U, 0)$. Consequently, in the absence of particle collisions, it is expected that the velocities of all the particles converge towards $(\pm U, 0)$, which is independent of their initial velocities and depends only on the wall velocity and the coefficients of restitution.

In the limit of small u_y, however, it cannot be assumed that particle-wall collisions are more frequent than particle-particle collisions. The frequency of

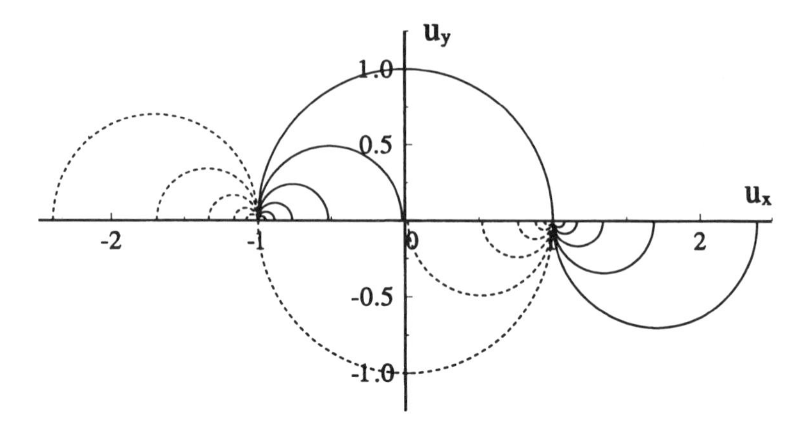

Fig. 2. The contours, C_i, of the particle velocities in the $u_x - u_y$ plane where the index, i, represents the number of times the particle has collided with the walls after it has acquired a velocity in the y direction due to a binary collision. The solid lines show the location of particles whose first collision is with the wall at $y = +(L/2)$ while the broken lines show the location of particles whose first collision is with the wall at $y = -(L/2)$. The coefficients of restitution, e_t and e_n, are both 0.7.

particle-wall collisions per unit length of the channel in the x direction scales as nru_y while that for particle-particle collisions is proportional to n^2r^2LU. This is because the difference in particle velocities scales as U for $u_y \ll U$. Therefore, the frequency of particle-particle collisions is the same as that of particle-wall collisions for $(u_y/U) \sim \epsilon$. To determine the effect of collisions to leading order in small ϵ, it is assumed that half of the particles have velocities $(U, 0)$ and the remaining half velocities $(-U, 0)$ prior to collision. Consider a collision between particle A with velocity $(U, 0)$ and particle B with velocity $(-U, 0)$. The velocity after collision is given by

$$\begin{aligned} u'_{Ax} &= -V\cos(2\theta) \; u'_{Ay} = -V\sin(2\theta)\,, \\ u'_{Bx} &= V\cos(2\theta) \quad u'_{By} = V\sin(2\theta)\,, \end{aligned} \tag{21}$$

where θ is the angle made by the line joining the centres of the particles at the point of collision with the x-axis. Therefore, binary collisions tend to transport particles onto a circle of radius, U, in velocity space as shown in Fig. 2. The subsequent collisions with the walls modify the velocity of the particles as indicated by Eq. (20), so that the velocity after i collisions, $(u_x^{(i)}, u_y^{(i)})$, is given by the parametric relations:

$$u_x^{(i)} + (1 + (-1)^{(i-1)} e_t^i) U = e_t^i U \cos(\chi) \quad \text{for } 0 < \chi < \pi$$
$$u_x^{(i)} - (1 + (-1)^{(i-1)} e_t^i) U = e_t^i U \cos(\chi) \quad \text{for } \pi < \chi < 2\pi \,. \tag{22}$$
$$u_y^{(i)} = e_n^i U \sin(\chi)$$

The above equations show that the particle positions are located along the ellipses C_i centred at $(\pm(1 + (-1)^{(i-1)} e_t^i) U, 0)$ with radii $e_t^i U$ and $e_n^i U$ lying along the x and y directions as shown in Fig. 2.

The distribution function along each of these contours is obtained by a flux balance in velocity space. The details of the calculation are not given here. The reader is referred to [8] for the details. The distribution function, $f_i(\chi)$, is defined such that $n f_i(\chi) d\chi$ is the number of particles in the differential angle, $d\chi$, about χ on the contour C_i. This is given by

$$f_i(\chi) = \frac{f_0(\chi)}{e_n^i} \prod_{j=1}^{i} \left[1 + \frac{2\epsilon}{(e_n)^j |\sin(\chi)|} \right]^{-1} \tag{23}$$

where $f_0(\chi)$, the distribution function after the binary collision, is

$$f_0(\chi) = \frac{\epsilon}{2|\sin(\chi)|} \left[\cos\left(\frac{\chi}{2} - \frac{\pi}{2} \right) \right] . \tag{24}$$

It can be easily shown that this distribution function is normalised:

$$\sum_{i=0}^{\infty} \int_0^{2\pi} d\chi f_i(\chi) = 1 \,. \tag{25}$$

The moments of the velocity distribution function can be easily calculated using the distribution function, Eq. (23). It is found that $\langle u_x^2 \rangle \to U^2$ and $\langle u_y^2 \rangle \sim V^2 \epsilon$ in the limit $\epsilon \to 0$. In addition, the cross-correlation, $\langle u_x u_y \rangle \sim U^2 \epsilon \log(\epsilon^{-1})$, and the shear stress decrease proportional to $\epsilon \log(\epsilon^{-1})$ in this limit. The qualitative behaviour of the velocity moments are the same for a three-dimensional suspension of elastic spheres as well as for suspensions of inelastic disks and spheres.

Conclusions

The derivation of the velocity distribution function for dilute particle suspensions in the kinetic and dissipative limits was discussed. In the kinetic limit, the distribution function is close to a Maxwell-Boltzmann distribution for a gas at equilibrium. For a sheared suspension, the asymptotic scheme for determining the distribution function is similar to that used in the Chapman-Enskog theory for dense gases. However, there is a minor difference: the "temperature" is not externally imposed, but is determined by a balance between the source of energy and dissipation due to shearing and inelastic collisions, respectively. For a bidisperse sedimenting suspension, the analysis is different from that used in the Chapman-Enskog theory due to a velocity-dependent drag force. The Boltzmann H-Theorem can be used to show that the magnitude of the fluctuating velocity scales as δU_m, where U_m is the mean velocity of the suspension. The difference in the mean velocity of the two species scales as $\delta^2 U_m$, where the small parameter $\delta \sim (\tau_c/\tau_v)^{1/2}$ with τ_c and τ_v being the time that elapsed between collisions and the viscous relaxation, respectively. Therefore, it can be inferred from the Boltzmann H-Theorem that the fluctuating velocity is small compared to the mean velocity of the suspension.

The distribution function in the dissipative limit is very different from the MB distribution. For a bidisperse sedimenting suspension, the distribution function is non-zero only in a finite region of the velocity space. It also has a divergence at the terminal velocities of the two species. The distribution function is highly anisotropic and the mean square velocity in the vertical direction is four times that in the horizontal direction. In addition, it is highly skewed and the skewness increases proportional to $(\tau_v/\tau_c)^{-1/2}$ in the limit, $\tau_v \ll \tau_c$.

For a sheared suspension of two dimensional disks, the dissipative limit corresponds to the regime $\epsilon \equiv (nrL) \ll 1$, where n is the particle number density, r is the particle radius and L is the width of the channel. In this limit, the frequency of particle-wall collisions is large compared to that of particle-particle collisions. The distribution function is sharply peaked around $(u_x, u_y) = (\pm U, 0)$ and is non-zero only along certain contours in the velocity space. This is shown in Fig. 2. The distribution function is highly anisotropic and the mean square velocity normal to the walls is $O(\epsilon)$ smaller than that in the flow direction. The cross-correlation, $\langle u_x u_y \rangle$, which is proportional to the shear stress, is $O(\epsilon \log(\epsilon^{-1}))$ smaller than the mean square velocity in the flow direction.

The above studies indicate that the distribution function in the kinetic limit is close to a Maxwell-Boltzmann distribution for a hard sphere gas. It can be determined using a perturbation analysis in which the MB distribution is the leading approximation. The distribution function in the dissipative limit is very different from the MB distribution and the analytical technique used is specific to the system under consideration.

Discussion

J. R. A. Pearson To what physical systems and/or phenomena do your approximate theories apply and provide physical insight?

V. Kumaran The theories derived here apply to the rapid flows of suspensions of particles in a gas, such as shear flows or settling suspensions in the kinetic and dissipative limits. In real systems, the flow may be in either of these limits, or in the intermediate regime. If the flow is in either of these limits, the theory applies without modifications. If it is in the intermediate regime, approximate distribution functions, such as the one devised by Kumaran, Tsao and Koch [9] can be used. The present analysis is useful for devising these approximation distribution functions since it provides the limiting behaviour to which any valid solution should converge.

In real systems, different points in the flow have different parameter values. In these cases, it would be necessary to use different distribution functions at different points in the flow to get a complete description. In this sense, the present description has an advantage over continuum descriptions. This is because in the latter, the same description is used throughout the flow even though the flow conditions could be very different.

M. J. Adams Could your method be adapted to describe the behaviour of fluidised beds which show complex behaviour such as the formation of bubbles?

V. Kumaran This analysis cannot be easily adapted to gas fluidised beds due to the complexity of the interaction between the gas and the fluid. The simple Stokes law for the interaction between the particles and the gas would not suffice. A more complete description of the gas-particle interaction at high Reynolds number in dense suspensions would be necessary. In addition, the assumption of molecular chaos would not be a good one for a fluidised bed where the particle density is quite high. However, the present analysis could be used to describe a vibrated fluidised bed where the fluidisation is due to the vibration of the bottom surface of the bed. In this case, the dynamics of the particles can be described by simple laws. It has also been experimentally observed that the density of the suspension is low enough to justify the use of the present theories for dilute suspensions.

M. E. Cates The limit $\epsilon \ll 1$ corresponds to the usual definition of Knudsen flow in gases for which the viscosity is independent of density. Your results are different. Why?

V. Kumaran The difference is in the boundary conditions used for the shear flow. For a gas sheared between two surfaces, the size of the molecules is small compared to that of the surface roughness on the surfaces. Therefore, the stochastic Maxwell boundary condition is used. This is because it is assumed that a fraction of the molecules incident on a surface are reflected elastically while the rest are reflected with a random velocity chosen so that the average temperature of the reflected molecules is equal to the temperature of the surface. In the present system, the size of the particles is large compared to the size of the surface roughness. Hence deterministic boundary conditions are used. This leads to a difference in the behaviour of the two systems.

J. Goddard Tsao and Koch [10] had recently performed an analysis of the shear flow of a particle suspension in which they reported the existence of two states of suspension, that is, an "ignited" and "collapsed" state. How are these states related to the asymptotic limits you talked about?

V. Kumaran The analysis of Tsao and Koch was for the shear flow of a suspension of particles in a gas where it is subjected to a shear flow. The "collapsed" state corresponds to a dilute suspension where most of the particles travel along the streamlines. Collisions also occur due to the relative velocity between particles travelling on different streamlines separated by a distance less than the particle diameter. The analysis of the collapsed state resembles closely the analysis for the dissipative limit of a bidisperse particle suspension discussed here, although the mechanism that induces particle collisions is different. The analysis of the ignited state is very similar to that for the kinetic limit of a bidisperse suspension. Therefore, the "collapsed" and "ignited" states correspond to the dissipative and kinetic limits.

M. Lal Can simulation methods, such as the lattice Boltzmann simulation, be used for these suspensions?

V. Kumaran If one is interested in simulating the behaviour of the particles using some simple assumptions (such as Stokes law) about their interaction with gas, it is easier to use discrete particle simulation procedures such as molecular dynamics or event-driven simulation. If one is interested in treating exactly the complex interaction between them, a technique like the lattice Boltzmann simulation would then be useful.

References

1. Liboff, R. L. (1990) *Kinetic Theory*. New Jersey: Prentice Hall
2. Chapman, S. & Cowling, T. G. (1970) *The Mathematical Theory of Non-uniform Gases*. Cambridge: Cambridge University Press
3. Kumaran, V. & Koch, D. L. (1993) Properties of a bidisperse gas — solid suspension. Part 1. Collision time small compared to viscous relaxation time. *J. Fluid Mech.* **247**, 623–641
4. Jenkins, J. T. (1997) Balance laws and constitutive relations for rapid flows of granular materials. In *Constitutive Models of Deformation*, ed. J. Chandra & R. P. Srivatsav. SIAM
5. Lun, C. K. K., Savage, S. B., Jeffrey, D. J. & Chepurniy, N. (1984) Kinetic theories for granular flow: inelastic particles in Couette flow and slightly inelastic particles in a general flow field. *J. Fluid Mech.* **140**, 223–256
6. Jenkins, J. T. & Savage, S. B. (1983) A theory for the rapid flow of identical, smooth, nearly elastic spherical particles. *J. Fluid Mech.* **130**, 187–202
7. Kumaran, V. & Koch, D. L. (1993) Properties of a bidisperse gas — solid suspension. Part 2. Viscous relaxation time small compared to collision time. *J. Fluid Mech.* **247**, 643–660
8. Kumaran, V. (1997) Velocity distribution function for a dilute granular material in shear flow. *J. Fluid Mech.*, in press
9. Kumaran, V., Tsao, H.-K. & Koch, D. L. (1993) Velocity distribution functions for a bidisperse sedimenting particle — gas suspension. *Int. J. Multiphase Flow* **19**, 665–681
10. Tsao, H.-K. & Koch, D. L. (1995) Simple shear flows of dilute gas — solid suspensions. *J. Fluid Mech.* **296**, 211–245

Dynamics of Complex Fluids, pp. 280–285
ed. M. J. Adams, R. A. Mashelkar, J. R. A. Pearson & A. R. Rennie
Imperial College Press–The Royal Society, 1998

Chapter 19

Migrational Instabilities in Particle Suspensions

J. D. GODDARD
*Department of Applied Mechanics and Engineering Sciences,
University of California, San Diego,
La Jolla, CA 92093-0411, USA
E-mail: jgoddard@ucsd.edu*

This work deals with an instability arising from the shear-induced migration of particles in dense suspensions coupled with a dependence of viscosity on particle concentration. The analysis summarised here treats the inertialess ($\mathrm{Re} = 0$) linear stability of homogeneous simple shear flows with a rheological model of the general variety proposed by Leighton and Acrivos (1987) for Stokesian suspensions. Depending on the importance of shear-induced migration relative to concentration-driven diffusion, this model admits short-wave instability arising from wave-vector stretching by the base flow and evolving into particle-depleted shear bands. Moreover, this instability in the time-dependent problem corresponds to a loss of ellipticity in the associated static problem ($\mathrm{Re} = 0, \mathrm{Pe} = 0$). While the isotropic version of the Leighton-Acrivos model is found to be stable with their experimentally determined parameter values for simple shear, Nott and Brady (1994) have suggested thatthe stable model may not always provide a good quantitative description of particle clustering in the core of pipe flow. The purpose of this note is to point out that an appropriate variant on the above model would allow such clustering to occur as an instability leading to a two-phase bifurcation in the base flow.

Introduction

The general theme of the present work is *material instability*, whose recent history, from the 1950s onwards, lies mainly in solid mechanics. In this setting, various strain-softening effects give rise to strain localisation in the form of "shear bands" or "necks". By comparison, there is a much smaller literature on related instabilities in fluid-like materials, such as non-monotone viscous stress in particulate suspensions and polymeric fluids or dissipative clustering in rapid granular flows.

A recent report [1] offers a brief review of a fairly extensive literature whose common thread is the "short wavelength" instability of homogeneous deformations and the associated change of type in the underlying field equations, such as theloss of static ellipticity or dynamic hyperbolicity. The following analysis deals with a form of material instability in which a scalar transport process plays a crucial rôle, a possibility which, in the present setting, is suggested by Pearson [2]. A number of prior, related studies involve the effects of heat conduction on thermally softening media, where mechanical dissipation (having no direct counterpart here) gives rise to phenomena such as bifurcation in liquid flows or "adiabatic shear bands" in solids [1].

Model and Analysis

With the neutrally buoyant "Stokesian" suspension model of Leighton and Acrivos [3] (*cf.* Phillips *et al.* [4] and Phan-Thien & Zang [5]) serving as a prototype, we consider an elementary constitutive model in which the stress and particle flux relative to the fluid are given, respectively, by

$$\mathbf{T} = 2\eta(\phi)\mathbf{D} - p\mathbf{1} \tag{1}$$

and

$$\mathbf{j} = -\{\kappa(\phi,\dot{\gamma})\nabla\phi + \nu(\phi,\dot{\gamma})\nabla\dot{\gamma}\}\,, \tag{2}$$

where ϕ denotes the particle volume fraction,

$$\mathbf{D} := \frac{1}{2}\{\nabla\mathbf{v} + (\nabla\mathbf{v})^T\}\,, \quad \dot{\gamma} := \{2\mathrm{tr}\,(\mathbf{D}^2)\}^{\frac{1}{2}}\,, \tag{3}$$

and $\mathbf{v}$ is mixture velocity. The quasi-static mechanics, $(\mathrm{Re} = 0, \mathrm{Pe} \geq 0)$, is governed by the above constitutive equations, the balances

$$\nabla\cdot\mathbf{T} = \nabla p \quad \text{with } \nabla\cdot\mathbf{v} = 0\,, \tag{4}$$

and

$$\partial_t\phi + \mathbf{v}\cdot\nabla\phi + \nabla\cdot\mathbf{j} = 0\,, \tag{5}$$

and suitable boundary conditions (compatibility with the base state and/or regularity of perturbed states in unbounded domains).

For planar infinitesimal perturbations $\mathbf{v}^{(1)}, \phi^{(1)}$ of the uniform base state:

$$v_x^{(0)} = \dot{\gamma}^{(0)} y \quad \text{and } \{v_y^{(0)}, v_z^{(0)}, \nabla\dot{\gamma}^{(0)}, \nabla\phi^{(0)}\} = 0 \,. \tag{6}$$

The standard stream-function *cum* Fourier representation in the xy plane, $\Psi(k_x, k_y, t), : \Phi(k_x, k_y, t)$, must satisfy

(7)

and

$$\partial_t \Phi - k_x \partial_{k_y} \Phi = -k^2 \{\nu(k_x^2 - k_y^2)\Psi + \kappa\Phi\} \,, \tag{8}$$

where we let

$$k^2 = k_x^2 + k_y^2 \quad \text{and } \alpha = \left(\dot{\gamma}\frac{d}{d\phi}\log\eta\right)^{(0)} \tag{9}$$

and adopt the change of notation,

$$\dot{\gamma}^{(0)} t \to t \,, \quad \frac{\kappa^{(0)}}{\dot{\gamma}^{(0)}} \to \kappa \,, \quad \frac{\nu^{(0)}}{\dot{\gamma}^{(0)}} \to \nu \,. \tag{10}$$

The stream function, Ψ, is readily eliminated from Eqs. (7) and (8) which are easily seen to be elliptic for $\kappa > \nu\alpha$ and otherwise hyperbolic with possible singular surfaces.

Wave-Vector Shearing

With Ψ eliminated, Eqs. (7) and (8) become

$$\partial_t \Phi - k_x \partial_{k_y} \Phi = \sigma\{\mathbf{k}\}\Phi \,, \tag{11}$$

where

$$\sigma\{\mathbf{k}\} = -k^2 \left\{\kappa - \frac{\nu\alpha}{k^4}(k_x^2 - k_y^2)^2\right\} \,. \tag{12}$$

Were it not for the term, $k_x \partial_{k_y}$, representing wave-vector shearing (Thomson [6], Savage [7], Wang *et al.* [8]), an effect that becomes dominant for large time, the preceding relation would represent a dispersion relation identical with that given by the 1-D analysis in the concluding section of Nott and Brady [9].

We note that wave-vector stretching in an arbitrary isochoric homogeneous flow with $\nabla \mathbf{v} = \mathbf{L}^T$ = const. and $\mathrm{tr}\{\mathbf{L}\} \equiv \nabla \cdot \mathbf{v} = 0$ can be concisely represented by the respective transformations between the material ("embedded") co-ordinates, $(\hat{\mathbf{x}}, \hat{\mathbf{k}})$, in physical space and "reciprocal" space:

$$\hat{\mathbf{x}} = \mathbf{F}^{-1}\mathbf{x}\,, \quad \text{with } (\partial_t)_{\hat{\mathbf{x}}} = (\partial_t)_{\mathbf{x}} + (\mathbf{L}\mathbf{x}) \cdot \partial_{\mathbf{x}} \tag{13}$$

and

$$\hat{\mathbf{k}} = \mathbf{F}^T\mathbf{k}\,, \quad \text{with } (\partial_t)_{\hat{\mathbf{k}}} = (\partial_t)_{\mathbf{k}} - (\mathbf{L}^T\mathbf{k}) \cdot \partial_{\mathbf{k}}\,, \tag{14}$$

where

$$\mathbf{F}(t) := (\partial_{\hat{\mathbf{x}}}\mathbf{x})^T = e^{\mathbf{L}t} \tag{15}$$

is the physical space deformation gradient.

For the simple shear at hand, the transformation in Eq. (14) gives $\mathbf{k}(\hat{\mathbf{k}}, t)$ as

$$k_x \equiv \hat{k}_x \quad \text{and } k_y = \hat{k}_y - \hat{k}_x t \tag{16}$$

and converts Eq. (11) into the o.d.e.,

$$\frac{d}{dt} \log \Phi = \sigma\{\mathbf{k}(\hat{\mathbf{k}}, t)\}\,, \tag{17}$$

where $\sigma\{\mathbf{k}\}$ is given by Eq. (12).

While Eq. (17) can be integrated subject to Eq. (16), asymptotic stability is determined by its limiting form for large t. There are also two cases to consider, $\hat{k}_x \neq 0$ and $\hat{k}_x = 0$, for which, respectively, one easily finds from Eqs. (16) and (17) that

$$\log \Phi \sim (\nu\alpha - \kappa)\frac{\hat{k}_x^2 t^3}{3} \sim (\nu\alpha - \kappa)\frac{k_y^2 t}{3} \tag{18}$$

and

$$\log \Phi \equiv (\nu\alpha - \kappa) k_y^2 t \tag{19}$$

for $t \to \infty$. Thus, $\kappa < \nu\alpha$ implies short wavelength instability in the form of a ("Kelvin mode") shear-band structure, with $k_y >> k_x$ representing particle-depleted strata lying perpendicular to the y-direction.

A Special Case

For the particle distribution in a fully developed rectilinear shear flow, several authors [3–5] employ Eq. (1) with the 1-D form for the particle flux,

$$j_y = -\phi^2 \left\{ K_1 \frac{d\dot{\gamma}}{dy} + K_2 \alpha \frac{d\phi}{dy} \right\} , \tag{20}$$

where

$$\alpha = \dot{\gamma} \frac{d}{d\phi} \log \eta \geq 0 \tag{21}$$

and the coefficients, $K_1 \geq 0$ and $K_2 \geq 0$, are constants (proportional to particle diameter squared).

With Eq. (2) as the obvious extension of Eq. (20) to 3D, which neglects shear-induced anisotropy, one can easily deduce fromthe preceding analysis that linear stability against planar disturbances requires that

$$K_2 > K_1 , \tag{22}$$

independently of the particular form for $\eta(\phi)$.

Equation (22) is clearly satisfied by the empirically assigned values of Leighton and Acrivos [3], for which $K_2 \approx 2K_1$ over a relatively broad range of ϕ. On the other hand, Nott and Brady [9] make the interesting observation that the stable form of Eq. (20) gives a cusped particle concentration profile in pipe flow which may not represent a particularly good quantitative description of the dense "plug flow" clustering of particles. While they propose to replace Eq. (20) by a more elaborate constitutive theory, a plausible alternative is obtained by merely allowing a variation with ϕ of the coefficients K_1 and K_2 in Eq. (2) to produce instability and a bifurcated two-phase "core annular" structure near the state of densest packing.

In summary, the above particle migration model can exhibit instability in simple shear, with the dominant linear mode being particle-depleted shear bands. Hence, the unstable model offers an alternative description of particle clustering in pipe flow. Whatever the physical validity of the model, the above analysis serves to illustrate an interesting mathematical connection between the loss of static ellipticity and time-dependent instability.

Acknowledgements

Partial support from the US National Aeronautics and Space Administration (Grant NAG 3-1888), the US National Science Foundation (Grant CTS

9510121) and the 1996 Programme on the Dynamics of Complex Fluids in the Isaac Newton Institute, Cambridge University, UK, is gratefully acknowledged.

Discussion

J. R. A. Pearson What is the the relevance of the analysis to "slugging" instabilities (which have been observed in two-phase flows)?

J. D. Goddard The author would venture to say that this type of instability, with $k_x \neq 0$, cannot be explained by the above type of linear stability analysis at $\mathrm{Re} = 0$.

References

1. Goddard, J. D. (ed., 1996) *Workshop on Material Instabilities, Report No. 95–20.* San Diego: Institute for Mechanics and Materials, University of California
2. Pearson, J. R. A. (1994) Flow curves with a maximum. *J. Rheology* **38**, 309
3. Leighton, D. T. & Acrivos, A. (1987) The shear-induced migration of particles in concentrated suspensions. *J. Fluid Mech.* **181**, 415
4. Phillips, R. J., *et al.* (1991) A constitutive equation for concentrated suspensions that accounts for shear-induced particle migration. *Phys. Fluids A* **4**, 30
5. Phan-Thien, N. & Fang, Z. (1996) Entrance length and pulsatile flows of a model concentrated suspension. *J. Rheology* **40**, 521
6. Thomson, W. (Lord Kelvin, 1887) Stability of fluid motion. Rectilineal motion of viscous fluid between two parallel planes. *Phil. Mag.* **24**, 188
7. Savage, S. B. (1992) Instability of unbounded uniform granular shear flow. *J. Fluid Mech.* **241**, 109
8. Wang, C.-H., Jackson, R. & Sundaresan, S. (1996) Stability of bounded rapid shear flows of a granularmaterial. *J. Fluid Mech.* **308**, 31
9. Nott, P. R. & Brady, J. F. (1994) Pressure-driven flow of suspensions: simulation and theory. *J. Fluid Mech.* **275**, 157

Dynamics of Complex Fluids, pp. 286–300
ed. M. J. Adams, R. A. Mashelkar, J. R. A. Pearson & A. R. Rennie
Imperial College Press–The Royal Society, 1998

Chapter 20

Review of Chaotic Behaviour of Suspensions of Slender Rods in Simple Shear Flow

K. S. KUMAR, S. SAVITHRI AND T. R. RAMAMOHAN*
Computational Materials Science, Unit I,
Regional Research Laboratory (CSIR),
Thiruvananthapuram - 695 019, India
**E-mail: ram@csrrltrd.ren.nic.in*

We report an overview of our research on the dynamics and rheology of periodically forced slender bodies in simple shear flow. The problem considered is the simplest case of a class of problems that do not appear to have received any attention in the chaos literature. We demonstrate that the rheological parameters of such suspensions evolve chaotically as well as the possibility of novel rheological behaviour, otherwise unobtainable, through control of chaos. These observations suggest the possible use of these results in the development of computer-controlled intelligent rheological behaviour of such suspensions.

Introduction

The study of the motion of small particles in Newtonian fluids has relevance in many disparate fields, ranging from soil mechanics to the micro-electronics industry [1]. These technological applications have motivated many groups to work on the dynamics and rheology of suspensions of small particles. Bretherton [2], Anczurowski & Mason [3], Leal & Hinch [4], Altan *et al.* [5] and Macmillan [6] are some of the authors who have analysed the dynamics of particles in a variety of linear flows and the effects of the orientation of these particles on the properties of the suspension. The evolution of particles in time-dependent flows has been analysed by Szeri *et al.* [7]. The analysis of the rheology of suspensions of small dipolar particles under the action of sinusoidal external forces has many practical applications, such as magneto-fluidisation [8], the magnetostriction of ferromagnetic particle suspensions [9], the rheological properties of ferromagnetic colloids [10] and the characterisation of magnetorheological suspensions [11–15]. Some of the above authors have examined rotating external fields which can be considered as a superposition of two orthogonal alternating fields.

The introduction of chaos theory has changed the dynamical system scenario appreciably as it is shown that even systems perceived to be simple can exhibit very complex behaviour. The idea of controlling chaos has further emphasised the importance of chaos theory by providing a new technique to obtain system behaviour which had heretofore been impossible. However, the possibility of the existence of chaos in suspension rheology has only been demonstrated recently [16]. Ramamohan *et al.* [17] have demonstrated that the dynamics of periodically forced slender bodies in linear flows can lead to chaotic orientation distributions. The effect of the chaotic orientation of particles on the rheology of the suspension has been analysed by Kumar and Ramamohan [16]. They have considered the rheology of dilute suspensions of periodically forced slender rods in simple shear flow in the presence of weak or negligible Brownian forces. They have also developed a novel technique for calculating the orientation distribution function of the particles and noted the weakness of the method proposed by Strand [18]. Strand [18] obtained the orientation distribution function from the diffusion equation by an orthogonal expansion in higher harmonics. This is not valid when the orientation of the particles evolves chaotically. Moreover, the diffusion equation is linear in the orientation distribution function. Therefore, the linearity of the diffusion equation precludes the possibility of chaotic solutions when it is solved with a continuous distribution as the initial condition. The method of Kumar and Ramamohan [16] involves obtaining the orientation distribution function by integrating the governing differential equations [17] of a set of particles distributed uniformly in phase space initially, and by taking the probability of a particle oriented in a particular direction at a given time as the fraction of the set of particles oriented in that direction at that instant. Since this numerical integration is done with single precision, the round-off errors in the computation which contribute a white noise term to the equations can be considered weak Brownian motion. This would be equivalent to solving the governing equations using the Langevin approach. This method is extended to delta (aligned) initial distributions in Kumar *et al.* [19]. In a delta initial distribution, all particles are aligned in a particular direction initially.

Recently, control of chaos has been pursued actively because of its potential importance in practical applications. The flexibility of forcing the system variables to oscillate in a pre-targetted orbit to achieve a desired purpose is worth noting. An algorithm for the control of chaos for chaotic rheological parameters has been proposed in Kumar *et al.* [25]. This algorithm is comparatively easy to implement in experimental set-ups and needs almost no information about the system.

Theory

Dynamics

The fundamental equations of the dynamics of slender rods in simple shear flow have been developed by Ramamohan *et al.* (1994) following the analysis of Berry and Russel [26]. The undisturbed velocity profile is chosen as

$$\mathbf{v}_\circ = \dot{\gamma} y \, \mathbf{i} \,, \tag{1}$$

where $\dot{\gamma}$ is the shear rate, y is the y co-ordinate and $\mathbf{i}$ is the unit vector in the direction of the x-axis. Typical second-phase fibres in composites and polymeric solutions can be modelled as slender rigid rods to an excellent approximation since their aspect ratio is generally greater than 50. The radii of the rods are assumed to be a and their length $2l$. All the quantities that appear in the analysis will be scaled after Berry and Russel [26]:

$$\begin{aligned} \text{length} &: \; l \,, \\ \text{velocity} &: \; \frac{l\dot{\gamma}}{\sqrt{2}} \,, \\ \text{force} &: \; \frac{8}{3\sqrt{2}} \pi \eta_s l^2 \dot{\gamma} \frac{1}{\ln(2r)} \,, \\ \text{torque} &: \; \frac{8}{3\sqrt{2}} \pi \eta_s l^3 \dot{\gamma} \frac{1}{\ln(2r)} \,, \\ \text{time} &: \; \frac{\sqrt{2}}{\dot{\gamma}} \,, \end{aligned} \tag{2}$$

where η_s is the viscosity of the liquid and r is the aspect ratio of the particle.

The rate of rotation of the fibre is given as

$$\dot{\mathbf{u}} = \frac{3}{2}\int_{-1}^{1} s\big[\mathbf{v}_\circ - \big(\mathbf{v}_\circ \cdot \mathbf{u}\big)\mathbf{u}\big]\,\mathrm{d}s + \mathbf{T} \times \mathbf{u} \tag{3}$$

by Berry and Russel [26]. Here s is the dimensionless position along the length of the fibre and $\mathbf{u}$ is the unit vector describing the orientation of the fibre. In Eq. (3), $\mathbf{T}$ is the net torque on the slender rod and is given by

$$\mathbf{T} = \int_{-1}^{1} s\big[\mathbf{u} \times \mathbf{f}(s)\big]\,\mathrm{d}s\,. \tag{4}$$

If, for example, we consider a situation where an electric or magnetic field is imposed on the rod, then $\mathbf{f}(s)$ will be a function of s alone. Hence

$$\int_{-1}^{1} s\mathbf{f}(s)\,\mathrm{d}s \tag{5}$$

can be assumed to be a vector of constant magnitude and direction. In such cases, the periodic torque induced on the fibre is given by $\mathbf{T} = \mathbf{u} \times \mathbf{k}\cos(wt)$, where

$$\mathbf{k} = \int_{-1}^{1} s\mathbf{f}(s)\,\mathrm{d}s\,.$$

$\mathbf{k}$ can be considered the orientation-independent part of the torque or its magnitude. It can also be considered a vector representing the interaction between the external field and a dipole either induced in the body or already present in it as, for example, a single domain magnetic particle. ω is the frequency of the periodic driver. Let k_1, k_2 and k_3 be the x, y and z components of $\mathbf{k}$.

In a spherical co-ordinate system, the evolution equations in Eq. (3) can be written as

$$\begin{aligned}
\dot{\theta} &= \sqrt{2}\sin\theta\,\cos\theta\,\sin\phi\,\cos\phi \\
&\quad + [k_1\cos\theta\,\cos\phi + k_2\cos\theta\sin\phi - k_3\sin\theta]\cos(wt)\,, \\
\dot{\phi} &= -\sqrt{2}\sin^2\phi + \left[-k_1\frac{\sin\phi}{\sin\theta} + k_2\frac{\cos\phi}{\sin\theta}\right] cos(wt)\,,
\end{aligned} \tag{6}$$

with $|\theta|$, $|\phi| \le \pi/2$. The singularity for $\dot{\phi}$ can be removed by taking $\phi_1 = \phi\sin\theta$. For a more detailed discussion of the method of solution of Eq. (6), the reader is referred to [17].

Rheology

The scaled normal stress differences and apparent viscosities for dilute suspensions of slender rods in simple shear flow have been given by Kumar and Ramamohan [16] based on the theory of Strand and Kim [24].

The orientation distribution function, $\psi(\mathbf{u},t)\,d\underset{\sim}{u}$, is defined as the probability that one particle is oriented in the solid angle, $\mathbf{u}+d\mathbf{u}$, about $\mathbf{u}$ at time t. The orientation average of a quantity B (say) is given by

$$\langle B\rangle = \int_0^{2\pi}\int_0^{\pi} B(\theta,\phi)\,\psi(\theta,\phi,t)\,\sin\theta\,d\theta\,d\phi\,, \tag{7}$$

where the orientation vector, $\mathbf{u}$, has been parametrised in spherical coordinates. The total stress tensor given by Strand and Kim [24] is

$$\begin{aligned}\boldsymbol{\sigma} = {}& -p\boldsymbol{\delta} + 2\eta_s\mathbf{E} + 2\eta_s\Phi \\ & \left\{2A_H\mathbf{E}\colon\langle\mathbf{uuuu}\rangle + 2B_H\left(\mathbf{E}\cdot\langle\mathbf{uu}\rangle + \langle\mathbf{uu}\rangle\cdot\mathbf{E}\frac{2}{3}\boldsymbol{\delta}\mathbf{E}\colon\langle\mathbf{uu}\rangle\right)\right. \\ & + C_H\mathbf{E} + F_H D_r\left(\langle\mathbf{uu}\rangle - \frac{1}{3}\boldsymbol{\delta}\right) + 3D_0\frac{m}{kT} \\ & \left.\times\left[\frac{(1-C)}{2}\langle\mathbf{uH}_\perp\rangle - \frac{(1+C)}{2}\langle\mathbf{H}_\perp\mathbf{u}\rangle\right]\right\},\end{aligned} \tag{8}$$

where p is the pressure, $\boldsymbol{\delta}$ the identity tensor, η_s the viscosity of the solvent, $\mathbf{E}$ the rate of strain tensor for the imposed flow field, Φ the volume concentration of particles, D_r the rotary diffusivity, D_0 the rotary diffusivity of a sphere of volume equal to that of the particle, $\mathbf{u}$ the unit orientation vector, $\mathbf{m}$ the magnetic or electric dipole moment of the particle, and C the shape factor, $(r^2-1)/(r^2+1)$, given as a function of aspect ratio, r. The other coefficients, A_H, B_H, C_H and F_H, are functions of the shape of the particle. $\mathbf{H}_\perp \equiv \mathbf{H}\cdot(\boldsymbol{\delta}-\mathbf{uu})$, where $\mathbf{H}$ is the external force field. The angle brackets indicate average over orientation.

The magnetic or electric dipole $\underset{\sim}{m}$ is assumed to be oriented parallel to the particle axis, $\mathbf{m}=m\mathbf{u}$. Now $m\mathbf{H}=\mathbf{k}$. The periodic torque induced on the fibre is given by

$$\mathbf{T} = \mathbf{m}\times\mathbf{H}\cos(\omega t) = m\mathbf{u}\times\mathbf{H}\cos(\omega t) = \mathbf{u}\times m\mathbf{H}\cos(\omega t)\,.$$

The scaled first apparent viscosity can be written as follows:

$$
\begin{aligned}
[\eta_1] &= \lim_{\substack{\Phi\to 0\\ r\to\infty}} \left(\frac{(\sigma_{xy}-\eta_s\dot{\gamma})100B_H}{\Phi\eta_s\dot{\gamma}}\right)\\
&= 75\langle\sin^4\theta\,\sin^2 2\phi\rangle\\
&\quad -300\sqrt{2}k_1\cos(\omega t)(\langle\sin\theta\,\sin\phi\rangle-\langle\sin^3\theta\,\sin\phi\rangle+\langle\sin^3\theta\,\sin^3\phi\rangle)\\
&\quad -300\sqrt{2}k_2\cos(\omega t)(\langle\sin^3\theta\,\cos^3\phi\rangle-\langle\sin^3\theta\,\cos\phi\rangle)\\
&\quad -300\sqrt{2}k_3\cos(\omega t)\left(\frac{1}{2}\langle\cos^3\theta\,\sin 2\phi\rangle-\frac{1}{2}\langle\cos\theta\,\sin 2\phi\rangle\right). \qquad (9)
\end{aligned}
$$

The other apparent viscosity,

$$
[\eta_2] = \lim_{\substack{\Phi\to 0\\ r\to\infty}} \left(\frac{(\sigma_{yx}-\eta_s\dot{\gamma})100B_H}{\Phi\eta_s\dot{\gamma}}\right), \qquad (10)
$$

and the normal stress differences,

$$
[\tau_1] = \lim_{\substack{\Phi\to 0\\ r\to\infty}} \left(\frac{(\sigma_{xx}-\sigma_{zz})100B_H}{\Phi\eta_s\dot{\gamma}}\right), \quad [\tau_2] = \lim_{\substack{\Phi\to 0\\ r\to\infty}} \left(\frac{(\sigma_{yy}-\sigma_{zz})100B_H}{\Phi\eta_s\dot{\gamma}}\right), \qquad (11)
$$

can be written similarly. We have scaled all the coefficients, A_H, B_H and C_H, with respect to $100B_H$ since A_H tends asymptotically to infinity and B_H tends to zero for slender rods of aspect ratio with r tending to infinity. However, A_HB_H tends to $3/4$ as r tends to infinity.

We have included a factor of 100 in the scaling since without this factor our scaled values of the various rheological parameters were of the order 10^{-3} in the chaotic regime and 10^{-7} in the non-chaotic regime. However, the range of the chaotic oscillations is well within the experimentally accessible range when these scaled values are converted to dimensional values. This is because the scaled values of the rheological parameters of the suspension in the absence of external forces is of the order 10^{-8}, and these values are easily measured experimentally.

A comparison of the magnitude of the external force used in our analysis with that of Strand's [18] is given in [16].

The method followed by Kumar and Ramamohan [16] for obtaining the various orientation averages is based on a brute force evaluation of the orientation distribution function from the evolution equations for an initially uniformly delta-distributed ensemble of orientation vectors. The ranges of θ and ϕ are converted into $[-\pi/2, \pi/2]$ by a linear transformation. The phase space is divided into 81 bins of equal size by dividing the ranges of both θ and ϕ into nine equal sub-intervals. The calculations were started off from a uniform initial distribution by taking one particle in each bin. The evolution of each of these particles is computed by a standard Runge-Kutta method with adaptive stepsize [27] in single precision. A discrete approximation to the orientation distribution function is calculated by taking the fraction of particles in each bin. We note that the round-off error in the integrator (in single precision) contributes a white noise term which can be considered weak or negligible Brownian motion.

The above method has been extended to delta (aligned) initial distributions [19]. A delta (aligned) initial distribution can be realised by imposing a strong external force at the start of the computation. An initial delta distribution remains one throughout in the presence of weak or negligible Brownian motion. In a delta distribution, we consider all the particles to be aligned in a particular direction initially and, since each particle evolves according to the evolution equations, their orientations are the same at any time. This implies that we need only consider a representative particle for the computation of the corresponding rheological parameters. The orientation of a single particle has been shown to behave chaotically. Hence the rheological parameters computed with this representative particle as a basis will also evolve chaotically. We note that the particle evolution trajectory forms a characteristic curve of the corresponding diffusion equation. This is especially clear in the case of the delta distribution.

Control of Chaos Algorithm

Control of chaos is implemented by applying a constant force at periodic intervals. A constant external force, other than the periodic force, is applied for the duration of one period, $(2\pi/\tilde{\omega})$, after every $(n-1)$ periods. In order to accomplish this, we modify the dynamical system as follows:

$$\begin{aligned}\dot{\theta} &= \sqrt{2}\sin\theta\,\cos\theta\sin\phi\,\cos\phi + [(k_1\cos(\tilde{\omega}t) + k_1')\cos\theta\,\cos\phi \\ &\quad + (k_2\cos(\tilde{\omega}t) + k_2')\cos\theta\sin\phi - (k_3\cos(\tilde{\omega}t) + k_3')\sin\theta]\,, \\ \dot{\phi} &= -\sqrt{2}\sin^2\phi + \left[-(k_1\cos(\tilde{\omega}t) + k_1')\frac{\sin\phi}{\sin\theta} + (k_2\cos(\tilde{\omega}t) + k_2')\frac{\cos\phi}{\sin\theta}\right]\,,\end{aligned} \quad (12)$$

where k_1', k_2' and k_3' are the x, y and z components of the scaled constant force. Accordingly, we modify the rheological parameters:

$$\begin{aligned}
[\eta_1] &= \lim_{\substack{\Phi\to 0\\ r\to\infty}} \left(\frac{(\sigma_{xy}-\eta_s\dot{\gamma})100B_H}{\Phi\eta_s\dot{\gamma}}\right)\\
&= 75\langle\sin^4\theta\,\sin^2 2\phi\rangle\\
&\quad - 300\sqrt{2}(k_1\cos(\tilde{\omega}t)+k_1')(\langle\sin\theta\,\sin\phi\rangle - \langle\sin^3\theta\,\sin\phi\rangle + \langle\sin^3\theta\,\sin^3\phi\rangle)\\
&\quad - 300\sqrt{2}(k_2\cos(\tilde{\omega}t)+k_2')(\langle\sin^3\theta\,\cos^3\phi\rangle - \langle\sin^3\theta\,\cos\phi\rangle)\\
&\quad - 300\sqrt{2}(k_3\cos(\tilde{\omega}t)+k_3')\left(\frac{1}{2}\langle\cos^3\theta\,\sin 2\phi\rangle - \frac{1}{2}\langle\cos\theta\,\sin 2\phi\rangle\right). \qquad (13)
\end{aligned}$$

Similarly, the other rheological parameters, $[\eta_2]$, $[\tau_1]$ and $[\tau_2]$, can be modified.

It may be noted that when $k_1' = k_2' = k_3' = 0$, the system evolves without any modifications.

A constant external force is applied throughout the ith period, where i is a multiple of n. That is, we set

$$k_2' = \begin{cases} k_2^0, & \text{if } \; i \equiv 0 \pmod{n} \\ 0, & \text{if } \; i \not\equiv 0 \pmod{n} \end{cases} \qquad (14)$$

for every period i, where k_2^0 is a non-zero quantity. This means that the system evolves without any modifications for periods which are not multiples of n. The choice of n depends on the period-m solution we want to stabilise. The integer n can be either m or a divisor of m, depending on the magnitude of k_2^0. In our calculations, we kept $k_1' = k_3' = 0$ throughout but, in general, k_1' and k_3' can also be set to non-zero values in the same manner as k_2'.

Results and Discussion

A novel technique to calculate the orientation distribution function was proposed in [16]. This technique focuses on the trajectories followed by a set of particles initially oriented along particular directions rather than on the evolution of an initial continuous distribution. We feel that this representation of the initial orientation distribution function may be more realistic in suspensions of macroscopic particles. The ranges of both the polar angle, ϕ, and the azimuthal angle, θ, are changed from $-\pi/2$ to $\pi/2$ by a linear transformation. The ranges are divided into nine equal sub-intervals which give rise to 81 bins.

In the case of a uniform initial distribution, we consider one particle in each of the 81 bins. The evolutions of these 81 particles are computed using Eq. (6) with a standard Runge-Kutta method with adaptive stepsize [27]. A discrete approximation to the orientation distribution function at any instant is thus obtained by calculating the fraction of the total number of particles in each bin at that instant. The number of bins is fixed at 81 since further sub-divisions lead to identical Poincaré sections and Lyapunov exponents.

The technique has been extended to initial delta (aligned) distributions in [19]. In an aligned initial distribution, all the particles are oriented in a particular direction initially. The orientations of these particles then evolve according to the evolution equations in Eq. (6). Since all the particles are assumed to start off with the same initial orientation, the averages required in the computations of the rheological parameters can be calculated on the basis of one particle initially oriented along the same direction. It may be noted that a delta distribution remains one throughout in the presence of weak or negligible Brownian motion.

It has been shown that the rheological parameters evolve chaotically when the orientation dynamics of the particles evolve chaotically. In the work

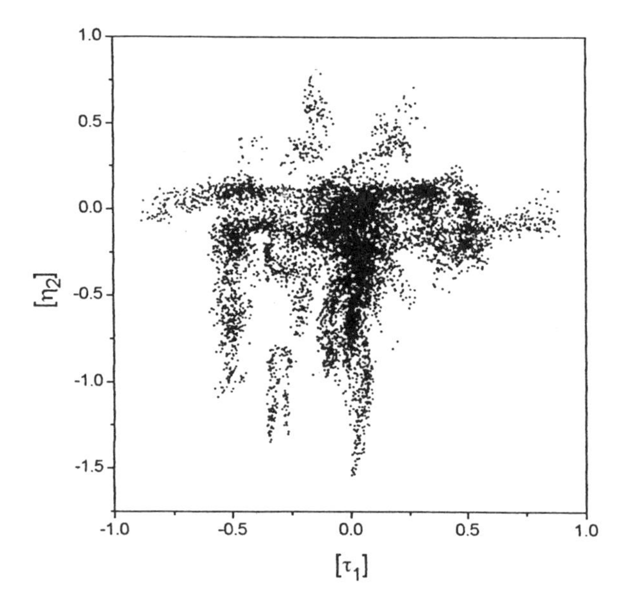

Fig. 1. Trajectory of $[\tau_1] \times [\eta_2]$ for $k_1 = k_3 = 0$, $k_2 = 0.20$ and $\tilde{\omega} = 1$.

reported earlier [19], we kept $k_1 = k_3 = 0$, $\tilde{\omega} = 1$ throughout and varied k_2 from zero to one in steps of 0.001 and computed the time series of $[\tau_1]$, $[\tau_2]$, $[\eta_1]$ and $[\eta_2]$. The maximum Lyapunov exponent of these time series was calculated using the algorithm of Kantz [28]. A trajectory plot of $[\tau_1] \times [\eta_2]$ for $k_1 = k_3 = 0$, $k_2 = 0.20$ and $\tilde{\omega} = 1$, which is chaotic, is given in Fig. 1.

For the uniform initial distribution, there appears to be three different chaotic regimes between $k_2 = 0.0$ and $k_2 = 1.0$. Chaos usually appears as an increase in the range of the rheological parameters. The range of the rheological parameters increase about 10,000 times in the chaotic regime compared to the non-chaotic regime for the uniform initial distribution. In the case of an aligned initial distribution, we found that the range of the rheological parameters increases by one more order of magnitude for the chaotic case.

For the case of an aligned initial distribution, there appear to co-exist three different attractors for the dynamics in the parameter range $0.005 \leq k_2 \leq 0.25$ and, correspondingly, three different types of behaviour for the rheological parameters. The three different attractors lead to three qualitatively different behaviours of the rheology. The basins of attraction of the three different attractors are intermingled so that the rheological parameters can be changed dramatically by a small change in the initial conditions. There also appears to be a boundary crisis around $k_2 = 0.24$, where the attractor collides with the basin of attraction of a stable torus and chaotic transients are found near this regime. For values between $k_2 = 0.37$ and $k_2 = 0.48$, we again encounter transient chaos and the co-existence of multiple attractors with what appears to be intermingled basins of attraction for the dynamics. The dynamics show very interesting behaviour in this regime. However, the rheological behaviour is qualitatively the same even though the dynamics exhibit a wide range of behaviour.

The applicability of the control of chaos algorithm discussed in the previous section has been demonstrated in [25]. The possibility of radically new behaviour of rheological parameters has also been demonstrated. In the computations, k_1' and k_3' were kept at zero throughout. Nevertheless, in general, they can also be assigned non-zero values.

Stabilising higher periods is also possible by using the control of chaos algorithm reported earlier [25]. In this system, once control is applied, the system stabilises rapidly even when it is started off from a uniform initial distribution. It is also possible to switch from one periodic solution to another. Furthermore, the period discussed can be pre-targetted to a large extent.

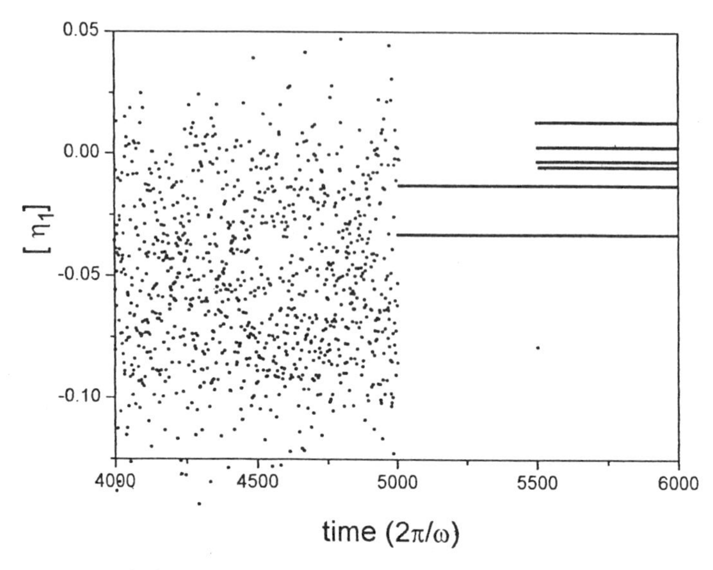

Fig. 2. Time series of $[\eta_1]$ controlled to a period-2 behaviour and then controlled to a period-6 behaviour for $k_1 = k_3 = 0$, $k_2 = 0.20$ and $\tilde{\omega} = 1$ using $k_2' = 0.085$ for both period-2 and period-6.

For $k_1 = k_3 = 0$, $k_2 = 0.20$, $\tilde{\omega} = 1$, the system behaves chaotically when started off from a uniform initial distribution. The time series of $[\eta_1]$, which is controlled to a period-2 behaviour and then to a period-6 behaviour, is given in Fig. 2. In this case, we set $k_1 = k_3 = 0$, $k_1' = k_3' = 0$, $\tilde{\omega} = 1$. A control force, $k_2' = 0.085$, is applied between the 5000th period and the 5500th period with $n = 2$. That is, a constant force, $k_2' = 0.085$, is applied throughout the ith period with $i \equiv 0(\mathrm{mod}2)$, or otherwise $k_2' = 0$, and the system then stabilises to period-2 behaviour. Similarly, we apply a constant force, $k_2' = 0.085$ with $n = 6$ between the 5500th period and the 6000th period, and the system stabilises to period-6 behaviour.

One of the interesting results of chaos control is the possibility of obtaining very complex, non-sinusoidal behaviour of the stabilised periodic solution. Figure 3 shows the time series of $[\eta_1]$ which is controlled to a period-9 behaviour. Another consequence of this control algorithm is the possibility of obtaining both positive and negative first and second normal stress differences. It has been reported that the first normal stress difference is practically always negative and numerically much larger than the second normal stress

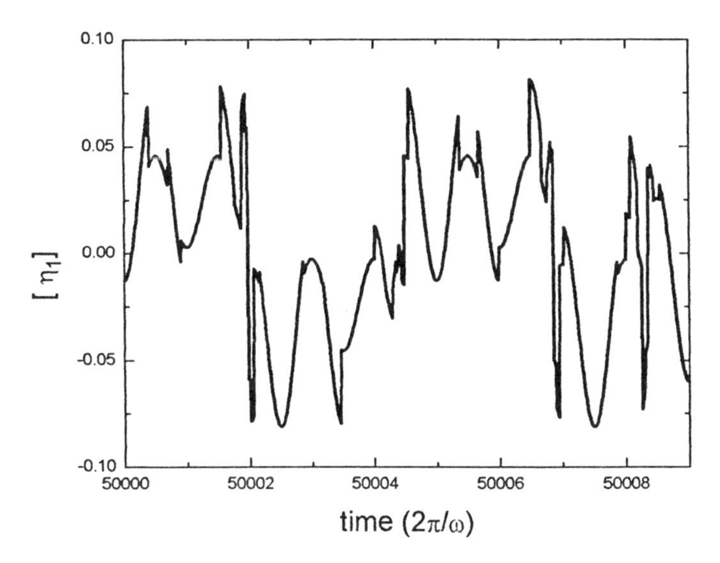

Fig. 3. Plot of time series of $[\eta_1]$ controlled to period-9 solution for $k_1 = k_3 = 0$, $k_2 = 0.20$, $k_2' = 0.1$ and $\tilde{\omega} = 1$. Note that this solution exhibits considerable complexity.

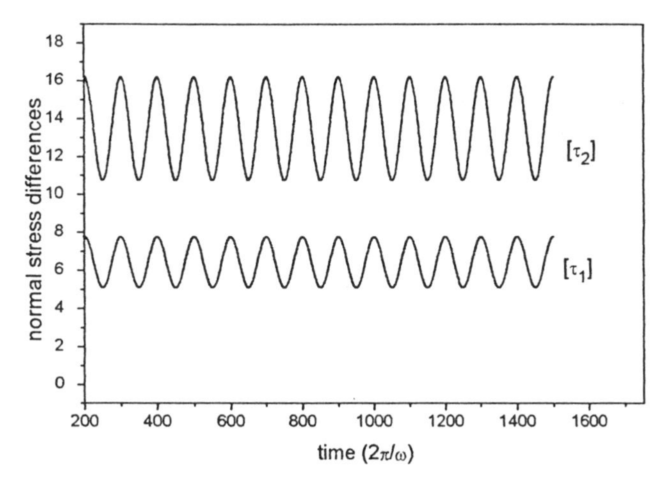

Fig. 4. Plot of controlled values of $[\tau_1]$ and $[\tau_2]$ versus time for $k_1 = k - 3 = 0$, $k_2 = 0.20$, $\tilde{\omega} = 1$ and $k_2' = 1.0$ applied at every period. Period stabilised is 1. Note that the values of $[\tau_1]$ vary over a positive range and that the values of $[\tau_2]$ are always greater in magnitude than the values of $[\tau_1]$.

difference for polymeric fluids [31]. Usually, the ranges of the first and second normal stress differences vary from negative to positive values. Therefore, it is theoretically possible to stabilise any one of these values by control of chaos. Figure 4 shows a stabilised period-1 orbit in which the first normal stress difference is positive and numerically less than the second normal stress difference. This observation may lead to fundamentally new behaviour in many standard rheological experiments.

The dynamics and rheology of periodically forced suspensions of slender rods in simple shear flow have been discussed. It has been demonstrated that chaotic solutions occur in certain parametric regimes. We have demonstrated that chaotic solutions lead to chaotic rheological parameters. Control of chaos is demonstrated to lead to novel rheological behaviour which is otherwise unobtainable. Furthermore, the system considered in this paper can be a truly mesoscopic system in which the chaotic fluctuations occur on a macroscopic scale. In the future, the effects of periodic forcing on suspensions of non-axisymmetric particles, including the effect of hydrodynamic interactions, will be considered. The results for these cases should be interesting since Leal and Hinch [29] have shown that deviation from axisymmetry can have major effects on suspension dynamics even in regular solution regime, and Pérez *et al.* [30] have shown that systems of weakly interacting chaotic oscillations can exhibit a wide range of behaviour.

Acknowledgements

The authors wish to thank Dr. A. D. Damodaran, Director of Regional Research Laboratory (CSIR), Thiruvananthapuram and Dr. B. C. Pai, Divisional Head for their constant encouragement. One of the authors (K. S. Kumar) wishes to acknowledge CSIR, India for financial support. Financial support from the Department of Science and Technology, Government of India, Grant No. SP/S2/R-04/93 is also gratefully acknowledged.

Discussion

S. Ramaswamy What would happen if we put in some noise by hand?

T. R. Ramamohan There is already some noise due to round-off errors.

E. J. Hinch When we focus on the rheological parameters which are averages of many particles, the sensitivity to initial conditions may not be there.

T. R. Ramamohan I think it is open to question. Experimentally periodic viscosities have been obtained for constant force fields, and hence it is likely that chaotic viscosities are possible. (Added after the discussion: In that sense, chaos is not possible in any system whatsoever because any observable quantity is an average over many molecules.)

M. E. Cates Perhaps the right experiment would be one that picks up the correlations of the orientations of individual particles like light scattering.

T. R. Ramamohan Possibly.

References

1. Happel, J. & Brenner, H. (1986) *Low Reynolds Number Hydrodynamics with Special Applications to Particulate Media.* Dordrecht: Martinus Nijhoff Publishers
2. Bretherton, F. P. (1962) The motion of rigid particles in a shear flow at low Reynolds number. *J. Fluid Mech.* **14**, 284–304
3. Anczurowski, E. & Mason, S. G. (1967) The kinetics of flowing dispersions. II. Equilibrium orientations of rods and discs. *J. Coll. Int. Sci.* **23**, 522–532
4. Leal, L. G. & Hinch, E. J. (1971) The effect of weak Brownian rotations on particles in shear flow. *J. Fluid Mech.* **46**, 685–703
5. Altan, M. C., Advani, S. G., Guceri, S. I. & Pipes, R. B. (1989) On the description of the orientation state for fibre suspensions in homogeneous flows. *J. Rheol.* **33**, 1129–1155
6. Macmillan, E. H. (1989) Slow flows of anisotropic fluids. *J. Rheol.* **33**, 1071–1105
7. Szeri, A. J., Milliken, W. J. & Leal, L. G. (1992) Rigid particles suspended in time-dependent flows: irregular versus regular motion, disorder versus order. *J. Fluid Mech.* **237**, 33–56
8. Buevich, Yu. A., Syutkin, S. V. & Tetyukhin, V. V. (1984) Theory of a developed magnetofluidised bed. *Magnetohydrodynamics* **20**, 333–339
9. Ignatenko, N. M., Melik-Gaikazyan, I. Ya., Polunin, V. M. & Tsebers, A. O. (1984) Excitation of ultrasonic vibrations in a suspension of uniaxial ferromagnetic magnetohydrodynamics. *Magnetohydrodynamics* **20**, 237–240
10. Tsebers, A. O. (1986) Numerical modelling of the dynamics of a drop of magnetisable liquid in constant and rotating magnetic fields. *Magnetohydrodynamics* **22**, 345–351
11. Cebers, A. (1993) Chaos in polarisation relaxation of a low-conducting suspension of anisotropic particles. *Magnetism and Magnetic Materials* **122**, 277–280
12. Cebers, A. (1993) Chaos: new trend of magnetic fluid research. *Magnetism and Magnetic Materials* **122**, 281–285
13. Kashevskii, BÉ. (1986) Torque and rotational hysteresis in a suspension of single-domain ferromagnetic particles. *Magnetohydrodynamics* **22**, 161–168
14. Petrikevich, A. V. & Raikher, Yu. L. (1984) Rheological characteristics of magnetic suspension in alternating magnetic field. *Magnetohydrodynamics* **20**, 122–127

15. Shul'man, Z. P., *et al.* (1986) Motion of an axisymmetric magnetically soft particle in hydrodynamic flow under the action of a strong rotating magnetic field. *Magnetohydrodynamics* **22**, 288–293
16. Kumar, K. S. & Ramamohan, T. R. (1995) Chaotic rheological parameters of periodically forced suspensions of slender rods in simple shear flow. *J. Rheol.* **39**, 1229–1241
17. Ramamohan, T. R., Savithri, S., Screenivasan, R. & Bhat, C. C. S. (1994) Chaotic dynamics of a periodically forced slender body in a simple shear flow. *Phys. Lett.* **A190**, 273–278
18. Strand, S. R. (1989) *Dynamic Rheological and Rheo-Optical Properties of Dilute Suspensions of Dipolar Brownian Particles.* PhD dissertation. Madison: University of Wisconsin
19. Kumar, K. S., Savithri, S. & Ramamohan, T. R. (1996) Chaotic dynamics and simple shear flow. *Jap. J. Appl. Phys.* **35**, 5901–5908
20. Brenner, H. (1970) Rheology of a dilute suspension of dipolar spherical particles in an external field. *J. Coll. Int. Sci.* **32**, 141–158
21. Brenner, H. (1974) Rheology of a dilute suspension of axisymmetric Brownian particles. *Int. J. Multiphase Flow* **1**, 195–341
22. Chaffey, C. E. & Mason, S. G. (1968) Particle behaviour in shear and electric fields V. Effect on suspension viscosity. *J. Coll. Int. Sci.* **27**, 115–126
23. Chaffey, C. E., *et al.* (1965) Particle motions in sheared suspensions, XVI. Orientations of rods and disks in hyperbolic and other flows. *Can. J. Phys.* **43**, 1269–1287
24. Strand, S. R. & Kim, S. (1992) Dynamics and rheology of a dilute suspension of dipolar non-spherical particles in an external field. Part 1. Steady-shear flows. *Rheol. Acta* **31**, 94–117
25. Kumar, K. S., *et al.* (1996) Novel rheological behaviour of suspensions of periodically forced slender rods through chaos control in a preliminary model. *J. Rheol.*, personal communication
26. Berry, D. H. & Russel, W. B. (1987) The rheology of dilute suspensions of slender rods in weak flows. *J. Fluid Mech.* **180**, 475–494
27. Press, W. H., Flannery, B. P., Teukolsky, S. & Vetterling, W. T. (1986) *Numerical Recipes: The Art of Scientific Computing.* Cambridge: Cambridge University Press
28. Kantz, H. (1994) A robust method to estimate the maximal Lyapunov exponent of a time series. *Phys. Lett.* **A185**, 77–87
29. Hinch, E. J. & Leal, L. G. (1979) Rotation of small non-axisymmetric particles in a simple shear flow. *J. Fluid Mech.* **92**, 591–608
30. Pérez, G., *et al.* (1995) In *Chaos in Mesoscopic Systems*, ed. G. Casati & H. A. Cerdeira, pp. 105–122. Singapore: World Scientific
31. Bird, R. B., Armstrong, R. C. & Hassager, O. (1987) *Dynamics of Polymeric Liquids, Vol. 1.* New York: John Wiley & Sons

Dynamics of Complex Fluids, pp. 301–314
ed. M. J. Adams, R. A. Mashelkar, J. R. A. Pearson & A. R. Rennie
Imperial College Press–The Royal Society, 1998

Chapter 21

Shear Thickening in Model Concentrated Colloids: The Importance of Particle Surfaces and Order Change Through Thickening

J. R. MELROSE[1*], J. H. VAN VLIET[1] AND R. C. BALL[2]
(1) Polymers and Colloids and (2) Theory of Condensed Matter Groups, Cavendish Laboratory, Madingley Road, Cambridge CB3 OH3, UK
**E-mail: jrm23@phy.cam.ac.uk*

Shear thickening effects in particulate dispersion colloids are not yet understood. We argue that the origin of these effects lies in the divergent viscous interactions of close approaching particle surfaces in a hydrodynamic medium. How close particles get in flow under an applied bulk stress is controlled by the conservative and Brownian forces. We report simulations of a reduced model in which the hydrodynamics are restricted to a resistance matrix with just pair terms between nearest neighbours that include the divergence found via lubrication theory in narrow gaps. We report the interplay between thickening and ordering under flow and show how surface effects, such as those of polymer-coated particles, can affect the bulk rheology in thickening.

Introduction

Dense colloidal suspensions at sufficiently high shear rates exhibit thickening effects: a rise of viscosity with increasing shear rate. For volume fractions between 0.45 and 0.5, a gradual rise is often reported. However, at higher volume fractions, a sudden thickening can occur with a large jump in viscosity and erratic flow behaviour after thickening. Thickening occurs in many systems [1], including monodisperse colloids [2, 3] of spherical particles.

The micro-structural origins of shear thickening have not been resolved. Some works have seen it as a general feature of non-equilibrium phase diagrams [4] while others have viewed it as a mechanical instability of layered flow [5]. Some theories [2, 5] emphasise the role of conservative repulsive forces. Yet others have speculated on the role of Brownian forces [3]. Thickening has been observed in molecular dynamics simulations. Nevertheless, the key finding for colloids [6] was the observation of a gradual thickening effect in simulations with hydrodynamic interactions. Here we study thickening in a simple model with hydrodynamic, Brownian and conservative interactions. In particular, we will introduce a simple model for polymer-coated particles and show how this can give quantitative predictions that compare well with experiment.

The Simulation Model

Modelling the flow of particles concentrated in a hydrodynamic medium is a significant challenge. Schemes capable of high accuracy, either through higher order expansion [7, 8] or explicit numerical solution of discretised Stokes equations [9, 10], are still not sufficiently numerically efficient for a study of concentrated systems in shear flow. Either the periodic moving boundary conditions cannot be implemented [10], or the need to handle very narrow gaps and high moment interactions between close surfaces demands too high a refinement of the discretisation [9] or too high a truncation of a moment expansion [7, 8]. Less accurate methods have been proposed [6, 11]. The most studied algorithm [6] involves $O(N^3)$ inversions of matrices which limit studies to small systems ($N < 100$, where N is the number of particles) in two-dimensions. However, the results for Brownian spheres in three-dimensions have been reported recently. Dissipative particle dynamic [11] methods have problems of compressibility which are serious for narrow gaps.

We argue that to motivate progress on the understanding of colloid rheology, it is too limiting to demand high accuracy from the hydrodynamic model, but that it is important to go beyond free-draining or so-called mean field models [12] which grossly ignore the physical symmetries of hydrodynamic interactions altogether. It is pertinent to study reduced hydrodynamic models relevant to particular limits. In the limit of high shear rates and concentrated systems, the divergent hydrodynamic interactions due to the relative motion of the surfaces of close particles must dominate the resistance to particle motion. Indeed, Brady and co-workers [6] have already found that hydrodynamic clustering is determined by pair lubrication. Our algorithm identifies the nearest neighbours and forms a resistance tensor pair-wise out of terms including the divergence found in the lubrication approximation. Note that the truncation of this resistance tensor is, at worst, a truncation of $(1/r^3)$ terms: each row in the resistance matrix relates to the problem of one sphere moving in a fixed array, hence resulting in a velocity field which decays as $1/r^3$ [13]. It should be noted, however, that a short-range truncation of the resistance tensor does, on inversion, generate a many-body mobility tensor which has a long-range $1/r$ dependence. However, it is likely to be an inaccurate approximation to the full hydrodynamic matrices since it ignores local multi-body effects in which the flows within the narrow gaps are coupled to the flows in the local pore space around the particle. We suspect that the errors are of the order of terms in the log of the inverse gap, which suggest errors at the 50–20% level in hydrodynamic viscosity [14]. Numerically, the algorithm requires O(N) inversion of a sparse matrix. Others have made a similar approximation [15] but the scope of our results is unprecedented. Furthermore, and contrary to others [15], we set up the matrix $\mathbf{R}$ solely out of Galilean invariant terms (those that depend just on relative velocities of particles). In the absence of an explicit fluid velocity field in the simulation, we argue that it is physically incorrect to break Galilean invariance in the particle velocities by coupling them to a background flow through a set of drag forces [15, 12]. We drive shear flow through the boundary conditions and refer to this model as a *frame invariant pair drag* model.

Hydrodynamic Model

Between close surfaces, pair hydrodynamic interactions can be divided into squeeze modes, the leading singular interaction along the line of centres, and modes arising from the transverse motion of neighbours. The code includes terms for all these modes. The squeeze mode is the dominant term. It diverges as the inverse of the inter-particle gap. For a given applied force, the time scale for particle relaxation will diverge as the inverse of the inter-particle gaps. *It is this divergence of the inter-particle drag that plays a crucial role in thickening.* The hydrodynamic, colloidal and Brownian force and torque balance on each particle to give the equation of motion:

$$-\mathbf{R} \cdot \mathbf{V} + \mathbf{F}_C + \mathbf{F}_B = 0\,. \tag{1}$$

The hydrodynamic force and torques, $\mathbf{R} \cdot \mathbf{V}$, are related linearly to the 6N velocity/angular velocity vector, $\mathbf{V}$, by the resistance matrix, $\mathbf{R}$, formed from a sparse sum of pair terms between nearest neighbours. Flow is driven by Lee-Edwards' [16] periodic boundary conditions on the particle velocity field. At each time-step, Eq. (1) was solved for $\mathbf{V}$ by conjugate gradient techniques. Since fixed time-step computations can lead to overlap, we adopt a variable time-step such that the particles never overlap but can approach arbitrarily close. Given a set of velocities from solving Eq. (1), the time-step is chosen as the smaller of a default value such that the gap between the first pair of colliding particles is reduced to a pre-set factor of its current value.

The Behaviour of Hard Spheres and Just Hydrodynamics

The bare model of hard spheres with only hydrodynamic interactions is a logical place to begin the study. However, we have shown that this model is pathological under shear flow: it does not have a steady state solution [17]. (Note that the model has only one time scale, that of the imposed shear rate, and it would therefore either have a single steady state solution or none at all). In our view, this represents an example of the ultimate thickened state. Clusters of particles form under shear down the compression axis. The divergent viscous interaction determines a strong coupling between the closeness of particle surfaces in a given cluster and the time scale for relaxation of that cluster under an applied force. For bare hard spheres, however, there are no mechanisms for relaxation and inter-particle gaps collapse to ever smaller values even at low strains. Farr [17] has developed a kinetic clustering theory of this effect by defining jamming in shear flow as the volume fraction above which clusters form a gel before they tumble. Note that the hard sphere model in shear flow represents a limiting pathology: an entirely different role to that which it has at rest.

Models for Particle Surfaces

Brownian and conservative forces provide mechanisms which can control the close approach of particles. These introduce additional time scales into the problem and a variety of steady states under varying shear rate. At high enough applied stress, these states enter a thickening regime in which particles approach ever closer with increasing stress. We can associate these with a time scale and a length scale characteristic of the gaps between close approaching particles under shear at high rates. In this paper, we consider Brownian hard spheres and spheres with a crude model for polymer-coated surfaces and Brownian forces.

For Brownian simulations, we generate random force and torque, $\mathbf{F}_B$, on each particle through a sum of weighted pair terms on each bond such that they obey the fluctuation-dissipation theorem: $\langle \mathbf{F}_B \mathbf{F}_B \rangle = 2kT\mathbf{R}$. Terms in the gradient of the diffusion tensor enter the first integral of the Langevin equation and we include these by using a time symmetric order dt^2 predictor-corrector algorithm for the particle trajectory. The Peclet number, Pe, sets the dimensionless shear rate — the ratio of the Brownian relaxation time to that of the shear rate (see Eq. (2)). For concentrated systems under shear, we also associate the Brownian forces with a length scale, δ_B, characteristic of close approach, by equating the Brownian forces between a pair of particles to that due to the shear stress, σ. This gives $kT/\delta_B \approx \sigma d^2$, where d is the diameter of the particle and hence the reduced length, δ_B/d, scales as the inverse stress,

$$\delta_B/d \approx kT/(\sigma d^3) \approx 1/(\eta_r \mathrm{Pe}) , \tag{2}$$

where η_r is the viscosity of the suspension relative to the fluid.

The second model is that of a polymer-coated particle. This interaction has both a conservative and a dissipative part. On the one hand, the osmotic pressure of the coat's polymers leads to a repulsion between particles on mutual compression of their coats. On the other hand, the porous coat acts to enhance the dissipation of fluid forced to flow through the coat under relative motion of the particles. The latter is a complex problem depending on the solvent polymer interactions, the pore space and concentration profile of the polymer coat and its dynamics. A model for the squeeze mode has been developed [18]. This model assumes a step function concentration profile for a coat of semi-dilute polymers. It also assumes the relative motions are such that the coats remain in their regime of linear response. Figure 1 shows the dissipative interaction law. Note that it has a greatly enhanced dissipation over that of the Reynolds law for hard spheres. However, under large compression, the coat has a weaker divergence than the inverse gap divergence of the Reynolds law. We have adopted the simplest model which includes just the squeeze mode of the polymer coat interaction with the shear friction left as for hard spheres. The model has three free parameters: the coat thickness, the strength of the conservative coat interaction (a lumped parameter depending on surface

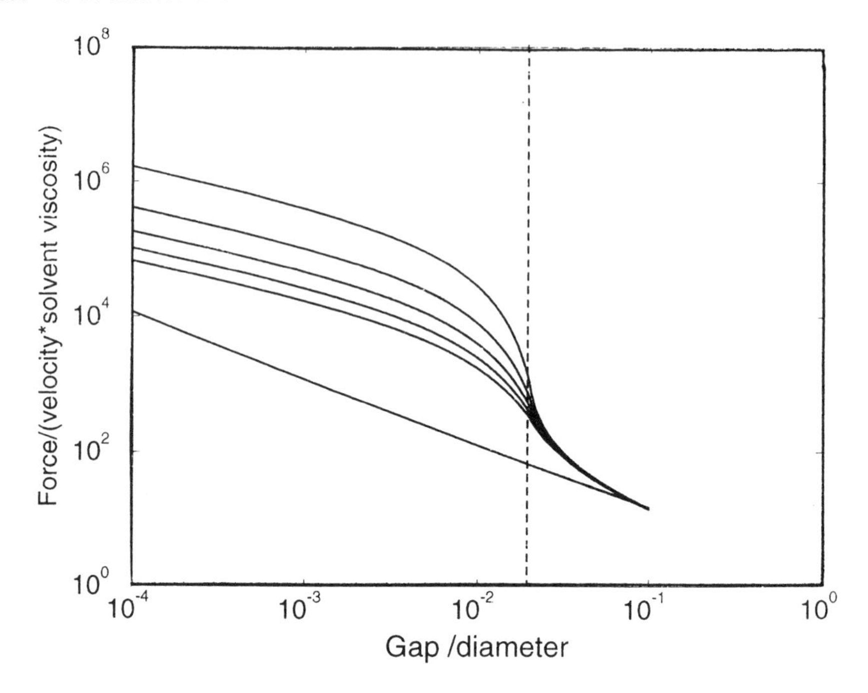

Fig. 1. The dissipative force law of [18] and the Reynolds lubrication law for relative motion normal to particle surfaces plotted against the gap between a pair of particles. The force is divided by the product of the relative velocity and the solvent viscosity. Curves top to bottom have coats 0.01 diameters thick with a porosity 0.001, 0.0008, 0.0006, 0.0004 and 0.0002 diameters. The thick line is the Reynolds law diverging as the inverse of the interparticle gap. The vertical line indicates the gap at contact of the coats.

coverage and the coat monomer size), and a nominal pore length scale which controls the strength of the dissipative interaction.

Results

We have presented the principal results elsewhere [19]. At high applied shear stresses, both the model Brownian spheres and polymer-coated spheres show shear thickening. The effect for Brownian spheres with hard sphere interactions is weaker than that of the polymer-coated spheres and experiments on polymer-coated particles [3]. We find, in general, that the inverse gap law of hard sphere Reynolds lubrication gives rise to just a logarithmic thickening. Physically reasonable parameters for the polymer coat model can give a fit to the experimental response. Figure 2 illustrates these results and includes some

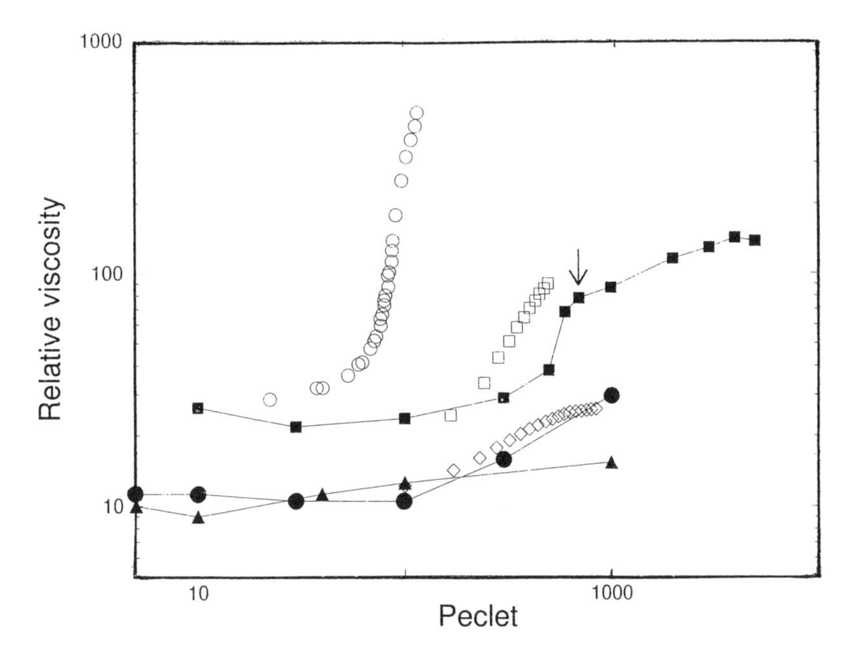

Fig. 2. Solid symbols are simulations of 50 spheres, hard spheres with Brownian forces alone at volume fraction 0.5 (triangles), and with coat forces of layer thickness 0.01, porosity 0.0005 and approximate coat strength $Q \sim 50kT/d$ at 0.5 volume fraction (circles) and 0.54 volume fraction (squares). Open symbols are experiments from [3] 690 nm polymethylmethacrylate particles with poly 12-hydroxy stearic acid polymer coats in didodecylphthalate at volume fractions 0.48, 0.51 and 0.55: diamonds, squares and circles, respectively.

new data at higher volume fractions. The shape of the thickening curve in the simulation at volume fraction 0.54 (solid squares), in particular the slower rate of thickening at the higher shear rates, correlates with the interaction law of Fig. 1 (see discussion below). We do find that at the highest rates, the thickening curve can flatten off into a plateau. While similar effects are seen in experiment [5], this prediction must be viewed with caution since in this regime, the model coats are highly compressed and the coat model may no longer be realistic. We will show elsewhere that the thickening curve is insensitive to system size. We have reproduced the same curve with N = 50, 200 and 700 simulations. Some configurations of the N = 200 runs are shown in Fig. 3 below.

It has long been speculated that shear thickening is an order-disorder transition [5]. Although our results suggest that order-disorder changes are not a

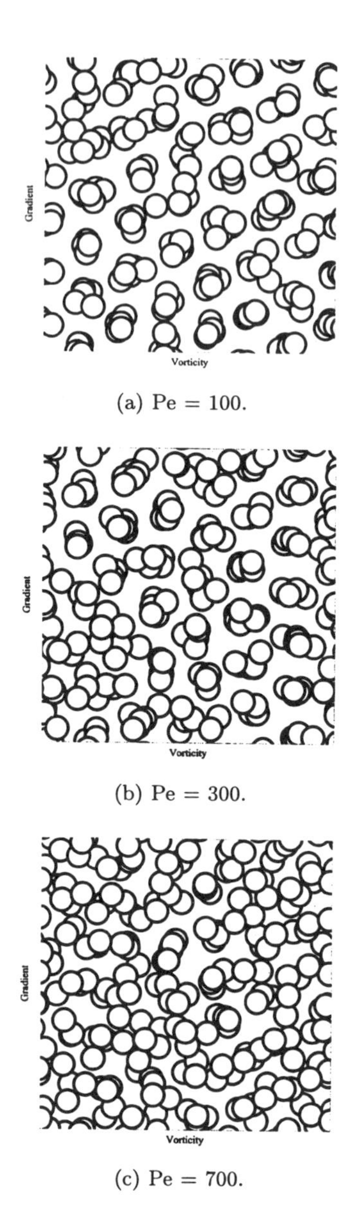

(a) Pe = 100.

(b) Pe = 300.

(c) Pe = 700.

Fig. 3. Some snapshot configurations of N = 200 particles at volume fraction 0.54 with model polymer coats under shear flow viewed along the flow direction in the shear gradient and vorticity plane. The parameters are those for solid squares in Fig. 2. The snapshots show the change of order as the shear rate is increased into the shear thickening regime (cf. Fig. 2). The particles are drawn half size.

necessary pre-condition for thickening, at the volume fractions of Fig. 2 the thickening regime does correspond to a change in order. This is illustrated in Fig. 3. The existence of ordering gives rise to a history dependence for the thickening transition: by allowing the system to form an ordered flowing phase at low shear rates and then gradually increasing the shear rate (as was done in Fig. 2), it can lead to a critical shear rate for the onset of thickening larger than that found by shearing configurations from rest at each shear rate.

Summary and Status of the Model

We have shown that model for colloids which undergoes shear thickening indicates that this effect is sensitive to the surface of the particle and is enhanced by mechanisms that increase the dissipative interaction between close surfaces. This suggests that the thickening effect is a probe of these interactions. The model does have limitations and we wish to discuss these here. A large portion of the real fluid flow and hydrodynamics in such systems were neglected in the model. However, we feel that this may not be significant in the shear thickening regime although this needs to be established by more complete computations. The model adopted for polymer-coated particles was simplified in many respects. There are many details of the interactions which we have ignored. The coat was assumed to have a step function concentration profile rather than a parabolic profile even though the coats in the experimental system we compared with ([3] and caption to Fig. 2) have chains of only a few monomers in length and would not be in the parabolic scaling regime. The coat was assumed to be in a regime of linear response under a given compression; local shear rates may bring the coats into a regime of non-linear response. Some theories predict an expansion of coats under strong flows. We plan to assess this question directly from the local motions in the simulation. The interaction law used has just squeeze motion. It ignored the relative shear motions of the coated surfaces (set in practice to those of hard spheres). In the experiments [3], thickening was promoted in the system by lowering the quality of the solvent for the coat. This will change the coat strength, its thickness and possibly roughen the surface. A key question is whether the polymer coats do inter-digitate on compression. If they do, the interaction law for the shear mode which we have neglected may be strongly modified.

Acknowledgements

We thank S. F. Edwards, A. Lips, R. Buscall, A. R. Rennie, S. M. Clarke, L. Silbert and, in particular, W. J. Frith and R. Farr for many useful discussions.

We thank the authors of [3] for allowing us to include some of their data. We thank the DTI/Colloid Technology project co-funded by the DTI, Unilever, Schlumberger, Zeneca and ICI for funding.

Discussion

M. E. Cates Can you clarify how the large Q limit (stiff coats) differs from the hard sphere Brownian case and whether the turning up in stress in the shear thickening regime is dissipative or elastic? For large Q, you have particles with a non-deformable porous outer layer, in which case it is hard to see how the dissipation could exceed that of hard spheres with the same outer radius. This is because a variational theorem states (with fixed boundaries) that the physical flow of the fluid is such as to minimise the dissipation. Hence if this is lowered by not having fluid flow within the coats, there will be no such flow.

J. R. Melrose The relation between particles with coats and hard spheres is subtle and requires further study. It is common in the literature to assume that coated spheres are equivalent to hard spheres with volume fraction set by the core plus coat diameter. However, properties will depend in a subtle way on the volume fraction and range of the coats. The fluctuation dissipation theorem determines that the Brownian forces act like a repulsion. At low shear rates, this will keep the particles apart. If the volume fraction is such that the typical particle gap exceeds the coat thickness, the equivalent hard sphere model will then work to the degree that the dissipative forces between particles appear to diverge at gaps of twice the thickness of the coat (c.f. Fig. 1). We have not yet included shear viscous forces modified by the coat. Until this is done, trustworthy quantitative answers cannot be given. If, however, the coats contact (and we find this is the case at 50% volume fractions for coat thickness 0.01 and even at Pe = 0.001), then the elastic coat forces will make a contribution to the thermodynamic component of the stress tensor. Whether or not this leads to a different form of shear thinning from that of Brownian spheres is unclear. Simple arguments by Brady suggest that both the elastic and Brownian contributions to the stress depend in the same way on the statistics of the matrix of contact vectors, $\mathbf{nn}$, ($\mathbf{n}$ is the normalised centre-to-centre vector of contacting neighbours).

The answer to the second issue is that shear thickening is the regime where the applied bulk stress is such that locally coats *do deform*. We monitored the distribution of inter-particle gaps between nearest neighbour particles. Figure 4

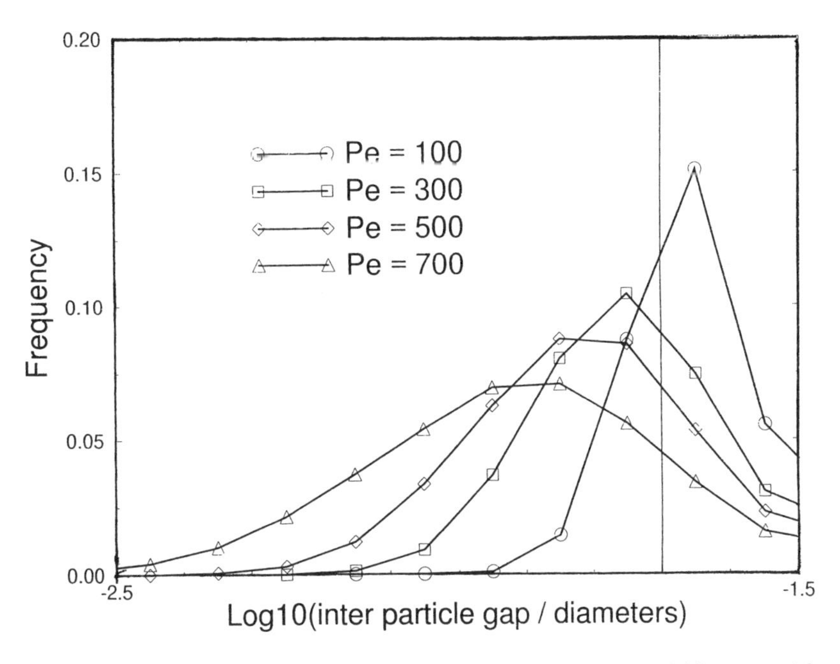

Fig. 4. The lower end of the distribution of gaps between nearest neighbour particles for the run of Fig. 2 plotted as solid squares. The frequency is the number of nearest neighbour pairs in a given bin range of inter-particle gap normalised by the total number of pairs. The vertical line shows the gap at contact of the coats. In the thickening regime (cf. Fig. 2), *circa* 10% of the coats are compressed.

shows data for one of the rheological plots of Fig. 2. As can be seen, the regime of thickening corresponds to the distribution of gaps by forming a tail at small gaps in which the polymer coats are compressed. (It is interesting to note that at Pe = 700, indicated by the arrow in Fig. 2, the peak in the distribution is approaching a gap of 10^{-2}, roughly half the compression of the 0.01 thick coats, and *circa* this region, the viscous interaction law turns over to an inverse square root of the gap divergence, cf. Fig. 1. It is suggestive that this is seen in the bulk rheology which lowers its rate of thickening at this shear rate.) The stress from the dissipative forces depends also on the relative velocity of the spheres which will be highly sensitive to the many-body structure, arguments based on isolated pairs may not be sufficient.

S. Ramaswamy If the suspensions were charge stabilised rather than polymer stabilised, what degree of shear thickening do you expect? Only logarithmic?

J. R. Melrose We would argue that plotted against the typical gaps at a given applied stress, the thickening would be logarithmic when the particles can be pushed against the repulsive interaction. Stronger elastic forces will not lead to a stronger thickening. Instead, they will delay the onset of thickening to a higher stress at which these forces can be compressed. The experimental evidence [2] suggests that the thickening in these systems is stronger than that of hard sphere dissipative forces. We must conclude, however, that in this case the near field viscous interaction laws are enhanced over that of hard spheres. Hence the physical mechanism of such is more mysterious than in the case of polymer coats. There is also the possibility of thickening associated with irreversible aggregation into the van der Waals minima.

M. R. Mackley Does clustering happen with or without Brownian motion? Does the structure relax upon the cessation of flow?

J. R. Melrose In the pure hydrodynamic case, hydrodynamic clustering is present. We are less certain of the degree of clustering in steady state thickening regimes. We are studying this at present. From the point of view of thickening, the Brownian forces act as the conservative forces controlling the close approach of particles. Structure will relax upon cessation of flow.

G. Marrucci In the case of sudden jamming, I would expect the reversing flow to destroy the gel. Is that correct?

R. C. Ball Simulations of hard spheres without Brownian motion, if not pushed to the limits of numerical precision, should indeed be time-reversible.

J. R. Melrose The percolation of the hydrodynamic clusters may be a breaking of the time reversal symmetry. In practice, the finite size of the periodic system will complicate the issue.

S. Ramaswamy Are the trajectories of the sheared particles highly sensitive to small changes in initial conditions?

J. R. Melrose We have not made a careful study of this. In the thickening regime, bulk results are sensitive to different starting states in that shear ordered states can be stable against thickening at shear rates where disorder initial states thicken. We think it incorrect to view the thickening regime as some sort of chaotic behaviour in the particle trajectories. Rather, the strong viscous interactions which take place as particles come close together lead to diverging relaxation times and a slowing down of the relative motion of close particles. The formation and rupturing of large clusters may, however, lead to a chaotic behaviour of bulk measurements.

M. Lal What is the fluctuation-dissipation equation corresponding to the relationship between the components of the mobility tensor (used in computing the Brownian terms) and the Brinkman parameter of the polymer layer between the particles?

J. R. Melrose We assume that the fluctuation theorem holds for the polymer coat. Hence Brownian forces generated are properly correlated with the dissipative law shown in Fig. 1.

M. J. Adams Could you speculate on how the introduction of elastohydrodynamic interactions would influence the results of your simulations?

J. R. Melrose The elasto-hydrodynamic coupling will introduce a non-linear dissipative interaction. We have coded a linearised version but have not yet performed a systematic study. There have been suggestions that elasto-hydrodynamic effects are related to shear thickening in polymer-coated systems. However, we find the arguments presented [20] inconclusive. The simpler, linear polymer interaction model used above does seem to predict the effect although it is clear to us that better theories need to be developed and these may involve elasto-hydrodynamic terms.

References

1. Barnes, H. A. (1989) Shear thickening (Dilatancy) in suspensions of non-aggregating solid particles dispersed in Newtonian liquids. *J. Rheol.* **33**, 329–366
2. Boersma, W. H., Laven, J. & Stein, H. H. (1990) Shear thickening (Dilatancy) in concentrated dispersions. *AIChE J.* **36**, 321–332; Chow, M. K. & Zukoski, C. F. (1995) Non-equilibrium behaviour of dense suspensions of uniform particles: volume fraction and size dependence. *J. Rheol.* **39**, 33–59
3. Frith, W. J., d'Haene, P., Buscall, R. & Mewis, J. (1995) Shear thickening in model suspensions of sterically stabilised particles. *J. Rheol.* **40**, 531–548
4. Woodcock, L. V. (1984) Origins of shear dilatancy and shear thickening phenomena. *Chem. Phys. Lett.* **111**, 455–459
5. Hoffman, R. L. (1974) Discontinuous and dilatant viscosity behaviour in concentrated suspensions 11. Theory and experimental tests. *J. Coll. and Interface. Sci.* **46**, 491–512
6. Bossis, G. & Brady, J. F. (1984) Dynamic simulation of sheared suspensions. *J. Chem. Phys.* **80**, 5141; Durlofsky, L., Brady, J. F. & Bossis, G. (1987) Dynamic simulation of hydrodynamically interacting particles. *J. Fluid. Mech.* **180**, 21–47; Phung, T. N., Brady, J. F. & Bossis, G. (1996) Stokesian dynamics simulation of Brownian suspensions. *J. Fluid. Mech.* 181–207

7. Cichocki, B., Felderhorf, B. U., Hinsen, K., Wajnryb, E. & Blawzdziewicz, J. (1994) Friction and mobility of many spheres in Stokes flow. *J. Chem. Phys.* **100**, 3780–3790
8. Ladd, A. J. (1990) Hydrodynamic transport coefficients of random dispersions of hard spheres *J. Chem. Phys.* **93**, 3484–3492
9. Kim, S. & Karrila, S. J. (1992) *Microhydrodynamics.* Butterworths
10. Ladd, A. J. (1994) Numerical simulations of particulate suspensions via discretized Boltzmann equation. Parts 1 & 2. *J. Fluid Mech.* **271**, 285–311
11. Koelman, J. M. V. A. & Hoogerbrugge, P. J. (1993) Dynamic simulations of hard-sphere suspensions under steady shear. *Europhys. Lett.* **21**, 363–368; Simulating microscopic hydrodynamic phenomena with dissipative particle dynamics (1992) *Europhys. Lett.* **19**, 155–160
12. Heyes, D. M. & Melrose, J. R. (1993) Brownian dynamics simulations of model hard-sphere suspensions. *J. Non-Newtonian Fluid Mech.* **46**, 1–28; Heyes, D. M. (1995) Mean-field hydrodynamics Brownian dynamics simulations of viscosity and self-diffusion of near-hard-sphere colloidal liquids. *J. Phys.: Cond. Matter* **7**, 8857–8865
13. Warren, P. B. (1996) Private communication to RCB
14. Farr, R., Melrose, J. R. & Ball, R. C. (1997) Unpublished
15. Frankel, N. A. & Acrivos, A. (1967) On the viscosity of a concentrated suspension of solid spheres. *Chem. Eng. Sci.* **22**, 847–853; Doi, M., Chen, D. & Saco, K. (1987) Simulations of concentrated colloids. In *Ordering and Organising in Ionic Solutions*, pp. 482. Singapore: World Scientific; Toivakka, M., Eklund, D. & Bousfield, D. W. (1995) Prediction of suspension viscoelasticity through particle motion modelling. *J. Non-Newtonian Fluid Mech.* **56**, 49–64
16. Lees, A. W. & Edwards, S. F. (1972) The computer study of transport processes under extreme conditions. *J. Phys.* **C5**, 1921–1929
17. Melrose, J. R. & Ball, R. C. (1995) The pathological behaviour of sheared hard spheres with hydrodynamic interactions. *Europhys. Lett.* **32**, 535–540; Farr, R., Melrose, J. R. & Ball, R. C. (1997) A kinetic model of jamming in hard sphere flows with only hydrodynamic lubrication. *Phys. Rev. E* **55**, 7203–7211
18. Fredrickson, G. H. & Pincus, P. (1991) Drainage of compressed polymer layers: dynamics of a squeezed sponge. *Langmuir* **7**, 786–795; Potanin, A. A. & Russel, W. B. (1995) Hydrodynamic interaction of particles with grafted polymer brushes and applications to rheology of colloidal dispersions. *Phys. Rev. E* **52**, 730–737
19. Melrose, J. R., van Vliet, J. H. & Ball, R. C. (1996) Continuous shear thickening and colloid surfaces. *Phys. Rev. Lett.* **77**, 4660–4663
20. Sekimoto, K. & Leibler, L. (1993) A mechanism for shear thickening of polymer bearing surfaces: elasto-hydrodynamic coupling. *Europhys. Lett.* **23**, 113–117

Dynamics of Complex Fluids, pp. 315–329
ed. M. J. Adams, R. A. Mashelkar, J. R. A. Pearson & A. R. Rennie
Imperial College Press–The Royal Society, 1998

Chapter 22

Micro-Structure of Colloidal Dispersions Under Flow

STUART M. CLARKE AND ADRIAN R. RENNIE*
Polymers & Colloids Group, Cavendish Laboratory,
Madingley Road, Cambridge, CB3 OHE, UK
**E-mail: rennie@gordon.cryst.bbk.ac.uk*

This article describes studies of the micro-structure of sheared colloidal dispersions and how the results can be related to computer simulations. Particular emphasis is made of light scattering and small-angle neutron scattering (SANS) experiments and the comparison with Brownian dynamics simulations. The importance of measurements in all three directions of a flow field is demonstrated. Experimental constraints and prospects for new advances are mentioned. The need for the hydrodynamic interactions between particles to be taken into account when modelling colloidal dispersions is also demonstrated.

Introduction

The relationship of non-Newtonian rheology with the molecular and microscopic structure of fluids has attracted considerable attention. In order to relate phenomena such as shear thinning and thixotropy to physical principles, it is necessary to understand the response of the micro-structure to the applied stresses. Determination of the structure under conditions of applied

*Present address: Department of Crystallography, Birkbech College, London, WC1E 7HX, UK

stress has been an essential part of the validation of models for theoretical calculations and computer simulations. The present article will describe scattering experiments applied to colloidal dispersions. Explanation of the advantages of different experiments and methods of comparison with models will be provided. Work on polymers and micelles under flow or deformation will not be covered in this article although some of the considerations are similar.

Rheo-optical studies of materials have been reported for many years and recent scattering work has been reviewed [1, 2]. The general topic of optical rheometry, which includes measurements such as flow birefringence and microscopy, has been described in a recent book [3]. The study of structure in sheared systems is important in several fields, particularly polymers, surfactant micelles and mesophases which [4] have been documented extensively. In the area of particulate dispersions, several key issues are the subject of intense study. These include the mechanisms of shear thinning and shear thickening, and the rôle of walls and boundaries on flow behaviour. Inter-particle hydrodynamic interactions may be of major importance in these areas. It is hoped that an understanding of these properties on the basis of physical principles may lead to improvements in the application and formulation of dispersions. The experiments are aimed at testing key aspects of theory and models.

Description of Scattering Techniques

Scattering Principles

An experiment that measures the radiation scattered from a sample as a function of angle, energy and wavelength can determine both the time-averaged or static correlations within a sample and the fluctuations with time. The experiments to be described here will involve only the determination of structure. This corresponds to the condition of no energy loss or gain between the radiation and the sample on scattering. Such experiments, involving the diffraction of x-rays, light and neutrons, are well described in the literature [5]. The amplitude, A, of radiation scattered from each scattering centre (for example, atom or nucleus) within the sample must be added with allowance being given for the phase difference to yield a coherent sum. The intensity, I, is calculated as AA^* where * indicates a complex conjugate. This simple formula for the calculation of scattering patterns amounts to a Fourier transform of the correlation in the spatial distribution of scattering centres, and may be written formally as follows:

$$I = \sum_{ij} A_i A_j{}^* = \sum_{ij} f_i \exp(i\mathbf{Q}\mathbf{r}_i)\, f_j \exp(-i\mathbf{Q}\mathbf{r}_j)$$

$$= \sum_{ij} f_i f_j \exp(i\mathbf{Q}(\mathbf{r}_i - \mathbf{r}_j)) \,. \tag{1}$$

The subscripts i and j run over all scattering centres located at positions $\mathbf{r}$, f is the scattering amplitude which describes the probability of a scattering event and $\mathbf{Q}$ is the momentum transfer given by

$$\mathbf{Q} = \mathbf{k}_i - \mathbf{k}_2 = (4\pi/\lambda)\sin(\theta/2)\,, \tag{2}$$

where $\mathbf{k}_1$ and $\mathbf{k}_2$ are the vectors describing the incident and scattered waves, λ is the wavelength and θ is the scattering angle. The details of the calculation of scattering patterns for structures such as those in colloidal dispersions can be found in several texts [6, 7].

For colloidal dispersions, it is often convenient to consider the scattering as a product of three terms:

$$I(\mathbf{Q}) = cP(\mathbf{Q})S(\mathbf{Q})\,, \tag{3}$$

where P describes the correlations within the individual particles and is known as the form factor and S, the structure or interference factor, corresponds to the correlations between particles. The constant, c, will depend on the concentration of particles and the scattering lengths or amplitudes, f. Under conditions of flow, both P and S could change but the discussion here is restricted to particulate dispersions in which the particles themselves are undeformable. Changes in intensity can be ascribed to different correlations between particles. Some care is needed, however, in the interpretation of changes in peak position as the form factor, $P(\mathbf{Q})$, may be very strongly modulated with various maxima and minima. The forms of $P(Q)$ for a sphere and $S(Q)$ for a hard sphere liquid, calculated according to the Ashcroft and Lekner [8] method, are shown in Fig. 1 as a function of Q, the amplitude of the vector $\mathbf{Q}$.

The simple description of scattering experiments implied in Eq. (3) may be somewhat modified by several factors. These include finite experimental resolution, polydispersity of particles which may affect both P and S, and nonlinear scattering behaviour such as that experienced when the scattering is large or there is extensive multiple scattering. While it may be possible to diminish some of these factors by appropriate choice of experiment or experimental conditions, in some circumstances it might be that the influence of, for

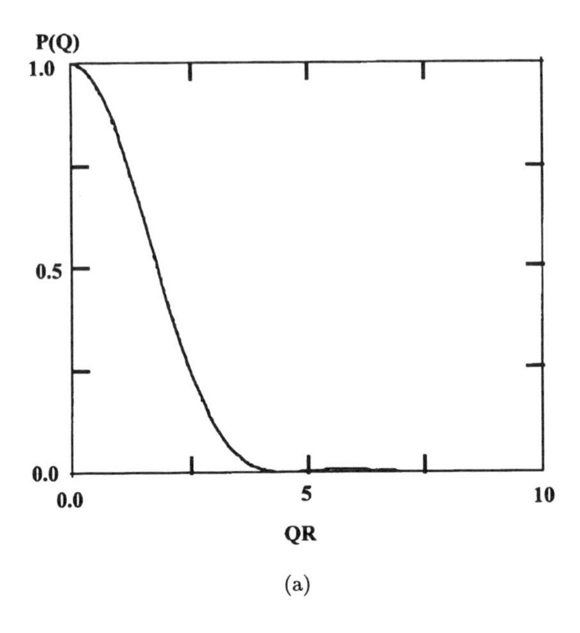

(a)

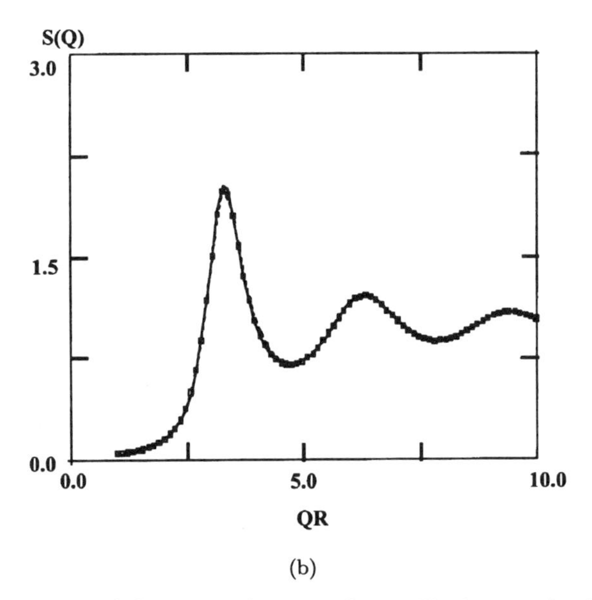

(b)

Fig. 1. (a) Form factor $P(Q)$ for monodisperse spheres. The function has been normalised to 1 at QR equal to zero. (b) Structure Factor $S(Q)$ for a "hard sphere" fluid at a volume fraction of 0.4 calculated according to Ashcroft and Lekner [8].

example, polydispersity may be an essential part of the investigation. Some of these factors will be discussed below as the different types of experiments are described.

Scattering experiments have proved particularly valuable in the determination of structure because they measure directly the pair correlations between particles. In this respect, they distinguish themselves from other experiments such as measurements of birefringence or dichroism, which are restricted to measurements of particular averages of orientation order parameters or structure. Direct microscopic imaging can be of help in understanding structure. In the case of liquids or other amorphous materials, it has been pointed out [9] that it can be extremely difficult to identify the difference in structure by direct observation, and that measurements of the correlation function are necessary.

X-ray, Neutron and Light Scattering

The choice of radiation for investigation may depend on several factors. The characteristic distance scales in the sample will force the choice of a combination of wavelength and range of scattering angles. However, a more important feature may be the contrast and absorption within the sample and sample containers which are described by the quantity f in Eq. (1). Some characteristic features of x-rays, neutrons and light are listed in Table 1.

Table 1. Properties of radiation used for scattering experiments.

	X-rays	**Neutrons**	**Light**
Wavelength	0.05–0.2 nm	0.02–3 nm	450–650 nm
Contrast	electron density	nuclear scattering (isotopic variation)	refractive index
Window or Cell Materials	beryllium, Li glass, thin plastic films	metals (e.g. Al, Nb, V), silica glass, Si crystals	glass, some plastics
Advantages	brightness of synchrotron	contrast variation	laboratory experiment
Disadvantages	cell construction difficult	expensive, low flux source	opaque and concentrated samples difficult

The nuclear scattering of neutrons can be used to particular advantage in multi-component fluids. If individual species are labelled, for example with the different isotopes of hydrogen, ^{1}H and ^{2}H, the location of specific species can be obtained. Even without isotopic contrast variation, the high transparency of samples and cell materials to neutrons may be important in performing experiments on concentrated dispersions or composite materials with a high loading of filler. A disadvantage of neutrons is the high cost and rarity of even relatively low flux sources and the absence of optical components such as lenses to focus beams or mirrors to deflect them through large angles.

The low flux of neutron sources often necessitates measurements at fairly poor resolution in ΔQ in order to obtain reasonable counting statistics. This may cause particular problems in separating the components P and S in Eq. 3. The resolution will convolute with the scattering from the sample in the measured intensity. The form of Eq. 3 might suggest that separate measurements of $P(Q)$ in a dilute or non-interacting dispersion would permit $S(Q)$ to be obtained by division. If both measurements are convoluted with a resolution function, this simple procedure may give highly erroneous results if either P or S contain sharp maxima or minima in the same range of Q. This is likely to be the case for concentrated dispersions of monodisperse spheres.

Light scattering suffers from a drawback: the relatively small differences in refractive index commonly found between materials can give rise to a large amount of scattering. In an extreme case, samples will multiply scatter to the extent that they are completely opaque. Under such circumstances, only observation of surface structure is possible. A careful choice of particles and dispersion media does, however, allow high optical transparency of concentrated dispersions. Although the choice of materials is restricted by the requirement to maintain colloidal stability, some important systems are accessible. Fluoroalkane particles which are charge-stabilised in water have been investigated in this way [10], and aqueous dispersions of polyethyl acrylate under shear have also been studied [11]. Sterically stabilised dispersions, which may be taken as a closer approximation to hard spheres, are simpler to prepare in index-matched fluids as a wide range of organic solvents is available. A considerable amount of work has been performed on PMMA latices [12, 13] as well as silica particles [14]. The requirement to approximately match the refractive indices in order to observe the bulk of a sample usually diminishes the problem of non-linear scattering (Mie theory, see [15]), which is associated with particles of different refractive index, n, on the scale $\Delta n\, R > 1$, to an insignificant level.

X-rays have proved useful in certain studies where the micro-focus and high brightness capabilities of synchrotron radiation are used to probe surface structures and bulk structures under shear [16]. Such experiments have proved difficult with laboratory sources.

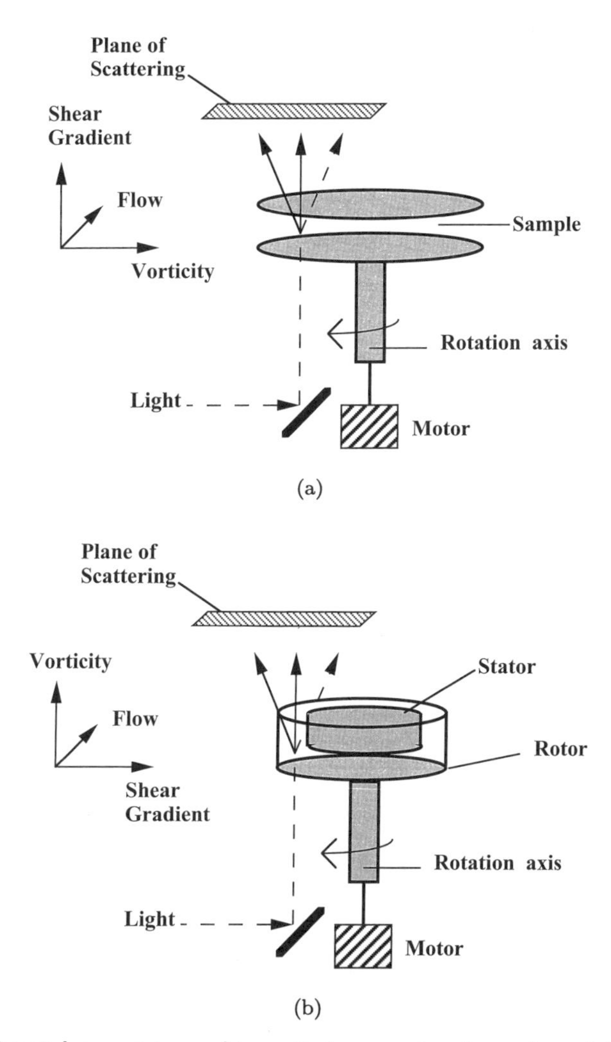

Fig. 2. Experimental geometries used in scattering experiments on sheared dispersions. (a) Disc-disc and (b) Couette cells used in light scattering experiments [13]. A Couette cell used in SANS experiments is shown in (c). The different positions of the incident beam with respect to the cell in attempts to observe all principal directions of flow are shown in (d).

Experimental Geometry

The geometry of the flow cell and the scattering experiment is of considerable importance. In order to obtain the full details of the structure, scattering data must be collected in all directions. The principal axes of flow are shown in Fig. 2 with respect to the orientation of flow cells used in light scattering and SANS experiments. The use of the two cells, shown in Figs. 2(a)

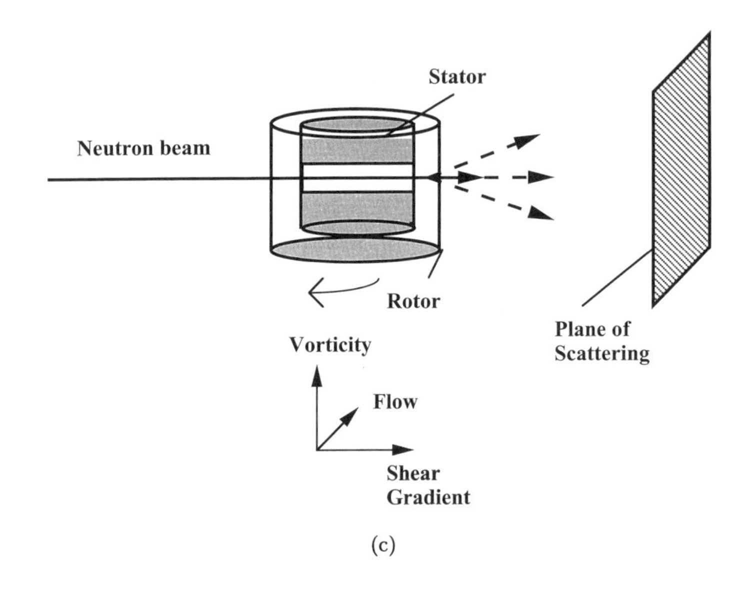

(c)

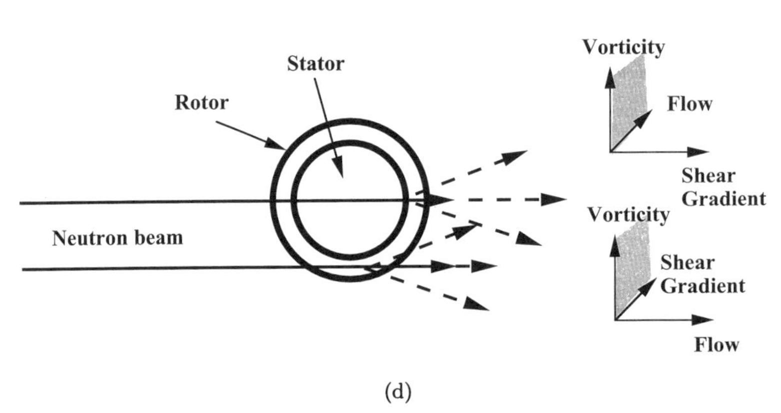

(d)

Fig. 2. (*Continued*)

2(b), permits observations with light along all three principal axes. Neutron scattering experiments are more difficult. This is because the mirrors to deflect the beam through large angles are not obtainable and most SANS instruments have been built on a horizontal axis. The usual SANS geometry is a Couette cell as shown in Fig. 2(c). Some measurements can be made through the edge of a cell (see Fig. 2(d)) if the cell is offset from the axis of the beam. Such data is often unsatisfactory as both the finite lateral extent of the cell and the beam (usually of order cm) lead to a spread of orientations and a sample thickness that is far from optimum, thereby causing multiple scattering.

The radiation used in an experiment may have a further consequence in that the measured scattering may not always lie on a plane formed from the principal axes of the flow geometry. This will often occur if the wavelength of the radiation becomes comparable to the dimensions investigated and the scattering angles are no longer small. Under these circumstances, considerable care may be required in making simple interpretations of scattering patterns. The region of scattering space over which counts have been integrated may be of considerable importance in determining the details of crystalline structures [17, 18].

Comparison with Theory and Simulation

The comparison of experimental scattering data with models and computer simulations has, until recently, primarily been of a qualitative rather than a quantitative nature. The need to make numerical comparisons of the extent and position of correlations between particles to distinguish different behaviour has been discussed elsewhere [19]. It has been pointed out [13] that the micro-structural changes associated with shear thinning in a dispersion of spherical particles may be rather small. A change of inter-particle spacing in the flow direction of just a few per cent may be associated with a reduction in viscosity of two orders of magnitude. Clearly, any experimental investigation or computer model will require data to a high precision if such structural changes are to be resolved. In practice, this poses stringent requirements on computer simulations as the number of particles and "box" size is often limited. Further, the need for adequate statistics in simulations cannot be neglected as the comparison with measured structures must be made with averages over a significant number of *independent* configurations. This can be difficult to achieve if realistic models are to be computed.

The magnitude of correlations between particles is as important as the position of the peaks in the correlation functions. In fluids, the position of the nearest correlations is often well-defined. It is the differences in the correlation between the second and third nearest neighbours that distinguishes amorphous structures from more regular crystalline arrangements. As the changes in rheological behaviour that are of interest in complex fluids are often associated with such transitions, it is important that such measurements can be made. In general, experimental methods are in a better position to distinguish the subtle changes in structure than many of the computer simulations which are still limited by the computing resources available. It is, however, only recently that numerical data from scattering experiments has been compared in detail with theory or simulations [19].

There are several advantages in making the comparison of models with experimental scattering patterns rather than the inverse procedure of deconvoluting the scattering data to obtain a particle correlation function which can be transformed to a real space particle distribution function. Most data sets are poorly terminated at both small and large Q. The deconvolution of resolution and particle polydispersity is then difficult. In contrast, the calculation of a scattering intensity from any known model configuration of particles follows in a straightforward manner from the theory outlined above. It is easy to include polydispersity in $P(Q)$ and make allowances for the known instrumental resolution.

Experiments on Colloidal Dispersions under Flow

The features of scattering experiments and the type of data that can be obtained on sheared dispersions will be illustrated in an example drawn from previous work. This has been chosen to show the behaviour of a dispersion of particles with almost "hard sphere" interactions obtained by steric stabilisation of latices. The example illustrates the use of light scattering. It is useful to describe briefly the interactions between colloidal particles before describing the experimental details. A systematic description of colloidal stability will not be given but can be found in textbooks [9].

Sterically stabilised dispersions of particles are often considered to behave approximately as hard spheres. The particles are provided with a surface layer of polymer or surfactant at low volume fraction which has a favourable interaction with the dispersion medium. The free energy associated with this interaction, which is largely entropic, amounts to a repulsive potential between the particles and is sufficient to overcome the van der Waals attraction. As the range of this repulsive interaction is limited to the physical extent of the adsorbed or grafted molecules, the potential is short-range and, for large particles, may approach that of ideal hard spheres. The data for such dispersions has therefore been used for comparison with theoretical models and computer simulations of hard spheres. In practice, several workers have shown that for small particles, the potential deviates significantly from that of hard spheres. The second principal method of establishing colloidal stability is to exploit charge on the particles. The interaction, even when screened by some of the counter-ions, is sufficient to provide a long-range repulsion which, when combined with the van der Waals attraction, provides the well-known DLVO potential [9].

Data will be reported on samples at different shear rates. For the purpose of comparison, it is convenient to use the reduced variable known as the Peclet number, Pe. This corresponds to a ratio of the energy of the shear field per particle to the thermal energy and is defined as

$$\mathrm{Pe} = 6\pi\, \gamma\, \eta\, a^3 / 8\, k_B\, T\,, \tag{4}$$

where γ is the strain rate, η is the viscosity, a is the particle diameter, k_B is Boltzmann's constant and T is the absolute temperature.

Shear of Sterically Stabilised Dispersions

Light scattering experiments have been made on PMMA particles (diameter 1.5 μm) stabilised with poly 12-hydroxy stearic acid in mixtures of decahydronaphthalene/tetrahydronaphthalene chosen to approximately match the refractive index ($n = 1.51$) of the particles. The apparatus [13] and preparation of the materials is described in detail elsewhere. The density difference between the particles is sufficient to produce significant sedimentation but at high volume fractions and shear rates, this was not apparent. The effects of sedimentation are discussed elsewhere [17]. Data is shown in Fig. 3 for this sample at rest and under shear in two different geometries at a Peclet number of 70.

The data in the flow-vorticity plane shows characteristics which have been mentioned previously [13, 20]. There is an increase in the inter-particle spacing in the flow direction and an increase in intensity corresponding to a greater correlation. The data in the other scattering plane is perhaps more interesting.

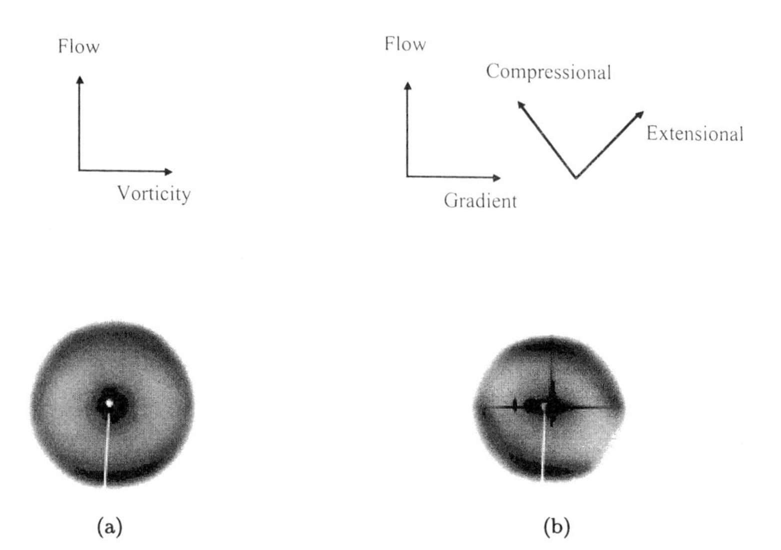

Fig. 3. Scattering from PMMA dispersion in decahydronaphthalene/tetrahydronaphthalene at 40% w/w. (a) In disc/disc geometry at Peclet number 70. (b) In Couette geometry at Peclet number 70. The principal axes of the shear flow are shown in the figure.

This flow-gradient plane contains the extensional and compressional directions of the flow field. There is some distortion of the scattering pattern due to the passage of light through a non-flat meniscus. However, a clear orientation of the scattering corresponding to an increase in separation of about 10% along the extensional axis is visible. The intensity along the compressional axis is reduced by about 40% which may be interpreted as a reduction in the spatial correlations of particles in this direction. There is no evidence of significant change in average particle separation along the compressional axis.

The results of SANS experiments [20] have been presented previously as a schematic drawing of particle arrangements. The dilation in the flow direction has been clearly identified by several authors. Scattering data that comes from a single plane limits the interpretation to a sketch of average positions in a single plane of real space. The combination of data from two orthogonal scattering planes allows us to draw schematic arrangements of particles in two planes. This provides information about projections down all principal axes of flow. This is shown in Fig. 4. The extension and compression in the shear-gradient flow plane are clearly visible in the data of Fig. 3 and have been included in this schematic drawing.

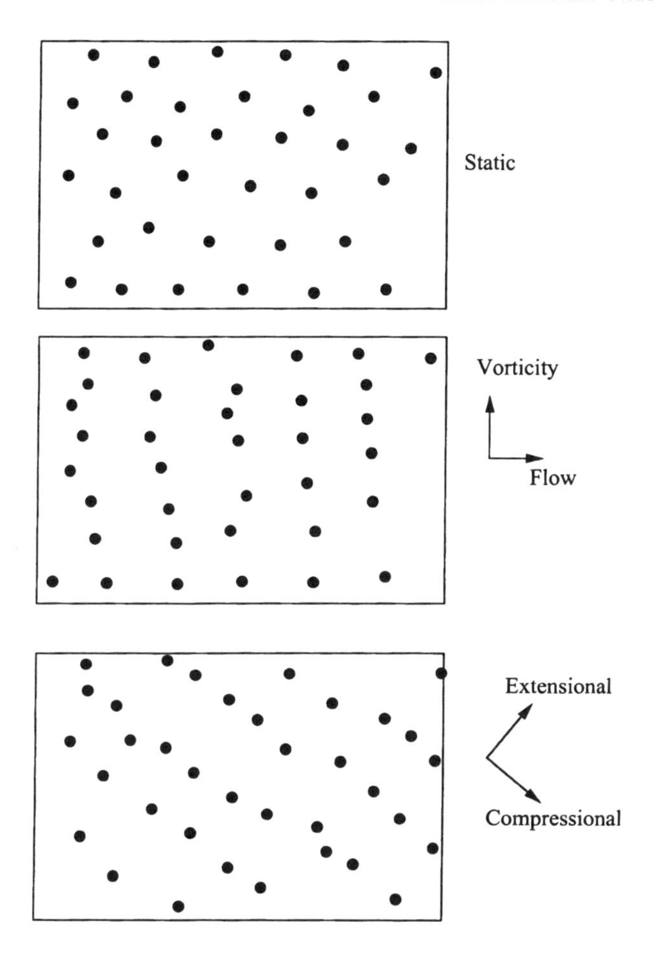

Fig. 4. Schematic diagram showing the distortion in particle positions for dispersions under flow based on the scattering data in Fig. 3.

Conclusions

This work has shown the importance of experimental geometries which provide information on the structure along all three principal flow directions. Light scattering has proved particularly convenient in this respect as beams can be deflected in any orientation. Other techniques such as SANS, however, could provide some similar information. The results have shown that the structures measured for sterically stabilised dispersions are only in reasonable agreement with Brownian dynamics models used in computer simulation if hydrodynamic interactions between particles are taken into account. Although the changes in nearest neigbour separations that occur during shear thinning are small, changes in the intensity corresponding to correlations between particles can be relatively large. The magnitude of these changes has implications on the requirements for the size of computer simulations. The benefits of obtaining data from scattering experiments, which is averaged over many particle configurations, are also demonstrated.

Acknowledgements

We are grateful to the UK EPSRC, the ECC International plc and the DTI programme in Colloid Technology (with Unilever, ICI, Schlumberger and Zeneca) for support of this work.

References

1. Rennie, A. R. & Clarke, S. M. (1995) Scattering by complex fluids under shear. *Current Opinion in Colloid & Interface Science* **1**, 34–38
2. Brady, J. F. (1995) Model hard-sphere dispersions-statistical-mechanical theory, simulations and experiments. *Current Opinion in Colloid & Interface Science* **1**, 472–480
3. Fuller, G. C. (1995) *Optical Rheometry of Complex Fluids.* New York: Oxford University Press
4. ACS Symposium Series (1994) Volume 578
5. Hukins, D. W. L. (1981) *X-Ray Diffraction by Disordered and Ordered Systems.* Oxford: Pergamon Press
6. Glatter, O. & Kratky, O. (ed.) (1982) *Small Angle X-ray Scattering.* London: Academic Press
7. Brumberger, H. (ed.) (1995) *Modern Aspects of Small-Angle Scattering.* Dordrecht: Kluwer
8. Ashcroft, N. W. & Lekner, J. (1966) Structure and resistivity of liquid metals. *Phys. Rev.* **145**, 83–90
9. Hunter, R. J. (1987) *Colloid Science, Vols. 1 & 2.* Oxford: Oxford Univ. Press
10. Ottewill, R. H. & Williams, N. St. J. (1987) Study of particle motion in concentrated dispersions by tracer diffusion. *Nature* **325**, 232–234
11. Sakabe, H. (1995) *The Structure and Rheology of Strongly Interacting Suspensions.* PhD thesis, Bristol
12. Pusey, P. N., van Megen, W., Bartlett, P., Ackerson, B. J., Rarity, J. G. & Underwood, S. M. (1989) Structure of crystals of hard colloidal spheres. *Phys. Rev. Lett.* **63**, 2753–2756
13. Clarke, S. M., Ottewill, R. H. & Rennie, A. R. (1995) Light scattering studies of dispersions under shear. *Adv. Coll. Interf. Sci.* **60**, 95–118

14. Yan, Y. D., Dhont, J. K. G., Smits, C. & Lekkerkerker, H. N. W. (1994) Oscillatory shear-induced order in non-aqueous dispersions of charged colloidal spheres. *Physica A* **202**, 68–80
15. van de Hulst, H. C. (1981) *Light Scattering by Small Particles.* New York: Dover
16. Idziak, S. H. J., Safinya, C. R., Sirota, E. B., Bruinsma, R. F., Liang, K. S. & Israelachvili, J. N. (1994) Structure of complex fluids under flow and confinement, x-ray couette shear cell and the x-ray surface force apparatus. *ACS Symposium Series* **578**, 288–299
17. Clarke, S. M. & Rennie, A. R. (1997) Ordering and structure at interfaces of colloidal dispersions under flow. *Faraday Discussions* **104**, 49–63
18. Clarke, S. M., Rennie, A. R. & Ottewill, R. H. The stacking of hexagonal layers of colloidal particles: study by small-angle neutron diffraction. *Langmuir* **13**, 1964–1969
19. Clarke, S. M., Melrose, J. R., Rennie, A. R., Heyes, D. M. & Mitchell, P. (1998) The Influence of Structure on the Rheology of Colloidal Dispersions: A Comparison of Scattering Experiments and Computer Simulation. In preparation
20. Lindner, P., Markovic, I., Oberthür, R. C., Rennie, A. R. & Ottewill, R. H. (1988) Small-angle neutron scattering studies of polymer latices under sheared conditions. *Prog. Coll. Pol. Sci.* **76**, 47–50

Dynamics of Complex Fluids, pp. 330–337
ed. M. J. Adams, R. A. Mashelkar, J. R. A. Pearson & A. R. Rennie
Imperial College Press–The Royal Society, 1998

Chapter 23

Measurement of the Flow Alignment of Clay Dispersions by Neutron Diffraction

A. B. D. BROWN, S. M. CLARKE AND A. R. RENNIE*
Polymers & Colloids Group, Cavendish Laboratory,
Madingley Road, Cambridge, CB3 OHE, UK
**E-mail: rennie@gordon.cryst.bbk.ac.uk*

A diffraction method to investigate the alignment of crystalline colloidal particles is described. The alignment of plate-like particles of kaolinite is studied at various concentrations and flow conditions. The concentration is varied from a volume fraction of 0.025 to 0.20. The study is made under both laminar and turbulent flow conditions in pipes and near baffles. Order parameters describing the distribution of particle orientations under flow are calculated.

Introduction

The study of order in sheared particulate dispersions has been the subject of considerable research [1, 2]. Information about the structure and orientational order in such systems will give an insight into their non-Newtonian flow behaviour. Control of particle alignment under flow should allow control of the rheological properties during processing and influence properties of the final products. For example, the application of coatings to paper usually involves a mixed colloidal dispersion containing a high volume fraction of anisotropic

*Present address: Department of Crystallography, Birkbech College, London, WC1 7HX, UK

particles applied under conditions of very high shear. The alignment of the particles during the application of the coat affects the appearance and properties of the finished product.

In this work, a technique to study the orientation of anisotropic colloidal particles is introduced and illustrated with a study of kaolinite suspensions under conditions of both laminar and turbulent pipe flow. The volume fraction and shear rate dependence of the alignment are described. The orientation distribution of a number of particles can be described by the even terms in a Legendre polynomial series. The coefficients of these terms are known as "order parameters". The technique presented here allows the full orientational distribution to be determined from which, theoretically, all terms of the series can be calculated. In comparison, other methods of measuring orientational order, such as dichroism and birefringence, provide only the first non-trivial order parameter due to the averaging inherent in these techniques.

Experimental Technique

Principles

A crystalline plate of clay will only diffract if it is oriented with respect to the incident beam and the detector to meet the Bragg condition. With many particles diffracting incoherently with respect to each other, the total intensity is the sum of intensities diffracted by each particle. Thus, the intensity of a given diffraction peak from a polycrystalline sample is proportional to the number of plates in the orientation for diffraction. By rotating the sample, the number of particles at each orientation can be measured and hence the complete angular orientation distribution can be obtained. Diffraction methods have been used in the past to study texture and preferred orientation in a variety of solid samples [3]. More recently, the technique has been applied to the study of aligned colloidal dispersions [4].

Apparatus

Neutron diffraction experiments were made with the diffractometer D1B at the high flux reactor of the ILL, Grenoble, France [5]. The experimental arrangement shown in Fig. 1 was described previously [4]. A syringe pump was used to force the dispersion down fused silica tubes at various flow rates ranging between 0.07 and 7.0 ml s^{-1}. Control signals from the pump were used to inhibit counting when the flow direction was changed to ensure that the data was collected under conditions of constant flow rate. The data represents an

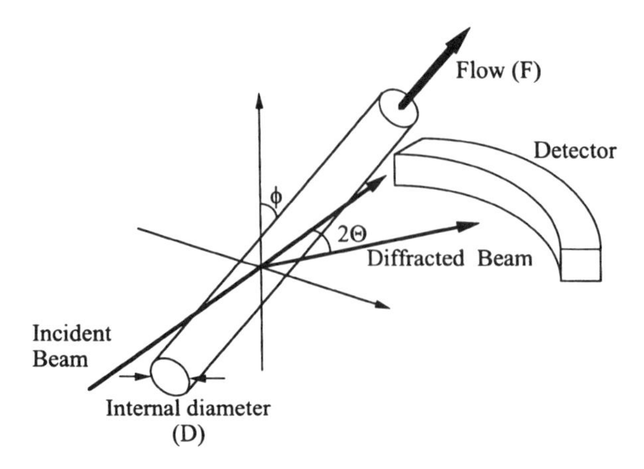

Fig. 1. Schematic illustration of the experimental geometry. The sample can flow along the pipe which is mounted on an Euler cradle in the neutron beam. The linear multi-detector is fixed and collects data over a range of scattering angles from five to 85 degrees in the horizontal plane. The sample pipe is rotated in a plane perpendicular to the beam to explore different orientations of the clay.

average over both flow directions, which are identical within the symmetry of pipe flow. Laminar flow was achieved using a uniform circular cross-section pipe (5 mm diameter). Turbulence was achieved by passing the dispersion through a baffle (2 mm diameter) placed in a tube (10 mm diameter).

Sample

Kaolinite forms as plate-like particles [6]. The particles used in this work had an average aspect ratio of about 30 and a size, measured by a Sedigraph, of 0.3 μm. It also had a polydispersity of ± 0.3 μm with a significant fraction at larger sizes. A sample with a small particle size was chosen so that sedimentation was reduced and thus the Brownian motion would be more significant at accessible shear rates.

The edges of kaolinite plates are positively charged while the faces are negative at neutral pH. Kaolinite was dispersed in D_2O and a low molecular weight polyacrylate, ($M_W \sim 3500$), was used as a stabiliser [7]. D_2O is used to reduce the incoherent background scattering that arises from hydrogen. Negatively charged polyacrylate stabilisers are expected to bind to positively charged edges of the plates, thereby neutralising the edge charge and preventing any associations that can arise from the edge to face attractions [8].

Kaolinite gives rise to very intense diffraction peaks, of the type $\{00\ell\}$, from the inter-planar spacing of the crystal. These diffraction peaks are related to the normal of the plate-like particles and are good indicators of particle orientation. The integrated intensities of the (001) and (002) peaks were used in the present experiments. The "in-plane" peaks, $\{hk0\}$, have relatively low intensity and were not observed except in concentrated and highly aligned samples. In general, the full three-dimensional orientation distribution of the particles' normals would require the full intensity distributions of the $\{00\ell\}$ reflections in three dimensions. However, the cylindrical geometry of the pipe imposes a symmetry on the system. Hence the full orientation distribution can be inferred from a suitably chosen one-dimensional data set. This data was collected by rotating the tube through ϕ from zero to 90 degrees in a plane perpendicular to the beam shown in Fig. 1. A powdered yttrium iron garnet powder was used to provide a uniform angular intensity distribution as calibration.

Results and Discussion

Laminar Flow

The intensity distribution at a range of flow rates is shown in Fig. 2. At rest, the intensity distribution is uniform with angle, indicating no preferred orientation of the particles. As the flow is increased, the intensity distributions show an increasing alignment of the particles with their normals perpendicular to the flow direction.

As can be seen from Fig. 2, a simple indicator of alignment is the intensity at $\phi = 0$. Figure 3 presents intensity at $\phi = 0$ as a function of flow rate. The principal trend in the data is a steady increase in intensity up to a flow rate of 3 ml s^{-1}, indicating an increase in alignment with increasing flow. Above this flow rate, a more complex flow regime is observed. The highest flow rate in this figure, 7.2 ml s^{-1}, corresponds to a Reynolds number of approximately 1,800. The flow in the pipe around this value may therefore be expected to exhibit turbulence.

The variation of intensity at $\phi = 0$ with volume fraction at three different flow rates is given in Fig. 4. It is evident that the alignment does not simply increase with volume fraction. Instead, it increases up to 0.14 and then decreases again as the volume fraction was increased up to 0.20 at all the flow rates investigated. The reduction in alignment at higher volume fractions could

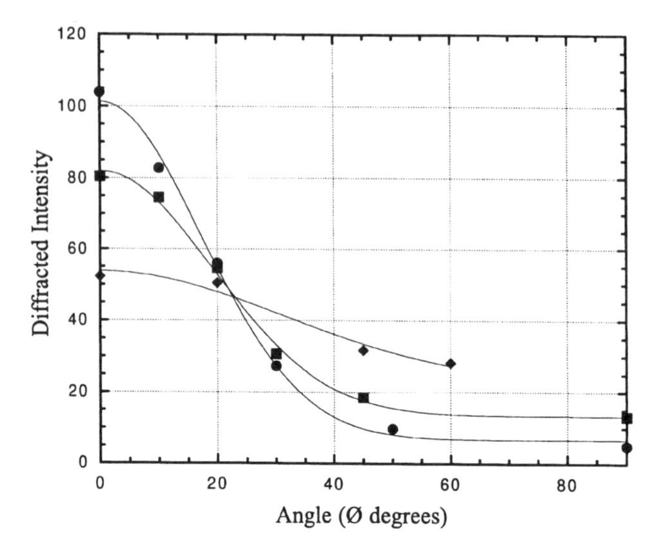

Fig. 2. Angular distribution of intensity at three flow rates. The data was measured at a weight fraction of 0.1 in a pipe of 5 mm diameter.
◆ 0.07 ml s^{-1} ■ 1.0 ml s^{-1} • 7.0 ml s^{-1}

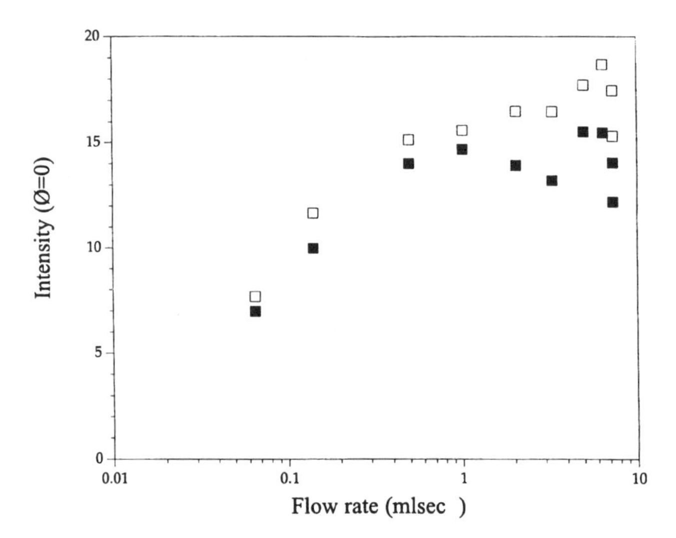

Fig. 3. Variation of alignment with flow rate. The intensity at zero degrees is taken as a measure of the order.
□ – (001) reflection ■ – (002) reflection

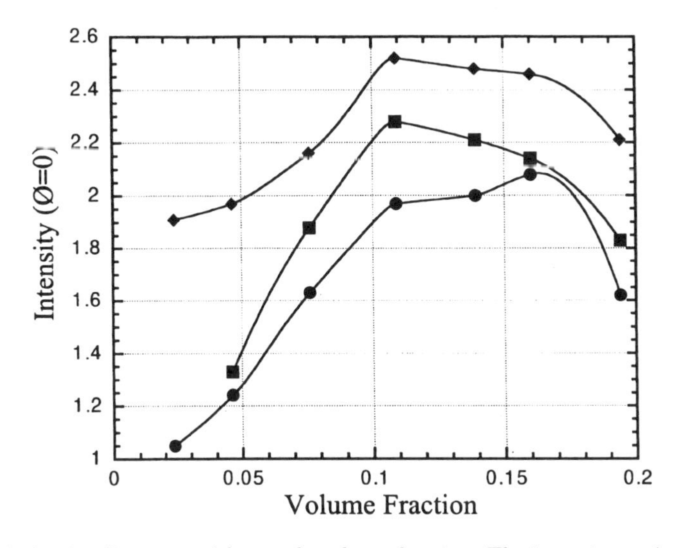

Fig. 4. Variation in alignment with sample volume fraction. The intensity at $\phi = 0$ is taken as a measure of the order. Measurements were taken from flow in a pipe of 5 mm diameter at three different flow rates: • 0.065 ml s^{-1} ■ 0.15 ml s^{-1} ♦ 1.1 ml s^{-1}.

Table 1. The order parameters calculated from the data displayed in Fig. 2.

Flow rate (ml s^{-1})	**7.0**	**1.0**	**0.07**
P_2 (Order Parameter)	−0.32	−0.27	−0.13
P_4	0.10	0.04	0.01

be due to shear-thinning, which restricts the shear to a smaller volume at the edge of the tube, or the formation of tumbling domains.

As discussed above, the extent of alignment can be described in terms of Legendre polynomials. Table 1 gives the first two even terms of this distribution for the data given in Fig. 2. The indicator of platelet orientation has been taken here to be the normal to the plates. The complete alignment of the normals in the flow direction corresponds to a P_2 of 1.0 while that which is perpendicular to the flow direction yields P_2 of $-1/2$ and P_4 of 0.375, and an isotropic distribution of particles gives a P_2 and P_4 of zero.

The negative values of P_2 and positive values of P_4 in Table 1 clearly indicates the preferred orientation of the platelets with their normals perpendicular to the flow direction.

Turbulent Flow

The sample at a volume fraction of 0.12 was forced at 2.5 ml s^{-1} through the baffle. The effect of turbulence was assessed by determining the extent of orientational order at a number of points along the tube. From the position of the baffle going downstream, the amount of orientational order in the tube was seen to decrease to a minimum at about 30 mm from the baffle. The order then increased towards the laminar flow value at larger distances from the baffle. The alignment at the minimum point was found to be approximately a third of that for laminar flow. These measurements were in keeping with visual observations as turbulence could be seen at the surface of the tube 30 mm from the baffle.

Conclusions

The results presented in this work clearly demonstrate that diffraction techniques can be used to measure the orientational order in concentrated dispersions of anisotropic particles. Data can be obtained over a wide range of sample volume fractions, flow rates and geometries. The extent of alignment is found to increase with flow although it is still only partial even in the most aligned case. Turbulent flow does significantly reduce the extent of alignment.

The cylindrical pipe flow geometry is convenient. However, the flow profile in a pipe is complex and the data presented here is an average over this flow profile. The cylindrical symmetry of pipe flow also means that there is no way to differentiate between the alignment in the vorticity and shear-gradient directions. In order to make measurements on well-defined, uniform shear fields, a plate-plate cell is being developed.

Acknowledgements

We are grateful to the UK EPSRC for a studentship for ABDB, and ECC International plc for the provision of samples and technical assistance.

References

1. Ackerson, B. J. (1996) Neutron, x-ray and light scattering rheo-optical techniques. *Curr. Opin. Coll. Interf. Sci.* **1**, 450–453
2. Rennie, A. R. & Clarke, S. M. (1996) Scattering by complex fluids under shear. *Curr. Opin. Coll. Interf. Sci.* **1**, 34–38
3. Subasilar, B. (1986) Effects of deformation zone geometry on particle alignment in the extrusion of clay. *Br. Ceram. Trans. J.* **85**, 49–51

4. Clarke, S. M., Rennie, A. R. & Convert, P. (1996) A diffraction technique to investigate the orientational alignment of anisotropic particles: studies of clay under flow. *Europhys. Letts.* **35**, 233–238
5. Ibel, K. (ed., 1994) *Guide to Neutron Research Facilities at the ILL, Institut Laue-Langevin*, pp. 46. B.P. 156, F-38042 Grenoble Cedex 9, France
6. Brindley, G. W. & Brown, G. (ed., 1980) *Crystal Structures of Clay Minerals and their X-ray Identification*, pp. 84. London: Mineralogical Society
7. Dispex N40 supplied by Allied Colloids Ltd
8. van Olphen, H. (1963) *An Introduction to Clay Colloid Chemistry*, pp. 90–93 & 168–170. London: John Wiley

Dynamics of Complex Fluids, pp. 338–344
ed. M. J. Adams, R. A. Mashelkar, J. R. A. Pearson & A. R. Rennie
Imperial College Press–The Royal Society, 1998

Chapter 24

The Rheology and Micro-Structure of Equine Blood

R. M. de Roeck* and M. R. Mackley
Dept. of Chemical Engineering, University of Cambridge, Pembroke St., Cambridge, CB2 3RA
**E-mail: rmdr2@com.ac.uk*

This paper reports the results of preliminary investigations that have been undertaken to establish a quantitative correlation between the rheology and micro-structure of equine blood. The long-term aim is to derive a quantitative and mechanistic model of structure formation in whole blood that can be used to predict rheological behaviour under complex flow conditions. The model must account for shear-thinning, viscoelastic and thixotropic behaviour. Initial trials of the Maxwell model have been made. Equine blood was the chosen system for investigation due to its strong aggregating tendency.

Introduction

The field of haemorheology has been active since the early 1960s following the advent of suitably accurate viscometers. Though generally accepted as a medical diagnostic aid, haemorheology also poses a challenge to the engineering community. Despite being a field that is open to new theory and conjecture, relatively little attention has been focused on obtaining a full rheological characterisation of blood.

The works of Thurston [1] and Chien *et al.* [2] have documented shear-thinning and viscoelastic characteristics. Furthermore, Schmid-Schönbein *et al.* [3] and Copley *et al.* [4] have performed microscopic observations of blood under shear and reported the presence of shear-induced aggregates. However, only *qualitative* correlations have been made between the observed micro-structure and the reported rheology. Consequently, this study's long-term objective aims to develop a *quantitative* model to help address some of the fundamental issues related to an understanding and prediction of blood behaviour under flow.

Experimental Techniques

The experiments in the current study are performed using heparinised blood (50 i.u./ml) from a healthy thoroughbred gelding. Samples are used within six hours of withdrawal. When not in use, the blood is kept still at room temperature. Before experimentation, the blood is hand-agitated for two to three minutes to re-suspend the cells prior to further dispersion protocols within the apparatus. All experiments are undertaken at $38.5°C \pm 1°C$.

Rheological characterisation is performed using a Rheometrics RDSII controlled strain rheometer in a 50 mm diameter parallel plate geometry. The sample is loaded between the plates, which are then set to a gap width of 150 μm. All blood samples were subjected to a pre-shear rate of 1000 s^{-1} for 30 seconds immediately before experimentation.

Strain and frequency sweeps were performed to examine the viscoelastic response of whole blood. A further steady-shear investigation was also made to determine shear-thinning behaviour and any thixotropic effects.

Microscopic observations of blood under controlled shear conditions were performed using the Cambridge Shear System (CSS450). The system employs a similar configuration used in the RDSII except that the plates are transparent, which allow the use of an Olympus BH2 microscope. A CCD camera is mounted on the microscope column to enable the use of a video recorder and an image capture system. The gap setting used in all CSS450 experiments was 50 μm due to the opacity of blood.

Results

Experiments were performed on whole blood of haematocrit (phase volume) 36%. The strain sweep data (Fig. 1) demonstrate a region of linear viscoelastic

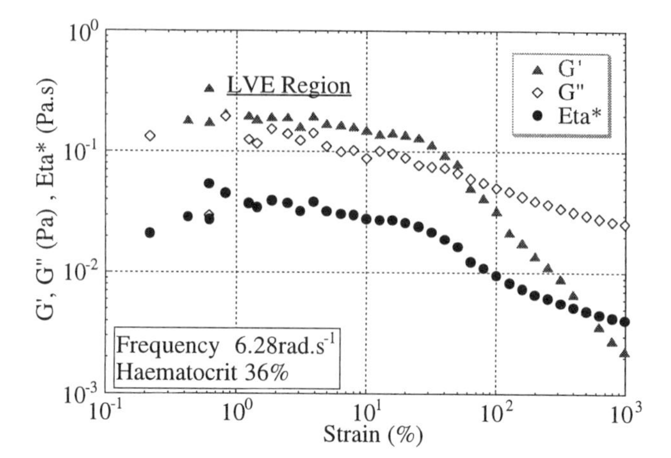

Fig. 1. Strain Sweep.

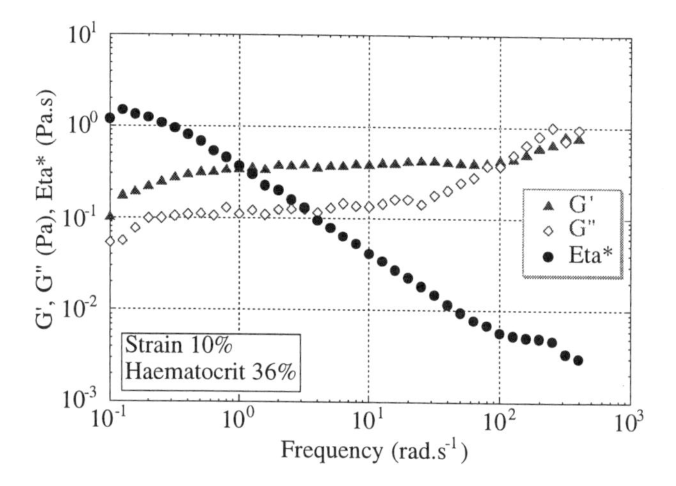

Fig. 2. Frequency Sweep.

behaviour up to a critical strain of about 20%. Of particular interest are the comparable magnitudes of the elastic (G') and viscous (G'') moduli in this region followed by a very steep reduction of the elastic component after the critical strain is reached. It is very likely that this is a function of long-range ordering present at small strain which breaks up as the strain increases.

The frequency sweep (Fig. 2) shows a pronounced shear-thinning in the complex viscosity, (η^*), reaching above two orders of magnitude. However, an apparent region of moduli frequency independence is also observed, which is uncharacteristic of Maxwellian materials. It should also be noted that a frequency sweep performed at this particular strain (10%) yields a result in which G' is greater than G'' in the low frequency range.

A further investigation was made on the steady-shear behaviour of blood. A rate sweep can be seen in Fig. 3 which demonstrates shear-thinning characteristics. One should also note that in the shear rate region of 2 s^{-1} to 100 s^{-1}, a deviation from an otherwise smooth curve exists. The steady decrease in apparent viscosity stops abruptly at 3 s^{-1} and is seen to rise and fall twice before a further steady decrease is seen after 100 s^{-1}. Consequently, steady-shear experiments was performed at different shear rates to look into the effects of thixotropy and to determine the validity of the observed point of inflection.

At very low shear rates, the blood quickly establishes a high viscosity (Fig. 4). As the applied shear rate is increased (1–3 s^{-1}), the viscosity value decreases, accompanied by a greater degree of thixotropy. As the shear rate is increased further (5 s^{-1}), thixotropy is still a prominent feature but there is now an increase in the apparent viscosity. At higher shear rates (10 s^{-1}), the time dependency is reduced to almost zero. However, an increase in the viscosity is seen at 7 s^{-1} before it finally decreased to the high shear rate

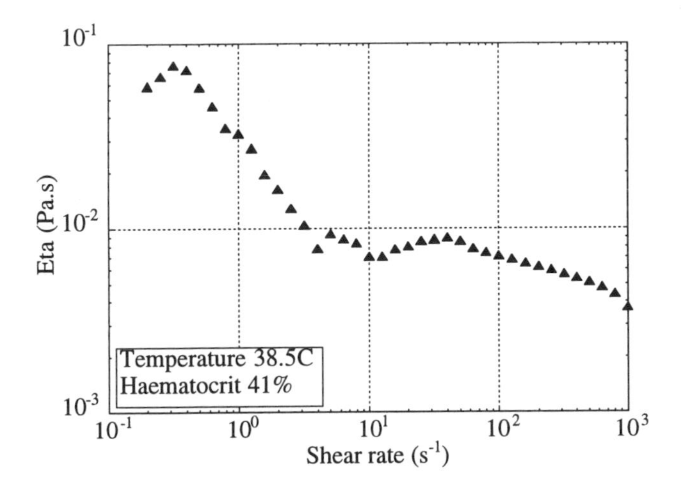

Fig. 3. Steady Shear Rate Sweep.

colloidal viscosity (25 s^{-1}), which is consistent with the corresponding value seen in Fig. 3. These results offer a reasonable confirmation of the observed anomaly.

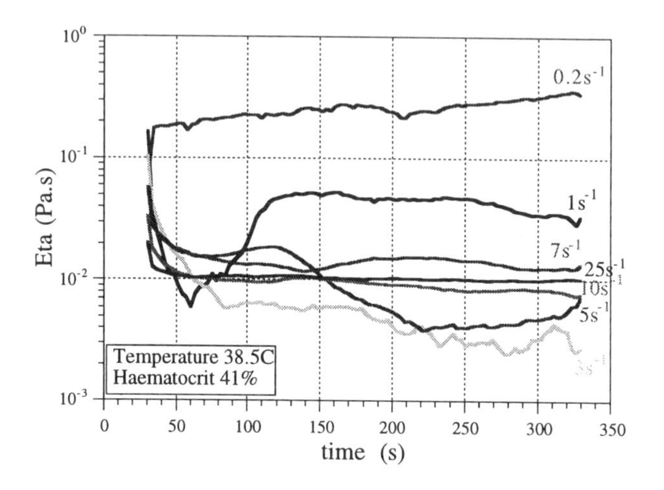

Fig. 4. Steady Shear Time Sweep.

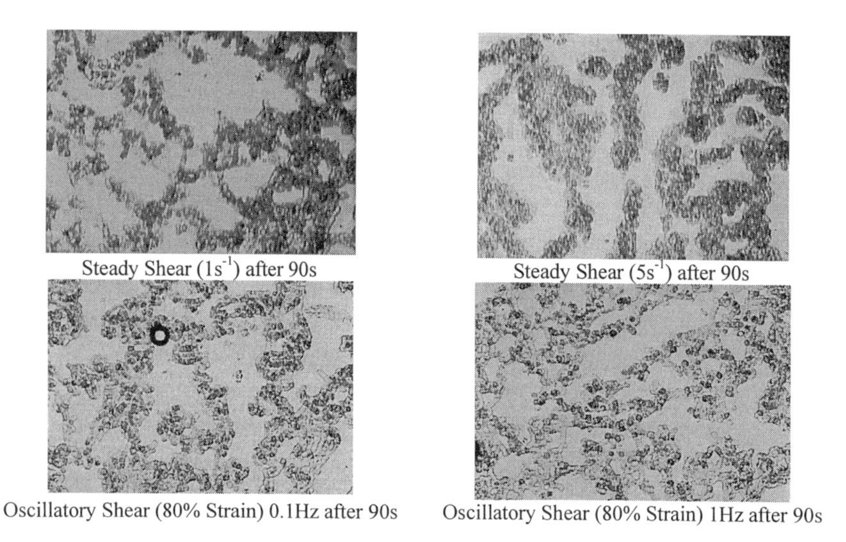

Fig. 5. Examples of the structure types observed on the CSS450 under various imposed flow conditions.

The analysis of structure formation is performed directly from video footage. Hence it is difficult to present a convincing argument on paper. Figure 5 represents a small selection of video-captured images which demonstrate the types of micro-structures observed. More information can be provided by contacting the authors at the address given.

It is seen that a number of different aggregate phases can be induced as a function of both steady and oscillatory flow conditions. In these experiments, aggregation was particularly noticeable under steady-shear conditions in the range 0.1–10 s^{-1} and under oscillatory conditions in the range 0.1–1 Hz. It was noticed that the oscillatory conditions were also more conducive to the formation of a network-type structure.

The images seen during the frequency sweep at 10% strain (not shown) demonstrate the formation of structure, although there does not appear to be a great distinction between the induced structures at the different frequencies. The images for the 80% strain case, however, showed much more distinct boundaries between the flow-induced structures. Distinct network structures are seen at frequencies of 0.1 Hz and 1 Hz whereas at10 Hz, no long-range extended structure is seen.

Summary

Cross-referencing the steady-shear rheological experiments (Figs. 3 and 4) with the video images, it can be seen that the shear rate region spanning the observed anomaly corresponds to images depicting an extended structure. The degree of aggregation seen at 0.1 s^{-1} corresponds to a high viscosity value of around 60 mPa.s. A large degree of structure formation is seen at 1 s^{-1} accompanied by a reduction in viscosity. This might be explained in terms of the formation of a cell-free plasma layer as the large aggregates sediment out.

Between 2 s^{-1} and 50 s^{-1}, different types of structure are formed, mainly as orientated cigar-shaped clumps that decrease in size as the shear rate increases. A dynamic aggregate coalescence/break-up mechanism was also observed, probably contributing to the anomalous effect.

The effect of oscillatory conditions can be seen to induce a network-type structure in the blood sample. It is very likely that this is responsible for the enhanced elastic component observed in dynamic experiments. The effect of performing frequency sweeps at different strains has also been seen to yield markedly different results, even at strains within the classical linear viscoelastic region observed in the strain sweep (Fig. 1). With reference to the video footage, it can be seen that the type of structure formed is very much dependent on the imposed strain. Efforts are currently being made to quantify the structure types as well as to determine the exact controlling conditions.

The protocols used so far to determine the rheological characterisation of blood have been based on fitting the resultant data to the Maxwell model. Although originally designed for viscoelastic polymer systems, Mackley *et al.* [5] have shown that the Maxwell model, in conjunction with a Wagner-type damping function [6], can also be used for other industrial fluids. However, it was found that this model could not fully reflect the observed rheology for the blood system, and therefore alone would not be suitable. A future model, based perhaps on the Maxwell model, will have to include parameters that account for viscoelasticity, thixotropy, the various types of structures being observed, and their rheological consequences.

References

1. Thurston, G. B. (1970) The viscoelasticity of blood and plasma during coagulation in circular tubes. In *Proceedings of the Sixth Conference of the European Society for Micro-circulation*, ed. S. Karger Basel, pp. 12–15. Aalborg
2. Chien, S., King, R. G., Skalak, R., Usami, S. & Copley, A. L. (1975) Viscoelastic properties of human blood and red cell suspensions. *Biorheology* **12**, 341–346
3. Schmid-Schönbein, H., Gosen, J. V., Heinich, L., Klose, H. J. & Volger, E. (1973) A counter-rotating "rheoscope chamber" for the study of the micro-rheology of blood cell aggregation by microscopic observation and micro-photometry. *Microvascular Research* **6**, 366–376
4. Copley, A. L., King, R. G., Chien, S., Usami, S. & Skalak, R. (1975) Microscopic observations of viscoelasticity of human blood in steady and oscillatory shear. *Biorheology* **12**, 257–263
5. Mackley, M. R., Marshall, R. T. J., Smeulders, B. A. F. & Zhao, F. D. (1994) The rheological characterisation of polymeric and colloidal fluids. *Chem. Eng. Sci.* **49**, 2551–2565
6. Wagner, M. H. (1976) Analysis of time-dependent non-linear stress growth data for shear and elongational flow of a low density branched polyethylene melt. *Rheol. Acta* **15**, 136–142

Part Four
Viscoplasticity

Introduction 347

Chapter 25. Crucial Elements of Yield Stress Fluid Rheology 351
J. M. Piau

Chapter 26. Are Plug-Flow Regions Possible in Fluids Exhibiting a Yield Stress? 372
M. M. Denn

Chapter 27. Non-Viscometric Flow of Viscoplastic Materials: Squeeze Flow 379
C. J. Lawrence and G. M. Corfield

Chapter 28. The Wall Yield of Rate-Dependent Materials 394
M. J. Adams, B. J. Briscoe, G. M. Corfield and C. J. Lawrence

Chapter 29. Strain Localisation During the Axisymmetric Squeeze Flow of a Paste 399
M. J. Adams, B. J. Briscoe, D. Kothari and C. J. Lawrence

Chapter 30. Viscoplastic Approaches in Forming Processes: Phenomenological and Computational Aspects 405
D. Perić and D. R. J. Owen

Chapter 31. Computation of Large-Scale Viscoplastic Flows of Frictional Geotechnical Materials 425
I. M. Smith

Chapter 32. Modelling of Liquefaction and Flow of Water Saturated Soil 446
F. Molenkamp, A. J. Choobbasti and A. A. R. Heshmati

Chapter 33. Analysis of Behaviour of Sand at Very Large Deformation 469
A. J. Choobbasti and F. Molenkamp

Introduction

The first viscoplastic models were the Bingham fluid, which is still commonly used by rheologists, and extensions of classical plasticity theory used to describe metalworking and soil mechanics. The contributions in this section were selected to emphasise both the similarities and differences between the two approaches. Viscoplasticity is a phenomenological description of materials that show a static yield surface which increases in size with increasing imposed rates of strain. While many materials such as hot metals exhibit this type of behaviour, the viscoplastic materials of most interest to rheologists are concentrated particulate dispersions. Given the enormous commercial value of these dispersions, a robust constitutive description of the bulk and at boundary walls has become increasingly important to process engineers who wish to exploit the growth in computer hardware and the numerical codes capable of optimising manufacturing operations. The processing environment is extremely demanding in terms of the complexity of the static and moving geometric boundary conditions, and the strains which may be large, transient and cyclic.

This section begins with a paper by Piau who provides a general introduction to viscoplasticity or "yield stress fluids", as they are usually termed by rheologists. A practical description of these fluids is offered in terms of the materials that may be processed (e.g., those that may be pumped, spread or extruded). However, they retain their geometric form after a cessation of the imposed strain. The idea of "rigid-viscoplasticity" is introduced, which is an approximation involving the neglect of the elastic strains and is applicable at large strains. The elasto-viscoplastic description draws upon classical yield criteria. However, most concentrated particulate suspensions show viscoelastic behaviour at stresses less than those required for yielding and this behaviour is difficult to formulate within a viscoplastic constitutive relationship. The resulting creep response is perhaps a factor in the continuing debate on the existence of a true yield stress. Piau also considers other facets that make the study of yield stress fluids demanding. These include localisation and thixotropy which fall under the general headings of path and time dependence. Other manifestations of this complexity include the various forms of strain hardening and pressure dependent yield criteria exhibited by powders and soils. Piau also employs the notion of a "gel", which may be translated as a material having a network structure in the static state that "fails" under the yield condition. The delayed recovery of such a structure invariably means that the viscoplasticity of particulate systems is coupled with thixotropic behaviour.

Because of the nature of the materials, the experimental techniques for characterising constitutive behaviour are rather different in the various disciplines. However, uniaxial compression is common in rheology, metalworking and soil mechanics, being termed as "squeeze flow", "upsetting" and "unconfined compression" respectively. As Piau points out, there are few simple flows for viscoplastic fluids and, in the case of squeeze flow, a simple velocity field is only developed for a frictionless wall boundary condition when a radial extension occurs. Denn reviews the history of squeeze flow, particularly with respect to the lubrication solution, with no-slip boundary conditions, which has proven to be controversial since its publication by Covey and Stanmore in 1981. As first pointed out by Lipscomb and Denn in 1984, the problem resolves itself to the non-existence of yield surfaces. In this context, Denn points out that the key factor is the finite height of real specimens. We know from the application of slip line field theory, using the cycloidal approximation, that a pair of central wedges is formed which become infinitely small as the ratio of the height to the width tends to zero.

Lawrence and Corfield re-evaluate the lubrication solution and formally show the equivalence between the solid and fluid formulation of the problem. While this has been widely appreciated in the engineering field, it is less familiar to rheologists. The conventional lubrication solution leads to a central unyielded plug flow region which extends radially and is hence kinematically inconsistent. Lawrence and Corfield show that, by re-scaling arguments, the strain rate in the "plug" flow region is small but finite so that the yield criterion is satisfied. Furthermore, this avoids the need to introduce more complex viscous descriptions such as the biviscosity relationship discussed by Denn.

Wall slip is a common feature of particulate dispersions and, starting with Sherwood and Durban in 1996, there has been a recent interest in deriving the effect of such boundary conditions on squeeze flow. Experimental means of obtaining the behaviour at walls are, of course, critical in any simulation scheme. Adams *et al.* consider such an analysis for the simple Tresca and more realistic Herschel-Bulkley type wall boundary conditions. The Tresca condition is commonly employed in metalworking and corresponds to the wall shear stress, being some fraction of the bulk value. The proportionality constant is sometimes considered a contamination (for example, oil and oxides) factor. However, the second wall condition appears to be more appropriate for pastes which show a velocity dependent wall slip. This is exemplified by the good agreement between measured wall radial pressure distributions and those calculated using this condition.

The following paper by Adams *et al.* also considers squeeze flow but from an experimental viewpoint. It shows that by varying the thermal, and hence frictional conditions, at the platen walls, it is possible to generate a wide range of displacement fields including strain localisation. From a processing perspective, this may be an unwanted phenomenon due to the generation of weak planes. The paper concludes with a summary of some of the key theoretical and computational work which attempts to understand this complex area.

Finite element analysis (FEA) is probably the most straightforward tool for examining shear banding, particularly with the advances made in contact-friction algorithms, enhanced element technology and adaptive meshing. Peric and Owen describe these advances which have become available in the last decade and also consider those that have been made in elasto-plastic formulations. Early FEA was most suited to linear engineering problems with strains limited by the Lagrangian framework. However, with many of the limitations removed, such as the ability to impose large strains and automatically remesh in regions of high stress gradients, this technique is almost certain to find applications in rheology and process engineering. For example, the Lagrangian framework is much more appropriate for free surface flows than classical CFD schemes based on an Eulerian framework.

The last three papers are provided by civil engineers and consider geotechnical problems from the continuum viewpoint. While it is debatable whether geotechnical flows are strictly fluid flows, rheologists are increasingly required to characterise difficult soft or wet solids as intermediate or final product forms of the type described by Smith. He describes the application of FEA but, unlike previous papers, a non-associated flow rule is invoked to account for the non-coaxiality of the stress and strain rate vectors which arises in compressible/dilational materials. Dilation is a critical factor in soil mechanics often because the primary grains or sub-unit agglomerates of clay particles are relatively large and angular. The much simpler case of associated flow is only valid when the angles of internal friction and dilation are equal. Smith introduces the idea of an "effective" stress, which is the result of subtracting the pore water stress from the total stress, and the use of the triaxial test method for characterising materials over a range of stress paths. As in the previous paper, mesh refinement is crucial for FEA to deal with large strains including those that occur in regions of localisation.

The triaxial tests described by Smith are carried out in an "undrained" state, namely, the pore water is unable to escape during the compression cycles. This theme is continued in the paper by Molenkamp *et al.* who consider liquefaction in which a high pore water pressure causes a loss of stiffness. In some instances, this may be sufficiently severe so that the soil behaves like a fluid. The constitutive modelling is different from that described previously involving the viscoplastic stress overshoot model. Here, a combination of a linear viscous and a complex elasto-plastic model is employed. As in all soil mechanics models, this requires a relatively large number of material parameters which necessitates the use of triaxial testing in order to impose complex stress paths. Here again, the model is implemented using FEA and the approach is illustrated for drained deviatoric loading in triaxial compression.

The final paper by Choobbasti and Molenkamp continues the theme of compressible granular materials. The interest here is the ability to describe such material at large deformations in the ultimate state. The flow and fracture of "wet" solids (cohesive powders) are gaining increasing commercial importance and the formulation described has enormous potential in this area.

In conclusion, this section has considered viscoplasticity in the broadest possible context starting from the classical rheological approach and, in the final papers, the plasticity based mechanical engineering and geotechnical approaches. It is clear that considerable progress has been made in the phenomenological or constitutive description and numerical modelling of the deformation and flow of these complex materials. One of the most difficult aspects remains the determination of realistic wall boundary conditions and the associated material parameters. With hindsight, perhaps a greater and more useful focus on this aspect could have been included. We would expect that improved microscopic descriptions, both from a theoretical and numerical basis, would be one of the major developments in the near future.

Dynamics of Complex Fluids, pp. 351–371
ed. M. J. Adams, R. A. Mashelkar, J. R. A. Pearson & A. R. Rennie
Imperial College Press–The Royal Society, 1998

Chapter 25

Crucial Elements of Yield Stress Fluid Rheology

J. M. PIAU
Laboratoire de Rhéologie, BP 53,*
Domaine Universitaire, 38041 Grenoble Cedex 9, France
E-mail: jmpiau@ujf_grenoble.fr

The question of the existence and nature of yield stress fluids, as well as that of thixotropy, has been raised. Positive and proper answers can be given by reviewing the preferred vocabulary and the contributions to this topic from different fields of science and technology. A major difficulty is that most flows of yield stress fluids are complex, even when the simplest boundary conditions, such as those found in a rheometer, are used. It thus appears of interest to list the phenomena at hand and to review the know-how in yield stress fluid rheometry. Complex boundary conditions allow yield surfaces and dead zones to be observed experimentally or simulated. It will be shown that the treatment of field equations is not generally a mere extrapolation of classical fluid mechanics developments. In particular, normal stresses deserve special attention. Applications will be given to illustrate these points for several colloidal systems or suspensions.

Introduction

The French Rheology Group gives the following definition of rheology in documents published since 1994 (see, for instance, each issue of *Les Cahiers de Rhéologie*):

*University of Grenoble (UJF and INPG) and CNRS (UMR 5520)

(i) Rheology (from the Greek *rheo* "to flow", the suffix *logy* indicating a science) is the science of flowing matter, of the stresses which must be applied, and of the structural changes which result.
(ii) Simply deformed matter is a particular case. Phenomena which belong to Newtonian fluid mechanics, to Hookean elasticity, or correspond simply to the flow of electrons are limiting cases.
(iii) Rheology is universal. It can be met in all the domains of human activity as well as in natural phenomena.
(iv) Rheology is multi-disciplinary. Its basic tools are found in mechanics, physics, chemistry, mathematics and biology. It is useful to each of these fields.

Multi-disciplinarity, as well as the growing number of researchers from very different fields who become involved, indicate that some open rheology meetings may have points in common with the tower of Babel. As long as rheology has not been studied in some depth in a laboratory, rheology may be the name of the room where the most expensive rheometer purchased recently from a rheology company is turned on to deliver what is then called the rheology of the sample, as if it was simply synonymous with viscosity. At the other extreme, rheologists have yet to discover many facets and to agree on many new concepts in this science, where complexity has to be faced and at least progressively mastered. All these difficulties apply to, in particular, the case of yield stress fluid rheology.

Several papers have been published on the possible existence and definition of a yield stress fluid. They have been reviewed in recent letters [12, 41]. However, a different and more positive point of view will be offered in the second section of the present paper. Comments will be given on the meaning words may take, depending on the scientific field, some of the phenomena and difficulties at hand will be pointed out, and the interest of interfacial properties in yield stress rheology will be underlined.

Big changes, which appear with respect to rheometry of yield stress fluids as compared to polymer rheometry, will be explained in the third section. Material samples will be shown to react in a sophisticated way to imposed boundary conditions. Indications will also be given on new and old techniques in shear rheometry.

Some of the specific features of Bingham fluid mechanics will be shown in the last section of this paper on complex flows without considering slip at the wall.

Yield Stress Fluids Rheology

Rheology draws the attention of a range of specialists to the pertinent experimental windows, and to the appropriate theoretical and numerical tools for solving the problem of flowing material at hand. They all participate in defining yield stress fluid rheology.

The Case of Newtonian Fluids

Several centuries after the first impressions of elasticity and viscosity were reported, applied mathematicians have defined clearly Newtonian fluids and elastic solids. A "Newtonian fluid" has very little in common with the writings of Newton. It is the composite name usually given to a mathematical set of four linear constitutive equations for stress tensor, heat flux vector, free energy and entropy. This set of black-box type equations is presently written in applied mathematics and engineering in order to build a mathematical model in Euclidean vector space. In addition, it is based on the equations for conservation principles, the equations for changes in parameters in constitutive equations with thermodynamic state, and on a series of initial and boundary conditions. Engineers hope that this complete mathematical model can be worked on and show some usefulness with respect to a practical situation where a particular species of matter is stressed in a particular experiment (or set of experiments). However, nobody can actually buy any Newtonian fluid in any shop. This kind of substance does not exist anywhere. Moreover, given any material sample under chosen conditions of temperature and pressure, it is always possible to prove that its behaviour cannot be represented by a Newtonian mathematical model in well-chosen processes or experiments. It is clear that no one-to-one identification is possible between mathematical models and matter.

Constitutive Equations for a (Matter-Experiment) Couple

Any constitutive equation is written in the hope that it can be used to model a (matter-experiment) couple reasonably well at least. "Matter" means that a chemical species or formula is specified. One consequence is that a mesoscopic structure is fixed with particular length, time, and other scales. "Experiment" tells exactly what process is applied to this matter, the shape and length, time, stress, scales of the time, and space boundaries. Inside the set of equations defining the mathematical model, the particular experiment being studied is specified through limiting and boundary conditions. However, this particular experiment also has a deep influence on any relevant constitutive equations. For instance, it is common to hear that some oil samples are not compressible, though this does not describe the intrinsic macroscopic properties of the oil at all. One must refer, at the same time, to an oil and to a particular set of experiments to know whether the incompressibility constitutive equation is useful in engineering. For instance, for a pressure transient in pipe flow, wave propagation need not be considered for short pipes and low oil compressibility. Hence the validity of the incompressibility constitutive equation depends on the length of the pipe used in the experiment as well as on the physical properties of the oil. This shows that engineers must select constitutive equations considering both the material and the process.

Yield Stress Fluids

Though it was identified more recently in science, the situation of yield stress fluid rheology can be described in the same way. "Yield stress fluids" is a shortened expression, commonly used in fluids engineering, which loosely designates materials (and the dedicated scientific activity) that can be pumped, spread and extruded, that is, processed without definitely loosing their gelled appearance. Conditions exist where they are seen to be strangely able to resist gravity, or surface tension forces, despite evidence showing a low viscosity. In fact, the main point to consider for a clear definition is, again, that this activity is linked to a series of typical mathematical models for stress constitutive equations introduced in plasticity.

The main mathematical idea of a plasticity definition is to specify a smooth convex yield surface in stress space with a constitutive equation inside the surface (elastic, for instance), and another different constitutive equation outside (viscoelastoplastic, for instance). Strain-hardening and anisotropy can also be accommodated. The yield surface in stress space introduces a characteristic stress parameter, or yield stress value, in addition to other characteristic time and viscosity values which may be needed. Constitutive equations written in plasticity show some usefulness for some (matter-experiment) couples.

Very simple rigid-viscoplastic mathematical models written by Bingham [5], Oldroyd [29–30] and Prager [38] form the main reference for the young field of yield stress fluid mechanics. Much more is clearly needed in the future, and several other constitutive equations have been introduced recently to take into account possibly more complex or realistic behaviour [14, 15, 45]. Nevertheless, all the different aspects of yield stress fluid mechanics and physics also need to be developed.

As with Newtonian fluids, a sample that is a Bingham fluid does not exist. Moreover, given any gel, it is always possible to prove that some aspects of its behaviour cannot be represented by a Bingham mathematical model in well-chosen processes or experiments. It is again clear that no one-to-one identification is possible between mathematical models and matter. In addition, models with a yield stress may prove useless for a gel at high stress values where a Newtonian model can prove to be adequate in some circumstances.

Instead of yield stress fluids, or plasticity, some researchers prefer to use the term physical gels, meaning that mesoscopic links exist between elements, thus creating a percolating network within matter. These links, being weak in some sense at least, are not irreversibly broken as would be the case with covalent links. They can reform after they have been broken by flow conditions (gelatine is not such a physical gel). Phase diagrams are introduced to identify the physico-chemical conditions of the existence of the gel.

A combination of macroscopic, as well as mesoscopic, experiments and material physico-chemical descriptions assists considerably in deciding whether rigid-viscoplastic mathematical models are physically relevant for a material in a process. It is most important that they all quantify phenomena in the correct experimental window pertinent for the yield stress phenomena.

As an example, let us consider concentrated dispersions of nanometric silica particles in water stabilised by long-range inter-particle forces [31]. They behave as soft solids. Under low stress, they show viscoelastic behaviour with viscous creep over long time scales. A shear stress plateau appears at large shear rate values (Fig. 1). This results from a localisation of shear in a fractured region of the sample. Water is released into a fracture which allows the solid regions to flow past each other. At rest, the solid regions re-adsorb this water and heal the fractures. On the basis of the repulsive forces between spherical silica particles, as well as on flow visualisations, one is clearly dissuaded from modelling this flowing suspension considering that it behaves in the bulk as a Bingham yield stress fluid.

Yield stress fluids introduce a change with respect to the usual objectives of rheology. The reason is that, in addition to bulk properties, it becomes necessary to study friction laws for slip at the wall and fractured zones. A rheology of interfaces needs to be developed.

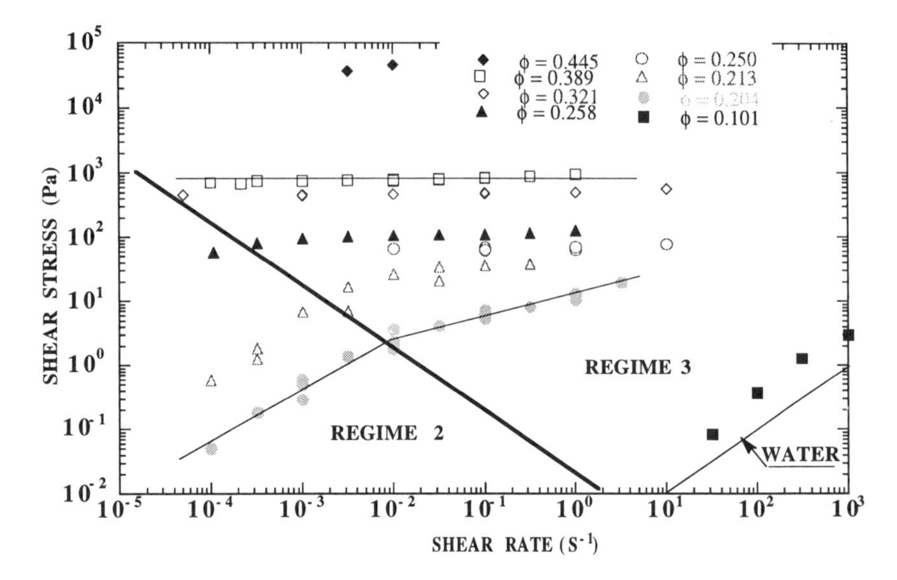

Fig. 1. The shear stress plateau represents fracture properties of the silica suspension [31].

Some data have recently been obtained on polymer slip at the wall for high wall shear stress levels. Yield stress fluid slip is quite another problem, as it is expected to be most noticeable for small wall stress values.

A sample which looks like an elastic solid, at least for some experimental conditions where it is not stressed too much, cannot be expected to respect the classical fluid mechanics condition of adherence to the wall. When a sample is made of elements which are big enough with respect to wall roughness, the same problem arises from the impossibility of placing a particle inertia centre at or inside the wall, and from the corresponding concentration gradient. In summary, the many details on the chemical formulation of the material studied, of the wall surface properties and roughness, can have a major influence on the friction law for slip at the wall.

The point can be illustrated with an example taken from the field of uncured rubber extrusion. A percolating network of carbon black particles and adsorbed polymer chains has been evidenced. The yield stress is also high. Such a mixture slips along smooth walls at low stress levels. Figure 2 shows data obtained for pure high molecular weight, EPDM, which slips at high shear stress values and follows a nonlinear law. One of the well-known procedures of rubber-making is to add a non-compatible lubricant. It migrates to the wall and

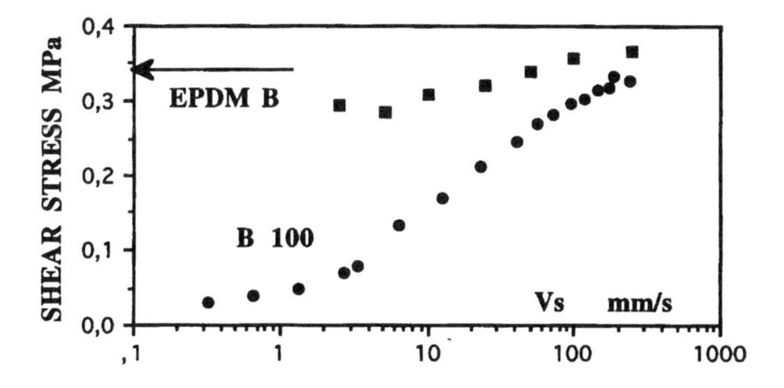

Fig. 2. Friction laws measured for a pure EPDM and for an uncured rubber formulated with the same EPDM [26].

helps the yield stress fluid to slip as easily as possible regardless of the stress level while changing the friction law from a nonlinear one, which generates flow instabilities [32], to a monotonic increasing law [26].

Thixotropy

Thixotropy may also be observed in yield stress fluid mechanics. It sometimes shares with yield stress fluids the undue privilege of being considered an obsolete and unnecessary concept not worthy of further consideration [25]. Thixotropy is introduced in engineering to state that the flow properties (like the apparent viscosity) of a substance change with time as it is stressed in a process, whereas the same initial properties are recovered after the substance has been left at rest for a while. Some very common paints are indeed sold for their thixotropic character which is so evident that no doubts have been cast on the concept.

Along the same lines, thixotropy can be linked to mathematical models with structural parameters and evolution equations for these parameters [9, 25]. Thixotropic models are intended to be more general than those used in nonlinear viscoelasticity. Thixotropy corresponds to the tendency of matter to exhibit a change with time in its reference properties (such as reference viscosity, yield value, time scale, elasticity modulus, loss modulus, ..., or conductivity) when stresses are applied and to recover progressively these reference properties when stresses are suppressed. Hence thixotropic effects are more general than elastic effects. The latter are defined as an instantaneous tendency of matter to recover its reference configuration when stresses are suppressed. The existence of matter reference properties do not necessarily involve the existence of a reference configuration. Elastic effects are not necessarily thixotropic in nature, but they are not excluded from thixotropy.

From the mesoscopic point of view, structural information is emerging to show that collective structural changes and rearrangements of the elements (molecules, particles, aggregates), which constitute the material sample, are a common characteristic of thixotropic systems. Collective rearrangements need a long time to be completed after the previously stressed sample is maintained macroscopically at rest. This is in contrast to the fast local changes and rearrangements at the level of elastic individual elements, such as may be considered in polymer melts.

A model 2D example can be considered. If some parsley or basil seeds are dropped at random on the surface of a glass of water, under the influence of capillary forces the seeds move across the surface and may come into close contact with each other. They aggregate and rearrange themselves very progressively and collectively over the entire field which they cover. This takes a long time when compared to the time needed for two seeds to assemble over comparable distances. The structure they build at rest becomes stronger and stronger until a limiting result is reached. The final structure may be made up of lines of seeds when the seed concentration is low. It fills the space available more densely when the concentration increases, but with numerous voids. The physical model so obtained shows many of the characteristics typical of yield stress fluid mechanics discussed in the third section of this paper. It also shows strong thixotropy. When stressed, it may slip easily at the wall and fracture. When small stresses are applied by blowing air on the surface, the percolated physical model seems to be rigid or elastic. It appears to be made up of large-scale interconnected aggregates. When larger stresses are applied by blowing stronger or with a fork, broken links between the main aggregates and free space available allow the sytem to flow. However, the aggregates are larger than individual seeds. Hence any study of an equivalent continuum will involve the consideraton of a very big sample as it must be larger than the aggregates. Very large stresses allow individual seeds to separate and the physical model probably approximates a Newtonian system. The dimensions of the structure decrease and a smaller sample than before seems to be required. When macroscopically at rest, the system again progressively builds an aggregated and rearranged structure under the influence of attracting

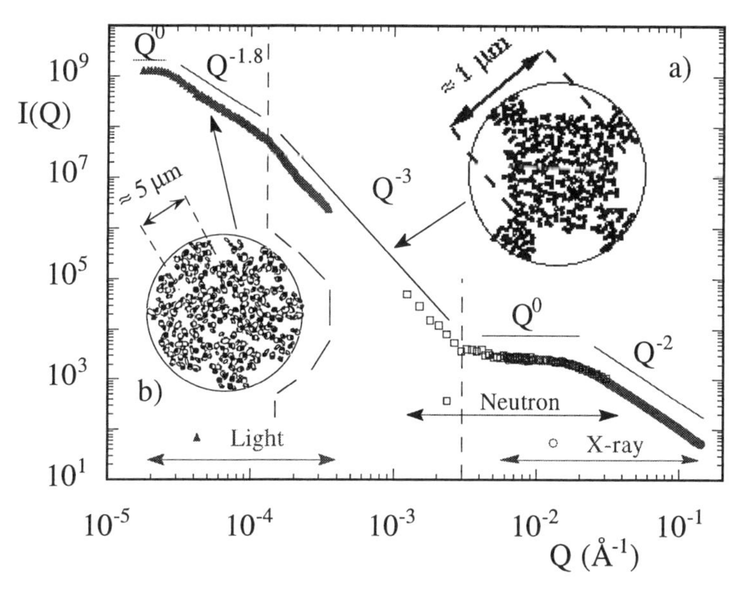

Fig. 3. Small-angle x-ray and neutron scattering, and static light-scattering from aqueous suspensions of Laponite at rest, at volume fraction 1.6% [NaCl] = 10^{-3} M, pH = 9.5. (a) Dense micron-sized aggregates. (b) Mass fractal made of alternance of aggregates and voids [37].

capillary forces. It may be noticed that the seeds can be introduced in many different ways. Thus, reference properties depend not only on concentration, but also on the preparation process. For instance, in the absence of any equivalent to Brownian motion, compact crystals could be built by placing each seed. Many colloidal systems with dominant attraction forces between particles may show similarities to this 2D system.

This is the case with clay (Laponite) suspensions, the micro-structure of which has recently been revealed using a combination of several radiation sources [36, 37]. The structure of these nanometric size clay suspensions has a characteristic length scale of the order of a micron (Fig. 3).

The Complexity Observed for Shear Rheometry

Boundary Conditions

For polymer solutions and melts, which do not change phase in the field, homogeneous velocity fields can be obtained experimentally so that the flow and stress fields in the bulk can be controlled from some non-deformable solid walls at the boundaries of the sample.

However, when applied to yield stress fluids, the same techniques generate flow fields which may be very complex indeed. This complexity is briefly described below together with some of the know-how which has been developed to maintain, extend and examine the existence of homogeneous velocity fields. No data can be reported convincingly without referring explicitly to this know-how.

Short List of Difficulties and Know-how

Though it is most often focused on shearing, considerable advances have been made with regards to yield stress fluid rheometry [27].

The data obtained allow constitutive equation parameters to be calculated and comparisons to be made between macroscopic properties and microscopic physical measurements. This gives an indication of the relevance of considering a yield stress.

However, in addition to all the usual considerations, numerous difficulties must be overcome in the shear rheometry of yield stress fluids:

(i) Many fluids are biphasic or pluriphasic media. Phases may be made to flow as a single continuum or separately, depending on boundary conditions. For instance, it is possible to drain, under gravity, the oil of a lubricating grease plug maintained inside a tube by a porous plate and to replace this oil with another oil or solvent. In this experiment, grease is no longer a yield stress fluid as it may be when it is pumped. Its soap skeleton is a rigid porous media adjacent to the porous plate and oil flows through it following a Darcy law. Other examples are given by clay for tile-making, ceramic slip or propellant, which, when forced in a die, may show solvent migration towards the walls.

(ii) A procedure is needed to generate a well-defined intial state of the samples before transient tests are performed. This can be achieved by systematically pre-shearing the samples.

(iii) The fluid is expected to slip at the wall. This important point has been introduced above and will be considered separately here after.

(iv) Sample edge effects may be stronger than for polymers.

(v) The sample may fracture in the bulk and when this occurs, data must either be discarded or interpreted on the basis of additional information given by a close observation of the flow field.

(vi) Controlled stress rheometry opens the question of effective control of slow motions and low stresses.

(vii) Many materials contain solvents or components which evaporate significantly under room conditions during the tests. This problem can be dealt with quite easily. Different kinds of solvent traps, sponges filled with solvent, or vapor generators can be adapted on the tools.

(viii) Sedimentation (or creaming) of particles is very important for suspensions of dense (or light) solid particles. This results rapidly in the appearance of a supernatant layer at the top of the sample and an increased concentration at the bottom. It may also result in the chaining of particles in the bulk of the sample. For several years, we have used tools adapted to such a situation (Figs. 4 and 5). Made of grillage, they are placed in the bulk of the suspension, a zone where concentration does not change too much. They do not prevent particles from migrating through their surfaces and they also mitigate slip efficiently due to their intrinsic roughness. Several practical combinations can be built from grillage tools [16] to some kind of Mexican spurs [43]. In addition, they can be designed as cone-plate or plate-plate tools. If an increase in the torque value is needed, they can also be superimposed like the tools in Maxwell's rheometer for gases.

(ix) The migration of particles during rheometer filling, the centrifugal forces taking place during the test, and the stresses gradients inside the rheometer between the gap and end zones can strongly affect Couette devices.

(x) The particle diameters may be so large, with respect to the gap, that continuum hypotheses are not respected, or they may interact with the apex of the cone in a cone-plate device. It is clear that big rheometers are needed for materials with large heterogeneities [10].

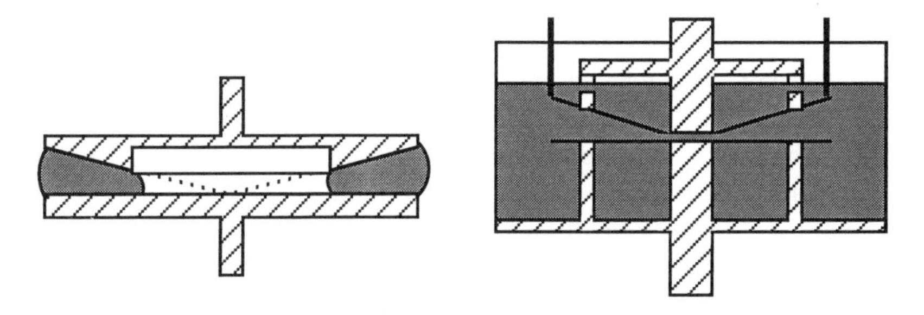

Fig. 4. Truncated cone (on the left), grillage tools (on the right).

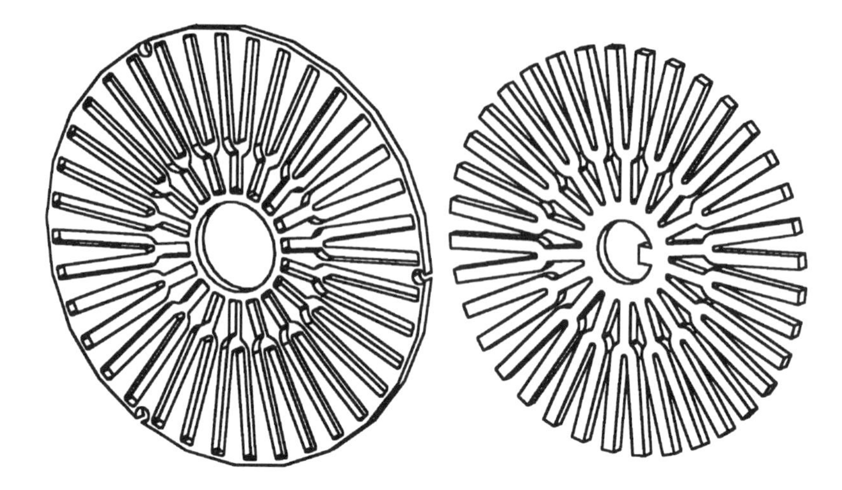

Fig. 5. Mexican spur tools: tool on the left is fixed at its periphery, tool on the right is fixed on its axis.

When problems with cone apex occur, plate-plate tools can be used instead of cone-plates. It is also wise to use severely truncated cones (Fig. 4) or cones made only of an annulus with a hole in the axis region up to half a diameter in width [7].

Instruments for Empirical Tests

Instruments have been used in industry for a long time to perform empirical tests and characterise pastes and gels [6]. Even when they are apparently more sophisticated than a Ford cup, Zahn viscometer or slump test, they measure parameters that may be poorly defined. However, from practical experience, they have been found to be easy to perform, and are rapid, useful and inexpensive. These tests are thus of great interest. In addition, they offer solutions which can be adapted, at least in part, in rheometry and in evaluating the pertinence of introducing a yield stress concept.

The vane-test is a safe classical method from soil mechanics. In this method, the slip at the wall is reduced. It can be performed easily on many materials with commercial rheometers equipped with simple adapted tools.

Bostwick or Adams consistometers can be used to measure the limits a sample flows to, under gravity, before it stops due to the yield value.

Penetrometers indicate penetration in a given time for different types of cones and needles designed to compensate, at least in part, slip.

All these instruments are to be used with a well-prepared sample. In the field of lubricating greases, standards exist on the preparation of samples using a worker. Workers have proved to be instruments of permanent usefulness as samples are prepared correctly with many different materials.

Mitigation of Slip at the Wall

Some particles or fluids may develop controlled interaction forces with the wall to reduce slip effects or to make them disappear in the range studied. This is the case when silicone is stuck at a wall using a primaire [22], or when some grease is rubbed manually on the wall prior to the introduction of the grease sample itself [34].

Slip can be mitigated using tools with rough surfaces [23] regardless of the physico-chemistry of the interface. Grooved tools or sand-blasted tools are less efficient than fakir net type surfaces. This is well-understood in vane-test crosses or striated Mooney rheometer tools for rubber.

Efficiently rough surfaces can be prepared either by gluing abrasive paper, abrasive powder or threading capillary tubes. In the case of big rheometers, more impressive roughnesses are prepared by welding big metal bars at the surface of the tools.

For free surface flows, many hydraulics solutions can be adapted, including laying a metal or plastic grillage sheet along the bed of the channel.

Many rheometer users do not dare (or are not allowed) to consider the rheometer they use as an ordinary apparatus which has to be handled, completed and modified on several points. But rheometers were mainly designed for polymers and they are often not equipped with respect to colloidal dispersions and suspensions. A number of equipment is required to be attached permanently to the rheometer in the field of yield stress fluids: a video camera and monitor, fine insoluble powders with good optical contrast [39], a variety of rough tools, sample preparation instruments and solvent traps.

The incidence of slip can be illustrated with the data obtained for human blood by using a Contraves Low Shear 40 rheometer equipped with a Couette cell (see Fig. 6). Human heritrocites have a diameter of about eight microns and can aggregate at rest. Standard smooth tools, and the same tools roughened using 56 micron silicium carbide particles, have been used successively. The large differences obtained are due to heritrocites slipping on standard tools. Rough tools allow heritrocite interactions to be studied [35].

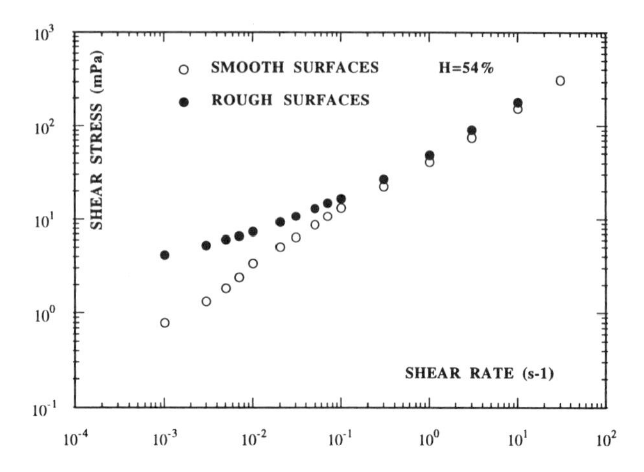

Fig. 6. Shear stress as a function of shear rate for red cells suspended in plasma for smooth and rough surfaced geometries at temperature 25°C. Cell concentration is 54% [35].

Bingham Fluid Complex Flows

In spite of the drastic simplifications introduced in order to obtain a Bingham fluid model, it remains nonlinear and a theoretical analysis of the mechanics of Bingham fluids is still difficult [20, 33, 46]. Great care needs to be exercised in computing flow fields [1, 4, 28].

Simple extensions of classical fluid mechanics methods may prove inadequate. The postulated yield surfaces introduced in [8, 11, 12, 17, 21, 40, 42, 44] are incompatible with the field equations.

Several experimental techniques must be carefully mastered prior to performing significant experiments with yield stress fluids. Fluids should be kept free from trapped gas bubbles. This is possible when a general pressurisation pot is installed on pipe and pump circuitry [19], or when free surface flows are organised with shallow deaerating zones and velocities are kept to a minimum [2] whenever they do not have to be large. Static pressure measurements, through measuring the deformation of very thin walled tubes, have also proved to be a useful technique [3, 19].

Yield Surfaces

Yield surfaces have mainly been studied on theoretical grounds since they are more difficult to determine experimentally.

The limits of plug flows in fully developed pipe flows, or uniform open channel flows, have been identified and methods do exist to calculate their shapes for complex cross-sections [18].

A sheared boundary layer has been shown to appear along the wall or surface of slender obstacles (Oldroyd's moving knife and jet problems [30]). It is separated from the unyielded fluid by an outer yield surface. A similar separating outer yield surface appears for blunt obstacles such as spheres [4].

Stagnation points seem to lead generally to the appearence of dead zones which have been computed for the squeeze film problem [28] and sphere creeping motion [4], and have been observed experimentally at the downstream end of the annular vortex in an abrupt divergent [13, 19, 24].

Dead zones have also been computed and observed experimentally in the corners of abrupt convergents and divergents or holes in the wall [1, 3, 19, 24].

A small velocity gradient indicates Bingham fluid flows and the second stress invariant is larger than the yield value. For confined flow (such as pipe, lubricating, squeezing and radial flow) and free surface flow singularities, non-zero elongation velocity gradients appear either along the axis or at the free surface. Hence the yield surface of a rigid Bingham fluid cannot pass continuously from one side to another in the singular section. When such a singularity is placed between two long cylindrical sections, the upstream and downstream plug flow zones are not connected. Numerous papers have overlooked this point.

Normalised Equations in a Long Domain and Dimensionless Parameters

Assuming that fluid motion is isochoric, the constitutive equations for a Bingham fluid are then written using two scalar material parameters, s and μ, corresponding to the yield stress value and the Newtonian viscosity observed at high shear rates, respectively.

The fluid domain boundary in space is assumed to be much longer than it is thick. Thus, only two characteristic length scales, L in the direction of its length and h, which is a measure of its thickness, have to be introduced to describe its shape.

It is reasonable to reduce the set of scalar equations produced hereafter to the two components of momentum along the x and y axes only.

A set of dimensionless variables is chosen so that they are normalised: X, Y and T are space and time U_0 is the reference velocity and the parameter, ε, is the ratio of h/L. U, V are the two normalised components of velocity. The normalised component of the strain rate intensity in the wall region is ζ.

In regions close to a free surface or an axis of symmetry, ζ should be changed to $\zeta\varepsilon$ for a correct estimation of the strain rate.

In terms of the dimensionless variables, the full equations of motion for 2D flows are written in wall regions [33]:

$$\begin{aligned}\dot{U} + P_{,X} &= \frac{1}{\varepsilon \mathrm{Fr}^2} \sin \alpha + \frac{1}{\mathrm{Re}} \left\{2\varepsilon^2 U_{,XX} + \left[U_{,Y} + \varepsilon^2 V_{,X}\right]_{,Y}\right\} \\ &\quad + \frac{\mathrm{Od}\,\varepsilon}{\mathrm{Re}} \left\{2\left(\zeta^{-1} U_{,X}\right)_{,X} + \left(\zeta^{-1}\left[\varepsilon^{-2} U_{,Y} + V_{,X}\right]\right)_{,Y}\right\} \\ &\quad \times \{\text{i.e.}\quad t_{xx,x} \quad \text{and} \quad t_{xy,y} \quad \text{yield terms}\}, \end{aligned} \tag{1}$$

$$\begin{aligned}\varepsilon^2 \dot{V} + P_{,Y} &= \frac{1}{\mathrm{Fr}^2} \cos \alpha + \frac{\varepsilon^2}{\mathrm{Re}} \left\{\left[U_{,Y} + \varepsilon^2 V_{,X}\right]_{,X} + 2\, V_{,YY}\right\} \\ &\quad + \frac{\mathrm{Od}\,\varepsilon}{\mathrm{Re}} \left\{\left(\zeta^{-1}\left[U_{,Y} + \varepsilon^2 V_{,X}\right]\right)_{,X} + 2\left(\zeta^{-1} V_{,Y}\right)_{,Y}\right\} \\ &\quad \times \{\text{i.e.}\quad t_{xy,x} \quad \text{and} \quad t_{yy,y} \quad \text{yield terms}\}, \end{aligned} \tag{2}$$

where α is the angle at which the plane Oxz is inclined to the horizontal, and the y-axis is located in the same vertical plane as x. P is pressure normalised by inertia.

The coefficients in Eqs. (1) and (2) are elementary functions of parameter ε, and of a list of dimensionless numbers which can be named after Reynolds, Froude and Oldroyd:

$$\mathrm{Re} = \frac{\rho U_0(\varepsilon L)}{k} \varepsilon = \varepsilon.\mathrm{Rh}; \quad \mathrm{Fr}^2 = \frac{{U_0}^2}{g\varepsilon L}; \quad \mathrm{Od} = \frac{s\,\varepsilon}{k}\left(\frac{\varepsilon L}{U_0}\right). \tag{3}$$

Definitions (3) simplify a presentation and discussion of the different terms in the equations.

Moreover, pressure is not necessarily limited to being normalised by inertia. Note that if liquid consistency or gravity are used to normalise pressure, then instead of inertia, the quantities multiplying the normalised pressure gradient terms in Eqs. (1) and (2) will be $1/\mathrm{Re}$ and $1/\mathrm{Fr}^2$, respectively, and not 1. Naturally, pressure gradient should then be grouped together with viscous terms or gravity terms instead of being written together with inertia. This was done in Eqs. (1) and (2).

The Importance of Normal Stress Derivatives

Two terms appear in Eq. (1) which are multiplied by Od ε/Re. The first one on the left comes from $t_{xx,x}$ and the second from $t_{xy,y}$. In Eq. (2), the term which is multiplied by Od ε/Re comes from $t_{yy,y}$. Since all three terms have the same order of magnitude, it would be irrelevant to retain only one of them (derived from $t_{xy,y}$) while neglecting the remaining two (derived from normal stress derivatives) without a good reason.

Considered together, it is apparent that in Eqs. (1) and (2) the inertia, pressure, gravity, viscous and yield stress terms are of the order of 1, 1, $1/\mathrm{Fr}^2$, 1/Re and Od ε/Re, respectively. Far from the walls, the yield stress terms become larger and their order of magnitude is Od/Re.

As an illustration, the flow being studied takes place along an inclined plane and the x-axis is located there [33]. The ordinate, y, is now the distance to this plane. At the free surface $y = h(x)$, the atmospheric pressure boundary condition used is $p = 0$. Thus, neglecting surface tension effects, the normal stress, Σ_{yy}, at the free surface must be zero.

From the analysis given above, it is possible to evaluate, due to normal extra-stresses, the relative influence of the yield stress value on the flow field by comparing $\cos\alpha/\mathrm{Fr}^2$ and Od/Re in Eq. (2), which are adapted to regions of high apparent viscosity far from the walls. It appears that

$$\frac{\mathrm{Fr}^2\ \mathrm{Od}}{\mathrm{Re}} = \frac{s}{\rho g h}, \tag{4}$$

that is, the yield stress value must be compared to the hydrostatic pressure level created by gravity.

It is possible to show that the yield stress value has a profound influence on the structure of the equations and the solution to practical problems such as steady varied flows in a channel [33]. Calculating the critical depth using classical hydraulics formulae yields a value of fluid depth for which the flow is still supercritical.

Experimental data seem to confirm these results. In addition, they show that a yield stress fluid can fracture into visible blocks at the free surface. This proves that it is under large tension stresses [2, 33].

When looking at confined low Reynolds number lubrication type flow, created either by motion of the walls or by a pressure gradient, pressure is normalised with liquid consistency. Normalised pressure gradient appears multiplied by 1/Re in Eqs. (1) and (2). The relative influence, due to normal extra-stresses, of the yield stress value on the flow field is simply given by Oldroyd number,

$$\mathrm{Od} = \frac{s}{\Delta p}\,, \tag{5}$$

that is, the yield stress value, s, has to be compared to the pressure level, Δp, created by viscous effects in the lubricated system.

Conclusion

Yield stress fluid and thixotropic fluid rheology are emerging fields. There is no point in questioning their existence. They should be worked on more intensively due to the numerous potential applications concerning materials and processes, and the many fundamental, theoretical and experimental problems that have to be addressed. Appropriate definitions can be suggested for yield stress and thixotropic fluids.

The experimental and theoretical techniques used so far need to be revisited carefully. Interfacial rheology also has to be considered simultaneously with more traditional bulk rheology.

The multi-disciplinary approach, typical of rheology, permits a much deeper understanding of the field and appears clearly as a key to scientific progress.

Acknowledgements

Exchanges with several colleagues and students in the field covered were helpful. Their names appear in the list of references. Particular mention should be made of Dr. A. Magnin's contributions.

Discussion

M. Renardy In your lecture, you claimed that it is not possible to buy a Newtonian fluid in any shop. However, in a shop here in Cambridge, I found a fluid manufactured in California which is clearly labelled "Newtonian".

J. M. Piau This wine may behave in such a way that a Newtonian model is appropriate. Ultimately, it depends not only on the model but also on what you are doing with it. Wine is a very particular fluid. When in a glass, it can even make tears and its wetting properties are interesting to observe.

V. M. Naik How trustworthy are the solutions for flow of yield stress exhibiting fluids through complex geometries, as obtained by you using

Bingham/H-B constitutive models, when one cannot predict phenomena such as flow localisation observed in simple affine geometry for representative materials? Are we not missing something when we disregard or ignore the past history and resultant micro-mechanical structure of material under investigation?

J. M. Piau Solutions for flows of Bingham fluids have been found for several confined flow geometries and seem to model correctly some experimental observations for yield stress fluids. However, it is true that more theoretical and experimental studies have to be completed, and new models may be required. Flow localisation, as you mentioned, seems to be predicted when Bingham models with concentration dependant parameters are introduced. Hence correct theoretical solutions for simple Bingham constitutive models can at least be considered as basic reference solutions which it is very important to provide.

References

1. Abdali, S. S., Mitsoulis, E. & Markatos, N. C. (1992) Entry and exit flows of Bingham fluids. *J. Rheol.* **36**, 389–407
2. Ayadi, A. (1996) *Ecoulement à surface libre des boues argileuses.* PhD Thesis, Université Joseph Fourier, Grenoble
3. Belhadri, M. (1996) *Ecoulements de fluides à seuil au travers de singularités convergentes et divergentes.* PhD Thesis, INPG, Grenoble
4. Beris, A. N., Tsamopoulos, J. A., Armstrong, R. C. & Brown, R. A. (1985) Creeping motion of a sphere through a Bingham plastic. *J. Fluid Mech.* **158**, 219–244
5. Bingham, E. C. (1922) *Fluidity and Plasticity.* New York: McGraw-Hill
6. Bourne, M. C. (1982) *Food Texture and Viscosity: Concept and Measurement.* New York: Academic Press
7. Coussot, P. (1992) *Rhéologie des boues et laves torrentielles. Etude de dispersions et suspensions concentrées.* PhD Thesis, INPG, Grenoble
8. Coussot, P. (1994) Steady laminar flow of concentrated mud suspensions in open channels. *J. of Hydraulic Research* **32**, 535–559
9. Coussot, P., Leonov, A. & Piau, J. M. (1993) Rheology of concentrated dispersed systems in a low molecular weight matrix. *J. Non-Newtonian Fluid Mech.* **46**, 179–217
10. Coussot, P. & Piau, J. M. (1995) A large-scale field coaxial cylinder rheometer for the study of the rheology of natural coarse suspensions. *J. Rheol.* **39**, 105–124
11. Covey, G. H. & Stanmore, B. R. (1981) Use of the parallel-plate plastometer for the characterisation of viscous fluids with a yield stress. *J. Non-Newtonian Fluid Mech.* **8**, 249–260

12. Dai, G. & Bird, R. B. (1981) Radial flow of a Bingham fluid between two fixed circular disks. *J. Non-Newtonian Fluid Mech.* **8**, 349–355
13. De Kee, D. & Chan Man Fong, C. F. (1993) A true yield stress? *J. Rheol.* **37**, 775–776
14. Doremus, P. & Piau, J. M. (1991) Constitutive equation of a yield stress fluid based on the network theory. In *Continuum Models and Discrete Systems*, ed. G. Maugin, pp. 20–29. Harlow: Longman
15. Doremus, P. & Piau, J. M. (1992) A generalised Maxwell model for yield stress fluid. *C. R. Acad. Sci.* **315-II**, 123–127
16. Ducerf, S. (1995) *Etude des propriétés rhéométriques de suspensions aqueuses minérales microniques denses, en fonction de leurs propriétés strucuturales.* PhD Thesis, INPG, Grenoble
17. Gupta, R. C. (1995) Herschel-Bulkley fluid flow development in a channel. *Polym. Plast. Technol. Eng.* **34**, 475–492
18. Huilgol, R. R. & Panizza, M. P. (1955) On the determination of the plug flow region in Bingham fluids through the application of variational inequalities. *J. Non-Newtonian Fluid Mech.* **58**, 207–217
19. Kouamela, N. D. (1991) *Etude expérimentale des écoulements en charge de fluides à seuil.* PhD Thesis, INPG, Grenoble
20. Lipscomb, G. G. & Denn, M. M. (1984) Flow of Bingham fluids in complex geometries. *J. Non-Newtonian Fluid Mech.* **14**, 337–346
21. Liu, K. F. & Mei, C. C. (1989) Slow spreading of a sheet of Bingham fluid on an inclined plane. *J. Fluid Mech.* **207**, 505–529
22. Magnin, A. & Piau, J. M. (1987) Shear rheometry of fluids with a yield stress. *J. Non-Newtonian Fluid Mech.* **23**, 91–106
23. Magnin, A. & Piau, J. M. (1990) Cone-and-plate rheometry of yield stress fluids. Study of an aqueous gel. *J. Non-Newtonian Fluid Mech.* **36**, 85–108
24. Magnin, A. & Piau, J. M. (1992) Flow of yield stress fluids through a sudden change of section. In *Theoretical and Applied Rheology*, ed. P. Moldenaers & R. Keunings, pp. 195–197. Amsterdam: Elsevier
25. Mewis, J. (1994) The thixotropic approach of time dependency in rheology. *Cah. Rhéol.* **13**, 2–9
26. Mezry, A. (1995) *Loi de glissement de caoutchoucs crus.* PhD Thesis, INPG, Grenoble
27. Nguyen, Q. D. & Boger, D. V. (1992) Measuring the flow properties of yield stress fluids. *Ann. Rev. Fluid Mech.* **24**, 47–88
28. O'Donovan, E. J. & Tanner, R. I. (1984) Numerical study of the Bingham squeeze film problem. *J. Non-Newtonian Fluid Mech.* **15**, 75–83
29. Oldroyd, J. G. (1947) A rational formulation of the equations of plastic flow for a Bingham solid. *Proc. Camb. Philos. Soc.* **43**, 100–105
30. Oldroyd, J. G. (1947) Two-dimensional plastic flow of a Bingham solid. A plastic boundary layer theory for slow motion. *Proc. Camb. Philos. Soc.* **43**, 383–395
31. Persello, J., Magnin, A., Chang, J., Piau, J.-M. & Cabane, B. (1994) Flow of colloidal aqueous silica dispersions. *J. Rheol.* **38**, 1845–1870

32. Piau, J. M., El Kissi, N., Toussaint, F. & Mezghani, A. (1995) Distorsions of polymer melt extrudates and their elimination using slippery surfaces. *Rheol. Acta.* **34**, 40–57
33. Piau, J. M. (1996) Flow of a yield stress fluid in a long domain. Application to flow on an inclined plane. *J. Rheol.* **40**, 711–723
34. Piau, J. M. (1989) Rhéométrie des graisses lubrifiantes. Unpublished report
35. Picart, C., Piau, J. M., Galliard, H. & Carpentier, P. H. (1996) Human blood shear yield stress and its hematocrit dependence. Submitted
36. Pignon, F., Piau, J. M. & Magnin, A. (1996) Structure and pertinent length scale of a yield stress colloidal discotic clay suspension. *Phys. Rev. Lett.* 4857–4860
37. Pignon, F., Magnin, A., Piau, J.-M., Cabane, B., Lindner, P. & Diat, O. (1996) A yield stress thixotropic clay suspension: investigation of structure by light, neutron, and x-ray scattering. Submitted
38. Prager, W. (1961) *Introduction to Mechanics of Continua.* Boston: Ginn and Company
39. Sacchettini, M., Magnin, A., Piau, J.-M. & Pierrard, J.-M. (1985) Caractérisation d'une graisse lubrifiante en écoulements viscosimétriques transitoires. *J. Theoretical and Applied Mech. Special Issue*, 165–199
40. Sherwood, J. D., Meeten, G. H., Farrow, C. A. & Alderman, N. J. (1991) Squeeze-film rheometry of non-uniform mudcakes. *J. Non-Newtonian Fluid Mech.* **39**, 311–334
41. Spaans, R. D. & Williams, M. C. (1995) At last, a true liquid-phase yield stress. *J. Rheol.* **39**, 241–246
42. Tichy, J. (1991) Hydrodynamic lubrication theory for the Bingham plastic flow model. *J. of Rheol.* **35**, 477–496
43. Vlachou, P. R. (1996) *Rhéologie de suspensions concentrées. Application aux laitiers de ciment pétrolier.* PhD Thesis, INPG, Grenoble
44. Wada, S., Hayashi, H. & Haga, K. (1973) Behaviour of a Bingham solid in hydrodynamic lubrication. *Bull. J. S. M. E.* **16**, n° 92, 422–431 and 432–440
45. White, J. L. (1981) Approximate constitutive equations for slow flow of rigid plastic viscoelastic fluids. *J. Non-Newtonian Fluid Mech.* **8**, 195–202
46. Wilson, S. D. R. (1993) Sqeezing flow of a yield stress fluid in a wedge of slowly-varying angle. *J. Non-Newtonian Fluid Mech.* **50**, 45–63

Dynamics of Complex Fluids, pp. 372–378
ed. M. J. Adams, R. A. Mashelkar, J. R. A. Pearson & A. R. Rennie
Imperial College Press–The Royal Society, 1998

Chapter 26

Are Plug-Flow Regions Possible in Fluids Exhibiting a Yield Stress?

MORTON M. DENN
Department of Chemical Engineering, University of California, Berkeley,
and
Materials Sciences Division, Lawrence Berkeley National Laboratory, Berkeley, CA 94720-1462 USA
E-mail: Denn@Chem.Berkeley.ED

Lubrication theory leads to the prediction of unyielded regions in squeeze flow of fluids exhibiting a yield stress, whereas a straightforward kinematical argument establishes that such regions cannot exist. Unyielded regions are also observed in finite element calculations. The apparent contradiction between the finite element solutions and the kinematical argument may arise because the latter is based on flow in an infinite geometry while the finite element grid is finite. In analogy to the squeeze flow of a Newtonian fluid with a transverse velocity gradient, it may be that the presence of unyielded regions depends on a dimensionless group which is the product of the Bingham number (which approaches infinity) and the square of the gap-to-radius ratio (which approaches zero). It is therefore necessary to determine the limits of the kinematical argument and to establish conditions under which approximations analogous to classical boundary layer and the lubrication theory are valid for yield stress fluids.

Introduction

Fluids exhibiting a yield stress are often described in simple shear by the Bingham equation,

$$\tau_{yx} = \tau_y \, sgn(dv_x/dy) + \eta_p \, dv_x/dy \, , \quad |\tau_{yx}| \geq \tau_y \, , \tag{1}$$

$$dv_x/dy = 0 \, , \quad |\tau_{yx}| < \tau_y \, . \tag{2}$$

Here, the velocity has a component only in the x-direction with a gradient in the y-direction, and τ_{yx} is the shear stress. The model has two parameters: the yield stress, τ_y, below which flow cannot occur, and the "plastic viscosity", η_p. (The latter nomenclature is unfortunate since viscosity is the chord from the origin to a point on the flow curve, and not the local slope of the flow curve. Hence, despite having a constant "plastic viscosity", the Bingham material is highly shear-thinning.) The Herschel–Bulkley material is of the form of Eqs. (1) and (2) but η_p has a power-law dependence on $|dv_x/dy|$. It is a characteristic of one-dimensional flows of Bingham-like materials that there is a yielded region near no-slip boundaries. However, there may be an unyielded plug where the fluid flows as a solid body and is not deformed. Continuity of velocity and shear stress are required at the interface between the yielded and unyielded regions.

Prager [1] and Oldroyd [2] have described properly invariant equations for a liquid with yield stress. The material is assumed to be elastic prior to yield. The yield surface is also assumed to satisfy a von Mises criterion, in which case the stress satisfies the equations

$$\boldsymbol{\tau} = \left[\tau_y / \left| \left(\frac{1}{2} \mathbf{d} : \mathbf{d} \right)^{\frac{1}{2}} \right| + \eta_p \right] \mathbf{d} \, , \quad \frac{1}{2} \boldsymbol{\tau} : \boldsymbol{\tau} \geq {\tau_y}^2 \, , \tag{3}$$

$$\boldsymbol{\tau} = \mu \mathbf{e} \, , \quad \frac{1}{2} \boldsymbol{\tau} : \boldsymbol{\tau} < {\tau_y}^2 \, . \tag{4}$$

$\boldsymbol{\tau}$ is the extra stress tensor, $\mathbf{e}$ the strain tensor, and $\mathbf{d}$ the deformation rate tensor. μ is an elastic modulus, and is usually taken to be infinite. In that case, the equation for the unyielded region is

$$\mathbf{d} = 0 \, , \quad \frac{1}{2} \boldsymbol{\tau} : \boldsymbol{\tau} < {\tau_y}^2 \, . \tag{5}$$

Equations (3) and (5) are the properly invariant generalisation of Eqs. (1) and (2). For the Bingham model, η_p is a constant while in the Herschel–Bulkley model, η_p is a power-law function of $|(\frac{1}{2} \mathbf{d} : \mathbf{d})^{\frac{1}{2}}|$. There are few solutions to this system with flows in complex geometries because of the difficulty of the computational problem associated with the unknown location of the yield surfaces. Some notable examples include the numerical solution of flow about a sphere by Beris and co-workers [3] and the asymptotic solutions by Craster [4]. Variational inequalities have recently been used to obtain approximate solutions in complex flows by Huilgol and Panizza [5] and Zwick and co-workers [6].

Lubrication Theory

A number of authors have used lubrication theory to obtain approximate solutions for the non-viscometric flow of Bingham materials [7–9]. In particular, Dai and Bird [7] analysed the radial flow between two fixed circular disks, and Covey and Stanmore [8] the squeeze flow between approaching disks. In both cases, the solutions included unyielded regions with position-dependent cross-sections. Lipscomb and Denn [10] pointed out through a strictly kinematical argument that the latter two solutions cannot be correct. This is because it is impossible to satisfy velocity and stress continuity at a surface enclosing a solid body with a changing cross-section. This argument has been reiterated more recently by Piau [11] and demonstrated analytically for channel flow by Huilgol [12]. The same critique is applicable to inertial boundary layer analyses of developing flow in a closed conduit [13].

Lipscomb and Denn [10] presented an analytical argument showing that no portion of the flow field in a squeeze flow can yield until yielding occurs everywhere. Their argument, which is based on allowing the elastic modulus in Eq. (4) to become arbitrarily large and hence keeping all deformations arbitrarily small, came under attack at the forum on *Dynamics of Complex Fluids.* We shall return to this point subsequently. (An alternative analysis, in which the no-slip condition is relaxed at the wall, is given by Sherwood and Durbin [14].)

Two-Viscosity Approximation

A common approach to avoiding the problem of the unknown yield surfaces has been to approximate the Bingham material by a purely viscous fluid with a continuous viscosity function. One example is the "two-viscosity" approximation,

$$\boldsymbol{\tau} = \left[\tau_y(1-\eta_1/\eta_0)\Big/\left|\left(\frac{1}{2}\mathbf{d}:\mathbf{d}\right)^{\frac{1}{2}}\right| + \eta_1\right]\mathbf{d}, \quad \frac{1}{2}\boldsymbol{\tau}:\boldsymbol{\tau} \geq {\tau_y}^2, \tag{6}$$

$$\boldsymbol{\tau} = \eta_0 \mathbf{d}, \quad \frac{1}{2}\boldsymbol{\tau} : \boldsymbol{\tau} < {\tau_y}^2, \tag{7}$$

with $\eta_0 >> \eta_1$. (For other purely viscous approximations, see [15, 16].) Analyses of the squeeze flow with a two-viscosity model are found in [10, 17–19]. A fundamental question in the use of such approximations is whether the correct Bingham limit is approached as $\epsilon = \eta_1/\eta_0 \to 0$. Wilson [19] has shown that the two-viscosity solution for a squeeze flow with a gap-to-radius ratio δ approaches that of the Bingham material only if $\delta/\epsilon << 1$. Since ϵ must approach zero, this implies that δ must approach zero even faster. We can therefore conclude that in general the Bingham limit of a continuous viscosity approximation is singular, and any purely viscous approximation is suspect unless convergence has been rigorously demonstrated.

Finite Geometry

O'Donovan and Tanner [18] reported an apparently convergent numerical solution of squeeze flow of a Bingham fluid showing an unyielded region on the centre axis (as well as at the outer edge); this seems to contradict the Lipscomb–Denn analysis. (While the calculation was initially carried out for a two-viscosity model, O'Donovan and Tanner estimated the locations of yield surfaces and showed that their result was consistent with a calculation in which the unyielded regions were replaced by rigid solids.) More recently, Adams and co-workers [20] have reported both finite element calculations and experiments showing an unyielded region on the axis in a squeeze flow. This apparent paradox may be rationalised by considering the similar problem of squeeze flow of a Newtonian fluid with a transverse viscosity gradient symmetric about the centre plane. An exact analytical solution to the latter problem is available for squeezing between infinite disks; a primary result is that there is a single maximum in the radial velocity profile located at the centre plane regardless of the ratio of maximum-to-minimum viscosity in the gap. This observation is inconsistent with the intuitive and experimentally observed result that low viscosity fluid near the disks can flow preferentially around a nearly undeformed high viscosity core. This results in an off-centre velocity maximum in the low viscosity region. The latter result is found in numerical solutions for disks of finite radius. Lee and co-workers [21] showed that the flow regime for finite disks depends on the value of a dimensionless group:

$$S = \eta_{\max} H^2 / \eta_{\min} R^2, \tag{8}$$

where H is the gap spacing, R the radius, and $\eta_{\max}$ and $\eta_{\min}$ the high and low viscosities in the core and outer regions, respectively. For finite $\eta_{\max}/\eta_{\min}$ but $H/R \to 0$, there is parallel squeezing with the velocity maximum at the centre plane. For finite H/R but $\eta_{\max}/\eta_{\min} \to \infty$, the velocity maximum occurs in the low viscosity fluid and the central core may be taken as rigid. This corresponds to a case in which low viscosity fluid can flow out preferentially from the space between the finite disks. If both limits ($H/R \to 0$ and $\eta_{\max}/\eta_{\min} \to \infty$) occur simultaneously with S remaining finite, there is a transition between the two modes of behaviour at a value of S of order unity.

It is likely that the computed evidence for an unyielded region in squeeze flow of a Bingham fluid is a consequence of the finite geometry, and there is no real conflict with the conclusion of Lipscomb and Denn [10], which assumes flow between infinite disks. The usual dimensionless group introduced for flow of yield stress fluids is the *Bingham number*,

$$Bi = \tau_y H/\eta_p V \, , \tag{9}$$

where H is a characteristic dimension and V a characteristic velocity. Yield behaviour is important when Bi is large, and this is clearly the relevant group for squeeze flow when the disks are infinite and there is a single characteristic length. For finite disks, however, the ratio H/R must be relevant. The minimum force required to move a plate of radius R scales as $\tau_y R^2$, while the force required to squeeze the plates when the yield stress is negligible scales as $\eta_p V R^4/H^3$ [22]. The ratio of these forces thus leads to a dimensional group $Bi(H/R)^2$. For $Bi \to \infty$ with finite H/R we can expect an unyielded region, whereas for $H/R \to 0$ with finite Bi we can expect yielding everywhere before any flow occurs. The intermediate case in which the limits $Bi \to \infty$ and $H/R \to 0$ are approached simultaneously, with a finite value of the product $Bi(H/R)^2$, is the situation of practical interest and it is likely that there will be a transition between the two types of limiting behaviour. (The *Plasticity number*, $Bi(H/R)$, arises naturally in the variational analysis of Zwick and co-workers [6]. It may be that this is the relevant group defining the flow regimes for finite disks. The physical significance is less obvious than that of the group introduced here. The idea that the finite geometry might be the source of the apparent conflict between the Lipscomb–Denn analysis and the finite element solutions arose in informal discussions after the close of the formal session at the Forum on *Dynamics of Complex Fluids.*)

Conclusion

Unresolved theoretical issues regarding the flow of materials exhibiting a yield stress, centering around the question of the existence and location of unyielded regions in complex (possibly time-dependent) geometries, remain. The finite boundary, which is an essential component in experiments and finite element calculations but generally neglected in analytical approximations, seems to be a critical factor in rationalising the apparent contradiction between theoretical arguments which require yielding everywhere in squeeze flows and computational results showing unyielded regions. It is therefore necessary to determine the limits of the kinematical argument used by Lipscomb and Denn [10], and to establish conditions under which approximations analogous to classical boundary layer and lubrication theory are valid for yield stress fluids. A recent paper by Piau [11] is a step in this direction.

Acknowledgement

This work was supported by the Director, Office of Energy Research, Office of Basic Energy Sciences, Materials Science Division of the US Department of Energy under Contract No. DE-AC073-76SF00098.

References

1. Prager, W. (1961) *Introduction to Continuum Mechanics*, pp. 136–153. Boston: Ginn
2. Oldroyd, J. (1947) A rational formulation of the equations of plastic flow for a Bingham solid. *Proc. Camb. Philos. Soc.* **43**, 100–105; Two-dimensional plastic flow of a Bingham solid. A plastic boundary-layer theory for slow motion. *Proc. Camb. Philos. Soc.* **43**, 383–395
3. Beris, A. N., Tsamopoulos, J. A., Brown, R. A. & Armstrong, R. C. (1984) Creeping flow around a sphere in a Bingham plastic. *J. Fluid Mech.* **158**, 141–172
4. Craster, R. V. (1995) Solutions for Herschel–Bulkley flows. *Quart. J. Mech. Appl. Math.* **48**, 343–374
5. Huilgol, R. R. & Panizza, M. P. (1995) On the determination of the plug flow region in Bingham fluids through the application of variational inequalities. *J. Non-Newtonian Fluid Mech.* **58**, 207–217
6. Zwick, K. J., Ayyaswamy, P. S. & Cohen, I. M. (1996) Variational analysis of the squeezing of a yield stress fluid. *J. Non-Newtonian Fluid Mech.* **63**, 179–199
7. Dai, G. & Bird, R. B. (1981) Radial flow of a Bingham fluid between two fixed circular disks. *J. Non-Newtonian Fluid Mech.* **8**, 349–355

8. Covey, G. H. & Stanmore, B. R. (1981) Use of the parallel-plate plastometer for the characterisation of viscous fluids with a yield stress. *J. Non-Newtonian Fluid Mech.* **8**, 249–260
9. Tichy, J. A. (1991) Hydrodynamic lubrication theory for the Bingham plastic flow model. *J. Rheol.* **35**, 477–496
10. Lipscomb, G. G. & Denn, M. M. (1984) Flow of Bingham fluids in complex geometries. *J. Non-Newtonian Fluid Mech.* **14**, 337–346
11. Piau, J. M. (1996) Flow of a yield stress fluid in a long domain. Application to flow on an inclined plane. *J. Rheol.* **40**, 711–723
12. Huilgol, R. J. (1994) Personal communication
13. Gupta, R. C. (1995) Developing Bingham fluid flow in a channel. *Math. Comput. Modeling* **21**, 21-28
14. Sherwood, J. D. & Durban, D. (1996) Squeeze flow of a power-law viscoplastic solid. *J. Non-Newtonian Fluid Mech.* **62**, 35–54
15. Papanastasiou, T. C. (1987) Flow of materials with yield. *J. Rheol.* **31**, 385–404
16. Taylor, A. J. & Wilson, S. D. R. (1997) Conduit flow of an incompressible, yield stress fluid. *J. Rheol.* **41**, 93–102
17. Gartling, D. K. & Phan-Thien, N. (1984) A numerical solution of a plastic fluid in a parallel-plate plastometer. *J. Non-Newtonian Fluid Mech.* **14**, 347–360
18. O'Donovan, E. J. & Tanner, R. I. (1984) Numerical study of the Bingham squeeze film problem. *J. Non-Newtonian Fluid Mech.* **15**, 75–83
19. Wilson, S. D. R. (1993) Squeezing flow of a Bingham material. *J. Non-Newtonian Fluid Mech.* **47**, 211-219
20. Adams, M. J., Aydin, I., Briscoe, B. J. & Sinha, S. K. (1997) A finite element analysis of the squeeze flow of an elasto-viscoplastic paste material. *J. Non-Newtonian Fluid Mech.* **71**, 41–57
21. Lee, S. J., Denn, M. M., Crochet, M. J. & Metzner, A. B. (1982) Compressive flow between parallel disks. I. Newtonian fluid with a transverse viscosity gradient. *J. Non-Newtonian Fluid Mech.* **10**, 3–30
22. Denn, M. M. (1980) *Process Fluid Mechanics*, pp. 255–261. Englewood Cliffs, N.J.: Prentice-Hall

Dynamics of Complex Fluids, pp. 379–393
ed. M. J. Adams, R. A. Mashelkar, J. R. A. Pearson & A. R. Rennie
Imperial College Press–The Royal Society, 1998

Chapter 27

Non-Viscometric Flow of Viscoplastic Materials: Squeeze Flow

C. J. LAWRENCE* AND G. M. CORFIELD

Department of Chemical Engineering & Chemical Technology,
Imperial College of Science, Technology & Medicine,
London SW7 2BY, UK
**E-mail: c.lawrence@ic.ac.uk*

It is widely known that a naive application of lubrication theory to the flow of viscoplastic materials can lead to a kinematic inconsistency if the flow is not fully developed. An example considered here is the squeeze flow in a parallel plate plastometer. The lubrication scaling is treated explicitly and a general form of the lubrication solution is presented. The naive solution has a region of plug flow which is apparently unyielded, but which is actually stretching as it flows radially. The inconsistency is resolved by introducing the appropriate scales for the inner region; it is shown that the strain rate in the plug region is actually small but finite and the yield criterion is satisfied. Thus, the lubrication solution provides a satisfactory leading-order solution despite the apparent inconsistency. The scaling analysis also indicates circumstances when the lubrication approach is insufficient.

Introduction

Plasticity analyses have traditionally been the province of solid mechanics, and have been developed extensively for metals and soils [1]. In contrast, fluid mechanics analyses have been largely concerned with viscous materials with relatively little attention to plastic flow [2]. Indeed, it is widely held that "fluid-like" materials may not have a true yield stress [3], but that yield-stress constitutive rules may provide convenient and useful means of modelling a broad class of materials. This class includes some polymer melts and solutions as well as micro-structured materials such as powders, suspensions, emulsions, foams and pastes. There has been a recent resurgence of interest in the flow of such materials in process engineering. This is because the interaction of micro-structure and rheology is important to the quality and performance of products such as foods and household products, while other products such as ceramics are processed via soft-solid intermediate states.

Lubrication theory provides a very useful means of studying the flow of viscous materials in slowly-varying or high aspect-ratio geometries, and it is natural to seek to extend the theory for application to yield-stress materials. The exact solutions for viscometric flows of plastic materials indicate that regions of unyielded material exist where the shear stress does not exceed the yield value [4]. These regions must be either stationary or move as a rigid body. Similar regions are found in lubrication solutions [5, 6] but Lipscomb and Denn [7] demonstrated conclusively that such solutions are inconsistent. In the squeeze flow example, the "unyielded region" is flowing radially outwards and is necessarily stretching as it does so.

For a plastic material, the viscosity is effectively infinite for stresses below the yield stress. Lipscomb and Denn showed that for a bi-viscosity material, which has a large but finite viscosity for small stress, the lubrication approach gives a good approximate solution. Indeed, experimental work [8, 9] has shown that the lubrication solution gives a very useful approximation for the total load measured in the squeeze flow experiment. The biviscosity model has formed the basis of numerical approaches [10] but Wilson [11] has shown that the lubrication scaling breaks down if the low-shear viscosity is too high.

Lipscomb and Denn [7] went further and argued that there could be no unyielded regions in complex internal flows. This argument appears to be contradicted by numerical solutions [9] which show conical unyielded zones at the two stagnation points and small rings of unyielded material in the fountain flow region at the edge of the sample. Beris *et al.* [12] found similar conical unyielded regions at the front and rear stagnation points in the flow of a Bingham material over a sphere. It is likely that the flow in the vicinity of a stagnation point at a no-slip boundary has a universal behaviour regardless of the remainder of the flow field. Thus, it seems that unyielded regions are possible, although they seem to occur only in regions where the flow is not gradually varying and the lubrication approximation does not hold.

In the mechanics of solids, the squeeze flow experiment is known as upsetting; there may be significant slip between the specimen and the boundary, and the interface is often lubricated. A series of solutions was given by Collins and Meguid [13] who considered the two-dimensional version of the flow and included various constitutive models. This approach has recently been adapted and applied to a soft solid material [14, 15]. Sherwood and Durban [16] have put forward an essentially similar approach based on fluid mechanics principles for the treatment of flows where wall slip is dominant.

Since Lipscomb and Denn's publication in 1983, much has been made of the inadequacy of standard fluid mechanics approaches, in particular lubrication theory, in describing non-viscometric flows of materials that possess a yield stress. In this paper, the simple squeeze flow geometry is again considered and the question of the sufficiency of the naive lubrication solution is revisited. It is first shown explicitly that the standard fluid mechanics approach may be recovered from a general solid mechanics plasticity approach. The equations are then scaled to obtain the lubrication solution. The unyielded plug flow region is predicted at leading order and a rescaling is used to show that this is the correct asymptotic solution. Finally, a number of cases are discussed in which the lubrication approach does fail.

Solid Mechanics and Fluid Mechanics Formulations

The formulation of constitutive equations for yield stress materials has evolved from different starting points and followed different paths in the disciplines of solid and fluid mechanics. The two approaches are necessarily consistent, as can be readily shown.

We define a scalar yield function, F, in stress space so that the isosurfaces of F are smooth and convex. F measures the distance from a point in stress space to the yield surface and is called the stress overshoot [17]. For $F < 0$, the material is unyielded and will be taken to be rigid. The surface $F = 0$ defines the plastic yield surface in stress space and when $F > 0$, the material undergoes plastic flow.

The plastic flow is often assumed to be associative. The rate of strain tensor is then orthogonal to the surfaces of constant F in stress space. This may be formalised as

$$e_{ij} = 2\Phi(F)\frac{\partial F}{\partial \sigma_{ij}}, \tag{1}$$

where $\Phi(F)$ is the flow curve that determines the strain rate for a given stress overshoot. Thus, $\Phi(F)$ is zero inside the yield locus, $(F < 0)$, and positive outside $(F > 0)$. A form commonly used for incompressible viscoplastic materials is the von-Mises yield function which is isotropic. This may be written as

$$F(\sigma) = k - \tau_0, \tag{2}$$

where τ_0 is the static (shear) yield stress and k is the magnitude of the deviatoric stress defined as follows. The deviatoric stress is $\mathbf{s} = \sigma + p\mathbf{I}$, where $p = -\frac{1}{3}\operatorname{tr}(\sigma)$ is the hydrodynamic pressure and σ is the stress tensor. The second invariant of $\mathbf{s}$ is $J_2 = \frac{1}{2}\,\mathbf{s}{:}\mathbf{s} = k^2$. In general, the yield function may also depend on the history of the deformation in a complex manner to account for strain hardening, strain softening, induced anisotropy or other phenomena. Recent advances have been summarised by Krempl and Gleason [18].

With the von-Mises yield function, the associated flow rule, Eq. (1), gives the rate of strain tensor, $\mathbf{e} = \Phi(F)\,\mathbf{s}/k$, and the second invariant of $\mathbf{e}$ is $I_2 = \frac{1}{2}\,\mathbf{e}:\mathbf{e} = \Phi^2(F)$. Hence we identify $\Phi(F)$ as the magnitude of the strain rate, $\Phi(F) = \dot{\gamma} = \sqrt{I_2}$. The stress-strain-rate relationship may then be written as

$$\begin{aligned} k &= \tau_0 + \Phi^{-1}(\dot{\gamma}) \qquad && k > \tau_0\,, \\ \dot{\gamma} &= 0 && k < \tau_0\,. \end{aligned} \tag{3}$$

This form, together with the flow rule,

$$\mathbf{e} = \mathbf{s}\dot{\gamma}/k\,, \tag{4}$$

is identical to the form used in non-Newtonian fluid mechanics [2]. Thus, we may determine the flow curve from standard viscometric measurements such as capillary rheometry. A useful model material is the Herschel–Bulkley fluid with

$$k = \tau_0 + \alpha\dot{\gamma}^n \quad k > \tau_0\,, \tag{3a}$$

which includes the Bingham $(n = 1)$, power-law $(\tau_0 = 0)$, Newtonian $(n = 1, \tau_0 = 0)$, and perfect plastic $(\alpha = 0)$ materials as special cases.

It is important to characterise the wall stress boundary condition at material interfaces. In the mechanics of metals and of compressible materials such as soils and powders, this is commonly treated using a Coulomb or Tresca boundary condition. The Coulomb condition relates the magnitude of the shear stress or wall friction, τ_w, to the (compressive) normal stress at the interface, σ_n, as $\tau_w \leq \mu\sigma_n$, where μ is the coefficient of friction. The Tresca condition relates the friction to the von-Mises stress magnitude as $\tau_w \leq mk$, where m is the friction factor. In each case, equality holds only when there is slip, or imminent slip, at the interface.

For the analysis of incompressible soft solids constrained by effectively rigid boundaries, a generalised Navier condition is better employed. This relates the wall shear stress to u_s, the magnitude of the slip velocity at the wall, and may be written in a form directly analogous to Eq. (3):

$$\begin{aligned} \tau_w &= \tau_{w0} + \phi^{-1}(u_s) \qquad && \tau_w > \tau_{w0}\,, \\ u_s &= 0 && \tau_w < \tau_{w0}\,, \end{aligned} \tag{5}$$

in which τ_{w0} is the wall yield stress, a threshold below which the wall slip velocity is zero. A useful model is the Herschel-Bulkley boundary condition which is analogous to the first part of Eq. (3):

$$\tau_w = \tau_{w0} + \alpha_w u_s^{n_w} \qquad \tau_w > \tau_{w0}\,. \tag{5a}$$

The first part of Eq. (5) includes no slip, power-law slip, (linear) Navier slip and frictionless conditions as special cases.

Squeeze Flow

In the upsetting or squeeze flow experiment, an incompressible material is squeezed between rigid platens as shown in Fig. 1. We assume axisymmetry

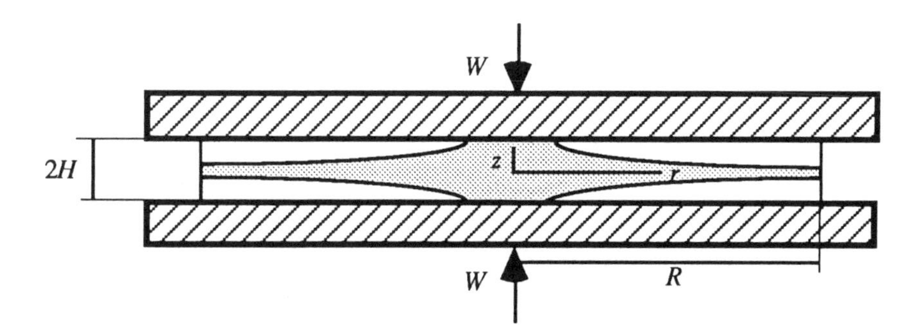

Fig. 1. The squeeze flow geometry. The specimen of radius, R, and thickness, $2H$, is reduced in thickness at a rate $2W$. According to the lubrication solution, the shaded region undergoes radial plug flow and yet is unyielded.

and use the cylindrical co-ordinates, (r, θ, z), with corresponding velocity components, $(u, 0, w)$. The velocity is prescribed at the upper and lower boundaries:

$$\text{at } z = \pm H \qquad w = -(\pm W)\,. \tag{6}$$

This generates a straining motion with non-zero components:

$$e_{rr} = 2u_{,r} \quad e_{\theta\theta} = 2\frac{u}{r} \quad e_{zz} = 2w_{,z} \quad e_{rz} = e_{zr} = u_{,z} + w_{,r}\,, \tag{7}$$

where a comma denotes differentiation with respect to the subscripted variable. The constraint of incompressibility requires that the rate of strain tensor is trace-free:

$$e_{rr} + e_{\theta\theta} + e_{zz} = 0\,. \tag{8}$$

The outer surface of the material is traction-free, so that at $r = R$,

$$s_{rr} = p \qquad s_{rz} = 0 \tag{9}$$

and the stress satisfies the quasi-static equilibrium equations:

$$p_{,r} = s_{zr,z} + s_{rr,r} + \frac{1}{r}(s_{rr} - s_{\theta\theta})\,, \tag{10}$$

$$p_{,z} = s_{zz,z} + s_{rz,r} + \frac{1}{r}\, s_{rz}\,. \tag{11}$$

For a specimen of small aspect ratio, $\delta = H/R \ll 1$, problems of this type may be considerably simplified by appropriate scaling [19].

Lubrication Solution

First we re-write the constitutive relation, Eq. (3), in an inverse form that is convenient for dimensional analysis as

$$\dot{\gamma} = \dot{\gamma}_0 g(k/\tau_0)\,, \tag{12}$$

where $\dot{\gamma}_0$ is a characteristic strain rate of the material, the yield stress τ_0 is the characteristic stress and $g(\cdot)$ is the dimensionless viscometric function. For the Herschel–Bulkley example in the first part of Eq. (3), we would have $\dot{\gamma}_0 = (\tau_0/\alpha)^{1/n}$ and $g(x) = (x-1)^{1/n}$ for $x \geq 1$ with $g(x) = 0$ for $x \leq 1$.

We use the natural scales of the problem, shown in Table 1, to render the equations dimensionless and retain the notation used above, though now for

Table 1. Scales used in the lubrication approximation.

	Scale	Variables	Scale	Variables
Lengths	H	z	$R = \delta^{-1}H$	r
Velocities	W	w	$\delta^{-1}W$	u
Strain rates	W/H	$e_{rr}, e_{\theta\theta}, e_{zz}$	$\delta^{-1}W/H$	$e_{zr}, e_{rz}, \dot{\gamma}$
Stresses	$\delta\tau_0$	$s_{rr}, s_{\theta\theta}, s_{zz}$	τ_0	s_{zr}, s_{rz}, k
Pressure	—	—	$\delta^{-1}\tau_0$	p

Table 2. Governing equations scaled for the lubrication approximation.

Strain rates	$e_{rr} = 2u,_r \quad e_{\theta\theta} = 2\dfrac{u}{r} \quad e_{zz} = 2w,_z \quad e_{rz} = e_{zr} = u,_z + \delta^2 w,_r$
Continuity	$e_{rr} + e_{\theta\theta} + e_{zz} = 0$
Equilibrium	$p,_r = s_{zr},_z + \delta^2 \left[s_{rr},_r + \dfrac{1}{r}(s_{rr} - s_{\theta\theta}) \right]$ $p,_z = \delta^2 \left[s_{zz},_z + s_{rz},_r + \dfrac{1}{r} s_{rz} \right]$
Flow	$\dot{\gamma} = \dfrac{1}{\omega} g(k)$
Associativity	$e_{ij} = \dfrac{\dot{\gamma}}{k} s_{ij}$
Strain rate	$\dot{\gamma}^2 = e_{rz}^2 + \dfrac{1}{2}\delta^2 \left[e_{rr}^2 + e_{\theta\theta}^2 + e_{zz}^2 \right]$
Flow stress	$k^2 = s_{rz}^2 + \dfrac{1}{2}\delta^2 \left[s_{rr}^2 + s_{\theta\theta}^2 + s_{zz}^2 \right]$

dimensionless quantities. The equations relevant to the current discussion have the dimensionless forms given in Table 2. The boundary conditions are not essential to the current discussion (though care must be exercised in some cases as discussed below) and are therefore omitted for brevity. Besides the aspect ratio $\delta = H/R$, the system is then characterised by the Weissenberg number, $\omega = WR/\dot{\gamma}_0 H^2$, which represents the ratio of the characteristic applied shear rate to the characteristic shear rate of the material.

In the lubrication approximation, we would neglect all terms of order δ^2 in the equations of Table 2 which leads to a straightforward solution. We use the caret to denote this naive lubrication solution. The pressure is a function of r only, $\hat{p}(r)$, the shear stress is linear in z, and the wall friction is simply related to the radial pressure gradient,

$$\hat{s}_{rz} = -\hat{\tau}_w z \qquad \hat{\tau}_w(r) = -\hat{p}'(r)\,. \tag{13}$$

Furthermore, only the shear components of the stress and strain rate are significant:

$$\hat{k} = |\hat{s}_{rz}| \qquad \hat{\dot{\gamma}} = |\hat{e}_{rz}|\,. \tag{14}$$

The strain-rate is given simply by the principal velocity gradient and is related to the stress by the one-dimensional constitutive relationship,

$$\hat{u}_{,z} = \hat{e}_{rz} = \pm\frac{1}{\omega}g(\hat{\tau}_w|z|)\,. \tag{15}$$

Equation (15) may be integrated directly to obtain the velocity profile. A further integration with a global mass balance gives the wall shear stress, and a third integration gives the pressure distribution.

Covey and Stanmore [6] applied the lubrication approach and obtained the solutions described above. Since $g(k)$ is zero for $k < 1$, the velocity profile has a characteristic plug flow region with $\hat{u} = \hat{u}_p(r)$ for $|z| < \hat{z}_p(r) = 1/\hat{\tau}_w$. Many observers [7, 9, 11] have pointed out that the material in this plug flow zone is unyielded since $k < 1$. This is inconsistent with the obvious fact that if u is non-zero, so is $e_{\theta\theta}$; then $\hat{\dot{\gamma}}$ is positive (but small) and the material must have yielded. Nevertheless, notwithstanding the extensive criticism, the solution given by Eqs. (13) and (15) is both kinematically and dynamically correct as will now be shown.

Self-Consistent Solution

In the region $|z| < z_p(r)$, the naive solution allows the flow stress, k, to fall below the yield value (unity) which is inadmissible. We must therefore rescale some of the variables to restore balance in the yield condition. The velocities of the naive solution are correct, but we must consider the next terms in an asymptotic expansion. The shear rate and the strain rate magnitude are much smaller, and the normal deviatoric stresses are correspondingly larger than for the naive solution, as shown in Table 3.

Table 3. Rescaling used in the plug flow region.

	Scale factor	Variables	Scale factor	Variables
Velocities	1	w	δ	$u - \hat{u}$
Strain rates	1	$e_{rr}, e_{\theta\theta}, e_{zz}$	δ	$e_{zr}, e_{rz}, \dot{\gamma}$
Stresses	δ^{-1}	$s_{rr}, s_{\theta\theta}, s_{zz}$	1	s_{zr}, s_{rz}, k

Table 4. Governing equations rescaled for the plug flow region.

Strain rates	$e_{rr} = \hat{e}_{rr} + 2\delta\tilde{u}_{,r} \quad e_{\theta\theta} = \hat{e}_{\theta\theta} + 2\delta\frac{\tilde{u}}{r} \quad e_{zz} = 2w_{,z}$ $\tilde{e}_{rz} = \tilde{e}_{zr} = \tilde{u}_{,z} + \delta w_{,r}$
Continuity	$e_{rr} + e_{\theta\theta} + e_{zz} = 0$
Equilibrium	$p_{,r} = s_{zr,z} + \delta\left[\tilde{s}_{rr,r} + \frac{1}{r}(\tilde{s}_{rr} - \tilde{s}_{\theta\theta})\right]$ $p_{,z} = \delta s_{zz,z} + \delta^2\left[s_{rz,r} + \frac{1}{r}s_{rz}\right]$
Flow	$\delta\tilde{\dot{\gamma}} = \frac{1}{\omega}g(k)$
Associativity	$\tilde{e}_{ij} = \frac{\tilde{\dot{\gamma}}}{k}s_{ij} \quad i \neq j \quad e_{ij} = \frac{\tilde{\dot{\gamma}}}{k}\tilde{s}_{ij} \quad i = j$
Strain rate	$\tilde{\dot{\gamma}}^2 = \tilde{e}_{rz}^2 + \frac{1}{2}\left(e_{rr}^2 + e_{\theta\theta}^2 + e_{zz}^2\right)$
Flow stress	$k^2 = \tilde{s}_{rz}^2 + \frac{1}{2}\left(\tilde{s}_{rr}^2 + \tilde{s}_{\theta\theta}^2 + \tilde{s}_{zz}^2\right)$

The rescaled quantities are denoted by a tilde and the equations in Table 2 are replaced by a slightly more complex system shown in Table 4. Note that the factor, δ, in the flow condition shows that $g(k)$ is small and hence k is just slightly greater than unity — the material is fully yielded.

The naive solution is still correct in the plug region in that the pressure is still a function of r only, $\hat{p}(r)$, and the shear stress is still linear in z, $\hat{s}_{rz} = -\hat{\tau}_w z$. However, all components of the stress and strain rate tensors are now significant. Thus, the procedure for calculating the velocity perturbation is less straightforward. The key equations are

$$\tilde{u}_{,z} = \tilde{e}_{rz} = \frac{\tilde{\dot{\gamma}}}{k} \quad \hat{s}_{rz} \tag{16}$$

and

$$\tilde{\dot{\gamma}} = 2\sqrt{\frac{\hat{u}_p'^2 + (\hat{u}_p/r)^2 + \hat{u}_p'\hat{u}_p/r}{1 - (\hat{\tau}_w z/k)^2}} = \frac{1}{\omega}\delta^{-1}g(k)\,. \tag{17}$$

Equation (17) may be solved for k and hence $\tilde{\dot{\gamma}}$. Hence Eq. (16) may be solved to find the velocity profile. Therefore, we may obtain a self-consistent solution for the plug flow region in which the material is fully yielded but the shear strain rate is very small. The naive lubrication theory solution is asymptotically correct and gives useful engineering approximations for the velocity profile, the wall shear stress, the pressure distribution and the total load. Equations that are analogous to Eq. (17) have been derived by Wilson [11] for the bi-viscosity model, and by Sherwood and Durban [16] for nearly homogeneous straining motions of the power-law fluid and rigid-plastic solid.

Results

Figure 2 shows the results of the naive lubrication calculation for a Herschel–Bulkley fluid with the no-slip boundary condition. Both the flow index and the Weissenberg number were taken to be 0.5. In this case, the plug flow region occupies more than half the flow domain and the velocity profiles are very flat.

Figures 3 and 4 show details of the results calculated for both the naive lubrication solution and the self-consistent solution. As for Fig. 2, a Herschel–Bulkley model has been used with flow index and Weissenberg number both set to 0.5, and the no-slip boundary condition has been applied. The aspect ratio has been set to 0.02 to demonstrate the asymptotic agreement of the solutions. Such a small value is unlikely to be met in practice. However,

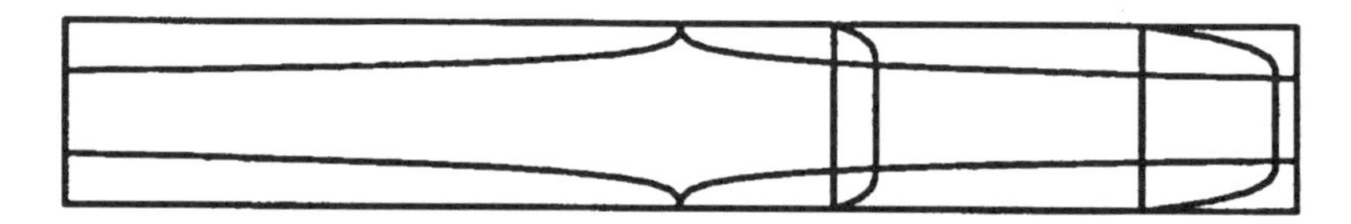

Fig. 2. The naive lubrication solution for a Herschel–Bulkley fluid with flow index and Weissenberg number both taken to be 0.5, showing the plug flow region and velocity profiles at $r = 0.25$ and 0.75.

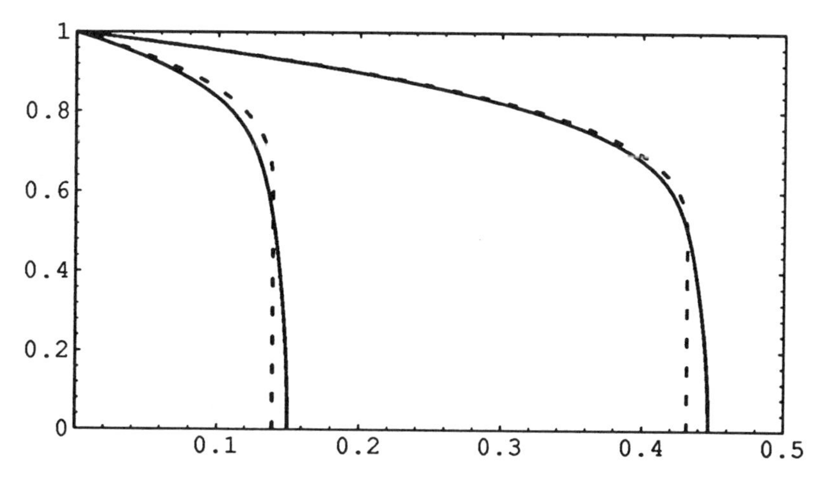

Fig. 3. The velocity profiles of the lubrication solution (...) and the self-consistent solution (___) at $r = 0.25$ and 0.75.

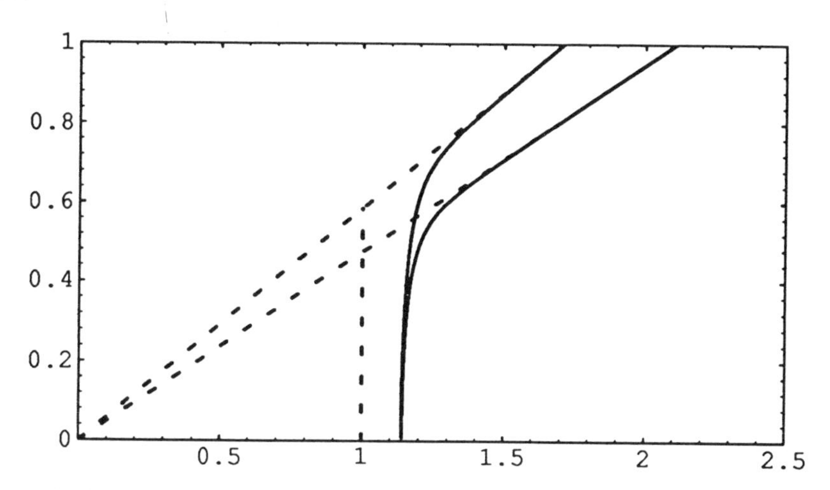

Fig. 4. Profiles of flow stress, k, calculated from the naive (...) and self-consistent (___) solutions.

similar profiles have been obtained for aspect ratios of 0.1 when more complex boundary conditions were used [15].

Figure 3 shows the velocity profiles at $r = 0.25$ and 0.5. The naive solution (dotted) shows the "unyielded" plug zone which gradually decreases in thickness with radius. The velocity profile is flat in this region whereas the self-consistent solution (solid line) has a smoothly rounded velocity profile.

The profile of the flow stress is shown in Fig. 4. The two sets of curves correspond to the velocity profiles of Fig. 3. The lubrication solution has a flow stress that is equal to the shear stress which is linear in z. The self-consistent solution, however, maintains a stress above the yield stress (unity) in the plug zone. The adjustment between the two regions is smooth but occurs over a short distance.

Further Discussion

The apparent inconsistency in the lubrication solution arises from the strict interpretation of the yield condition. The material is either yielded or not and there is no in-between state. This approach is perhaps over-zealous. It may be argued [3] that no material possesses a true yield stress. Rather, the yield stress provides a simple model of complex material behaviour. The lubrication approximation is also a simplifying assumption and the two approximations are not entirely consistent. Nevertheless, the results of the lubrication approximation are correct and it is unreasonable to insist that such results should be discarded. In fact, the lubrication approach provides a convenient, simple and *accurate* solution. This has been demonstrated here for the example of squeeze flow. However, other gradually varying flows are subject to a similar analysis. Another example is the spin coating process in which a thin layer of material flows off a rotating disk under the action of centrifugal force [19–21].

The biviscosity model has been introduced to avoid the problems with the yield stress [7, 11]. This is neither necessary nor, in general, desirable. The equations of the system are made more complex when the biviscosity model is used because of the introduction of another parameter — the viscosity ratio. This is an artificial parameter and the results of the analysis may depend on its value [11]. The yield stress model is actually simpler, though neither model is likely to be an exact representation of a real material. Since the properties of a real material are always measured over a finite range, the ultimate behaviour at vanishing stress must be inferred by extrapolation. The yield stress model is capable of giving perfectly good results with a smooth velocity profile and no inconsistencies.

It is important to note the limitations of the lubrication solution. In the squeeze flow problem studied here, the scaling is invalid near the origin of the co-ordinates, where there are stagnation points, and near the free surface, since the boundary condition, Eq. (9), cannot, in general, be satisfied. Numerical simulations [9] indicate that the lubrication solution actually fails in these regions. The flow near the free surface is a fountain flow similar to that in mould filling; the material barrels outwards, part of the free surface contacts the platen, and thin rings of unyielded material may occur in the vicinity of the contact lines. Furthermore, conical unyielded regions similar to those seen by Beris *et al.* [12] occur near the stagnation points. Nevertheless, when the aspect ratio is reasonably small, the lubrication solution gives a good approximation for most of the flow field, the pressure distribution and the total load.

There are a few known situations where the lubrication scaling fails severely. When the friction at the surfaces is small, the flow is nearly a homogeneous pure straining motion [16]. In this case, the lubrication solution predicts the correct plug velocity field but underestimates the pressure and load, which must be sufficient to make the material yield in compression [8]. A different problem arises when the Weissenberg number defined above is large. Then a distinguished limit may exist, so that the lubrication solution is incorrect [11]. Fortunately, squeeze flows are generally quite slow and thus this limit is unlikely to occur in practice. A third type of difficulty arises in heated compression, where the viscosity of the material near the wall may be much smaller than that in the core [22]. In this case, the material near the wall can squeeze out preferentially and the velocity profile has maxima near the walls and a local minimum on the central plane. The edge boundary conditions, Eq. (9), are very important in this case and the lubrication solution fails. Finally, yield stress materials are often subject to flow localisation; the flow may take a very different form, as in the slip line field solutions of classical plasticity [1], with multiple blocks of relatively undeformed material sliding over one another.

Acknowledgements

We gratefully acknowledge the contributions of members of the Particle Technology Research Programme at Imperial College: Professors M. J. Adams and B. J. Briscoe, Drs. I. Aydin and S. Sinha and D. C. Kothari and C. Fielenbach. Work on this topic has been supported by DTI, EPSRC, MAFF and Unilever. This version of the paper was prepared after the Forum and inevitably draws on material from the discussion that occurred there. We acknowledge specifically the insights of Professor M. M. Denn.

References

1. Hill, R. (1950) *The Mathematical Theory of Plasticity.* Oxford University Press
2. Bird, R. B., Armstrong, R. C. & Hassager, O. (1987) *Dynamics of Polymeric Liquids, Vol. 1,* 2nd ed. Wiley

3. Barnes, H. F. & Walters, K. D. (1985) The yield stress myth? *Rheol. Acta* **24**, 323–326
4. Craster, R. V. (1995) Solutions for Herschel–Bulkley flows. *Q. J. Mech. Appl. Math.* **48**, 343–374
5. Bird, R. B., Dai, G. C. & Yarusso, B. J. (1983) The rheology and flow of viscoplastic materials. *Rev. Chem. Eng.* **1**, 1–70
6. Covey, G. H. & Stanmore, B. R. (1981) Use of the parallel plate plastometer for the characterisation of viscous fluids with a yield stress. *J. Non-Newtonian Fluid Mech.* **8**, 249–260
7. Lipscomb, G. G. & Denn, M. M. (1984) Flow of Bingham fluids in complex geometries. *J. Non-Newtonian Fluid Mech.* **14**, 337–346
8. Adams, M. J., Edmondson, B., Caughey, D. G. & Yahya, R. (1993) An experimental and theoretical study of the squeeze-film deformation of elastoplastic fluids. *J. Non-Newtonian Fluid Mech.* **51**, 61–78
9. Adams, M. J., Aydin, I., Briscoe, B. J. & Sinha, S. K. (1997) A finite element analysis of the squeeze flow of an elasto-viscoplastic paste material. *J. Non-Newtonian Fluid Mech.* **71**, 41–57
10. Gartling, D. K. & Phan-Thien, N. (1984) A numerical simulation of a plastic fluid in a parallel-plate plastometer. *J. Non-Newtonian Fluid Mech.* **14**, 347–360
11. Wilson, S. D. R. (1993) Squeezing flow of a Bingham material. *J. Non-Newtonian Fluid Mech.* **47**, 211–219
12. Beris, A. N., Tsamopoulos, J. A., Brown, R. A. & Armstrong, R. C. (1984) Creeping flow around a sphere in a Bingham plastic. *J. Fluid Mech.* **158**, 141–172
13. Collins, I. F. & Meguid, S. A. (1977) On the influence of hardening and anisotropy on the plane strain compression of thin metal strip. *J. Appl. Mech.* **44**, 271–278
14. Adams, M. J., Briscoe, B. J., Corfield, G. M., Lawrence, C. J. & Papathanasiou, T. D. (1997) An analysis of the plane-strain compression of viscoplastic materials. *J. Appl. Mech.* **64**, 420–424
15. Corfield, G. M. (1996) *The constrained flow of pastes.* PhD Thesis. Imperial College, London University
16. Sherwood, J. D. & Durban, D. (1996) Squeeze flow of a power-law viscoplastic solid. *J. Non-Newtonian Fluid Mech.* **62**, 35–54
17. Perzyna, P. (1966) Fundamental problems in plasticity. *Adv. Appl. Mech.* **9**, 244–377
18. Krempl, E. & Gleason, J. M. (1996) Isotropic viscoplasticity theory based on overstress (VBO). The influence of the direction of the dynamic recovery term in the growth law of the equilibrium stress. *Int. J. Plasticity* **12**, 719–735
19. Lawrence, C. J. & Zhou, W. (1991) Spin coating of non-Newtonian fluids. *J. Non-Newtonian Fluid Mech.* **39**, 137–187
20. Jenekhe, S. A. & Schuldt, S. B. (1985) Flow and film thickness of Bingham plastic liquids on a rotating disk. *Chem. Eng. Commun.* **33**, 135–143

21. Burgess, S. L. & Wilson, S. D. R. (1996) Spin coating of a viscoplastic material. *Phys. Fluids* **8**, 2291–2297
22. Lee, S. J., Denn, M. M., Crochet, M. J. & Metzner, A. B. (1982) Compressive flow between parallel disks: I. Newtonian fluid with a transverse viscosity gradient. *J. Non-Newtonian Fluid Mech.* **10**, 3–30

Dynamics of Complex Fluids, pp. 394–398
ed. M. J. Adams, R. A. Mashelkar, J. R. A. Pearson & A. R. Rennie
Imperial College Press–The Royal Society, 1998

Chapter 28

The Wall Yield of Rate-Dependent Materials

M. J. ADAMS, B. J. BRISCOE, G. M. CORFIELD AND C. J. LAWRENCE*
Department of Chemical Engineering,
Imperial College of Science, Technology and Medicine,
London SW7 2AZ
**E-mail: c.lawrence@ic.ac.uk*

A theoretical analysis of the axisymmetric squeeze flow of a viscoplastic material is described. A comparison is made of the stress fields for two types of wall stress boundary condition: the Herschel–Bulkley and Tresca. Only the former prescribes a slip velocity dependence of the wall traction and this leads to major differences in the stress fields. The results of these analyses are compared with the radial wall pressure measurements obtained from the axisymmetric squeeze flow of an aqueous starch paste.

Recently, there has been an interest in the theoretical analysis of the squeeze flow of viscoplastic materials, such as soft solids and pastes under slip boundary conditions [1, 2]. These analyses employed the simple Tresca boundary conditions where the wall shear stress, τ_w, is given by

$$\tau_w = mk\,, \tag{1}$$

where $m(0 < m \leq 1)$ is termed the friction factor and k is the bulk shear yield stress. However, rheological studies of pastes [3] have revealed that the wall shear stress is a function of the slip velocity, U, so that the most general wall boundary condition (wbc) may be written in a form that is analogous to the Herschel–Bulkley (HB) relationship. Thus

$$\tau_w = \tau_{wy} + KU^n \, , \tag{2}$$

where τ_{wy} is the wall yield stress and K and n are material constants. The Navier condition is recovered when $\tau_{wy} = 0$. Here, we describe the results using an axisymmetric squeeze flow analysis with these wbcs, and based on an asymptotic perturbation of the radial velocity field and an associated viscoplastic flow rule given by Perzyna [4]. The analysis is analogous to that reported by Lawrence and Corfield [5] except that they considered the special case of stick boundary conditions.

Figures 1 and 3 show contour plots of the calculated shear stresses for the Tresca and HB wbcs, respectively

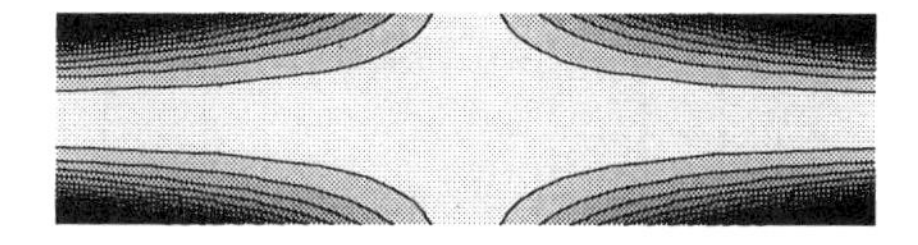

Fig. 1. A Contour plot (of an axial section in the plane of deformation) of the shear stresses generated by traction at the wall boundary between the squeeze flow platens illustrating the profile for the Herschel–Bulkley wall boundary condition (Eq. (1)). Darker shading depicts increasing shear stress.

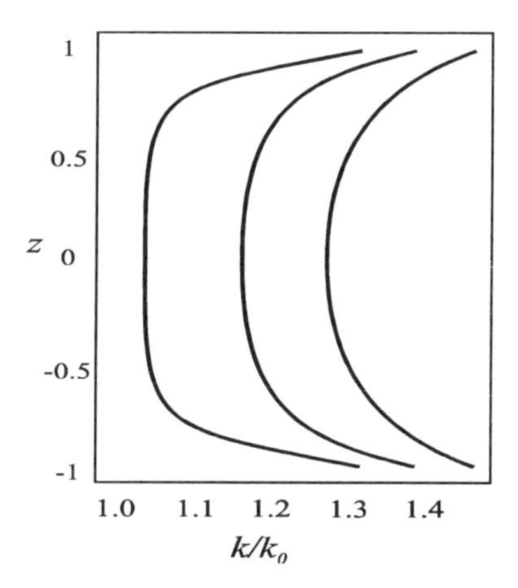

Fig. 2. The stress variance for the Herschel–Bulkley wall boundary condition is plotted for increasing platen velocities at $r = 0.5$; r is the radial co-ordinate in Fig. 1 where $r = 0$ is the central axis and $r = 1$ the free surface. The ordinate axis, z, represents the location between the platens where $z = 0$ is the material centre and $z = \pm 1$ the wall boundaries. The illustration shows the gradual transition from plug flow (left-hand trace) at lower platen velocities to a more rational distribution of the flow stresses at higher velocity (right-hand trace).

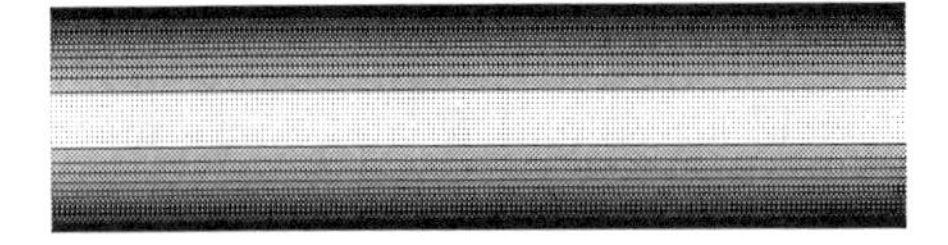

Fig. 3. A Contour plot of the flow stresses generated with the Tresca wall boundary condition. Darker shading depicts increasing shear stress.

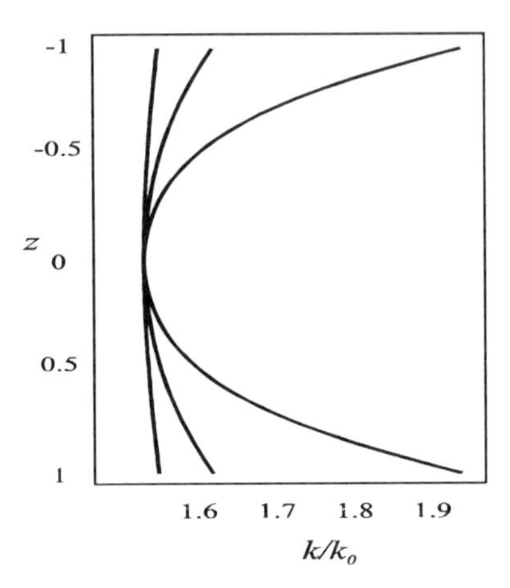

Fig. 4. For an arbitary platen velocity, the stress variance for the Tresca wall boundary condition is plotted for $m = 0.1$ (left-hand trace), 0.5 (middle trace) and 0.9 (right-hand trace), where $m = 0$ represents "slip" and $m = 1$ represents "stick". The ordinate axis, z, represents the thickness of the material between the platens where $z = 0$ is the material centre and $z = \pm 1$ are the wall boundaries. For $m = 0.9$, the stress variance increases significantly at the boundaries.

In the case of the Tresca wbc, the stress is invariant with respect to the radial co-ordinate for all values of the axial co-ordinate, including those at the wall where it so defined. However, for the HB wbc, the contour plot is a function of the radial co-ordinate and is similar to that obtained from the lubrication solution with a stick wbc [6, 7]. The corresponding stress variance (defined as the ratio of the current flow stress, k, to the static shear yield value, k_0) plots are given in Figs. 2 and 4.

An important point to note is that $k/k_o > 1$ for all values of the axial co-ordinate so that there are no unyielded regions as obtained from the low order lubrication solution (see [5]). For the Tresca wbc, the profile tends to that expected for plug flow as m tends to zero and becomes more fully developed

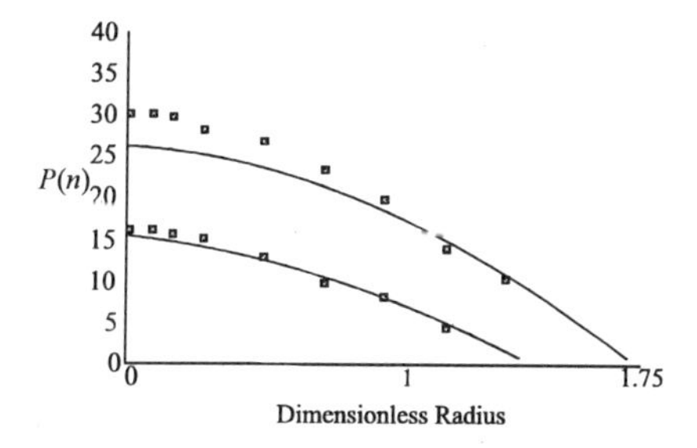

Fig. 5. The radial pressure distribution for a starch-based paste with 50% moisture content deformed to 55% and 65% of the original specimen height. The points are experimental data and the lines are derived from the model using a Herschel–Bulkley wall boundary condition.

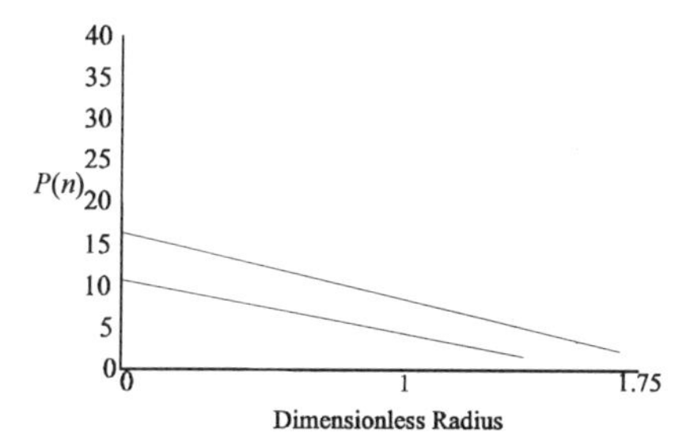

Fig. 6. The radial pressure distribution as predicted for an experiment identical to Fig. 5 employing the Tresca wall boundary condition. Most conspicuous is the linear profile of the prediction and the associated reduction in the pressure magnitude.

with increasing values of m. The contours are invariant with respect to the imposed platen velocity. However, the magnitude of the stress variance will increase proportionally with this velocity. In the case of the HB wbc, the flow becomes more fully developed with increasing values of the platen velocity but, at low velocities, there is a central core where the yield criterion is only just satisfied. This is reminiscent of the lubrication solution mentioned above.

Radial wall pressure distributions, (Pn), were obtained in the axisymmetric squeeze flow of an aqueous potato starch paste by measuring the normal stresses at discrete points along the platen wall (Fig. 5). Capillary rheometry established that this paste was viscoplastic and exhibited a HB wbc. The measured constitutive relationships were employed in the current analysis scheme and the calculated pressure distributions are shown in Fig. 5. Unlike the best fit of a Tresca wbc, Fig. 6, which leads to a linear distribution and which is associated with wall slip at all values of the radial co-ordinate, the agreement is excellent.

In conclusion, the current analysis using a HB wbc provides a realistic description of the squeeze flow of a typical paste where slip is associated with a minimum value of the wall shear stress.

References

1. Sherwood, J. D. & Durban, D. (1996) Squeeze flow of a power law fluid. *J. Non-Newtonian Fluid Mech.* **14**, 347–60
2. Adams, M. J., Briscoe, B. J., Corfield, G. M. & Lawrence, C. J. (1997) An analysis of the plane strain compression of viscoplastic materials. *J. Applied Mech. ASME* **64**, 420–424
3. Yilmazer, U. & Kalyon, D. M. (1989) Slip effects in capillary and parallel disk torsional flows of highly filled suspensions. *J. Rheology* **33** (8), 1197–1212
4. Perzyna, P. (1996) Fundamental problems in plasticity. *Adv. Appl. Mech.* **9**, 244–377
5. Lawrence, C. J. & Corfield, G. M. (1998) Non-viscometric flow of viscoplastic materials: squeeze flow. In *Dynamics of Complex Fluids*, ed. M. J. Adams, R. A. Mashelkar, J. R. A. Pearson & A. R. Rennie, pp. 379–393. London: Imperial College Press–The Royal Society
6. Covey, G. H. & Stanmore, B. R. (1981) Use of the parallel plate plastometer for the characterisation of viscous fluids with a yield stress. *J. Non-Newtonian Fluid Mech.* **8**, 249–260
7. Adams, M. J., Edmondson, B., Caughey, D. G. & Yahya, R. (1996) An experimental and theoretical study of the squeeze film deformation of elastoplastic fluids. *J. Non-Newtonian Fluid Mech.* **51**, 1225–1233

Dynamics of Complex Fluids, pp. 399–404
ed. M. J. Adams, R. A. Mashelkar, J. R. A. Pearson & A. R. Rennie
Imperial College Press–The Royal Society, 1998

Chapter 29

Strain Localisation During the Axisymmetric Squeeze Flow of a Paste

M. J. ADAMS, B. J. BRISCOE*, D. KOTHARI AND C. J. LAWRENCE

Department of Chemical Engineering,
Imperial College of Science, Technology & Medicine,
London, SW7 2BY
**E-mail: b.briscoe@ic.ac.uk*

An experimental study of the axisymmetric squeeze flow of a model paste is described. Inhomogeneous deformation was observed at the edge of some specimens in the form of barrelling or bollarding, and in the interior as localised shear bands. The nature of these flow instabilities was critically dependent on the wall boundary conditions which were varied by changing the temperature of the platens.

Introduction

Pastes are common intermediate products in processing industries such as those concerned with the production of foods and ceramics. They contain a high volume fraction of a particulate phase dispersed in a fluid. Their rheological behaviour is essentially elasto-viscoplastic and the associated material parameters may be obtained using nominally simple measurement geometries such as squeeze flow. The analysis of experimental data obtained from this method involves either numerical or analytical solutions although, for the latter, the elastic strains are invariably neglected [1]. These procedures have focused upon continuous velocity fields and the analytical solutions are valid for large values of the aspect ratio, R/H, where R and H are the radius and height of the specimen, respectively, and in the region, $0 < r < R$, where r is the radial co-ordinate. However, in any industrial forming operation involving such materials, it is possible for unwanted flow instabilities to occur which result in local strain concentrations. In squeeze flow, this may include shear banding or non-uniform geometric changes at the periphery of a specimen. Such phenomena are critically dependent on the wall stress boundary conditions. Here, we describe an experimental study aimed at exemplifying these instabilities using a model paste ("Plasticine") by varying the boundary conditions with heated platens.

Experiment and Results

Squeeze flow experiments were performed on cylindrical Plasticine specimens of 80 mm diameter and 20 mm height using an Instron universal testing machine (model 6022). Overhanging parallel and polished stainless steel platens were used for this purpose with the lower platen remaining stationary and the upper platen closing with a velocity of 10 mm/min. The temperature of the platens was controlled using electrical heaters. Each specimen was cut longitudinally through the centre and the exposed surfaces were marked with a 1 mm square grid. To minimise bulk heating the specimens were then reassembled and positioned on the lower platen immediately prior to compression.

Figure 1 shows photographs of the exteriors and interiors of the specimens for different platen temperatures after an imposed uniaxial strain of 50%. Multiple shear bands are most prominent in the temperature range 40–60°C with the displacement fields being continuous at 20°C and 80°C, and with that at 30°C corresponding approximately to uniform extension. At 20°C, barrelling may be observed while unambiguous bollarding occurs at 50°C and above. The bollarding is so severe at 80°C that longitudinal cracks are formed at the periphery. Barrelling is defined as a relative increase in the mid-plane specimen diameter while bollarding is defined as a relative increase in the values of the diameter adjacent to the platens. In all cases, the diametric changes are not symmetric about the mid-plane because of slight differences in the platen temperatures and hence the wall friction.

Discussion

The origins of barrelling and bollarding are well established. When the platen friction is high, or no-slip conditions prevail, the imposed strain is

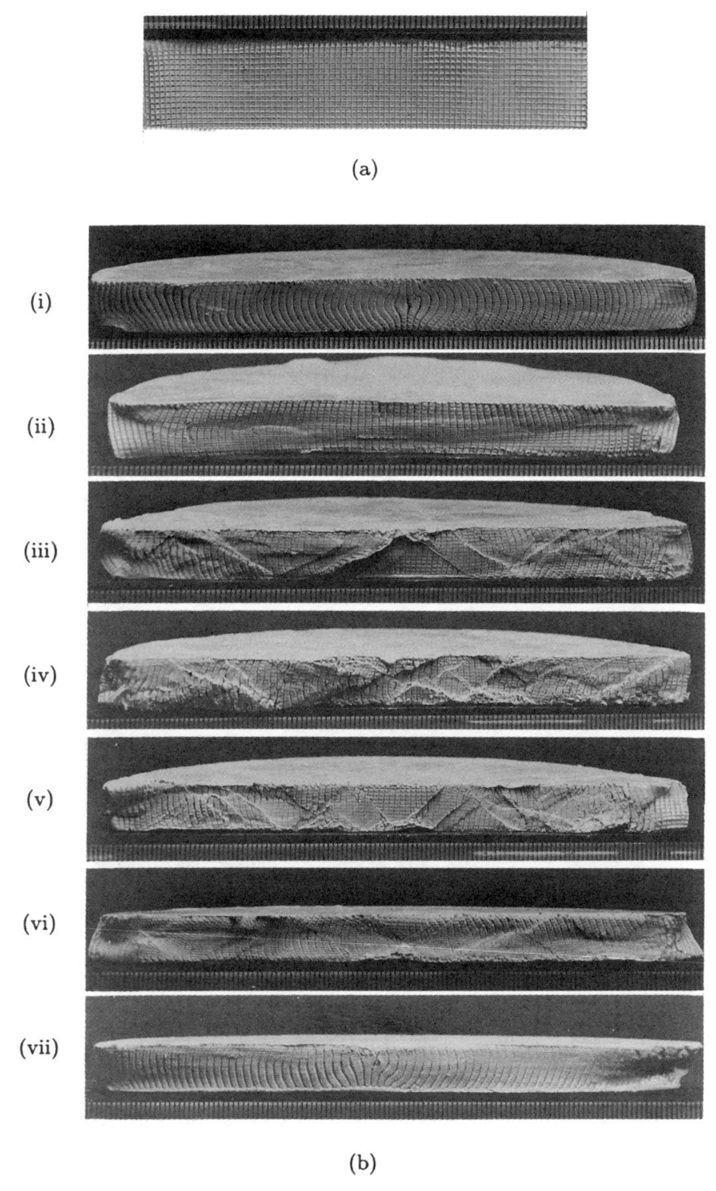

Fig. 1. Photographs of (a) the interior gridded surface of an uncompressed cylindrical "Plasticine" specimen, (b) and (c) the interior and exterior surfaces of specimens compressed by 50% with platen temperatures of (i) 20°C, (ii) 30°C, (iii) 40°C, (iv) 50°C, (v) 60°C, (vi) 70°C and (vii) 80°C.

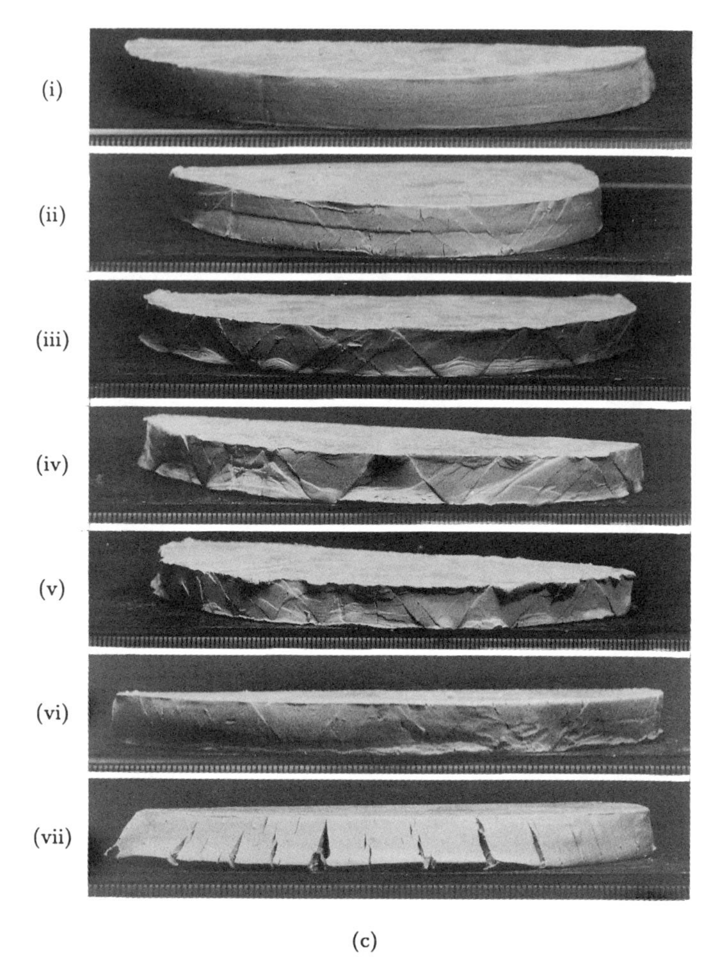

Fig. 1. (*Continued*)

accommodated by barrelling and folding of the peripheral surface onto the platens [2]. This corresponds to the lower platen temperatures. As the temperature of the platens is increased, the Plasticine softens locally and is preferentially "squeezed", thus causing a shear stress to act on the core which leads to bollarding [3]. In metalworking practice, an instability criterion has been developed that takes into account the influence of the initial thermal non-uniformities, strain rate and temperature dependence of the flow stress [4].

The prediction of the initiation, direction and thickness of shear bands and the post-localisation behaviour is considerably more difficult although significant progress has been made for individual bands. Shear banding is an ubiquitous phenomenon for plastically deformed ductile materials such as metals, polymers, rocks, powders and wet soils [5]. Lüders bands in some metals are perhaps the most well-known example. Moreover, Semiatin and Jonas [6] have published photographs of compressed metal specimens that are similar to those provided here. As in the case of crack propagation, micro-structural imperfections are believed to play a critical role in the initiation of shear bands due to their ability to concentrate the stress. This has been established quantitatively by using finite element analysis on a viscoplastic material in which multiple shear band formation was predicted albeit in the simpler case of plane strain [7]. A crucial factor in the formation of shear bands is the low rate of work hardening with the region of localisation being subject to work softening. The latter process may arise from dilation and void formation which are characteristic of particulate materials. The location of shear bands depends on the interaction of the wall boundary conditions with the micro-structure.

The orientation of the shear bands corresponds to the direction of the maximum shear flow, and hence may be predicted using the Mohr–Coulomb yield criterion [8]. However, this criterion describes only the instantaneous stress and is consequently unable to predict the onset of localisation. One approach is to determine when the governing field equations undergo a transition from elliptic to hyperbolic [9]. In the former case, the deformation is diffused while for the latter, the velocity characteristics are indicative of the large local strains associated with the formation of shear bands. The method of characteristics then provides an alternative basis for predicting the orientation of the shear bands.

References

1. Lawrence, C. J. & Corfield, G. M. (1998) Non-viscometric flow of viscoplastic materials: squeeze flow. In *Dynamics of Complex Fluids*, ed. M. J. Adams, R. A. Mashelkar, J. R. A. Pearson & A. R. Rennie, pp. 379–393. London: Imperial College Press–The Royal Society
2. Adams, M. J., Briscoe, B. J. & Kamjab, M. (1993) The deformation and flow of highly concentrated dispersions. *Adv. Coll. Int. Sci.* **44**, 143–182
3. Papathanasiou, A. C., Macosko, C. W. & Scriven, L. E. (1986) Analysis of lubricated squeezing flow. *Int. J. Number. Meth. Fluids* **6**, 819–839
4. Oh, S. I., Semiatin, S. L. & Jonas, J. J. (1992) An analysis of the isothermal hot compression test. *Metal. Trans.* **23A**, 963–975

5. Rice, J. R. (1976) The localisation of plastic deformation. In *Theoretical and Applied Mechanics*, ed. W. T. Koiter, pp. 207–220. North-Holland Publishing Co
6. Semiatin, S. L. & Jonas, J. J. (1984) In *Formability of Workability of Metals: Plastic Instability and Flow Localisation.* American Society of Metals
7. Belytschko, T., Chiang, H.-Y. & Plaskacz, E. (1994) High resolution two-dimensional shear band computations: imperfections and mesh dependence. *Comput. Methods Appl. Engrg.* **119**, 1–15
8. Lee, Y. K. & Ghosh, J. (1996) The significance of J_3 to the prediction of shear bands. *Int. J. Plasticity* **12**, 1179–1197
9. Lee, Y. K. (1989) Conditions for shear banding and material instability in finite elastoplastic deformation. *Int. J. Plasticity* **5**, 197–226

Dynamics of Complex Fluids, pp. 405–424
ed. M. J. Adams, R. A. Mashelkar, J. R. A. Pearson & A. R. Rennie
Imperial College Press–The Royal Society, 1998

Chapter 30

Viscoplastic Approaches in Forming Processes: Phenomenological and Computational Aspects

D. PERIĆ* AND D. R. J. OWEN
Department of Civil Engineering, University of Wales Swansea,
Swansea, SA2 8PP, United Kingdom
**E-mail: d.peric@swansea.ac.uk*

The main objective of this work is to discuss a sound continuum basis and associated numerical formulation of viscoplastic approaches used in simulation of forming processes. From the outset, a Lagrangian framework is adopted, as is often done in solid mechanics, and large deformations and material strains are accommodated within the described framework. Several phenomenological viscoplastic constitutive models are reviewed that are applicable for particular materials. Computational issues are discussed in the context of the finite element method which is formulated within a rigorous nonlinear framework. Within the computational scheme, a robust algorithm based on an operator split method (elastic predictor-plastic corrector) is used for numerical integration of the viscoplastic constitutive equations. Large deformations represent standard working conditions during forming operations, which causes continuous changes of the optimal finite element mesh configuration throughout the deformation process. Therefore, an adaptive finite element procedure is formulated and employed in numerical simulations. The finite element simulation of a series of numerical tests is carried out and the results are compared with available experimental evidence.

Introduction

The role of computational mechanics in the solution of nonlinear industrial problems is becoming increasingly established due to parallel advances in theoretical formulations, together with associated algorithmic implementation, and unabated developments in computer technology. Such simulation procedures, which are almost exclusively based on the finite element method, are now routinely employed in the modelling of processes such as metal forming operations. The acceptance of such approaches to industrial design is continually increasing, due to improved awareness, the availability of appropriate software and ever reducing computational costs. The strong impact on the computational community is evident from numerous conferences (see Shen and Dawson [1] for the most recent meeting).

Probably the most striking example of developments in this field are the strides made recently in the numerical solution of finite strain plasticity problems. The formulation of rigorous solution procedures has been the subject of intense debate over the last decade, and only recently has some consensus been reached on an appropriate constitutive theory based on tensorial state variables to provide a theoretical framework for the macroscopic description of a general elasto-plastic material at finite strains. In computational circles, effort has been directed at the formulation of algorithms for the integration of constitutive equations relying on operator split methodology. In particular, the concept of consistent linearisation has been introduced to provide quadratically convergent solution procedures. By employing logarithmic stretches as strain measures, a particularly simple model for large inelastic deformations at finite strains is recovered. In particular, the effects of finite strains appear only at the kinematic level, and the integration algorithms and corresponding consistent tangent operators for small strain situations can be directly employed.

For several areas of application, the inclusion of rate effects on material deformation is essential. Such problems include sheet stamping operations, in which rate effects can be significant, and hot forging processes in which the material deformation is highly strain rate dependent. For problems of this type, rate dependent behaviour can be included through the use of an elasto-viscoplastic model in place of the rate independent elasto-plastic approach. The formulation of solution procedures for finite strain elasto-viscoplastic problems follows that for the rate independent case, and the use of logarithmic strains as a kinematic variable offers similar benefits.

Practical applications involving both finite strains and inelastic material deformation often require numerical modelling of other physical phenomena. For example, since the development of finite strains is frequently a precursor to failure, the modelling of failure mechanisms such as ductile damaging may be necessary. The inclusion of damage effects within a finite strain elasto-plastic setting can again be accomplished through an operator split implementation based on consistent linearisation principles.

A further class of nonlinear problem for which considerable advances in numerical modelling have been made in recent years is that of contact-friction behaviour. Contact-friction phenomena arise in many important practical areas and numerical treatment in the past has, of necessity, relied on temperamental and poorly convergent algorithms. This situation has changed markedly with a recognition of the complete analogy that exists between contact-friction behaviour and the classical theory of elasto-plasticity. Thus, the operator split algorithms and consistent linearisation procedures developed for the latter case translate directly into contact-friction models that provide robust and rapidly convergent numerical solutions.

Despite such advances, most applications reported in the literature are still restricted to the generalised Amontons-Coulomb law of perfect friction. As pointed out by Curnier [2], such a simplified theory may represent only a limited range of tribological situations and the state conditions of surfaces in contact are influenced by a number of complex phenomena, such as wear, internal straining, lubrication and chemical reactions [3]. In situations involving large sliding distances, which are typical in deep drawing operations, the evolution of surface wear may become particularly important in the definition of frictional behaviour. In particular, for coated sheet metals, appreciable deviations from the standard Amontons-Coulomb model are observed in experiments. When subjected to high normal pressures and large sliding distances, the surface coating present in these materials may be worn away to expose the bulk metal, thereby causing a rapid increase in the friction coefficient. In these cases, numerical predictions with reasonable accuracy may demand a consideration of more realistic friction rules (we refer to [4] for recent contributions).

The demands of industry have also increased. Apart from technical requirements, which includes developments in material, structural and contact models, the finite element based simulation must also be appropriately integrated within a CAD/CAM environment to reduce both design to product time and minimise costs. In such a *decision support system*, a crucial link between numerical analysis and other phases of the design process is provided by *automated, adaptive finite element strategies.*

A general feature of finite element simulation of forming operations is that the distribution of the optimal mesh refinement changes continually throughout the forming process. Considerable benefits may accrue from an adaptive analysis in terms of robustness and efficiency. At the same time, error estimation procedures play a crucial role in quality assurance of the decision-making process by providing a reliable finite element solution.

This paper examines some of the above issues in detail and presents computational strategies suitable for the modelling of industrial scale forming problems. A set of numerical examples is provided to illustrate the effectiveness and robustness of the techniques developed.

Elasto-Viscoplastic Solids

The constitutive model for elasto-viscoplastic solids that has a wide range of applications may be represented in the following form:

$$\begin{aligned} f(\boldsymbol{\sigma},\bar{\sigma}) &= F(\boldsymbol{\sigma}) - \bar{\sigma}\,, \\ F(\boldsymbol{\sigma}) &= \sqrt{3J_2}\,, \\ \dot{\boldsymbol{\varepsilon}}^{vp} &= \gamma\langle\Phi(\bar{\sigma})\rangle\sqrt{\frac{3}{2}}\frac{\operatorname{dev}[\boldsymbol{\sigma}]}{\|\operatorname{dev}[\boldsymbol{\sigma}]\|}\,, \end{aligned} \tag{1}$$

in which $\boldsymbol{\sigma}$ denotes the stress tensor; J_2 is the second invariant of deviatoric stress tensor, $\operatorname{dev}[\boldsymbol{\sigma}]$; $\bar{\sigma}(\bar{\varepsilon}^p)$ is the uniaxial yield stress; $f(\boldsymbol{\sigma},\bar{\sigma})$ is the yield function; γ is the fluidity parameter; and Φ is the *viscoplastic flow potential.* The uniaxial stress is often expressed as a function of the equivalent plastic strain, $\bar{\sigma} := \bar{\sigma}(\bar{\varepsilon}^p)$, which indicates a degree of hardening. The viscoplastic model, Eq. (1), is supplemented by an elastic law,

$$\boldsymbol{\sigma} = \mathbf{C} : \boldsymbol{\varepsilon}^e\,, \tag{2}$$

where $\mathbf{C}$ is the elasticity tensor and $\boldsymbol{\varepsilon}^e$ denotes the elastic part of the total strain tensor, $\boldsymbol{\varepsilon}$. The notation, $\langle\bullet\rangle$, in Eq. (1) denotes the so-called ramp function defined as $\langle x\rangle = (x + \mid x \mid)/2$. Physically, the viscoplastic deformations will be present if the flow potential, $\Phi(\bar{\sigma})$, takes non-negative values. In practical metal forming operations under high temperature conditions, the effective stress (usually termed the flow stress) is also rate-dependent. With small modifications, the viscoplastic model, Eq. (1), is still suitable for simulation. Several empirical relations exist for flow stress prediction, which are based on experimental tests and are applicable to particular materials under specific conditions. The three most commonly employed expressions are summarised below:

(i) *Hajduk expression.* The flow stress, $\overline{\sigma}$, is assumed to be expressed as

$$\overline{\sigma} = K_{fo} K_T K_{\overline{\varepsilon}} K_{\dot{\overline{\varepsilon}}} \,, \tag{3}$$

where the three coefficients are functions of the form,

$$K_T = A_1 \exp[-m_1 T] \; ; \; K_{\overline{\varepsilon}} = A_2 \overline{\varepsilon}^{\,m_2} \; ; \; K_{\dot{\overline{\varepsilon}}} = A_3 \dot{\overline{\varepsilon}}^{\,m_3} \,, \tag{4}$$

where A_i and m_i are material constants determined from tests.

(ii) *Sellars-Tegart expression.* This is based on the following interpolation equation:

$$Z = \dot{\overline{\varepsilon}} \, \exp\left[\frac{Q}{R(T+273)}\right] = C[\sinh(\alpha\overline{\sigma})]^n \,, \tag{5}$$

where Z is the Zener-Hollomon parameter; Q is an activation energy usually independent of temperature and, in many cases, also independent of strain; R is the gas constant, 8.31 J/molK; T is the temperature in degrees Celsius; and C, α and n are material constants.

(iii) *ALSPEN expression* [5]. This expression is found to cover closely the properties of some aluminium alloys by fitting experimental curves in the form of

$$\overline{\sigma} = c(T)\,(\alpha + \alpha_0)^{n(T)}\, \overline{\varepsilon}^{\,m(T)} \,. \tag{6}$$

Coefficients $c(T)$, $n(T)$ and $m(T)$ are described in [5, 6]; α_0 is a constant taken to be 0.001; and $\mathrm{d}\,\alpha = \mathrm{d}\,\overline{\varepsilon}^{vp}$ for temperatures below the onset limit, $T_0 \approx 700$ K. Otherwise, $\mathrm{d}\,\alpha = 0$.

Since practically $\overline{\varepsilon} = \overline{\varepsilon}^{\,vp}$ and $\dot{\overline{\varepsilon}}^{\,vp} = \gamma\Phi$ for J_2 plasticity, parameters γ and Φ can be obtained, as shown in Box 1, for all three interpolation functions. In all three cases, the yield stress is taken to be zero. Hence the assumption is made that some part of the strain is always inelastic.

Finite Strain Plasticity

Multiplicative Decomposition

The main hypothesis underlying the approach employed for finite strain elastoplasticity is the multiplicative split of the deformation gradient into elastic and plastic parts, that is,

Box 1. Elasto-viscoplastic material model.

(i) Additive strain rate decomposition:

$$\boldsymbol{\varepsilon} = \boldsymbol{\varepsilon}^{e} + \boldsymbol{\varepsilon}^{vp}$$

(ii) Elastic response:

$$\dot{\boldsymbol{\varepsilon}}^{e} = \mathbf{C}^{-1} : \dot{\boldsymbol{\sigma}}$$

(iii) Flow rule:

$$\dot{\boldsymbol{\varepsilon}}^{vp} = \gamma \langle \Phi(\overline{\sigma}) \rangle \sqrt{3/2} \frac{\operatorname{dev}[\boldsymbol{\sigma}]}{\| \operatorname{dev}[\boldsymbol{\sigma}] \|}$$

(iv) γ and Φ from interpolation functions:

(1) Hajduk

$$\gamma = \left[\frac{1}{A_1 A_2 A_3} \exp[m_1 T] \overline{\varepsilon}^{vp(-m_2)} \right]^{(1/m_3)}$$

$$\Phi = \left(\frac{\overline{\sigma}}{K_{fo}} \right)^{(1/m_3)}$$

(2) Sellars-Tegart

$$\gamma = C \exp \left[\frac{-Q}{R(T + 273)} \right]$$

$$\Phi = [\sinh [\alpha \overline{\sigma}]]^{n}$$

(3) ALSPEN

$$\gamma = [c(T)]^{(-1/m(T))} [\alpha(\overline{\varepsilon}^{vp}) + \alpha_0]^{(-n(T)/m(T))}$$

$$\Phi = (\overline{\sigma})^{(1/m(T))}$$

$$\boldsymbol{F} := \boldsymbol{F}^{e} \boldsymbol{F}^{p} . \tag{7}$$

This assumption, first introduced by Lee [7], admits the existence of a local unstressed *intermediate configuration*. Due to its suitability for the computational treatment of finite strain elasto-plasticity, the hypothesis of

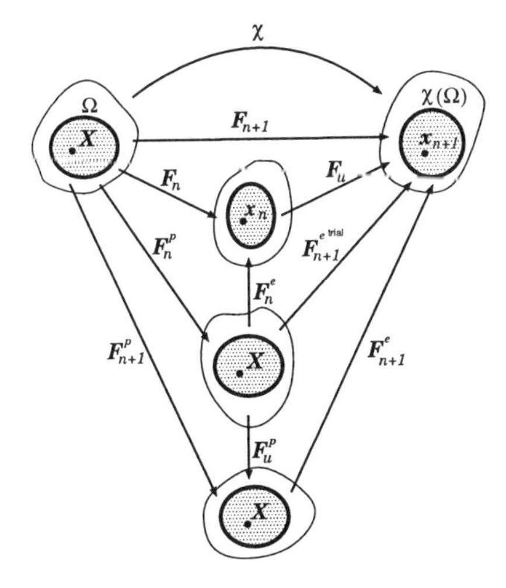

Fig. 1. Multiplicative decomposition of deformation gradient.

multiplicative decomposition is currently widely employed in the computational mechanics literature [8–10].

Stress Update for Finite Strains

In the context of finite element analysis of path dependent problems, the load path is followed incrementally and a numerical approximation to the material constitutive law is needed to update stresses as well as the internal variables of the problem within each increment. Given the values of the variables, $\{\boldsymbol{\sigma}_n, \boldsymbol{q}_n\}$, at the beginning of a generic increment, $[t_n, t_{n+1}]$, an algorithm for the integration of evolution equations is required to obtain the updated values, $\{\boldsymbol{\sigma}_{n+1}, \boldsymbol{q}_{n+1}\}$, at the end of the increment.

In the present work, the *backward Euler scheme* is employed for time integration and the *Newton–Raphson algorithm* is used in the solution of the resulting set of nonlinear equations. For the sake of convenience, the operations on the kinematic level of the algorithm for integration of the constitutive equations at finite strains are summarised in Box 2. The small strain stress update is then performed by employing the standard stress integration procedures which rely almost exclusively on the operator split methodology (elastic predictor-plastic corrector).

Box 2. Algorithm for integration of constitutive equation.

(i) For given displacement, $\boldsymbol{u}$, and increment of displacement, $\boldsymbol{\Delta u}$, evaluate total and incremental deformation gradient:

$$\boldsymbol{F}_{n+1} := \mathbf{1} + \mathrm{Grad}_{\varphi n+1}[\boldsymbol{u}_{n+1}]\,, \quad \boldsymbol{F}_{\boldsymbol{u}} := \mathbf{1} + \mathrm{Grad}_{\varphi n+1}[\boldsymbol{\Delta u}_{n+1}]\,.$$

(ii) Evaluate elastic trial deformation gradient and elastic trial Finger tensor:

$$\boldsymbol{b}^{e\,\mathrm{trial}}_{n+1} := \boldsymbol{F}_{\boldsymbol{u}}(\boldsymbol{b}^e_n)(\boldsymbol{F}_{\boldsymbol{u}})^T\,.$$

(iii) Compute eigenvalues (principal stretches, $\boldsymbol{\lambda}^{e\,\mathrm{trial}}$) and eigenvectors (rotation tensor, $\boldsymbol{R}^{e\,\mathrm{trial}}_{n+1}$) of elastic trial Finger tensor, $\boldsymbol{b}^{e\,\mathrm{trial}}_{n+1}$.

(iv) Evaluate elastic trial left strain tensor and its logarithmic strain measure:

$$\begin{aligned} \boldsymbol{V}^{e\,\mathrm{trial}}_{n+1} &:= (\boldsymbol{R}^{e\,\mathrm{trial}}_{n+1})^T(\boldsymbol{\lambda}^{e\,\mathrm{trial}})\boldsymbol{R}^{e\,\mathrm{trial}}_{n+1}\,, \\ \boldsymbol{\varepsilon}^{e\,\mathrm{trial}}_{n+1} &:= (\boldsymbol{R}^{e\,\mathrm{trial}}_{n+1})^T \ln[\boldsymbol{\lambda}^{e\,\mathrm{trial}}]\boldsymbol{R}^{e\,\mathrm{trial}}_{n+1}\,. \end{aligned}$$

(v) Perform stress updating procedure for small strain.

(vi) Update Cauchy stress and internal variable:

$$\begin{aligned} \boldsymbol{\sigma}_{n+1} &:= \det[\boldsymbol{F}_{n+1}]^{-1}\boldsymbol{\tau}_{n+1}\,, \\ \boldsymbol{V}^e_{n+1} &:= \exp[\boldsymbol{\varepsilon}^e_{n+1}]\,, \quad \boldsymbol{b}^e_{n+1} := (\boldsymbol{V}^e_{n+1})^2\,. \end{aligned}$$

Finite Element Technology

It is a well-known fact that the performance of low order kinematically based finite elements is extremely poor near the incompressible limit. Problems of elasto-plastic forming simulations under plastic dominant deformations and the assumption of isochoric plastic flow are included in this class of analysis. In such situations, spurious *locking* frequently occurs as a result of the inability of low order interpolation polynomials to adequately represent general volume preserving displacement fields. However, due to their simplicity, low order elements are often preferred in large-scale computations and several formulations have been proposed to allow their use near the incompressible limit. Within the context of geometrically linear theory, the class of assumed enhanced strain methods described by Simo and Rifai [11], which incorporates popular procedures such as the classical incompatible modes formulation [12] and *B-bar* methods [13], is well-established and is employed with success in a number of existing commercial finite element codes. In the geometrically nonlinear regime, however, the enforcement of incompressibility is substantially more demanding and the development of robust and efficient low order finite elements is by no means trivial. Different approaches have been introduced in the computational literature. The class of mixed variational methods developed by Simo *et al.* [14], the mixed u/p formulation proposed by Sussman and Bathe [15], the nonlinear B-bar methodology adopted by Moran *et al.* [16], and the family of enhanced elements of Simo and Armero [17] are particularly important. However, due to the occurrence of pathological hourglassing patterns, a serious limitation on the applicability of enhanced elements for the elasto-plastic finite strain case has been identified by de Souza Neto *et al.* [18]. An element which does not suffer from such drawbacks is the F-bar element presented in [19]. It is based on the concept of multiplicative deviatoric/volumetric split in conjuction with the replacement of the compatible deformation gradient with an assumed modified counterpart.

Adaptive Strategies for Industrial Forming Operations

The history-dependent nature of the process necessitates a transfer of all relevant problem variables from the old mesh to the new one as successive remeshing is applied during the process simulation. As the mesh is adapted, with respect to an appropriate error estimator, the solution procedure, in general, cannot be re-computed from the initial state, but has to be continued from the previous computed state. Hence some suitable means for transferring the state variables between meshes, or *transfer operators*, needs to be defined. A class of transfer operators for large strain elasto-plastic problems occurring in forming processes is defined in the next section.

Adaptive Solution Update

Error Indicators

The extension of the error estimation based on the *plastic dissipation functional* and the rate of *plastic work* described by Perić *et al.* [20] for large strain elasto-plasticity has been found to be the most appropriate solution to the adaptive solution of metal forming processes.

Mesh Regeneration

An unstructured meshing approach is used for mesh generation and subsequent mesh adaptation. The algorithm employed is based on the *Delaunay triangulation* technique, which is particularly suited to local mesh regeneration. An extension of the Delaunay scheme to quadrilateral elements in 2-D is also available and creates the possibility of employment of the low order elements described in the previous section. In 3-D, unstructured meshes based on tetrahedral elements are employed in numerical analyses since automatic meshing algorithms based on hexahedral elements are, at present, not available for arbitrary geometries.

Transfer Operations for Evolving Meshes

After creating a new mesh, the transfer of displacement and history–dependent variables from the old mesh to a new one is required. Several important aspects of the transfer operation have to be addressed [21, 22]:

(i) consistency with constitutive equations;
(ii) requirement of equilibrium (which is fundamental for implicit FE simulation);
(iii) compatibility of the history-dependent internal variables transfer with the displacement field on the new mesh;
(iv) compatibility with evolving boundary conditions; and
(v) minimisation of the numerical diffusion of transferred state fields.

To describe the transfer operation, let us define a state array, ${}^h\mathbf{\Lambda}_n = ({}^h\boldsymbol{u}_n, {}^h\boldsymbol{\varepsilon}_n, {}^h\boldsymbol{\sigma}_n, {}^h\boldsymbol{q}_n)$, where ${}^h\boldsymbol{u}_n, {}^h\boldsymbol{\varepsilon}_n, {}^h\boldsymbol{\sigma}_n, {}^h\boldsymbol{q}_n$ denote values of the displacement, strain tensor, stress tensor and a vector of internal variables at time t_n for the mesh, h. Furthermore, assume that the estimated error of the solution, ${}^h\mathbf{\Lambda}_n$, respects the prescribed criteria while these are violated by the solution, ${}^h\mathbf{\Lambda}_{n+1}$. In this case, a new mesh, $h+1$ is, generated and a new solution, ${}^{h+1}\mathbf{\Lambda}_{n+1}$, needs to be computed. As the backward Euler scheme is adopted, the internal variables, ${}^{h+1}\boldsymbol{q}_n$, for a new mesh, $h+1$, at time t_n need to be evaluated. In this way, the state, ${}^{h+1}\widetilde{\mathbf{\Lambda}}_n = (\bullet, \bullet, \bullet, {}^{h+1}\boldsymbol{q}_n)$, is constructed, where $\sim$ is used to denote a reduced state array. It should be noted that this state characterises the history

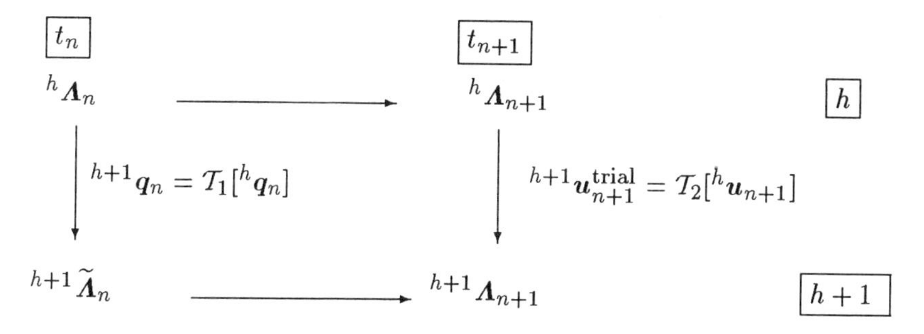

Fig. 2. Transfer operator diagram.

of the material and, in the case of a fully implicit scheme, provides sufficient information for the computation of a new solution, ${}^{h+1}\boldsymbol{\Lambda}_{n+1}$. Conceptually, Fig. 2 summarises a typical transfer operation that includes both the mapping of the internal variables and the mapping of the displacement field. The implementation of the given general transfer operation is performed in several steps and details are provided by Perić *et al.* [21] and Lee and Bathe [22].

Numerical Verification

To illustrate the computational strategy, its application to examples of industrial metal forming operations are presented in this section.

(i) *Prediction of tool wear for forging operations: A plane strain bulk forming of a crane hook.* A statistical investigation covering more than 100 forging geometries has categorised the principal causes of tool failure. Different types of wear (mainly at corners and roundings) are the cause for scrapping some 60% of the tools. Various types of crack formation account for approximately 25% and local plastic deformation, 5%. Consequently, the prediction of tool wear is a primary concern in the implementation of a preventive maintenance strategy within the forging industry.

Current approaches adopt a semi-empirical wear model which is incorporated into a finite element model to predict "instantaneous" tool wear. It must then be integrated to estimate the accumulated wear occurring over a larger number of forming cycles. Typical adhesive and abrasive wear models take the form of

$$W = K\,\frac{q \times s}{H}\,, \tag{8}$$

where W is the wear volume per unit area, q is the normal pressure, s the sliding length, H the surface hardness and K is a wear constant.

The local hardness is a function of temperature which necessitates a fully coupled thermo-mechanical approach to finite element modelling. During a drop forging cycle, where the cooler tool is in close contact with the hot billet, the surface temperature of the tool will rise rapidly due to heat transfer and the heat generated by friction. It is also necessary to estimate the variation of the temperature over the die surface during the forming cycle.

The sliding length, which is the amount of material passing a specific point on the die surface, is considered to be the most important parameter in the wear calculation. This is due to its strong influence on the temperature distribution generated through frictional heat.

The normal pressure, normalised by the yield stress, usually ranges from unity up to 3–4, with higher pressures occurring during the last stages of closed die forging operations or in open die forging with a narrow flash gap. An important aspect is the variation of the friction stress with the normal pressure. Usually, the law of constant friction is used. However, it is often advantageous to take the variation of the friction stress with the normal pressure into account by using the general friction model formulated by Bay and Wanheim [23].

Figure 3 illustrates the hot forging of a section of a crane hook. The material elastic constants are $E = 120\,000$ [N/mm^2] and $\nu = 0.3$. The viscoplastic flow is assumed to follow the Sellars-Tegart interpolation function, Eq. (5), with material coefficients $C = 1.2 \times 10^{12}$ [s^{-1}], $\alpha = 1.12 \times 10^{-2}$ [mm^2/N], $n = 3.5$ and $Q = 352\,690$ [J/mol]. The initial workpiece temperature is $T = 1100$ [°C] and a Coulomb frictional law is assumed with the coefficient of friction being $\mu = 0.3$.

The coupled thermo-mechanical analysis of the forming process is performed. Figures 3(b–d) show the temperature development in the tool over one forging cycle, which is primarily due to frictional sliding between the workpiece and tool. It is seen that the areas of greatest temperature increase coincide with regions of high curvature in the tool profile. This results in large normal pressures and, consequently, relatively large amounts of frictional work being dissipated as heat. By employing Eq. (8), the wear profile may easily be obtained which indicates that high wear coincides with regions of high temperature (low hardness) and normal pressure.

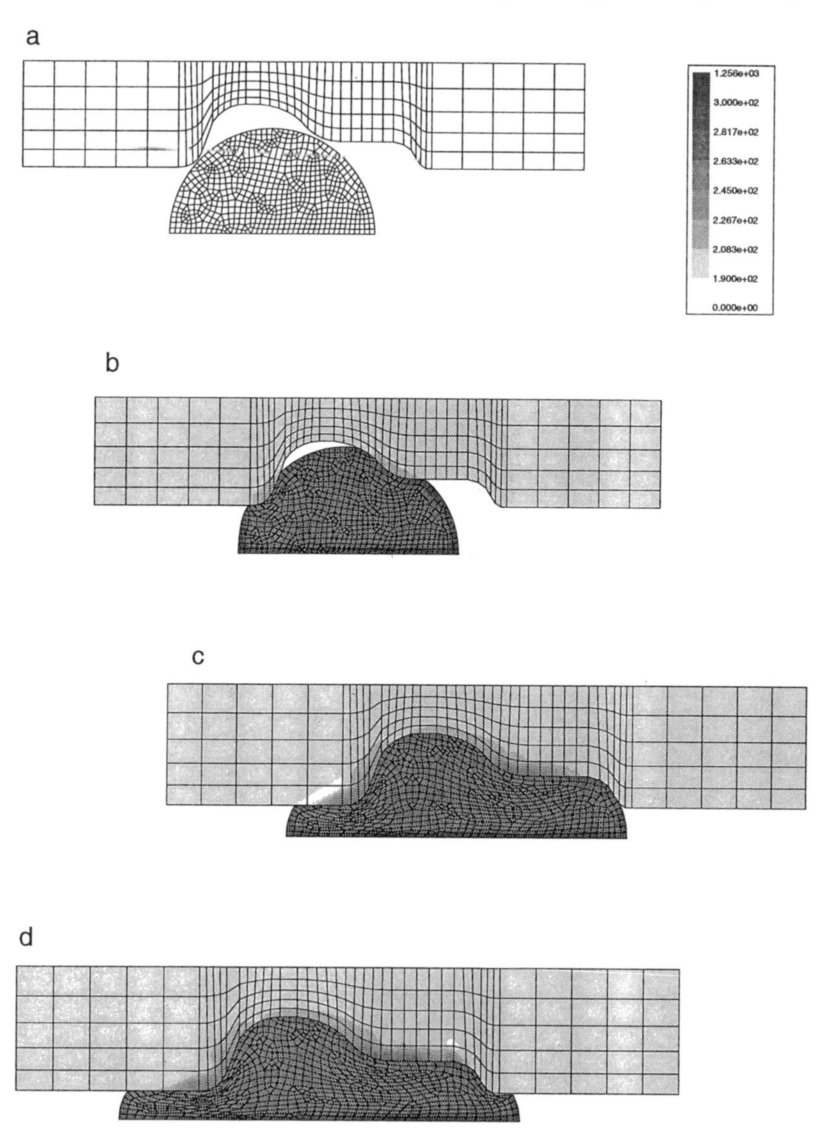

Fig. 3. A plane strain bulk forming of an industrial component. (a) Initial finite element meshes for tool and workpiece. (b–d) Deformed finite element meshes with temperature distributions at various stages of punch displacement.

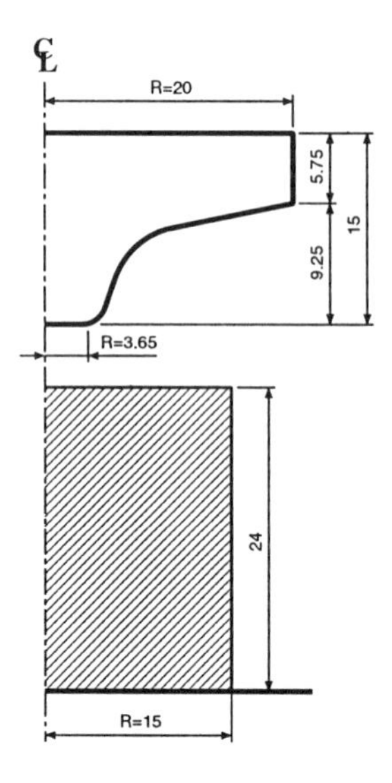

Fig. 4. Axisymmetric piercing. Geometry.

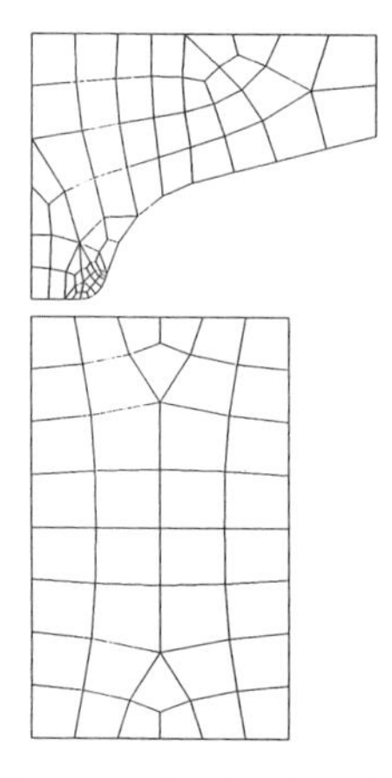

Fig. 5. Axisymmetric piercing. Initial mesh.

(ii) *Axisymmetric piercing.* The finite element simulation of the axisymmetric piercing of a cylindrical workpiece is presented. The geometry of the problem is shown in Fig. 4 and the initial mesh in Fig. 5. The workpiece is assumed to be made from an elastic-plastic material with Young's modulus, $E = 210$ [GPa], Poisson ratio, $\nu = 0.3$, yield stress, $\sigma = 100$ [MPa] and linear hardening with hardening modulus, $H = 900$ [MPa] while the punch is assumed to be rigid. Frictional contact between workpiece and tool is defined by a Coulomb law with the coefficient of friction being $\mu = 0.1$.

During analysis, an error indicator based on the rate of plastic work is used. The initial mesh consists of 101 quadrilateral elements and the final mesh contains 426 elements. Convergence of the finite element solution is established on the basis of the standard Euclidean norm of the out-of-balance forces with a tolerance of 10^{-3}. No difficulties related to the convergence have been observed during simulation despite frequent remeshings.

The distribution of the effective plastic strain on deformed meshes at various stages of the process is shown in Fig. 6. The deformed meshes show no hourglassing patterns, which is in agreement with analyses of a similar class of problems carried out in [19].

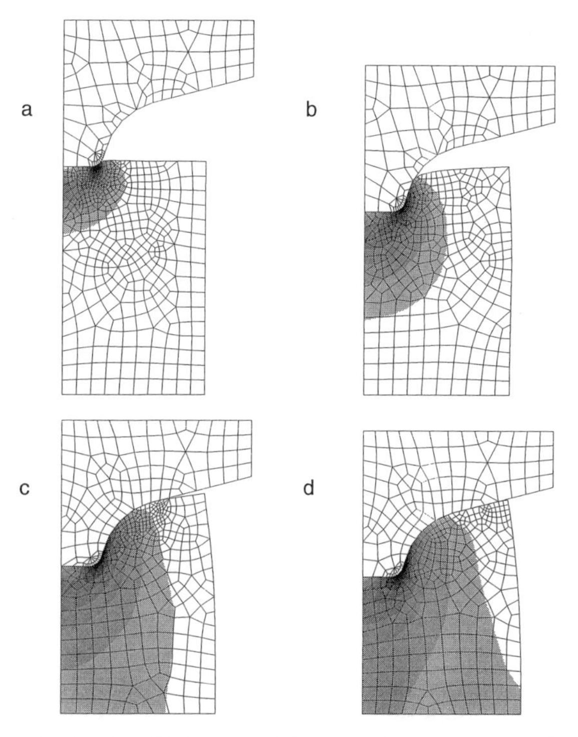

Fig. 6. Axisymmetric piercing. Evolution of effective plastic strain on adapted meshes. (a) U = 0.62, (b) U = 5.66, (c) U = 9.83 and (d) U = 11.00.

Conclusions

Some recent advances in the finite element analysis of metal forming problems have been reviewed, indicating the progress that has been made in both a theoretical understanding of inelastic material behaviour under finite strains and the associated numerical implementation. While the state of knowledge in some areas is relatively mature, further understanding and development is required in others. For example, issues related to the modelling of complex contact-friction phenomena are far from being settled, and a more comprehensive treatment of friction may necessitate the integration of micro-mechanical studies with computational approaches.

A reliable and efficiently automated, adaptive finite element strategy based on effective error estimates and automatic remeshing techniques for the FE analysis of forming operations has been described in this paper. Numerical examples are also provided to illustrate the effectiveness and robustness of the developed approach.

Although adaptive strategies are, at present, routinely performed for linear elliptic problems, their extension to nonlinear elliptic problems — in particular to metal forming problems where, typically, large inelastic deformations at finite strains are standard working conditions — is by no means trivial. Apart from the issues briefly mentioned in this paper, several important aspects of adaptive strategies related to nonlinear industrial applications which need further attention include the introduction of various types of error estimators and their comparative analysis, and the consideration of data transfer strategies.

The implementation of the procedures discussed in this paper will increasingly rely on advanced solution strategies and parallel processing concepts in order to provide solutions to industrial problems within acceptable computing time scales. It is expected that for general 3-D problems, which may arise in the simulation of forming processes, iterative methods combined with multi-grid solutions can potentially offer substantial savings in terms of CPU time and memory requirements in comparison to direct solution methods, thus offering a viable alternative to explicit solution strategies for large classes of problems.

Discussion

M. Lal A thorough validation of modelling codes is an essential step in the development process of the software fit for use in the studies of practical problems. A common approach for validating newly developed codes is to perform simulations on a number of "benchmark" models for which analytical results are available. Satisfactory agreement between the analytical and simulation results would ensure validity of the code. A second approach is to select computational models that would correspond closely with well-characterised materials on which reliable experimental data exist. Good agreement between the computed and experimental results would serve to establish the validity of the modelling methodology. What procedures did you pursue to validate your methodologies and codes? How does your model take into account the presence of interfacial regions in multi-phase systems, particularly when the thickness of these regions is appreciable?

D. R. J. Owen The testing and benchmarking of complex software is currently a subject of considerable interest. In view of the emergence of large software systems incorporating nonlinear material and other behaviours, and the continuing improvement in the performance/cost ratio of computers, the need to validate such codes is clear. One organisation which is dedicated to this objective is NAFEMS, whose primary activity is the development of benchmark tests for commercial and research finite element codes together with other related computational simulation procedures.

Whilst certain aspects of finite element codes can be rigorously validated, such as element performance and nonlinear return mapping algorithms, the development of adequate benchmarks which fully test the performance of large scale codes under general nonlinear material behaviour and phenomena such as frictional contact is still incomplete. At present, and for the foreseeable future, the role of code validation by comparison with experimental data is vital. Of course, our finite element codes are continuously validated against the available, and any newly-developed, benchmark tests and experimental data, both published and those made available to us by our industrial partners.

D. Perić It should be emphasised that the process of validation is not one-directional, that is, from experiment to numerical analysis. It has been accepted that numerical solutions can be used in the process of characterisation of complex materials. The whole field of inverse methods that is rapidly expanding is effectively combining numerical and experimental techniques in the description of complex phenomena.

J. Goddard To what extent are the computational methodologies dominated by material type — elastic versus plastic versus viscous — and (Mike Cates' concern) what about transitional regimes?

M. E. Cates There is clearly a continuum spectrum connecting elastic/plastic solids at one end and viscoelastic solids at the other. It is worrying that the algorithms used in the two limits are so completely different, especially if one wishes to model materials in the middle. Presumably, there are some materials where different algorithms will perform better in different flows for the same system.

D. Perić It is a fact that the mathematical model and its numerical formulation should reflect the essential aspects of the physical problem at hand. This seems to be the natural way of formulating numerical algorithms that will provide the optimal performance in terms of efficiency and robustness.

In that respect, the model for glass forming presented in the lecture is based on the assumption that elastic effects are negligible, and that the material is effectively modelled as a non-Newtonian fluid whereby viscosity is highly temperature-dependent. This is reflected in a numerical model which is based on the so-called flow formulation, which models glass forming within a simple Lagrangian framework for fluid flow and does not provide directly for elastic deformations. However, it has been recognised that for processing metals at room temperatures, elastic deformations are important. With respect to elastic springback which follows the end of the process and also, with respect to possible strain localisation where elastic deformations offer additional bifurcation paths, the inclusion of elasticity has proven to be essential. Therefore, the numerical model for these types of problems must incorporate elasto-plastic deformations at large strains. It is the task of the numerical analyst to formulate the associated numerical algorithms in the most appropriate and optimal way.

References

1. Shen, S.-F. & Dawson, P. R. (eds., 1995) *Proceedings of the Fifth International Conference on Numerical Methods in Industrial Forming processes.* Rotterdam: A. A. Balkema
2. Curnier, A. (1984) A theory of friction. *Int. J. Solids Struct.* **20**, 637–647
3. Oden, J. T. & Martins, J. A. C. (1985) Models and computational methods for dynamic friction phenomena. *Comp. Meth. Appl. Mech. Engng.* **52**, 527–634
4. de Souza Neto, E. A., Hashimoto, K., Perić, D. & Owen, D. R. J. (1996) A phenomenological model for frictional contact accounting for wear effects. *Phil. Trans. R. Soc.* **A354**, 819–843

5. Fjaer, H. G. & Mo, A. (1990) Alspen — a mathematical model for thermal stresses in direct chill casting of alluminium billets. *Metal. Trans.* **B21**, 1049–1060
6. Schönauer, M., Rodič, T. & Owen, D. R. J. (1993) Numerical modelling of thermomechanical processes related to bulk forming operations. *J. Phys. IV* **3**, 1199–1209
7. Lee, E. H. (1969) Elastic-plastic deformation at finite strains. *J. Appl. Mech.* **36**, 1–6
8. Eterovic, A. L. & Bathe, K.-J. (1990) A hyper-elastic based large strain elasto-plastic constitutive formulation with combined isotropic-kinematic hardening using the logarithmic stress and strain measures. *Int. J. Num. Meth. Engng.* **30**, 1099–1114
9. Simo, J. C. (1992) Algorithms for static and dynamic multiplicative plasticity that preserve the classical return mapping schemes of the infinitesimal theory. *Comp. Meth. Appl. Mech. Engng.* **99**, 61–112
10. Perić, D., Owen, D. R. J. & Honnor, M. E. (1992) A model for finite strain elasto-plasticity based on logarithmic strains: computational issues. *Comp. Meth. Appl. Mech Engng.* **94**, 35–61
11. Simo, J. C. & Rifai, S. (1990) A class of mixed assumed strain methods and the method of incompatible modes. *Int. J. Num. Meth. Engng.* **29**, 1595–1638
12. Taylor, R. L., Beresford, P. J. & Wilson, E. L. (1976) A non-conforming element for stress analysis. *Int. J. Num. Meth. Engng.* **10**, 1211–1219
13. Hughes, T. J. R. (1908) Generalisation of selective integration procedures to anisotropic and nonlinear media. *Int. J. Num. Meth. Engng.* **15**, 1413–1418
14. Simo, J. C., Taylor, R. L. & Pister, K. S. (1985) Variational and projection methods for the volume constraint in finite deformation elasto-plasticity. *Comp. Meth. Appl. Mech. Engng.* **51**, 177–208
15. Sussman, T. & Bathe, K.-J. (1987) A finite element formulation for nonlinear incompressible elastic and inelastic analysis. *Comp. Struct.* **26**, 357–409
16. Moran, B., Ortiz, M. & Shih, F. (1990) Formulation of implicit finite element methods for multiplicative finite deformation plasticity. *Int. J. Num. Meth. Engng.* **29**, 483–514
17. Simo, J. C. & Armero, F. (1992) Geometrically nonlinear enhanced strain mixed methods and the method of incompatible modes. *Int. J. Num. Meth. Engng.* **33**, 1413–1449
18. de Souza Neto, E. A., Perić, D., Huang, G. C. & Owen, D. R. J. (1995) Remarks on the stability of enhanced strain elements in finite elasticity and elastoplasticity. *Comm. Num. Meth. Engng.* **11**, 951–961
19. de Souza Neto, E. A., Perić, D., Dutko, M. & Owen, D. R. J. (1996) Design of simple low order finite elements for large strain analysis of nearly incompressible solids. *Int. J. Solids Struct.* **33**, 3277–3296
20. Perić, D., Yu, J. & Owen, D. R. J. (1994) On error estimates and adaptivity in elastoplastic solids: application to the numerical simulation of strain localisation in classical and cosserat continua. *Int. J. Num. Meth. Engng.* **37**, 1351–1379

21. Perić, D., Hochard Ch., Dutko, M. & Owen, D. R. J. (1996) Transfer operators for evolving meshes in small strain elasto–plasticity. *Comp. Meth. Appl. Mech. Engng.* **137**, 331–344
22. Lee, N.-S. & Bathe, K.-J. (1994) Error indicators and adaptive remeshing in large deformation finite element analysis. *Fin. Elem. Anal. Design* **16**, 99–139
23. Bay, N. & Wanheim, T. (1990) Contact phenomena under bulk plastic deformation conditions. In *Advanced Technology of Plasticity 1990, Proceedings of the Third International Conference on Technology of Plasticity*, pp. 1677–1691. The Japan Society for Technology of Plasticity

Dynamics of Complex Fluids, pp. 425–445
ed. M. J. Adams, R. A. Mashelkar, J. R. A. Pearson & A. R. Rennie
Imperial College Press–The Royal Society, 1998

Chapter 31

Computation of Large-Scale Viscoplastic Flows of Frictional Geotechnical Materials

I. M. Smith

School of Engineering, University of Manchester,
Manchester M13 9PL, UK
E-mail: ian.smith@man.ac.uk

Traditionally, the computation of geotechnical flows had to be restricted to small strain analyses due, initially, to a lack of comprehensive constitutive laws and later to computational inadequacies. More recently, constitutive behaviour has been clarified and two-dimensional problems solved although still at "small" strains and deformations. Currently, attention has shifted to the computation of large, three-dimensional problems involving significant deformations (strains of several hundred percent). Previous work will be reviewed briefly, as will constitutive relations. Computational issues will be dealt with at some length since algorithms had to be redesigned completely for current and future parallel computation. Two case histories illustrate the nature of current problems which are of practical interest. Excavations at CERN for collider extensions involve long-term creep of geomaterials while a waste tip in California failed in a complex manner that involves multiple mechanisms. Without these modern computational aids, the physical phenomena cannot be understood.

Viscoplasticity in Geomechanics

Geotechnical soil materials are "soft" solids in the sense of elastic moduli of up to a few MPa and shear strengths up to a few hundred kPa. They behave (usually) as frictional materials and are multi-phase with solid skeletons surrounded by water and/or gas. "Soft" rocks have moduli extending to a few GPa at strengths of some MPa. Although clearly viscoplastic in their response, geomaterials are often treated much more simply: as elastic solids for the purpose of deformation calculations and as rigid-plastic solids for ultimate load calculations. The exceptions are for very fast processes such as pile-driving, where penetration rates of a few metres per second are reached, and very slow processes such as creep around underground excavations.

Pile-driving has been analysed (one-dimensionally) as a viscoplastic process since the early 1950s. It was one of the first practical applications of digital computing [1]. The realistic analysis of creep in excavations and tunnels has depended on the development of complete constitutive relationships for geomaterials and their incorporation into computer codes, usually based on the finite element method. [2] gives a recent survey of such activity.

Yield and Viscoplastic Flow

For the purpose of this paper, elasto-visco-perfectly plastic behaviour is assumed to apply to geomaterials. The serious divergences between these assumptions and reality are discussed later (these apply to certain coupled "undrained" flows).

The underlying perfectly plastic flows take place when the Mohr-Coulomb yield (failure) criterion is reached. The Mohr-Coulomb surface in stress space is shown in Fig. 1. In the figure,

$$\theta = \text{Lode angle}, \tag{1}$$

$$\sigma_m = \frac{\sigma_1 + \sigma_2 + \sigma_3}{3}, \tag{2}$$

$$\overline{\sigma} = \frac{1}{\sqrt{2}}[(\sigma_1 - \sigma_2)^2 + (\sigma_2 - \sigma_3)^2 + (\sigma_3 - \sigma_1)^2]^{1/2}, \tag{3}$$

c, ϕ = 'cohesion' and internal angle of friction (material parameters), and the yield function may be expressed as

$$F_{mc} = \sigma_m \sin\phi + \frac{\overline{\sigma}}{\sqrt{3}}\left(\cos\theta - \frac{\sin\theta \sin\phi}{\sqrt{3}}\right) - c\,\cos\phi\,. \tag{4}$$

The nature of the viscoplastic flow is governed by a plastic potential function, Q, which in this simple model is assumed to take the same form as F with

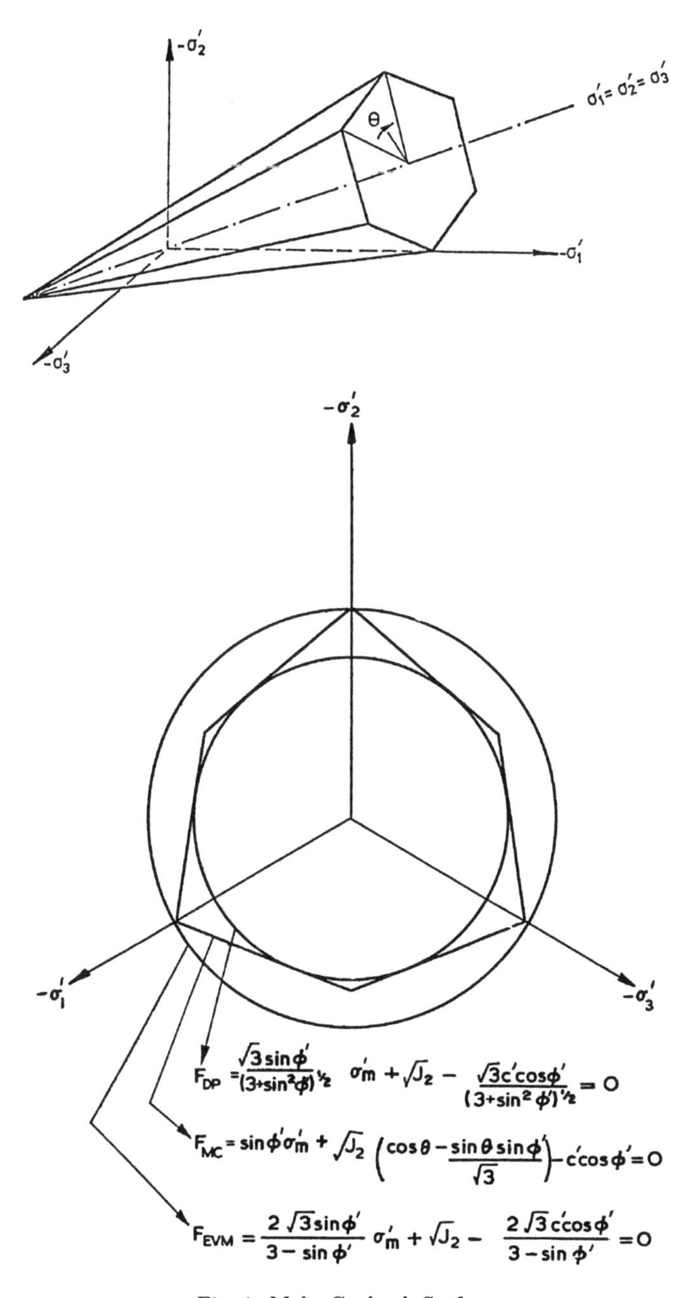

Fig. 1. Mohr-Coulomb Surface.

the dilation angle, ψ, in place of the angle of friction, ϕ. In general, ψ is assumed to lie between zero and ϕ. When $\psi = \phi$, flow is associated and the bound theorems of plasticity hold. However, the implied dilation rates are not observed in real geomaterials and computations are often performed with $\psi = 0$, that is, non-associated flow with no plastic volumetric strain. The general viscoplastic strain rate law may be written as

$$\dot{\boldsymbol{\epsilon}}^{vp} = \gamma\, f\left(\frac{F}{F_o}\right)\frac{\partial \mathbf{Q}}{\partial \boldsymbol{\sigma}}, \tag{5}$$

where γ is the material fluidity parameter and F_o non-dimensionalises F [3].

Various forms of the function, $f(\frac{F}{F_o})$, have been proposed [4]. For example,

$$f\left(\frac{F}{F_o}\right) = \frac{F}{F_o} \quad \text{(linear)}, \tag{6}$$

$$f\left(\frac{F}{F_o}\right) = \left(\frac{F}{F_o}\right)^n \quad \text{(power)}, \tag{7}$$

$$f\left(\frac{F}{F_o}\right) = \exp\left(\frac{F}{F_o}\right) - 1 \quad \text{(exponential)}. \tag{8}$$

In a computational algorithm, estimates of the accumulated viscoplastic strains at each step in time are made and converted into internal self-equilibrating stresses. Hence, this leads to self-equilibrating nodal loads in a finite element mesh. Over a time-step, the increment of viscoplastic strain is given by

$$\mathbf{d}\boldsymbol{\epsilon}^{vp} = \int \dot{\boldsymbol{\epsilon}}^{vp} dt. \tag{9}$$

In contrast to "initial stress" algorithms where the strain-rate integration is merely a numerical device, integration via Eqs. (5) and (9) has physical significance due to the material parameters γ and f. It is quite common for a simple forward Euler integration to be used, that is,

$$\mathbf{d}\dot{\boldsymbol{\epsilon}}^{vp} = \dot{\boldsymbol{\epsilon}}^{vp}\Delta t, \tag{10}$$

which, of course, has limited stability properties.

Algorithm for Coupled, Transient, Elasto-Viscoplastic Flow

A Structure Chart for the (finite element) algorithm is shown in Fig. 2 and the rheological model in Fig. 3. It can be seen that a complicated interaction between the consolidation (dissipative time-dependent) and viscoplastic time-dependent processes is, in principle, possible. The present algorithm envisages time scales for viscoplastic flow to be short (say, weeks) in comparison to dissipative time scales (say, months or years).

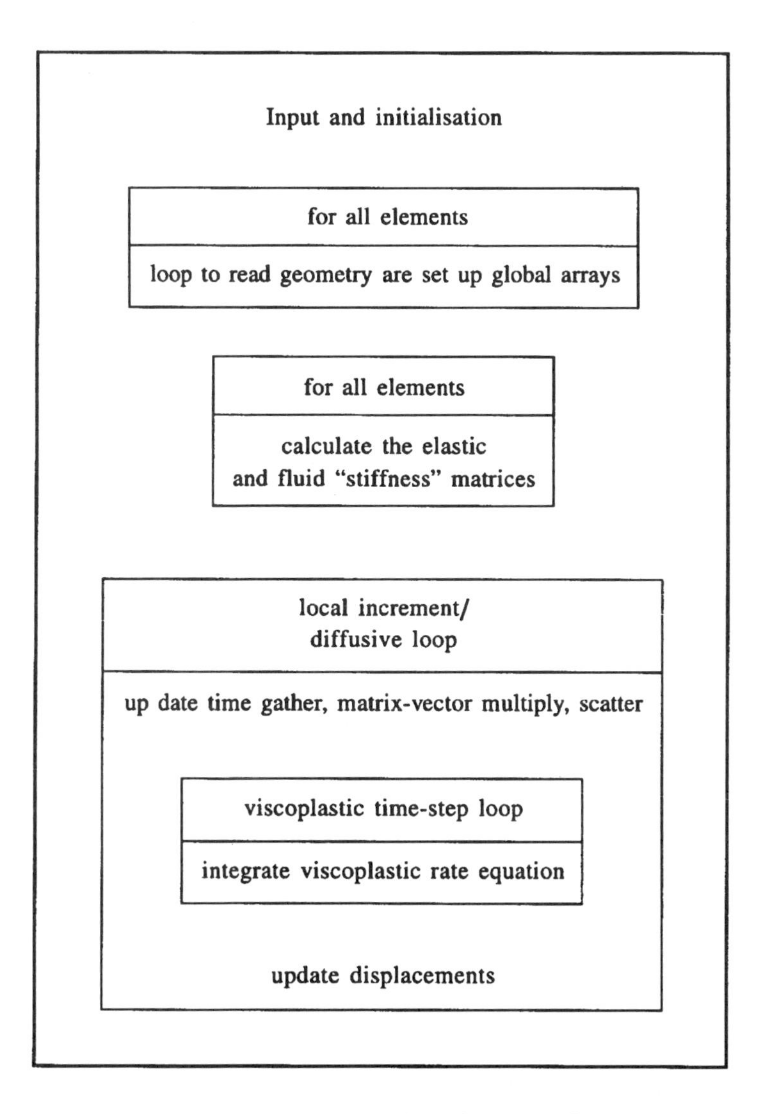

Fig. 2. Structure Chart for coupled biot elasto-viscoplastic process.

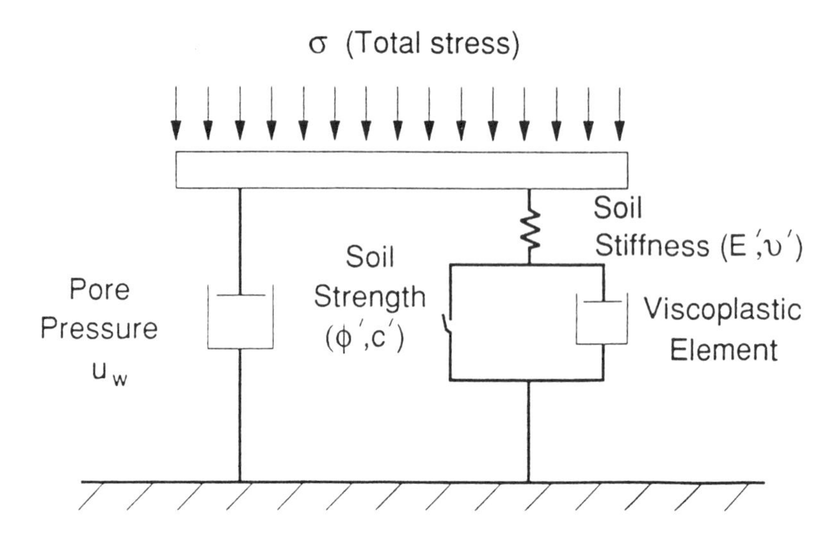

Fig. 3. Elasto-visco plastic rheological model.

If $\mathbf{r}$ are the displacements of the body shown in Fig. 3 and $\mathbf{u_w}$ the excess pore water pressures, the equations to be solved are

$$\mathbf{km}\,\mathbf{r} + \mathbf{c}\,\mathbf{u_w} = \mathbf{f} \quad \text{(equilibrium)}, \tag{11}$$

$$\mathbf{c}^T \frac{\mathbf{dr}}{\mathbf{dt}} - \mathbf{kp}\,\mathbf{u_w} = \mathbf{0} \quad \text{(continuity)}. \tag{12}$$

Here $\mathbf{km}$ is the elastic solid stiffness matrix and $\mathbf{kp}$ the fluid (Laplacian) "stiffness" matrix. The matrix, $\mathbf{c}$, is a coupling matrix and t refers to time. The vector, $\mathbf{f}$, contains all the forces in the system, generated externally and internally generated by the viscoplastic strains.

A simple implicit time-stepping method can be used to discretise the dissipation time-dependent equations, Eqs. (11) and (12), incrementally in the matrix form:

$$\begin{bmatrix} \mathbf{km} & \mathbf{c} \\ \mathbf{c}^T & -\Delta t\theta\mathbf{kp} \end{bmatrix} \begin{Bmatrix} \Delta\mathbf{r} \\ \Delta\mathbf{u_w} \end{Bmatrix} = \begin{Bmatrix} \Delta\mathbf{f} \\ \Delta t\mathbf{kpu_w} \end{Bmatrix}. \tag{13}$$

Note the negative diagonal terms implied by the $-\Delta t\theta\mathbf{kp}$ term. Indeed, as $\Delta t \to 0$, these equations are analogous to the Stokes equations for a fluid written in terms of velocities and pressures.

Constitutive Issues and Localisation

In coupled problems, the solid stresses which appear in the equilibrium equations are "effective" stresses, that is, those acting between the grains of the geomaterial. It is important to appreciate the limitations of the simple elasto-visco-perfectly plastic behaviour which has been assumed.

Figure 4 shows the effective stress paths that are typical of saturated geomaterials at different densities, tested in undrained (constant volume) triaxial tests. A loose geomaterial exhibits a stress path bending back towards the origin accompanied by large compressive values of u_w, the excess pore water pressure. The elastic effective stress path, however, climbs vertically to reach the Mohr-Coulomb surface and is therefore only appropriate for medium-dense materials. In general, more complicated multi-surface constitutive models [5] are necessary but these can, in principle, be analysed using the same techniques described here.

In the results which follow, fixed meshes (although with large displacements) have been used. Geomaterials do exhibit shear localisation and for an analysis of this, adaptive mesh refinement can be employed [6]. An example of such a mesh, adapted on the basis of gradients of plastic shear strain increment, is shown in Fig. 5.

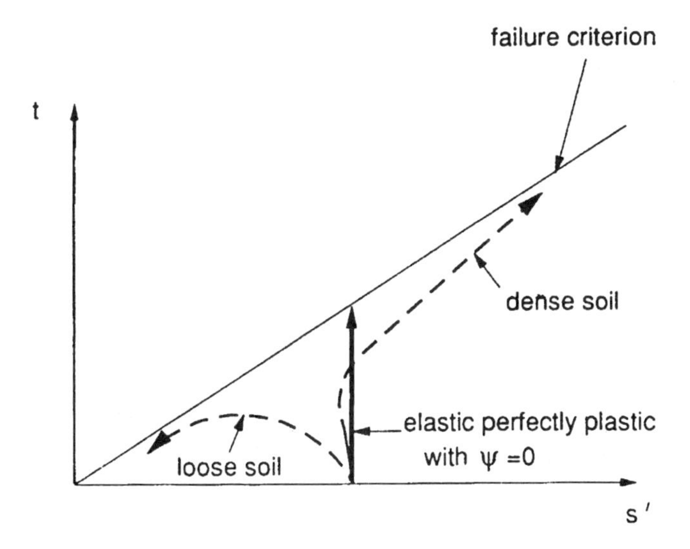

Fig. 4. Effective stress paths in undrained triaxial tests. s′ is related to σ_m and t to $\bar{\sigma}$.

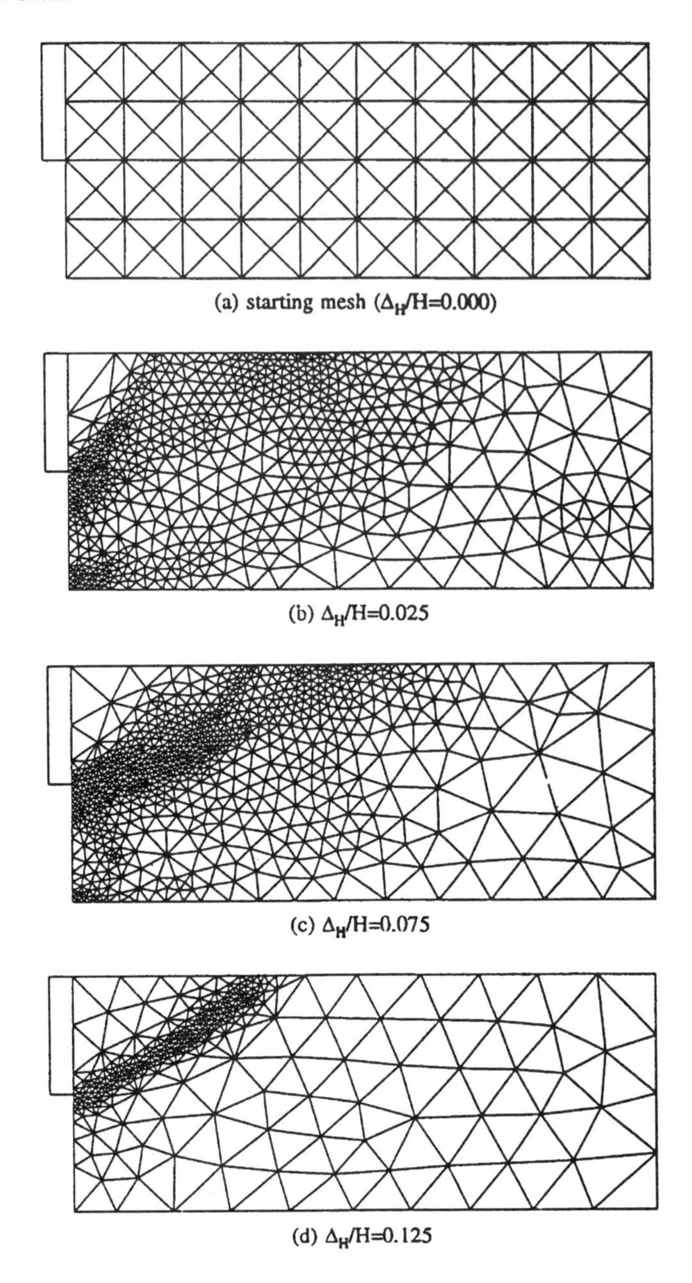

Fig. 5. Adaptively refined finite element meshes.

Analytical Details of Large Displacement Analysis

In the large displacement finite element analyses, the tangent stiffness matrix consists of two parts: a "geometrical" part which depends on the current spatial co-ordinates, and an "initial stress" part which depends on the current stresses. An updated Lagrangian approach has been adopted with the global stiffness matrix reformed at each load step.

The polar decomposition theorem permits the deformation gradient to be written uniquely as the product of a rotation (orthogonal) and a stretch matrix. A logarithmic measure of the principal values of the stretch matrix is used as the strain measure [7], and the conjugate stress is a rotated Jaumann stress. A simple hyper-elastic constitutive model based on the principal stretch is used so that stress can be derived directly from the strain energy function. The multiplicative decomposition allows the deformation gradient to be split into its elastic and plastic parts. The elasto-plastic rate equation can be integrated by Euler methods. The tangent stiffness matrix can be consistently formed to achieve better convergence rates.

Computational Issues

Equation (13) has to be solved over large meshes. For example, a subsequent case history will show that some flow phenomena just cannot be reduced to two dimensions, as has commonly been done. Whenever three-dimensional meshes become extensive, and possibly localisation is involved, the strategy of assembling millions of element equations like Eq. (13) into large global matrices becomes unattractive. Due to linearisation, Eq. (13) is a set of linear simultaneous equations to be solved at every time-step. In place of elimination strategies, like Gaussian elimination, we consider iterative strategies such as preconditioned conjugate gradients for equation solution. The basic algorithm for solving the system,

$$\mathbf{A}\mathbf{x} = \mathbf{b}\,, \tag{14}$$

can be expressed as

$$\mathbf{p}^0 = \mathbf{r}^0 = \mathbf{b} - \mathbf{A}\mathbf{x}^0\,, \tag{15}$$

where $\mathbf{r}^0$ is the "residual" or error for a first trial, $\mathbf{x}^0$, and then k steps of the process:

$$
\begin{aligned}
\mathbf{u}^k &= \mathbf{A}\mathbf{p}^k \\
\alpha^k &= \frac{(\mathbf{r}^k)^T \mathbf{r}^k}{(\mathbf{p}^k)^T \mathbf{u}^k} \\
\mathbf{x}^{k+1} &= \mathbf{x}^k + \alpha_k \mathbf{p}^k \\
\mathbf{r}^{k+1} &= \mathbf{r}^k - \alpha_k \mathbf{u}^k \\
\beta^k &= \frac{(\mathbf{r}^{k+1})^T \mathbf{r}^{k+1}}{(\mathbf{r}^k)^T \mathbf{r}^k} \\
\mathbf{p}^{k+1} &= \mathbf{r}^{k+1} + \beta_k \mathbf{r}^k
\end{aligned}
\tag{16}
$$

until the difference between $\mathbf{x}^{k+1}$ and $\mathbf{x}^k$ is "sufficiently" small. In the above, $\mathbf{u}, \mathbf{p}$ and $\mathbf{r}$ are vectors of length NEQ, the number of equations to be solved, while α and β are scalars.

The essential operations can be seen as a matrix-vector multiply followed by a series of whole-vector operations such as dot products. These have to be completed on every conjugate gradient iteration of every time-step, both viscoplastic and diffusive. Although this is computationally intensive, there is no need to assemble elements into global systems. Hence the iterative approach is attractive as it can be programmed for parallel computers. Typically, the author targets MIMD-type architectures by using a message-passing library MPI. On a scalar computer, over 90% of the time in the conjugate gradient algorithm is spent in the matrix-vector multiplication of Eq. (6). This vectorises very well. Even the most modest parallelisation leads to further considerable gains in this section of the code so that the vector operations section becomes much more significant [8].

Figure 6 shows the typical performance of this type of algorithm on a parallel computer, in this case for only 100,000 equations. The speed-up scales quite well up to 32 processors and the overall speed is disappointing. This is due mainly to the poor single-node performance of this particular computer. For several million equations, scalability extends to hundreds of processors. What these results show is that this kind of algorithm can make full use of parallel architectures and will speed up and scale well as single-node performance improves, for example, by using vector hardware in place of scalar for the nodes.

The "condition" (eigenvalue spectrum) of the left-hand side matrix in Eq. (13) is vital to the performance of the pcg algorithm. So far, the author

		Number of Processors					
		1	2	4	8	16	32
Total time	cpu	28.1	15.6	8.8	4.86	3.1	2.37
Host time	cpu		0.81	0.83	0.79	0.80	1.03
Serial time	cpu	0.67	0.66	0.70	0.65	0.63	0.86
Slave time	cpu		0.97	0.97	0.82	0.74	0.97
Common setup time	cpu		0.50	0.33	0.25	0.11	0.22
Iterations	cpu	27.4	14.1	7.4	3.73	2.11	1.11
	spdup	1.00	1.94	3.71	7.35	13.0	24.6
	Mflps	64	123	235	468	825	1562
	% Pk	43	41	39	39	34	33
Matrix multiply	cpu	22.9	11.8	6.3	3.06	1.68	0.87
	spdup	1.00	1.94	3.67	7.45	13.6	26.3
	Mflps	75	145	272	557	1014	1969
Vector operations	cpu	4.5	2.34	1.2	0.66	0.43	0.25
	spdup	1.00	1.94	3.75	6.9	10.70	18.4
	Mflps	9	17	36	63	97	170
Peak	Mflps	150	300	600	1200	2400	4800

Spdup - speedup from single node version
Mflps - Mflop rate

Fig. 6. Parallel performance of pcg algorithm.

has used simple diagonal preconditioning and pcg, although it is recognised (from experience with the Stokes equations, for example) that superior iterative algorithms for problems of this type may exist, based perhaps on the MINRES scheme [9].

Fortunately, in general, the larger the problem, the better it is for these techniques. Figure 7 shows the number of conjugate gradient iterations

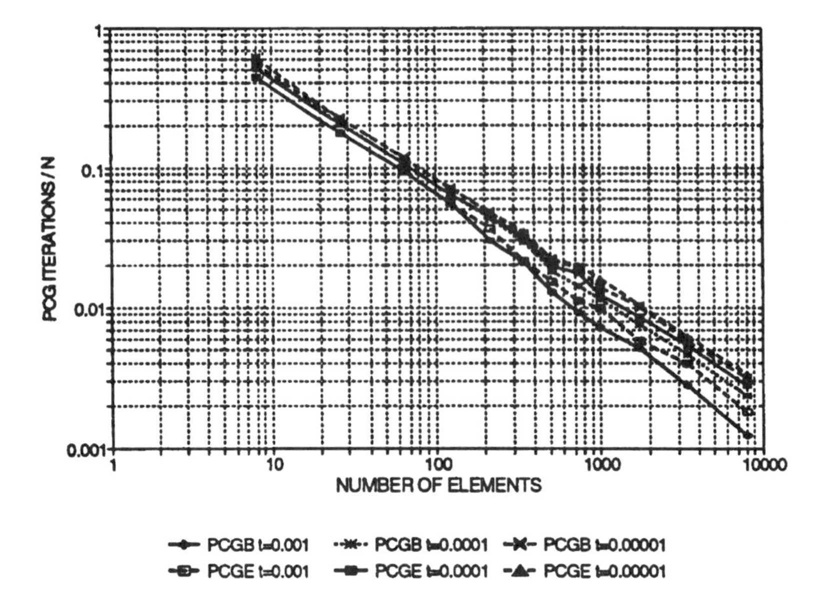

Fig. 7. PCG iterations as a function of number of finite elements.

expressed as a proportion of the number of equations dropping rapidly even as the number of equations to be solved reaches millions.

Case Study 1

L'Organisation Européenne pour la Recherche Nucléaire (CERN) proposes to upgrade its existing particle accelerator facilities. These upgrades will enable the site near Geneva, Switzerland, to be used for fundamental physics research into the next century.

The existing facilities for the Large Electron Positron (LEP) project are housed in a series of tunnels, shafts and caverns distributed around a 27 kilometre, quasi-circular Electron Beam Tunnel. Measurement stations are located strategically in caverns around the facilities' circumference. CERN is proposing an upgrade of these facilities to accommodate the Large Hadron Collider (LHC) project and is investigating the possibility of excavating additional underground openings at points on the circumference of the Beam Tunnel.

The proposed Point 1 facilities require new caverns to be excavated in very close proximity to the existing underground facilities, which comprise the

Beam Tunnel, two local enlargements to the Beam Tunnel (at a 14-metre span), a cavern offset from the Beam Tunnel (at a 20-metre span), twin 9-metre diameter shafts to the surface, and various tunnelled connections to the two enlargements.

The proposed layout for the LHC project comprises two new caverns with associated shafts and tunnels. The main cavern, proposed at 34 metres wide by 53 metres long, is designed to hold the LHC particle detector. The cavern will be positioned between the two existing shafts in such a way that it will incorporate the existing Beam Tunnel and intersects the end wall of the existing 20-metre span cavern. A second cavern, 20 metres wide by 62 metres long, will house computing facilities and is to be excavated at right angles to the side wall of the main cavern. In addition, two shafts, 30 metres and 12 metres in diameter, will intersect the main cavern roof. The size and position of these shafts will be dictated by the size of the equipment that must be lowered into the main cavern once it is completed.

Because the LHC project requires very large excavations in close proximity to existing structures, the preliminary design proposals and the feasibility of construction had to be assessed to judge the associated risks, the likely rock support requirements, and the optimum layout and excavation sequencing.

The study required a summary of anticipated geotechnical conditions from a review of existing data and a preliminary site-specific ground investigation. Due to the complex layout of the proposed excavations, both three-dimensional and two-dimensional finite element analyses were carried out in order to quantify the effects of construction on the existing structures.

The geotechnical data showed that the existing complex of tunnels, shafts and caverns at CERN was excavated partly within interbedded sandstones and marls of the Molasse rocks found in western Switzerland and partly within Jurassic Limestones. The studies showed that:

(i) the Molasse deposits comprise thickly to thinly interbedded sequences of arenaceous and argillaceous rocks;
(ii) the lithologies present vary from sandstones to marls with frequent intermediate material — described as sandy marl and marly sandstone;
(iii) the mode of deposition dictates that the thickness and, characteristically, the lateral persistency of the beds, vary; and
(iv) in general, the rocks are weak and soft, with design strengths and stiffnesses varying from between 22 MPa to 5 MPa, and 3 GPa to 0.5 GPa, respectively.

The three-dimensional analyses involved studying three different excavation sequences that assumed, in each case, elastic and elasto-plastic mass response to excavation both with and without concrete linings. The results confirmed that the design proposals for Point 1 are feasible but careful sequence excavation and support practices would be required.

Following the three-dimensional work, two-dimensional analyses were performed on selected representative sections to define rock support and excavation sequencing. The results from these analyses indicated that over-stressing of the rock mass in the very small shaft pillars above the main cavern crown and also beneath the invert are likely to cause problems during construction.

The recommendations for the proposed design include the pre-placement of support to the floor of the main cavern prior to excavation. Conventional top-down excavation of the caverns will begin after a reinforced concrete invert arch has been installed and anchored to the underlying rock. In addition, a primary support system has been proposed for the cavern and shaft system which comprises cable anchors, rock bolts and shotcrete placed in a staged manner. Figure 8 illustrates the typical stages of the construction option.

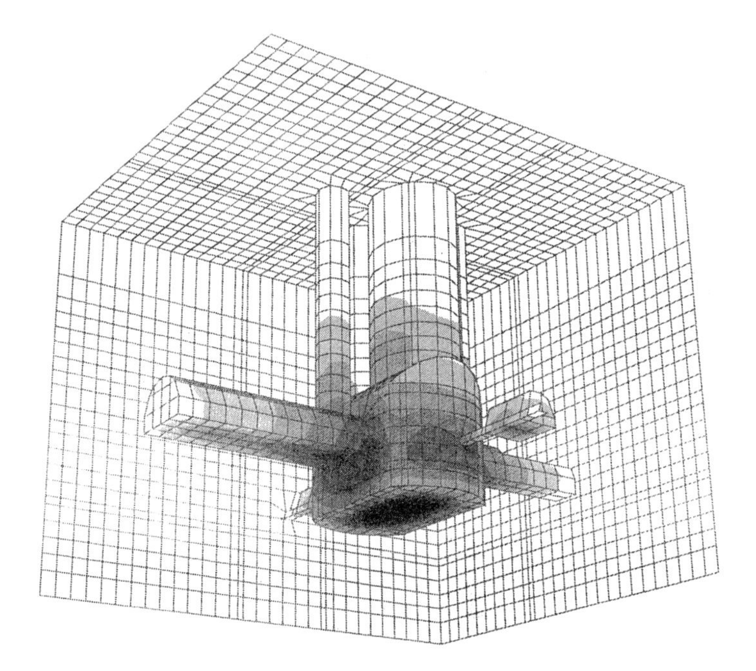

Fig. 8. Staged construction of the LHC project at CERN.

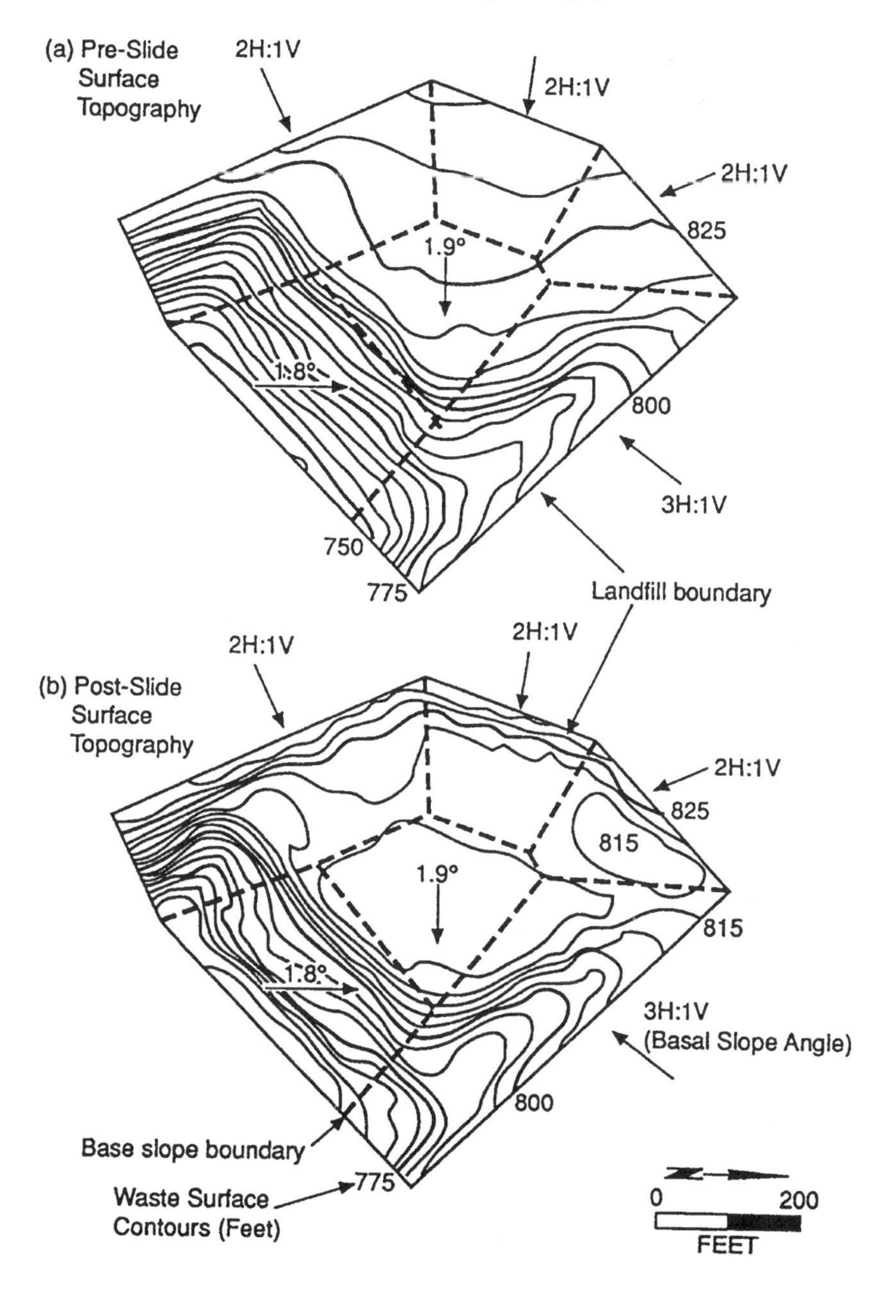

Fig. 9. Topography before and after Kettleman Hills collapse.

Case Study 2

In March 1988, there was a failure at Kettleman Hills, California, where a large landfill consisting of hazardous waste was being stored and treated before disposal [10–12]. Over half a million cubic metres of material had been placed at the time of failure and the waste had reached a height of 27 metres. The topography before and after the failure is shown in Fig. 9.

The waste had been stored in a bowl-shaped excavation with a relatively flat base and excavated side slopes of between 2:1 and 3:1. Hence it was geometrically unsymmetrical. Any failed flow would necessarily be three-dimensional in nature, involving (probably) the failure of the steepest side slopes, the less steep side slopes and finally, the whole mass funnelling towards the opening in the excavation at the south-east corner.

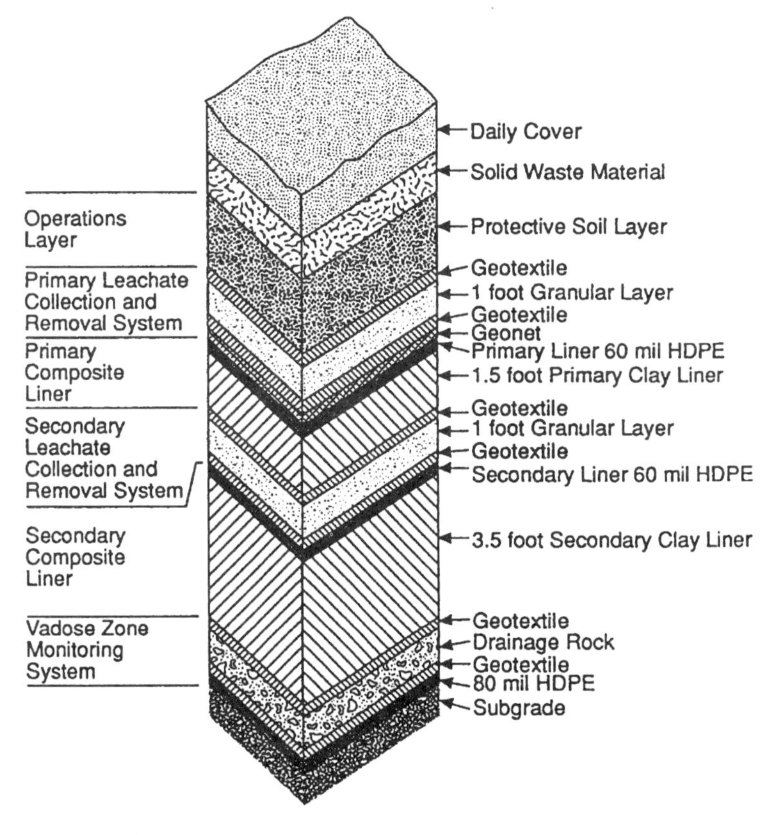

Fig. 10. Base liner details.

After failure, when at an unknown displacement of possibly 0.2 metres, the mass began to accelerate to its final location. Horizontal movements of 10 metres associated with vertical slumps of the order of 4 metres were also observed.

The containment liner system used on the base is shown in Fig. 10. Post-failure investigations showed that movements in the base took place via the material above the secondary clay liner moving en masse along the secondary clay/secondary liner interface. Indeed, the same conclusion could also be reached about the side slope movements. Figure 11 shows, in very general terms, the movements which could be inferred from the pre- and post-failure topographies together with post-failure site investigations.

Laboratory studies had indicated low residual angles of friction on the postulated failure surfaces (of the order of about eight degrees) and a general picture of the strengths of the components of fill is given in Table 1.

When it came to stability analyses, [10–12] analysed rigid plastic block sliding mechanisms in two and three dimensions. In particular, in the latter case, these are of dubious kinematic admissibility and in [11], it is conceded that

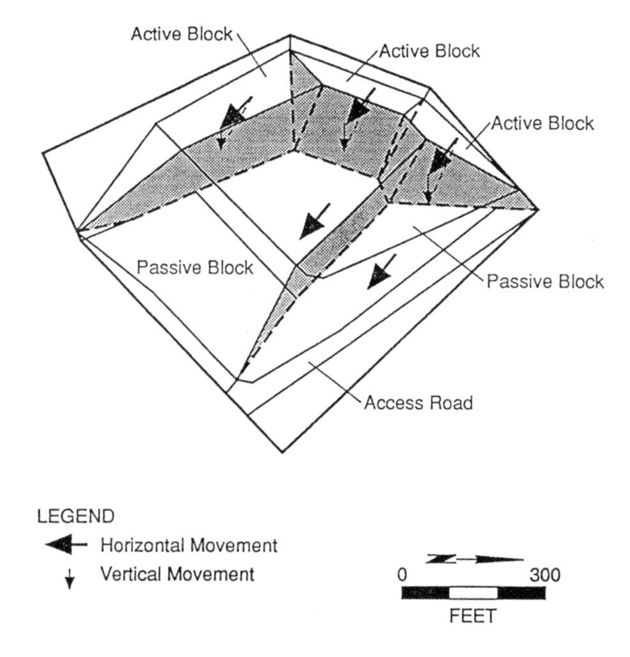

Fig. 11. General interpretation of failure movements.

Table 1. Strengths of components of fill.

Yield Criterion	von Mises			Mohr-Coulomb		
Material	Fill	Liner	Base Soil	Fill	Liner	Base Soil
Cohesion (kN/m^2) [11]	215.5	431	431	0	0	431
Cohesion (kN/m^2) [12]	167.5	335	335	0	0	335
Friction angle (°)	0	0	0	40	8	0
Young's Modulus (kN/m^2)	7200	7200	36000	7200	7200	36000

"three-dimensional analyses involve some degree of engineering judgement and probably have a level of uncertainty of about $\pm 10\%$". The present finite element computations are therefore possibly the first statically and kinematically admissible three-dimensional analyses of this failure.

The mesh (of 14-node cuboidal elements) is shown in Fig. 12 and the results of the analyses in Fig. 13. A comparison of Fig. 13 and Fig. 11 shows that the general disposition of movements has been reproduced correctly in the calculations. Meanwhile, "large" movements, taken to signify collapse or "failure", occurred when the weight of the fill material had its true value (17.4 kN/m^3). In the analyses, "gravity" was kept constant and the material strengths decreased until movements become "large". In the computations, very large strains (several hundred percent) occurred in the thin elements modelling the liner/clay interface. However, the movements computed were about one tenth of the post-failure movements on site (approximately 1m horizontal and 0.4 m crest slump). Nevertheless, the computations give a remarkably accurate simulation of what takes place in practice.

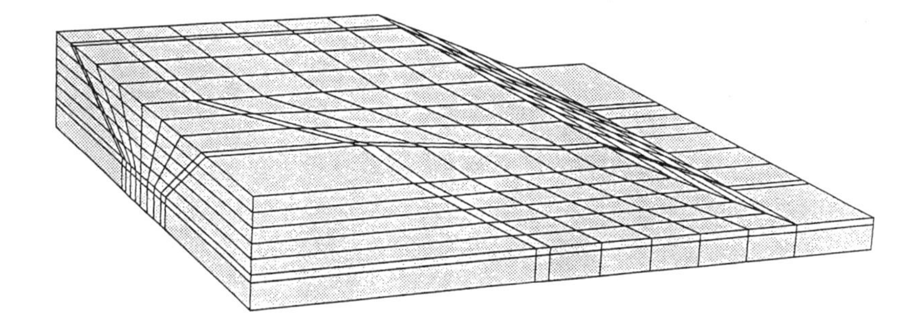

Fig. 12. Mesh of Kettleman.

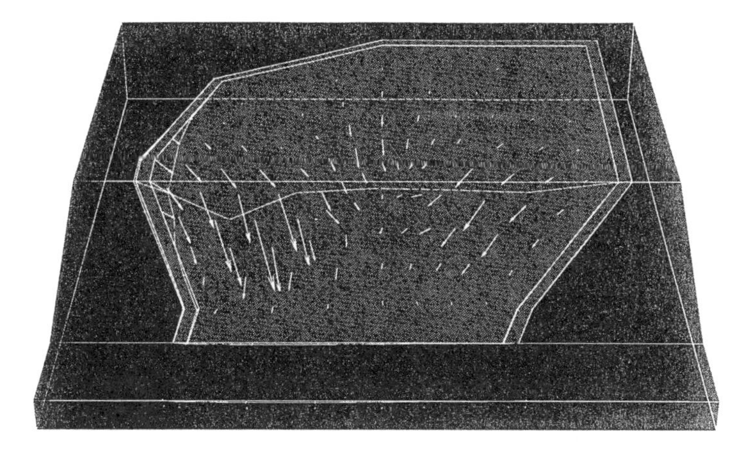

Fig. 13. General computation of failure at Kettleman.

Conclusions

Elasto-perfectly-viscoplastic flow behaviour of frictional geomaterials has been computed in practical situations. The computations are demanding but can be scaled to make use of parallel computing hardware. More work is necessary to devise optimal iterative algorithms for the non-positive definite equation systems which have to be solved. Visualisation techniques are essential for interpreting the results of the three-dimensional analyses.

Discussion

G. Marrucci Boundary conditions are at infinity conceptually. Is the finite size of the computational model found by trial and error?

I. M. Smith Essentially, yes. In seismic problems where waves originate in the far field, boundary conditions can impose significant difficulties. For static equilibrium, the theory of elasticity suggests that the far field feels disturbances from the local region being analysed. Fortunately, nonlinear, dissipative mechanisms such as plasticity have the effect of localising disturbances. Hence boundary conditions are less serious then one might fear. "Infinite" elements or similar devices can be used if boundaries do pose significant problems.

G. Marrucci Are you involved in the Pisa leaning tower problem?

I. M. Smith No. An international commission is studying the problem. There seem to have been about three "solutions" so far, but I don't think any of them have been successful.

M. E. Cates/J. Goddard There is clearly a continuous spectrum connecting elastic/plastic solids at one end and viscoelastic fluids at the other. It is worrying that the algorithms used in the two limits are so completely different, especially if one wishes to model materials in the middle. Presumably, there are some materials where different algorithms will perform better in different flows for the same system. To what extent are the computational methodologies dominated by material type, elastic vs. plastic vs. viscous, and what about transitional regimes?

I. M. Smith You are of course right that, in principle, there should be a continuous spectrum and that a unified method or range of methods might be conceivable. The crux seems to be the importance of shear and in geomaterials, for example, of shear-volume coupling. As we seek to solve larger and larger problems, there will always be the tendency to simplify equations wherever possible, leading to "horses for courses". As a simple example, frictional plasticity leads to quite a significant increase in computational complexity relative to non-frictional plasticity.

References

1. Smith, E. A. L. (1960) Pile driving analysis by the wave equation. *Proc ASCE, JSMFD* **86**, 35–61
2. Cristescu, N. D. & Gioda, G. (eds., 1994) Visco-plastic behaviour of geomaterials. *Courses and Lectures No. 350*, pp. 317. Springer-Verlag
3. Zienkiewicz, O. C. & Cormeau, I. C. (1974) Viscoplasticity, plasticity and creep in elastic solids, a unified approach. *IJNME* **8**, 821–845
4. Perzyna, P. (1986) Fundamental problems in viscoplasticity. *Advances in Applied Mechanics* **9**, 243–377
5. Molenkamp, F. (1981) Elasto-plastic double hardening model MONOT. *Report of Delft Soil Mechanics Laboratory*
6. Hicks, M. A. & Mar, A. (1994) A combined constitutive model-adaptive mesh formulation for the analysis of shear bands in soils. *Proc of the Third European Conf. Num. Meth. Geot. Eng.*, pp. 59–66. Manchester, UK
7. Peric, D. (1993) On a class of constitutive equations in viscoplasticity: Formulation and computational issues. *IJNME* **36**, 1365–1393
8. Smith, I. M. & Griffiths, D. V. (1996) Programming the Finite Element Method. 3rd edn., pp. 420. J. Wiley & Sons
9. Fischer, B. (1994) *MINRES implementations based on the Lanczos basis*

10. Mitchell, J. K., Seed, R. B. & Seed, H. B. (1990) Kettleman Hills waste landfill slope failure I: Linear system properties. *Proc ASCE J. Geot. Eng.* **116**(4), 647–668
11. Seed, R. B., Mitchell, J. K. & Seed, H. B. (1990) Kettleman Hills waste landfill slope failure II: Stability analyses. *Proc ASCE J. Geot. Eng.* **116** (4), 669–690
12. Byrne, R. J., Kendall, J. & Brown, S. (1992) Cause and mechanism of failure, Kettleman Hills landfill 13–19 Phase IA. *Slopes and Embankments, ASCE*, 1118–1215

Dynamics of Complex Fluids, pp. 446–468
ed. M. J. Adams, R. A. Mashelkar, J. R. A. Pearson & A. R. Rennie
Imperial College Press–The Royal Society, 1998

Chapter 32

Modelling of Liquefaction and Flow of Water Saturated Soil

F. MOLENKAMP*, A. J. CHOOBBASTI AND A. A. R. HESHMATI
University of Manchester Institute of Science and Technology,
Department of Civil and Structural Engineering,
Manchester, M60 1QD, England
**E-mail: frans.molenkamp@umist.ac.uk*

The numerical simulation of liquefaction and flow of water-saturated loose granular materials by means of a continuum approach involves an incremental formulation of large deformation based on polar decomposition of the deformation gradient. The formulation and its numerical implementation in a finite element programme are discussed. The continuum type of constitutive model describes the response of the granular skeleton to a small increment of deformation and rotation. The model is isotropic double hardening elasto-plastic with the mean effective stress, the density and the hardening parameters as state parameters. Special attention is given to the description of the ultimate state of deformation, which is reached during continuous flow. For undrained flow, the formulation enables the simulation of both pre-peak hardening and post-peak softening towards and inclusive of the ultimate state of continuous flow. The global viscous aspect of material behaviour observed is assumed to be a consequence of the induced shear of the pore fluid in between the moving grains and is accounted for by an additional linear viscosity. The structure of the numerical implementation of the constitutive model and the connection to the finite element programme are discussed briefly. The characteristics of the model are demonstrated by means of a simulation of a drained laboratory test used to measure the behaviour of soils.

Introduction

Liquefaction is the loss of stiffness of a water-saturated granular skeleton due to the generation of high pore pressure. This causes the mean effective stress to reduce significantly and the material behaviour to become liquid-like.

The first aspect of liquefaction [1, 2], often called "static liquefaction", involves the consequence of undrained instability of slopes. The unstable undrained soil flows away like a liquid, known as a "flow slide".

The second aspect of liquefaction [3], sometimes also called "initial liquefaction" or "cyclic softening", occurs as a consequence of alternating loading, for example, due to an earthquake. Shaking of level ground may lead to large horizontal displacement amplitudes when the vertical effective stress is reduced to zero by excess pore pressure generation. Structures resting on this liquefied soil may tilt and sink into the liquefied ground.

In sloping ground, the static stress includes a shear component on potential failure planes parallel to the ground surface. During the alternating loading of sloping ground by an earthquake, which leads to virtually undrained behaviour, the alternating component of the deformation, the so-called "cyclic softening", may increase while the average deformation may also build up gradually to unacceptable magnitude, known as "cyclic mobility". Eventually, a flow slide may occur if the remaining undrained strength becomes smaller than the static driving stress and the liquefied soil can be carried away over a large distance. However for very loose soils flow slides may also be induced by static causes, such as the addition of additional weight on top of a slope during construction, the erosion of the toe of a slope and a change in the ground water regime.

After having observed the consequences of flow slides of loose water-saturated fine sands, Casagrande [1] hypothesised the existence of a "critical void ratio" which distinguishes liquefiable states from non-liquefiable ones. This "critical state" would be reached during continuous deformation in simple shear at constant volume and at constant strain rate. Loose materials would contract monotonically during continuous shear to reach this "critical state" while dense materials would dilate to reach it.

When laboratory testing methods improved, it became clear [4] that the envelope of states of ultimate stress and density reached at this continuous deformation is not identical to the boundary between liquefiable and non-liquefiable states involved in the undrained instability of slopes, but that this "liquefaction" boundary is related in some way to that "ultimate state" envelope. Nevertheless, the word "critical state" [5] has become synonymous to "ultimate state of continuous deformation" or "steady state of deformation" [6].

To illustrate the observed behaviour in Fig. 1, the initial states of five very loose sand samples are indicated in terms of the mean effective stress, $p = (\sigma_a + 2\sigma_r)/3$, and the void ratio, e_v, together with the probable range of

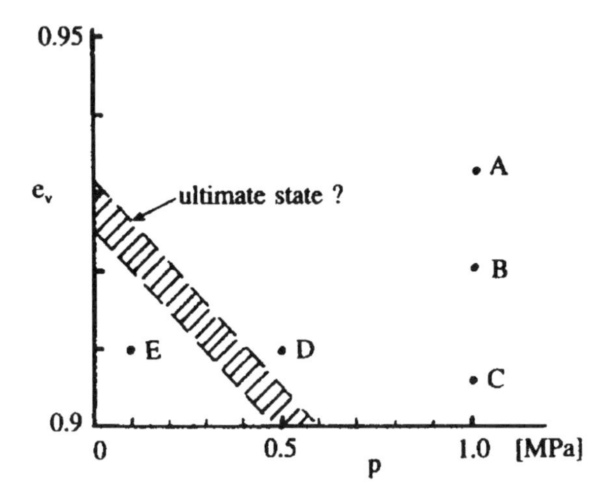

Fig. 1. Initial state of five sand samples in terms of mean effective stress p and void ratio e_v. The probable range of the ultimate state is also indicated.

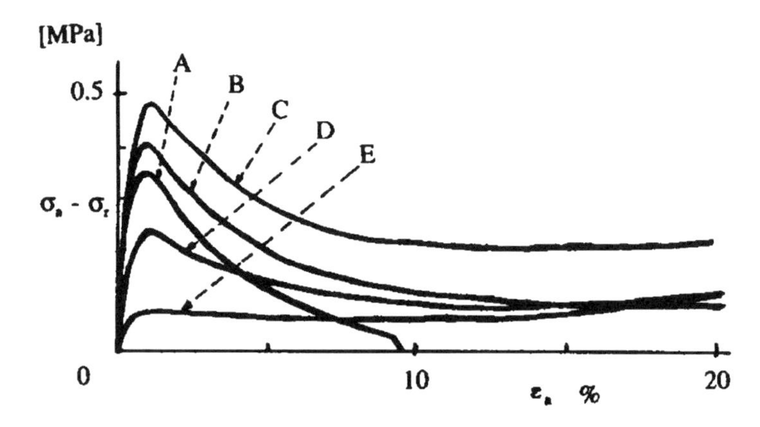

Fig. 2. Relation between deviatoric stress and axial strain of five undrained triaxial compression tests.

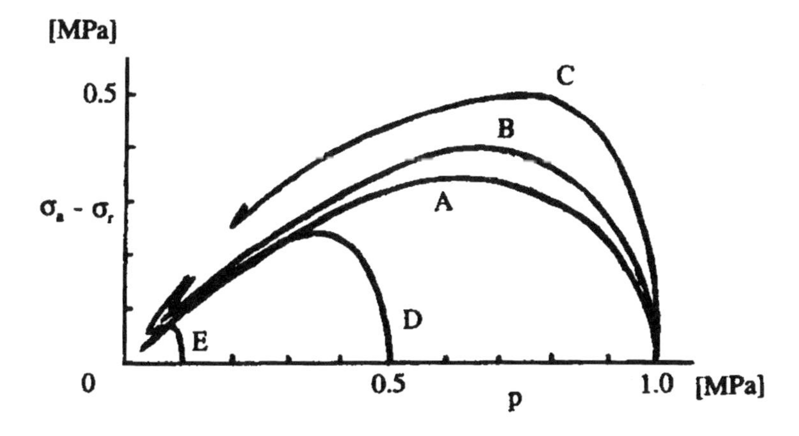

Fig. 3. Effective stress paths of five undrained triaxial compression tests.

the ultimate state based on the experimental data [7] shown in Figs. 2 and 3. Figure 2 shows the results in terms of the deviatoric stress, $q = \sigma_a - \sigma_r$, versus the axial strain, ε_a, for displacement-controlled undrained triaxial compression, and in Fig. 3 the corresponding effective stress paths in terms of p and q are shown.

The only sample that certainly liquefies is sample A, which has a void ratio $e_v = 0.933$ and an initial mean effective stress $p = 1$ MPa. Its post-peak deviatoric stress reaches even zero. Sample B, with an initial void ratio, $e_v = 0.921$, shows a monotonic post-peak decrease of the deviator up to the maximum observed strain of 20%. Due to the observed small strain range, it remains unclear whether or not it has reached the ultimate state. Sample C, with an initial void ratio, $e_v = 0.906$, also shows a post-peak softening, but this is followed by a post-peak hardening. Figure 3 shows that this post-peak hardening occurs together with an increase in the mean effective stress. Hence the post-peak behaviour is non-monotonic and the ultimate state will be reached at a mean effective stress larger than the one obtained at the end of the test if a much larger uniform deformation would be enforced. However, when enforcing larger axial compression in an undrained triaxial test, the sample usually fails due to the occurrence of a shear band. This is because at the applied rate of deformation, pore water movement will give rise to a locally drained behaviour. In such case, the ultimate density is only reached inside the shear band.

The minimum undrained post-peak resistance is often called "undrained residual strength" [7]. The state of deformation at the minimum undrained post-peak "residual" strength is also known as the "quasi-steady state" [7, 8].

Samples D and E, at a void ratio, $e_v = 0.910$, and initial mean effective stresses, $p = 0.5$ and 0.1 MPa, respectively, show similar behaviour as sample C, thus illustrating again non-monotonic behaviour in reaching the ultimate state.

The majority of observed flow slides do not involve sufficient deformation ($> 100\%$) in approaching the ultimate state. Furthermore, much smaller deformations ($< 20\%$) are usually already fatal to the functioning of most geotechnical structures. Hence in geotechnical engineering, the "quasi-steady state" is often more important than the "ultimate state".

Characteristics of Finite Element Programme for Flow of Liquefied Material

For finite element analyses of problems involving liquefaction and subsequent flow of liquefied granular material, not only must the occurrence of undrained instability of the equilibrium solution be analysed, but the post-instability behaviour must also be simulated. For this, the effects of the mass of the liquefied soil and the viscosity of the liquefied sand-water mixture must be taken into account. If the re-sedimentation had to be analysed as well, then the assumption of undrained behaviour could not be used. Instead, a consolidation analysis involving the relative motion of the pore water with respect to the soil skeleton would have to be considered. However, the latter falls outside the scope of this paper.

The effect of the mass of the soil can be taken into account in a dynamic analysis. This requires the usage of the mass matrix and nodal accelerations. The numerical solution can be obtained using direct integration in the time domain.

The effect of the apparent viscosity of the liquefied soil can be taken into account in an approximate way by adding to the effective stress according to the elasto-plastic model an additional viscous stress component. In its simplest form, this additional viscous stress can be assumed to be described by a linear viscous model which involves only a deviatoric stress. For the viscosity, the apparent viscosity of the liquefied sand-water mixture must be used. Furthermore, the pore pressure due to the undrained response needs to be added to the initial pore pressure.

The discretised form of the equation of motion at the new state can be expressed as the nodal unbalanced vector [9] for the kth node and the rth Cartesian direction, namely,

$$
\begin{aligned}
f^u_{rk} = &- \int_{V_0} B_{rkji} \left(R^{rT}_{ip}\, \sigma^0_{pq}\, R^r_{qj} + \Delta \overset{\circ}{\sigma}_{ij} \right) |\mathbf{H}| dV \\
&+ \int_{V_0} N_k\, \rho\, b_r\, |\mathbf{H}|\, dV - \int_{V_0} N_k\, \rho\, N_1\, \delta_{rs}\, |\mathbf{H}|\, dV\, \ddot{u}_{sl} \\
&+ \int_{S^*_{0u}} N_k \left(R^{rT}_{rp}\, \sigma^0_{pq}\, R^r_{qj} + \Delta\, \overset{\circ}{\sigma}_{rj} \right) n_j{}^n\, |\mathbf{G}|\, dS + \int_{S^*_{0T}} N_k\, \overline{T}_r\, |\mathbf{G}|\, dS\,,
\end{aligned}
\tag{1}
$$

in which V_0 is the volume at the current state; S^*_{0T} is the area with prescribed traction at the current state; S^*_{0u} is the area with prescribed displacements at the current state; N_k is the shape function of the kth node; B_{rkij} is the strain-nodal displacement matrix with i and j representing Cartesian directions; R^r_{qj} is the incremental rotation from the current state to the new state with q and j being Cartesian directions; σ^0_{pq} is the Cauchy stress at the current state; $\Delta\overset{\circ}{\sigma}_{ij}$ is the objective Cauchy stress increment; T_r is the prescribed traction on S_T in the new state; $|\mathbf{H}|$ is the determinant of the incremental deformation gradient between the current state and the new state; $|\mathbf{G}|$ is the ratio of the boundary areas in the new and current states; ρ is the specific mass in the new state; b_r is the body force per unit of volume in the new state; u_{sl} is the displacement of the lth node in the sth Cartesian direction; and n^n_j is the normal vector on the boundary in the new state.

Constitutive Modelling of Flow of Granular Materials

The modelling of liquefaction is achieved by combining a simple linear viscous model and a complex elasto-plastic model. The elasto-plastic double hardening model is an extension of an earlier version called Monot [10]. The extensions involve an ultimate state of continuous flow, a refinement for one-dimensional compression, and a proper numerical treatment of fluidisation. It involves six components, each applied to a different region of principal stress space as illustrated in Fig. 4, namely, the nonlinear elastic model 1, the elasto-plastic deviatoric model 2, the elasto-plastic compressive model 3, the vertex model 4 of the elasto-plastic deviatoric and compressive models, the elasto-purely

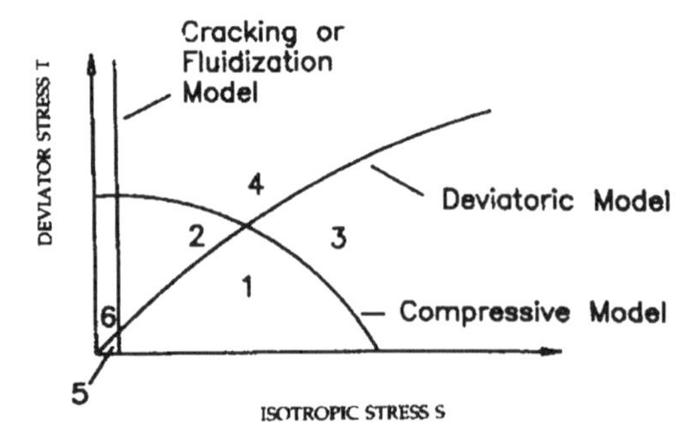

Fig. 4. Illustration of the components of the constitutive model.

plastic fluidisation model 5, and the vertex model 6 of the elasto-plastic deviatoric and fluidisation models.

Nonlinear Elastic Model

The nonlinear elastic component [11, 12] describes the behaviour of soils or any other granular material during small unloading and reloading. It is based on a complementary potential function assumed to be a function of both the isotropic stress, S, and the deviatoric stress level in terms of T/S, in which T is the stress deviator.

Plastic Compressive Model

The plastic compressive component is based on the relevant model component of [13]. This model has been extended [14] to improve the simulation of the K_0-effective stress path, in which for 1-dimensional compression of the material without any lateral strain a specific principal stress ratio is reached. The improvement of the plastic compressive model involves changing the sphere-shaped yield surface to an ellipsoid. The original spherical model has two parameters, B and B^p, which follow from a fit of the stress-plastic strain behaviour for isotropic loading. The improvement requires two additional parameters, namely, K_0 and α. The parameter K_0 is the principal stress ratio in one-dimensional compression and α allows a proper fit to the one-dimensional compressive strain to be attained.

The yield surface, F^c, and the identical plastic potential surface, G^c, for this model component is expressed as

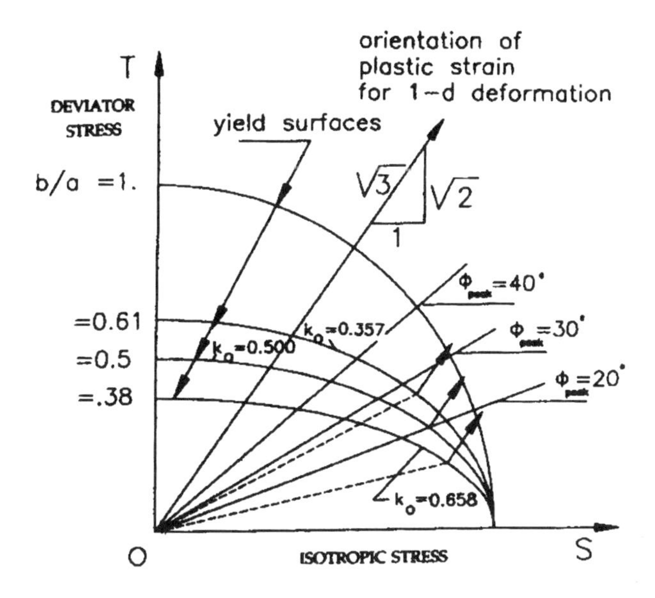

Fig. 5. Characteristic elliptical yield surfaces with the corresponding peak and K_0 mobilized friction levels.

$$F^c = G^c = \frac{S^2}{P_a^2} + \frac{T^2}{P_a^2}\left(\frac{1+2K_0}{1-K_0}\right) - \frac{f_c}{P_a^2} = 0\,, \tag{2}$$

in which f^c is the size of the compressive yield surface.

Figure 5 illustrates several associative surfaces in principal T-S space for several common values of friction angles. The corresponding K_0-stress paths are also shown. For $K_0 = 0$, the K_0-stress path coincides with the direction of the plastic strain path.

Plastic Deviatoric Model

The plastic deviatoric component is the extended isotropic version of the deviatoric component of the Alternat model [15]. It can simulate the ultimate state of deformation, in which for a deviatoric effective stress path the changes in the increment of the isotropic strain and deviatoric stress with increasing deviatoric strain become equal to zero.

In this deviatoric model, the yield surface for triaxial compressive stress states is expressed in terms of the shear stress level, $X = T/S$, as a function of the isotropic stress level, S/Pa, and the void ratio, e_v. It consists of a combination of two components, namely, function X_1 for low, pre-peak shear stress levels and function X_2 for large, post-peak shear stress levels, namely,

$$X = \frac{X_1 \, X_2}{(X_1^n + X_2^n)^{\frac{1}{n}}} \,. \tag{3}$$

This expression ensures that the lower one of X_1 and X_2 is approximated from below as the parameter, $n >> 1$. Function X_1 for the low, pre-peak range reads

$$X_1 = \frac{T}{S} = M^y \left\{ (1 + \nu^y) + \nu^y \cdot \exp\left(-\psi^y \frac{S}{P_a} \right) \right\} \exp(-\beta^y \, e_v) \,, \tag{4}$$

where ν^y, ψ^y and β^y are the material parameters and M^y, S/P_a and the void ratio, e_v, are state parameters. The post-peak surface is expressed by

$$X_2 = \frac{T}{S} = M^p \left\{ (1 + \nu^p) + \nu^p \cdot \exp\left(-\psi^p \frac{S}{P_a} \right) \right\} \exp(-\beta^p \, e_v) \,, \tag{5}$$

in which M^p, ν^p, ψ^p and β^p are material parameters. The parameters of the deviatoric yield surface are chosen such that a close agreement is obtained with experimental data [16]. By shifting this deviatoric yield surface along the isotropic axis towards the tensile part of the stress space, cohesive materials can be simulated.

The shape of the yield surface in the pi-plane [17] reads

$$F^d = \frac{I_1^3}{I_3} - 27 - f^d(X) = 0 \,, \tag{6}$$

in which I_1 and I_3 are the first and the third invariants of the stress and f^d represents a function of the shear stress level, $X = T/S$, as defined in Eq. (3).

After noting that for purely deviatoric loading in triaxial compression the state parmeter M^y depends on the plastic deviatoric strain e^p, thus $M^y = M^y[e^p]$, and after introducing the following shorthand expression for Eq. (4), namely $X_1 = M^y A$, then the yield surface of the deviatoric model can be expressed by

$$F[\sigma_{ij}, e_v, M^y] = \frac{X[f^d] X_2}{A(X_2^n - \{X[f^d]\}^n)^{\frac{1}{n}}} - M^y[e^p] = 0 \,, \tag{6a}$$

which is based on Eqs. (3)–(6) and in which the square brackets [ ] indicate the independent variables.

The concept of an "ultimate" state of stress and deformation has been included in the formulation by adding a surface of zero dilatancy for triaxial compression. The latter depends on both the void ratio and isotropic stress which are expressed by

$$X_\mu = \left(\frac{T}{S}\right)_\mu = C\left\{(1+\xi) - \xi\exp\left(-\theta\frac{S}{P_a}\right)\right\}\{1-\exp(-\zeta e_v)\}\,, \tag{7}$$

where C, ξ, ζ and θ are the parameters. The difference between the post-peak shear stress level, X_2, and this shear stress level, X_μ, at zero dilatancy decreases when the stress approaches the ultimate state. This difference becomes zero when that state is reached.

The definition of the plastic potential requires the dilatancy ratio, dv^p/de^p, which is the ratio of the rates of the plastic isotropic and deviatoric strains. This ratio is based on Rowe's stress-dilatancy relations [18, 19]. Having calculated the dilatancy ratio, $(dv^p/de^p)_c$, for triaxial compression and $(dv^p/de^p)_e$ for triaxial extension, the lowest Fourier function in terms of the Lode angle, θ, is used to calculate the dilatancy ratio, $(dv^p/de^p)_i$, for any other intermediate value, namely,

$$\left(\frac{dv^p}{de^p}\right)_i = \frac{1}{2}\left\{\left(\frac{dv^p}{de^p}\right)_c(1-\sin(3\theta)) + \left(\frac{dv^p}{de^p}\right)_e(1+\sin(3\theta))\right\}. \tag{8}$$

The deviatoric model describes the plastic strain in terms of the deviatoric strain in deviatoric loading, and a non-associated flow rule in which the dilatancy ratio is incorporated, namely,

$$d\varepsilon^p_{ij} = \frac{\partial G}{\partial\sigma_{ij}}\lambda = \left(\frac{dv^p}{de^p}\frac{1}{\sqrt{3}}\delta_{ij} + \frac{\partial G^d}{\partial\sigma_{ij}}\right)\lambda\,, \tag{9}$$

where the super index d indicates the deviatoric component.

The change in size of the plastic deviatoric yield surface is related to the plastic deviatoric strain for purely deviatoric loading only. Consequently, no hardening parameter can be formulated to facilitate this relation. The hardening modulus, H^d, is calculated by

$$\mathbf{H} = -\left|\frac{\partial G^d}{\partial\sigma_{ij}}\right|\left(\frac{dM^y}{de^p} - \frac{\partial F}{\partial e_v}\frac{de_v}{de^p}\right), \tag{10}$$

where e^p is the deviatoric strain in purely deviatoric loading in triaxial compression.

Fluidisation Model

This simple model component ensures a proper numerical treatment of fluidisation, being a case of approaching zero effective stress. It can also be used to simulate the phenomenon of cracking a cohesive material in tensile loading. The yield surface for the cracking and fluidisation model has been defined as

$$F^f = -S + C_s = 0\,, \tag{11}$$

where S is the isotropic stress and C_s is the isotropic strength measure. For the simulation of the fluidisation phenomenon of frictional material without cohesion, the value of the isotropic strength, C_s, should be taken very small, for example, 1 N/m^2, to prevent numerical errors from occurring at the apex of the yield surface of the deviatoric model. The flow rule for the cracking and fluidisation model is associative, meaning that the yield and potential surfaces are coincident, and the behaviour is purely plastic with hardening modulus, $H^f = 0$.

Numerical Implementation of Constitutive Model

In the finite element programme for the analysis of static instability and subsequent flow of an elasto-plastic solid, the initial state is first specified. The "Modified Newton-Raphson" iteration is then used for each load increment to calculate the new estimate of the nodal displacements followed by a corresponding estimate of the unbalanced vector.

To this end, from the calculated nodal displacements at each integration point the resulting increments of strain and rotation are derived. On the basis of these increments and the state parameters at the beginning of the increment, improved estimates of the resulting stress, σ, and the state parameters are calculated. For the calculation of this improved stress, σ, and the state parameters, a constitutive sub-routine is needed.

Major Constitutive Sub-Routines

The numerical implementation of the constitutive model in a finite element programme involves three basic sub-routines, namely, Monot_initial,

Monot_stiffness and Monot_step together with a set of lower level sub-routines, which are only called in those three.

Sub-routine Monot_initial checks that the material parameters are in range and that the initial stress state is acceptable. It calculates the state parameters if they are not specified by assuming normal consolidation. In this sub-routine, some important physical parameters such as the peak mobilised friction angle, the friction angle at zero dilatancy, the ultimate void ratio at the current mean stress and the ultimate mean pressure at the current void ratio, are also calculated.

The sub-routine Monot_stiffness composes the material tangent stiffness matrix for the above mentioned six regions in stress space. The elasto-plastic matrices at vertex points (models 4 and 6) are formed by repeated application of the sub-routine for the composition of an elasto-plasticity matrix.

The sub-routine Monot_step calculates the new stress and state parameters on the basis of the extended Monot model for given increments of strain and material rotation. In this calculation, it is assumed that the prescribed increments of the strain and the rotation are small and finite. The stress at the start of the increment is either inside the yield surface or in an acceptable range of deviation from the relevant yield surface.

With the initial assumption of elastic behaviour in the current increment, the location of the new stress is determined. This initial assumption of elastic behaviour is checked by comparing the location of the new yield surface with the location of the corresponding yield surface in stress space. If the new stress state is outside the corresponding single yield surface, then plasticity must have occurred involving this yield surface. If this new "elastic" stress is outside more than one yield surface, a more detailed analysis is then required to establish the active yield surfaces.

Control of Numerical Drift

In principle, in the numerical calculation of the new position of the stress, the small errors in the calculation of each individual increment may accumulate in subsequent increments. This accumulated error may cause that the new stress after the application of an increment of stress is not located in the acceptable range of deviation of the corresponding yield surface. Hence it may lead to a difference in the calculated stress-strain response.

To minimise the eventual accumulated error, two aspects should be distinguished. One aspect involves the magnitude of the applied increments of stress and strain. Considering the nonlinear nature of the models formulation, the stress increments should not be too large. The other aspect concerns an iterative correction procedure to be applied at the end of any strain increment to keep the relevant stresses on the corresponding yield surfaces. Consequently, the eventual errors in the plastic deformation will not accumulate severely.

It should be noted that the correction procedure depends strongly on the formulation of the different models. For the purely elasto-plastic fluidisation model, only the new stress needs to be corrected. For the elasto-plastic compressive model, corrections of both the stress and state parameter are needed. The correction procedure for the elasto-plastic deviatoric model cannot be applied in the context of the general elasto-plasticity theory since no hardening parameter could be defined for this model. The correction procedure requires the definition of an acceptable range of deviation of the stress from the relevant yield surface. The specification and the magnitude of the acceptable range of deviation plays an important role in the convergence criterion of the correctional iteration process.

Numerical Implementation of Incremental Large Deformation

The implementation of the large deformation theory for the analysis of the three-dimensional motion of materials, as expressed by Eq. (1), into a finite element computer programme requires the following additional components compared to a small deformation analysis.

Gradients of Incremental Displacement and Deformation

The incremental displacement gradient, $\mathbf{M}$, between the current state and the new state is needed at each integration point of the volume. This is to enable a calculation of the increments of strain and rotation which are required in calculating the new stress. The corresponding incremental deformation gradient, $\mathbf{H}$, and its determinant, $|\mathbf{H}|$, are needed for the integration over the current volume and for the area ratio, $|\mathbf{G}|$.

The calculation of the incremental displacement gradient, $\mathbf{M}$, involves the global derivatives of the shape functions, $\partial N_k/\partial x_j$, in the usual way and arranging them in a matrix by a sub-routine to facilitate multiplication by the incremental displacements. The incremental deformation gradient, $\mathbf{H}$, is obtained easily by adding the unit tensor to $\mathbf{M}$. Subsequently, the determinant, $|\mathbf{H}|$, is calculated.

Incremental Rotation and Stretch in New State

The incremental rotation, $\mathbf{R}^r$, between the current and new states is needed at each integration point to calculate the new stress. Unfortunately, the incremental rotation tensor, $\mathbf{R}^r$, and the new stretch tensor, $\mathbf{V}^n$, can only be solved iteratively if the stretch tensor, $\mathbf{V}$, at the current state and the incremental displacement gradient, $\mathbf{M}$, are known [20]. This iteration process has been arranged within a sub-routine which has $\mathbf{M}$ and $\mathbf{V}$ as intent in parameters, and $\mathbf{R}^r$ and $\mathbf{V}^n$ as intent out parameters.

Normal Vector on Plane Boundary Element

The unit normal vectors, $\mathbf{n}$, on each plane boundary element of the current state is needed to calculate the surface ratio, $|\mathbf{G}|$. This surface ratio, $|\mathbf{G}|$, is needed to calculate the boundary integrals in Eq. (1). The sub-routine developed for this purpose has as its input parameters the nodal co-ordinates and as output parameters, the area, dS, and the outward unit normal vector, $\mathbf{n}$, on the plane surface.

It should be noted that in Eq. (1), the normal vector, $\mathbf{n}^n$, occurs in the integral for the nodal loads on the boundary with prescribed displacement. However, this calculation does not occur in the Modified Newton Raphson iteration process to satisfy the equation of motion when the nodal displacements are used as variables which, like in this case, are prescribed. Nevertheless, for a back analysis of these nodal loads, they are needed. Because the nodal displacements are known at this stage, the same sub-routine can be used as mentioned above to calculate of the normal vector for the current state.

Ratio of Area of Surface Element at New and Current States

The ratio, $|\mathbf{G}|$, of the surface elements at the new and current states is needed at each integraetion point of each plane boundary element as expressed by Eq. (1). The sub-routine has as its input parameters the incremental deformation gradient matrix, $\mathbf{H}$, at an integration point of the plane boundary element, and the outward normal vector, **n**, of this element at this point at the current state. The calculated ratio, $|\mathbf{G}|$, is the only output parameter.

Numerical Simulations of Laboratory Tests

The properties of the constitutive model are illustrated by showing the calculated response for a characteristic drained triaxial test. In contrast to the usually small strains in such physical experiments in the numerical simulations, very large strains of up to 300% have been applied to clarify the modelled behaviour.

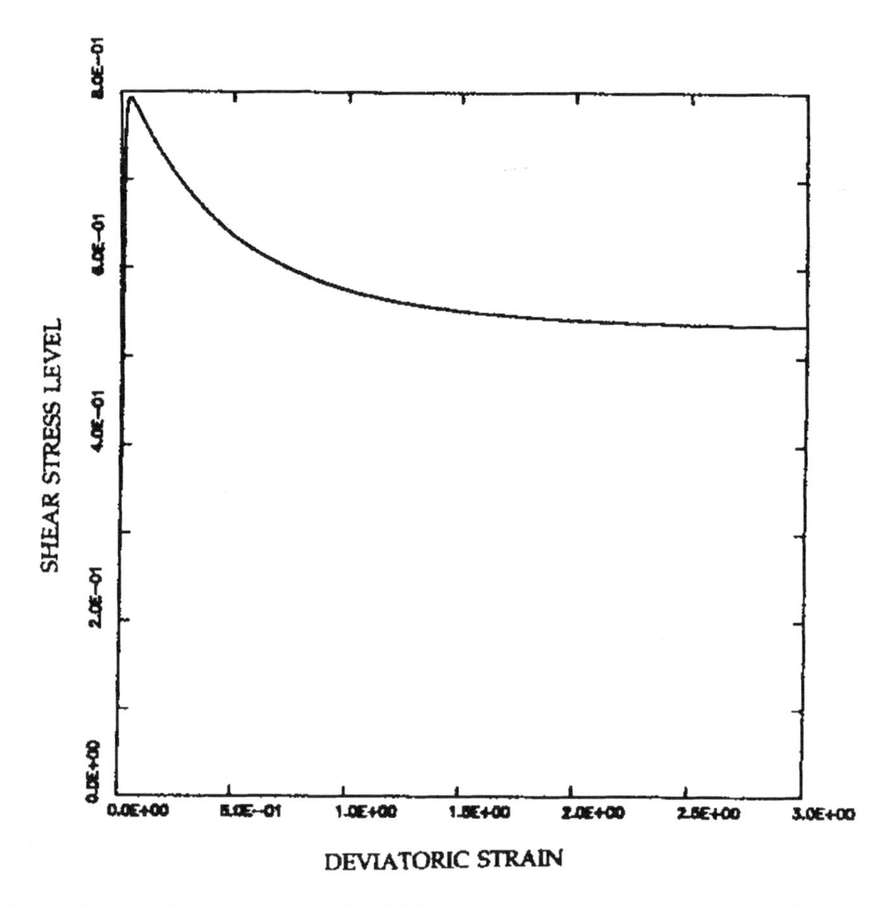

Fig. 6. Calculated deviatoric stress level T/S versus deviatoric strain for drained deviatoric loading in triaxial compression at constant isotropic stress.

The properties of the analytical formulation and its numerical implementation are demonstrated by calculating the response for a strain-controlled and purely drained deviatoric loading at a mean effective stress equal to the mean atmospheric pressure, and an initial void ratio of 0.4 for a dense sand [21].

The calculated relation between the deviatoric stress level, X, and the deviatoric strain, e, is shown in Fig. 6. The pre-peak strain hardening behaviour occurs in the small strain range. The post-peak behaviour shows a

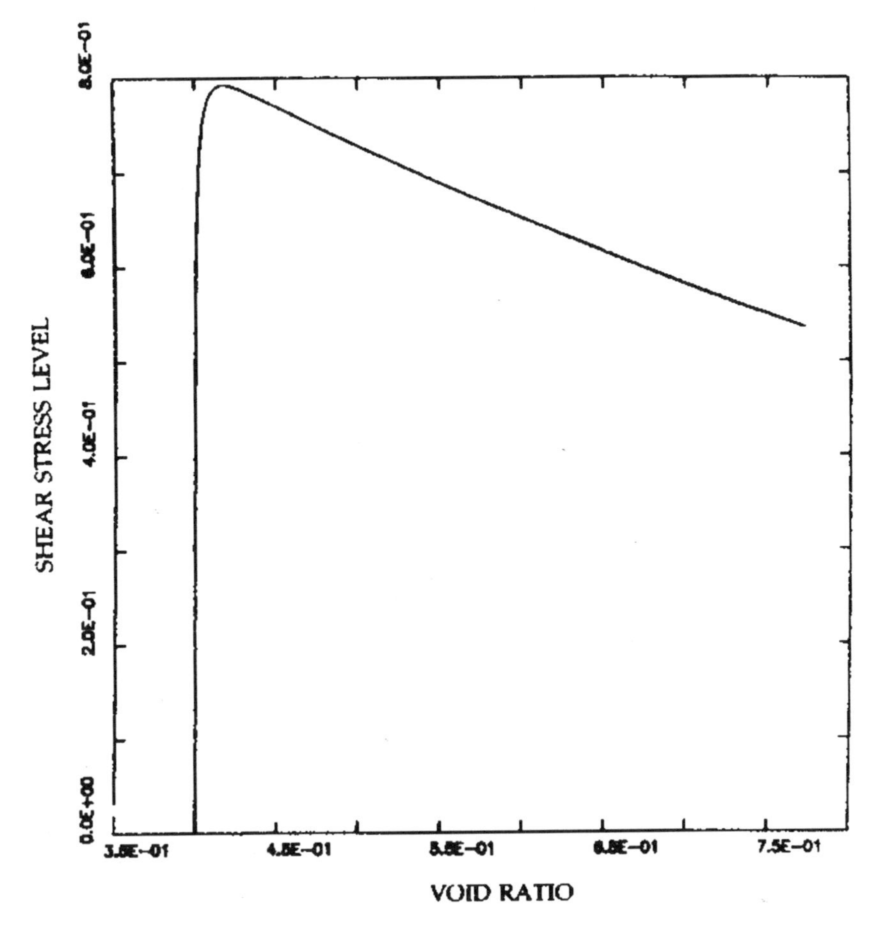

Fig. 7. Calculated deviatoric stress level T/S versus void ratio for drained deviatoric loading in triaxial compression at constant isotropic stress.

monotonic reduction of the deviatoric stress level towards the ultimate state of deformation which, at the enforced strain of 300%, has practically been reached.

The corresponding change in the void ratio, e_v, is illustrated in Fig. 7. Due to the selected parameters, the void ratio is shown to increase from 0.4 to about 0.77.

The surfaces of the corresponding post-peak deviatoric stress level, X_2, and the zero-dilatancy deviatoric stress level, X_μ, are illustrated in Fig. 8. It can be seen that both curves intersect at the ultimate state for the applied deviatoric strain of 300%.

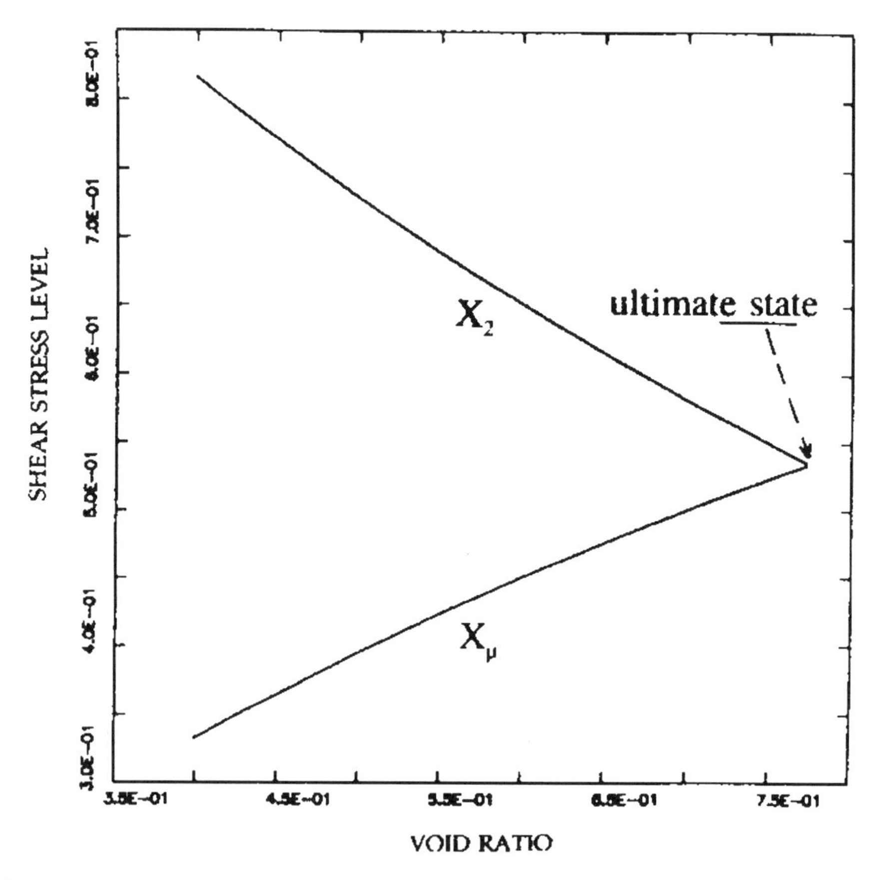

Fig. 8. Paths of the post-peak and zero dilatancy stress levels and void ratios for drained deviatoric loading in triaxial compression at constant isotropic stress.

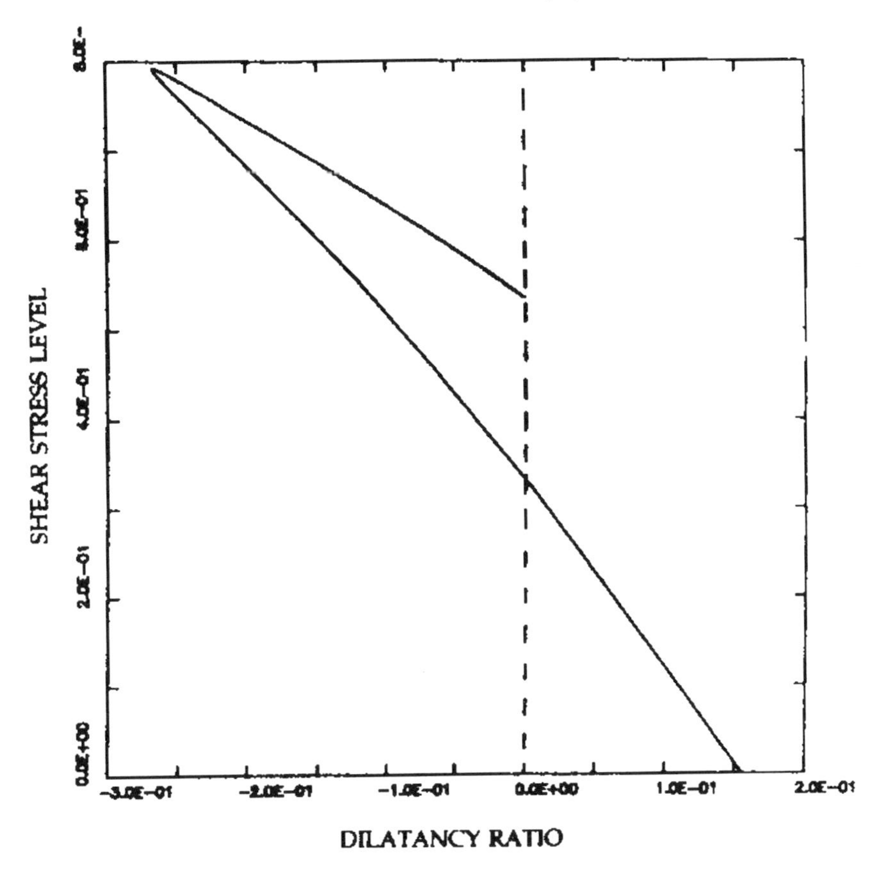

Fig. 9. Calculated dilatancy ratio as a function of the deviatoric stress level for drained deviatoric loading in triaxial compression at constant isotropic stress.

The resulting relation between the dilatancy ratio and the deviatoric stress level, X, is illustrated in Fig. 9. It shows that in the pre-peak range, contraction is followed by dilation in accordance with Rowe's stress-dilatancy theory [18, 19]. In the post-peak range, the dilatancy continues, but at an adjusted rate, until the ultimate state is reached.

Conclusions

Liquefaction and flow of sand have been shown to be a complex phenomenon. The undrained behaviour usually involves both a "quasi-steady state" with a minimum undrained post-peak resistance, and an "ultimate state" with the ultimate state resistance. For very loose sand, both states coincide at zero mean effective stress.

The extensions of the Monot model have been shown to involve a proper definition of the ultimate state of continuous flow. Attention has also been paid to modelling fluidisation and 1-dimensional compression.

The implementation of the incremental formulation of large deformation has been shown to involve four basic components, of which the one concerning the incremental rotation is the more complex one.

Discussion

J. D. Goddard These non-monotonic curves, experimental drop-offs, suggest failure by shear bands or something like this. Some researchers in soil mechanics have doubts about these curves because they are based on measurements on the boundaries of the specimens. Inside the specimens, non-homogeneous behaviour may be occurring. The peaks may be a consequence of such non-homogeneous internal behaviour. Could you clarify your feelings about these views?

F. Molenkamp The phenomenon of instability of material behaviour is a complex one. I analysed this [22] for the extreme cases of both drained and undrained behaviour by considering uniform deformation and shear banding, while applying some popular coaxial elasto-plastic models. From this kind of analysis, it can be understood that for drained behaviour, shear banding can occur before the peak, particularly for plane-strain deformation. In such case, the deformation at the peak will be non-uniform.

For contractant material and undrained behaviour, instability by uniform deformation was calculated for much smaller shear stress levels than those at which drained shear banding was calculated. The undrained instability by uniform deformation occurs at mobilised friction angles of the order of 10 to 20 degrees, while the drained shear banding occurs at mobilised friction angles of the order of 30 to 40 degrees. This suggested to me that the deformation at the undrained peak is still uniform. Recently Desrues *et al.* [23] published experimental data of the deformation of drained triaxial tests. They actually measured internal mechanisms involving localisation at peak level using tomography. However, it should be understood that in undrained triaxial tests, the peaks, as illustrated in Fig. 2, occur at much smaller mobilised shear stress levels and, consequently, the deformation will still be completely uniform.

J. D. Goddard It can be questioned whether these curves for undrained deformation are really stress-strain curves since uniform deformation of the specimen is required. In fact, the curves represent some macroscopic global force versus displacement relation. In principle, they are not really stress-strain curves in the sense that the specimens may be deforming non-uniformly.

F. Molenkamp It should be noted that at the undrained peak state, the effective stress is still at a rather low mobilised shear stress level at which no localisation should be expected yet. Localisation occurs later, when the stress has reached the climbing branch of the effective stress path, as illustrated in Fig. 3, where the effective shear stress level approaches the one describing drained shear banding. Recently, Vardoulakis [24] has investigated this aspect in more detail, by taking into account consolidation and non-coaxiality of the elasto-plasticity.

M. J. Adams Could you clarify your views about shear banding? Do you think that strain softening is possible without shear banding?

F. Molenkamp Yes. I do think that for undrained deformation strain softening is possible without shear banding.

J. D. Goddard Both Vardoulakis and Desrues indicate that the liquefaction process is not due to simple shear banding, but in fact occurs by a very complicated internal network of bands.

F. Molenkamp In those studies, drained shear banding was observed at the elevated mobilised shear stress levels while the undrained peaks occur at much lower shear stress levels. So I agree that localisation can be observed for stress states along the climbing branches of the effective stress paths. However, I do not expect localisation to occur at the undrained peak or at the kink in the stress path at the minimum isotropic effective stress. I believe that deformation at these states is still uniform.

J. D. Goddard That is where I am not sure about. Can you back up your view with experimental results, or is this just a theoretical or computational idea?

F. Molenkamp In the undrained triaxial tests I myself have performed, the deformation, as seen from the outside, is still completely uniform at both the undrained peak and at the kink in the effective stress path. Shear bands can be observed at a later stage when the effective stress is on the climbing branch of the stress path if small strain rates are applied.

M. J. Adams Could it be that the onset of the kink is an indicator of the beginning of localisation?

F. Molenkamp It is understood that at the kink in the stress path, the soil starts to dilate. It tends to increase in volume. With the pore water being unable to leave the specimen, pore suction starts to develop and the mean effective stress starts to increase. That causes the kink in the stress path. Localisation starts to develop later on when the stress is on the climbing branch and starts to approach the shear stress level at which drained shear banding can be expected. To become absolutely sure about this initiation of localisation, tomographic measurements on undrained specimens have to be done. However, considering the low strain rates applied in such undrained tests, I expect the behaviour to be very similar to that observed in drained tests.

V. Entov You mentioned that you simulate liquefaction by considering flow and sedimentation separately. What is the condition for the flow to stop?

F. Molenkamp As I mentioned, I excluded sedimentation from my considerations. I concentrated only on undrained behaviour.

V. Entov Even when considering only undrained behaviour, what is your assumption about the restoration of strength?

F. Molenkamp I believe that restoration of strength occurs because of resedimentation, which involves the redensification of the soil skeleton coupled with the redistribution of the pore water. This is contrary to the assumption of undrained behaviour as I have used, so I believe that I am not considering this restoration properly. To take resedimentation into account would require a complete two-phase medium model. I can only justify my current "over"-simplified assumptions by stating that with the remaining, still rather complicated, model, the severity of the consequences of liquified flow are overestimated, which for some practical situations at the current state of knowledge seems already useful.

M. J. Adams Could you clarify how you take into account rate effects?

F. Molenkamp In the described model for both drained and undrained soil behaviour, the major complexity concerns the elasto-plastic frictional contractive-dilative aspects of the behaviour. The viscous aspects of the behaviour have been treated very approximately only by just using linear viscosity with one parameter, in parallel with the complex frictional resistance.

References

1. Casagrande, A. (1936) Characteristics of cohesionless soil affecting the stability of slopes and earth fills. *J. of Boston Society of Civil Engrg*
2. Casagrande, A. (1975) Liquefaction and cyclic deformation of sand. A critical review. In *Proc. of the Fifth Pan-Amer. Conf. on Soil Mech. and Found. Engrg* **5**, 79–133. Buenos Aires, Argentina
3. Mogami, T. & Kubo, K. (1953) The behaviour of soil during vibration. *Proc. of the Third Int. Conf. Soil. Mech.* **1**, 152–153
4. Casagrande, A. & Watson, J. D. (1938) Compaction tests and critical density investigations of cohesionless materials for Franklin Falls dam. In *Appendix BII Report to the US Engineer Corps.* Boston
5. Schofield, A. N. & Wroth, C. P. (1968) *Critical State Soil Mechanics.* London: McGraw-Hill
6. Poulos, S. J., Castro, G. & France, J. W. (1985) Liquefaction evaluation procedure. *J. Geotechn. Engrn, ASCE* **111**, 772–792
7. Ishihara, K. (1993) Liquefaction and flow failures during earthquakes. *Geotechnique* **43** (3), 351–415
8. Castro, G. (1975) Liquefaction and cyclic mobility of saturated sands. *J. Geotechn Engrg, ASCE* **101** (6), 551–569
9. Molenkamp, F. & Choobbasti, A. J. (1995) Incremental analysis of large deformation using polar decomposition. In *Proc. of the Fifth Int. Symp. Numer. Models in Geomechanics*, pp. 375–380. Davos: Balkema Publications
10. Molenkamp, F. (1980) Elasto-plastic double hardening modelling Monot. *Report Laboratorium voor Grondmechanica*, Delft, CO-218595
11. Molenkamp, F. (1988) A simple model for isotropic nonlinear elasticity of frictional materials. *Int. J. Numer. Anal. Methods in Geomechanics* **12**, 457–475
12. Molenkamp, F. (1992) Application of nonlinear elastic model. *Int. J. for Numer. Anal. Methods in Geomechanics* **16**, 131–150
13. Lade, P. V. (1977) Elasto-plastic stress-strain theory for cohesionless soil with curved yield surfaces. *J. Solids Structures* **13**, 1019–1035
14. Choobbasti, A. J. & Molenkamp, F. (1995) Adaption of spherical cap model for simulation of K_0-state. In *Computational Mechanics in UK, Fourth ACME-UK*, pp. 77–80. Glasgow
15. Molenkamp, F. (1990) Reformulation of Alternat model to minimise numerical drift due to cyclic loading. *Report of the University of Manchester, Department of Engineering*
16. Tatsuoka, F. & Ishihara, K. (1974) *Yielding of Sand in Triaxial Compression, Soils and Foundations* **14** (2), 63–76
17. Lade, P. V. & Duncan, J. M. (1975) Elasto-plastic stress-strain theory for cohesionless soil. *J. Geot. Eng. Div., ASCE* **101**, 1037–1053
18. Rowe, P. W. (1962) The stress-dilatancy relation for static equilibrium of an assembly of particles in contact. *Proc. Royal Soc.* **269**, 500–527
19. Rowe, P. W. (1971) Theoretical meaning and observed value of deformation parameters for soil. In *Proc. Roscoe Mem. Symp.*, pp. 143–194. Cambridge

20. Molenkamp, F. (1986) Limits to the Jaumann stress rate. *Int. J. Numer. Anal. Methods Geomechanics.* **10**, 151–176
21. Choobbasti, A. J. (1997) *Numerical simulation of liquefaction.* PhD Thesis University of Manchester Institute of Science and Technology
22. Molenkamp, F. (1991) Material instability for drained and undrained behaviour. Part 1. Shear band generation. *Int. J. Num. and Anal. Methods in Geomechanics* **15**, 147–168; Part 2. Combined uniform deformation and shear band generation, **15**, 169–180
23. Desrues, J., Chambon, R., Mokni, M. & Mazerolle, F. (1996) Void ratio evolution inside shear bands in triaxial sand specimens studied by computed tomography. *Geotechnique* **46** (3), 529–546
24. Vardoulakis, I. (1996) Deformation of water-saturated sand. I. Uniform undrained deformation and shear banding. *Geotechnique* **46** (3), 441–456; Deformation of water-saturated sand. II. Effect of pore water flow and shear banding. *Geotechnique* **46** (3), 457–472

Dynamics of Complex Fluids, pp. 469–474
ed. M. J. Adams, R. A. Mashelkar, J. R. A. Pearson & A. R. Rennie
Imperial College Press–The Royal Society, 1998

Chapter 33

Analysis of Behaviour of Sand at Very Large Deformation

A. J. CHOOBBASTI AND F. MOLENKAMP*

University of Manchester Institute of Science and Technology,
Department of Civil and Structural Engineering,
Manchester, M60 1QD, England
**E-mail: frans.molenkamp@umist.ac.uk*

This paper presents the main features involved in the development of an isotropic elasto-plastic constitutive model to simulate the behaviour of sand towards a state of very large deformation.

Introduction

The elasto-plastic constitutive model is a refinement of the Monot model [1], enabling the simulation of very large deformations of sand. It consists of six components, namely, the nonlinear elastic model 1, the elasto-plastic deviatoric model 2, the elasto-plastic compressive model 3, the elasto-plastic deviatoric and compressive vertex model 4, the elastic-purely plastic fluidisation model 5, and the elasto-plastic deviatoric and fluidisation vertex model 6. Figure 1 illustrates the different components of this model.

The nonlinear elastic component [2], which describes the behaviour of soils during small unloading and reloading, is based on a potential complementary function. The plastic compressive components are based on the relevant model

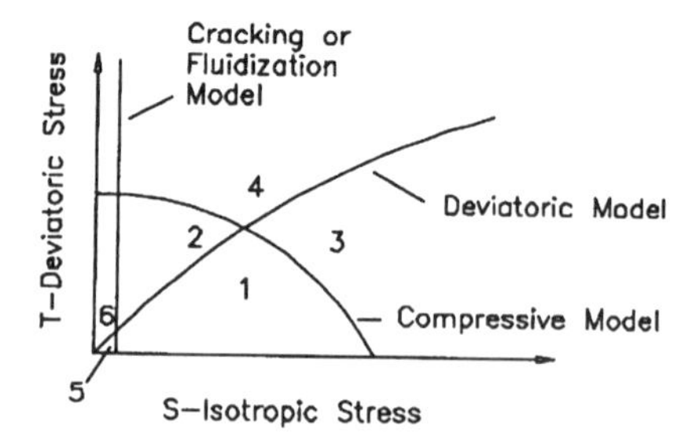

Fig. 1. Illustration of the components of the constitutive model.

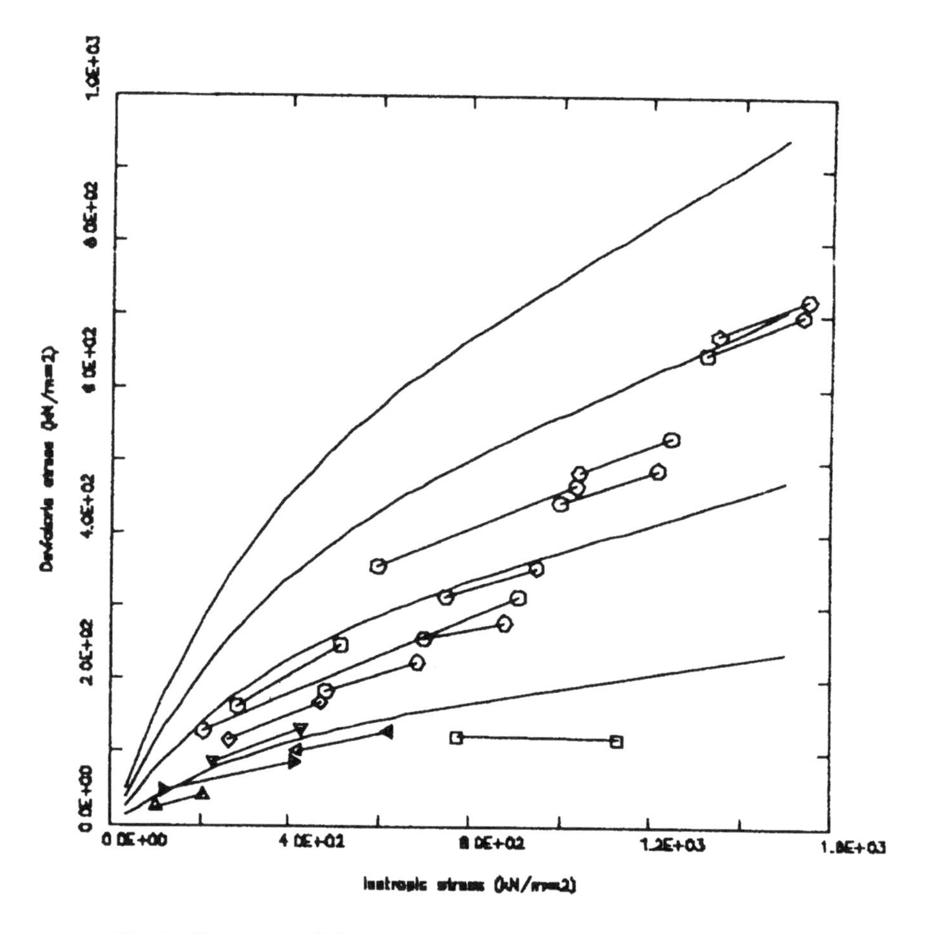

Fig. 2. Illustration of the fitting quality of the deviatoric yield surface.

component described in [3]. The components of this model have been extended by [4] to improve the simulation of the K_0-effective stress path where a specific principal stress ratio is reached for a vertical compression of the material without any horizontal flows. The plastic deviatoric component is a new version of the Alternat model [5] which was extended by [6] during the course of his PhD work. This new formulation enables a simulation of the "ultimate" state of deformation where, a condition of constant volume is reached for continuing flow. The cracking and fluidisation model component is an independent and simple but important model. It was added to Monot to [1] in order to simulate the behaviour of a cohesive material in tensile loading, when cracking is possible, and in order to simulate a cohesionless material in tensile loading at a very low isotropic effective stress level, and thus at a state of fluidisation. The vertex model components 4 and 6 involve aspects of two models. They are model 4, which involves aspects of models 2 and 3, and model 6, which involves those of models 2 and 5. The capabilities of the model to simulate the behaviour of material up to the "ultimate" state of deformation are then demonstrated. Special attention is paid to the responses of the deviatoric model. The functional form for this deviatoric yield surface is chosen such that a close agreement is obtained with the data from [7]. Figure 2 shows this consistency. A full description of the implementation of the above-mentioned constitutive model in a finite element programme is given by [6].

Properties and Performances of the Model

The capabilities of the model to simulate the behaviour of material for the pre-peak and post-peak strain ranges is illustrated in this section using strain controlled loading. The main motivation is the application of the model in analysing the behaviour of material up to the "ultimate" state of deformation using strain control. To this end, a purely deviatoric loading at an initial isotropic stress level of $S/P_a = \sqrt{3}$ and for a series of initial void ratios of $e_v = 0.4$ to 1.3 has been considered. Only the response of the deviatoric model is discussed here. Thus, during the process of deformation of each sample, the void ratio of the sample changes and the isotropic stress level remains the same.

The response of the soil model at large deformation is presented in Figs. 3 and 4. Typical relationships between the shear stress level, T/S, and the deviatoric plastic strain, e^p, of the material deforming at the "ultimate" state of deformation are shown in Fig. 3. This figure illustrates that for the selected

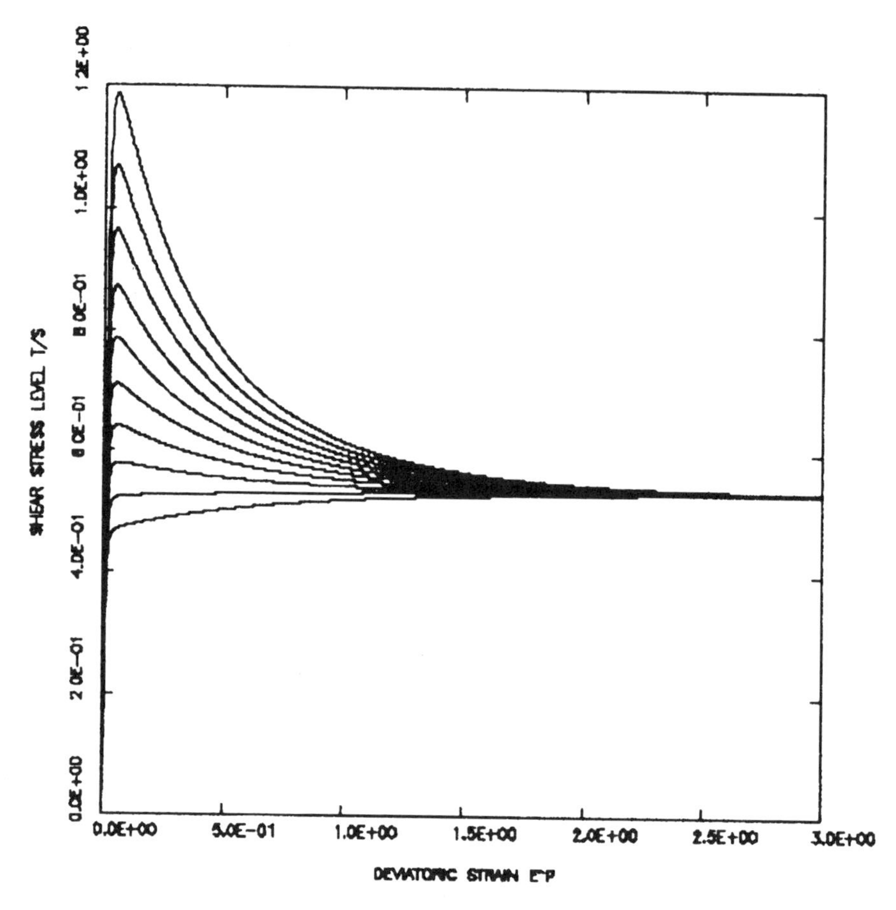

Fig. 3. Calculated deviatoric stress level $(T/S)_\mu$ versus deviatoric strain e^p.

set of material parameters, the "ultimate" state of deformation is reached at 300%. The many characteristics of this state can be seen from this figure immediately. Of particular importance is the loose sand which shows a monotonically increasing shear stress level, T/S, with increasing shear strain, e^p, until an "ultimate" shear strength $(T/S)_\mu$ is eventually reached. However, for dense sand, a peak strength is first reached with further shearing and increase in volume, the resistance drops and the same "ultimate" shear strength, $(T/S)_\mu$, is reached.

In Fig. 4, the applied deviatoric strain, e^p, is shown on the horizontal scale and the calculated void ratios on the vertical scale. The results shown in this

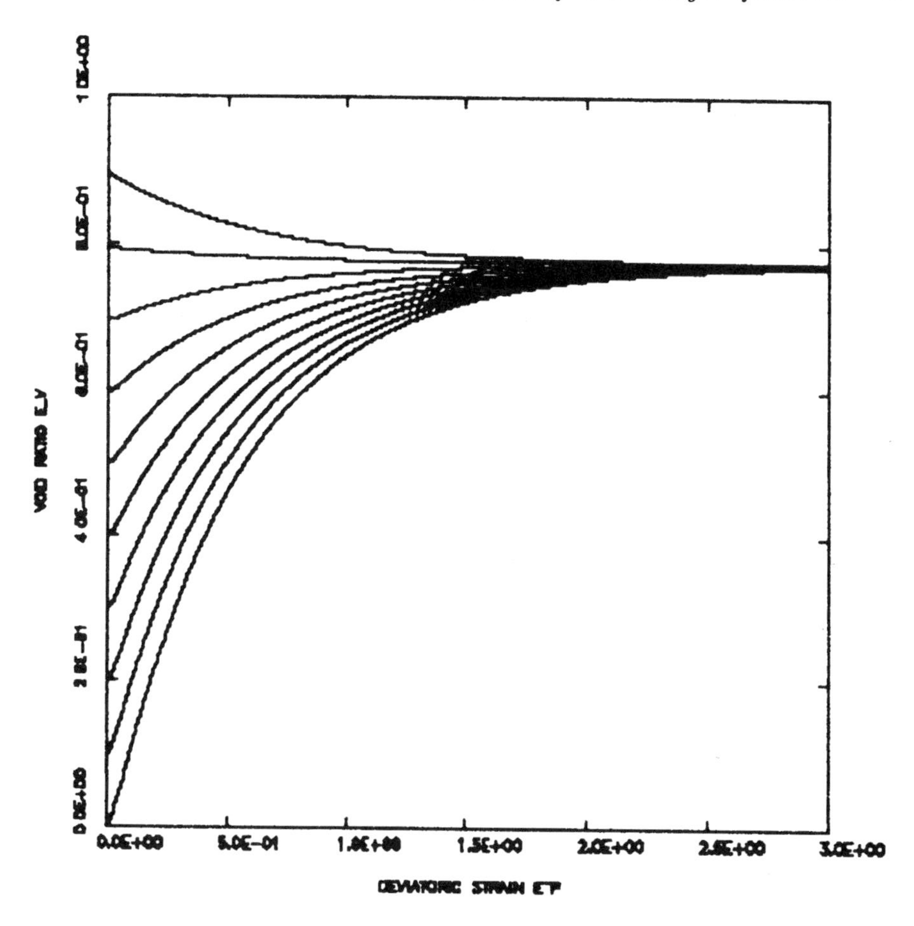

Fig. 4. Calculated void ratio versus deviatoric strain e^p.

figure are a visualisation of the way in which the void ratios of dense and loose sand samples change during continued uniform deviatoric deformation to reach the void ratio at the "ultimate" state. For dense sands, the corresponding curves reveal dilation. Of particular importance is that these results reveal the capability of the model to simulate contraction which usually occurs in experiments at the initial phase of the test. According to these results, a sample with an initial void ratio equal to the "ultimate" void ratio reveals an initial contraction followed by a dilation prior to reaching its "ultimate" void ratio at the "ultimate" state.

It should be noted that the responses of the model presented above are supported by recent experiments using tomographic measurements [8] where, the "ultimate" state was reached in drained deformation in shear bands.

Conclusions

The main features of the developed "ultimate" state constitutive model have been discussed. The application of the constitutive model to simulate the behaviour of soil at sufficiently large deviatoric strain of, for example, 300%, illustrates the appropriateness of the theoretical expressions and the validity of the numerical implementation.

References

1. Molenkamp, F. (1980) Elasto-plastic double hardening model Monot. Report Laboratorium voor Grondmechanica, Delft, Co-218595
2. Molenkamp, F. (1988) A simple model for isotropic nonlinear elasticity of frictional materials. *International Journal for Numerical and Analytical Methods in Geomechanics* **12**, 457–475
3. Lade, P. V. (1977) Elasto-plastic stress-strain theory for cohesionless soil with curved yield surfaces. *J. Solids Structures* **13**, 1019–1035
4. Choobbasti, A. J. & Molenkamp, F. (1996) Adaption of spherical cap model for simulation of K_0-state. *Fourth ACME-UK Conference.* Glasgow
5. Molenkamp, F. (1990) *Reformulation of Alternat Model to Minimise Numerical Drift Due to Cyclic Loading.* Report of the University of Manchester
6. Choobbasti, A. J. (1997) *Numerical Simulation of Liquefaction.* PhD Thesis, University of Manchester Institute of Science and Technology (UMIST)
7. Tatsuoka, F. & Ishihara, K. (1974) Yielding of sand in triaxial compression. *Soils and Foundations* **14** (2), 63–76
8. Desrues, J., Chambon, M., Mokni, M. & Mazerolle, F. (1996) *Geotechnique* **3**, 529–546

AUTHOR INDEX

Adams, M. J. 394, 399

Ball, R. C. 301
Briscoe, B. J. 394, 399
Brown, A. B. D. 330

Cates, M. E. 199
Choobbasti, A. J. 446, 469
Clarke, S. M. 315, 330
Corfield, G. M. 379, 394

de Roeck, R. M. 338
Denn, M. M. 372

Gilbert, D. 30
Goddard, J. D. 280

Halin, P. 88
Harden, J. L. 199
Heshmati, A. A. R. 446
Hulsen, M. A. 106

Ianniruberto, G. 176

Johner, A. 193

Keunings, R. 88
Kothari, D. 399
Kumar, K. S. 286
Kumaran, V. 263

Larson, R. G. 230
Laso, M. 73
Lawrence, C. J. 379, 394, 399
Lele, A. K. 131

Mackley, M. R. 30, 338
Marrucci, G. 176
Mashelkar, R. A. 131
McKinley, G. H. 6
Melrose, J. R. 301
Molenkamp, F. 446, 469
Müller, S. 188

Öttinger, H. C. 61
Owen, D. R. J. 405

Perić, D. 405
Piau, J. M. 351

Ramamohan, T. R. 286
Ramaswamy, S. 248
Ravi Prakash, J. 155
Renardy, M. 38
Rennie, A. R. 315, 330
Rutger, R. P. G. 30

Savithri, S. 286
Schmidt, C. 188
Siebert, H. 193
Smith, I. M. 425

Tanner, R. I. 47
Tildesley, D. J. 213

van den Brule, B. H. A. A. 106
van Heel, A. P. G. 106
van Vliet, J. H. 301

Wittner, J. 193

SUBJECT INDEX

adaptive meshing
 see computational techniques
adsorption 129
 adsorbed layer xii, 199, 210, 214
aggregates
 aggregated 343
algorithm 110, 303
 see also computational techniques
alignment
 see flow
anisotropy 80, 246, 330, 354
 Ashcroft and Lekner model 317
association
 hydrogen bonding 127, 133, 152, 239
atomistic simulation
 see computational techniques
AVSS simulation
 see computational techniques

bead-rod, bead-spring model
 see constitutive models
Bernoulli spiral
 see instability
bicontinuous gyroid
 see liquid crystal
bidisperse systems 267
 see also suspension, dispersion
birefringence
 flow induced 14, 316, 331
 stress 30
Bingham fluid
 see constitutive equations
Bingham number
 see dimensionless groups
biviscosity fluid
 see constitutive equations
block copolymer
 see polymer
blood
 equine 338
Boger fluid
 see fluid
bollarding
 see localisation
boundary lubrication
 see lubrication
boundary condition
 see wall boundary condition
boundary layer
 see stress
Brinkman equation
 see equation
Brownian dynamics simulation
 see computational techniques
Brownian motion 75, 94, 108, 252, 288, 302, 312, 332, 359
brush polymers
 see polymer

Capillary number
 see dimensionless groups
capillary rheometer
 see rheometer
chaos xii, 287, 312
Chapman-Enskog theory 276
charge stabilised dispersion
 see dispersion
Chilcott-Rallison model
 see constitutive models
clay
 see particle
coarse-grained simulation
 see computational techniques
coating
 film 148
collision
 inelastic 245, 263

colloid (*see* also dispersion)
  colloidal crystal 250
complex flow
  see flow
computational techniques x
  adaptive meshing 90, 101, 349, 413, 431
  atomistic 129, 214, 222
  AVSS 55
  Brownian dynamics 10, 62, 73, 89, 128, 160, 252, 302, 328
  coarse-grained 62, 91, 129, 156, 218, 222, 255
  CONNFFESSIT 4, 62, 74, 107
  dissipative particle dynamics 302
  DEVSS 112
  EVSS 55
  finite difference 74
  finite element analysis (FEA) xiii, 26, 52, 74, 90, 108, 349, 375, 403, 406, 412, 421, 428, 450, 471
  finite volume 4, 47, 58, 74
  Lagrangian meshing 90, 246, 349, 422, 433
  lattice Boltzmann 278
  molecular dynamics 70, 214
  Monte Carlo 68, 130, 195, 226, 230, 239
  SFEM 54
  SIMPLER 4, 53
  SIMPLEST 49
  stochastic meshing 74
  Stokesian dynamics 252, 302
  thermo-mechanical 416
cone-and-plate rheometer
  see rheometer
configuration 14, 91, 107
  distribution function 62, 77, 93, 109, 265
  entropy 3
  polymer 61, 75
conformational 62
CONNFFESSIT simulation
  see computational techniques
constitutive
  equations x, 7, 40, 48, 62, 74, 89, 353
    Bingham 347, 354, 364, 369, 372
    biviscosity 374, 380
    Hajduk 409
    Herschel-Bulkley 349, 369, 373, 382
    Newtonian 4, 49, 90, 353, 368, 375
    Sellars-Tegart 409
  models 3, 7, 91
    bead-rod, bead-spring 4, 62, 91, 133, 167, 219
    Chilcott-Rallison 21, 42
    Doi-Edwards 92, 128, 176, 185
    dumbell 4, 5, 74, 88
      FENE 4, 10, 75, 93
      FENE-P 4, 77, 93
      FENE-CR 22
      Hookean 67, 70, 75, 107, 156
    Giesekus 19
    Jeffreys
      convected 159
    Maxwell 8, 58, 107, 344
      upper-convected 3, 38, 49, 90, 127, 132
    Oldroyd B 4, 5, 8, 49, 78, 106, 127, 132
    Phan-Thien Tanner,(PTT) 4, 24, 38, 49
    Rouse 10, 127
    Zimm 7, 8, 10, 127, 159, 171, 205, 207

continuum
mechanics x, xii, 73, 85, 89, 107
corner
flow (*see* flow)
Couette cell
see rheometer
Cox-Merz rule 24, 128, 176
creep
see viscoelasticity, viscoplasticity
crosslink density
see network
cubic phase
see liquid crystal

Deborah number
see dimensionless groups
desorption 129, 200
deviatoric stress
see stress
dilation 349, 455, 473
angle 350, 428
dimensionless groups
Bingham 347, 376
Capillary 21, 149
Deborah 3, 6, 18, 54, 79, 115
Froude 366
Görtler 3
M (McKinley) 3, 17
Oldroyd 366
Peclet 249, 205, 260, 325
Reynolds 39, 115, 249, 258, 333, 366
Weissenberg 3, 18, 38, 47, 50, 58, 90, 94, 385
dispersion (*see* also colloid, suspension, polymer latex) 347
charge stabilised 320
colloidal 80, 245, 287, 301, 316, 330, 355
sterically stabilised 200, 320, 332
dissipation 222, 266, 281, 309

Doi-Edwards model
see constitutive models
dumbell model
see constitutive models

elasticity 260
elastic fluid (*see* fluid)
elastic solid ix, 352, 422, 426
elasto-plastic
see plastic
electrostatic force 217, 249
entanglement 137
renewal 177
Ericksen-Leslie-Parodi theory 189
equation
Brinkman 205, 313
Fokker-Planck 4, 63, 91, 109
Irving-Kirkwood 224
Langevin 4, 62, 91, 249, 288, 305
stochastic differential equation (*see* stochastic)
equine blood
see blood
EVSS simulation
see computational techniques
extensional flow
see flow
extrudate
instability (*see* instability)
swell 58

filament stretching (*see* rheometer)
finite element analysis
see computational techniques
finite volume analysis
see computational techniques
Flory theory 127, 135, 196
flow
around sphere 47
capillary (*see* rheometer)
complex x, xii, 3, 64, 80, 89, 91, 364

corner 38, 90
Couette (*see* rheometer)
eccentric cylinder 38, 54, 58, 95
extensional 3, 4, 6, 20, 185
fountain 380
granular 451
instability (*see* instability)
journal bearing 77, 82
laminar 333
oscillatory 9, 159, 343
particle alignment 330
past cylinder 52, 114
past plate 3
pipe 42, 284, 354, 365
plug 284, 365, 386
shear xiii, 3, 6, 20, 41, 68, 172, 181, 203, 210, 273, 278, 284, 288
slow 264
start-up 68, 78, 106
squeeze xiii, 348, 365, 374, 380, 394, 400
turbulent x, 259, 336
transient 26
viscometric 3, 39, 95, 128, 380
flow induced birefringence
see birefringence
fluctuation 67, 85, 160, 173, 196, 251, 313, 316
fluid
Bingham (*see* constitutive equations)
Boger 6, 91
elastic 8
flow
Herschel-Bulkley (*see* constitutive equations)
ideal elastic 9
Newtonian (*see* constitutive equations)
rigid-viscoplastic (*see* viscoplastic)
yield stress 347, 354, 364, 369, 380
fluidisation 456
bed 245, 248, 263, 277
Fokker-Planck equation
see equation
form factor 317
fountain flow
see flow
fractionation 200
friction (*see* also wall boundary condition) 129, 213, 222, 407, 421
internal angle 350, 457, 464
fracture
see localisation
Froude number
see dimensionless groups

gel
see network
Giesekus model
see constitutive models
Görtler number
see dimensionless groups
granular flow
see flow
granular temperature 270

Hajdnk equation
see constitutive equations
haemorheology 338
hard sphere particle
see particle
Herschel-Bulkley fluid
see constitutive equations
hexagonal phase
see liquid crystal
hydrodynamic lubrication
see lubrication

hydrodynamic interaction 10, 67, 128, 156, 171, 201, 210, 246, 249, 303, 312, 328
hydrogen bonding association
see association

inelastic collision
see collision
instability
Bernoulli spiral 17
elastic 16
extrudate 30
flow x, xii, 3, 6, 246
fluidised bed 251
migrational 280
spiral 22
Taylor-Couette 17

Jeffreys model
see constitutive models
journal bearing
see flow

kinetic theory 16, 61, 88, 100, 101, 107, 157, 265

lamellar phase
see liquid crystal
laminar flow
see flow
Langevin equation
see equation
Langmuir-Blodgett deposition 214
latex
see polymer
lattice Boltzmann simulation
see computational techniques
Leighton-Acrivos model 281
Lennard-Jones potential
see potential
light scattering 14, 246, 261, 316
liquefaction 447
liquid crystal ix, 128, 188
bicontinuous gyroid 230
cubic 230
hexagonal 190, 230, 241
lamellar 190, 230, 240
lyotropic 188, 231
thermotropic 188
localisation xiii, 431
barrelling 400
bollarding 400
fracture 350, 355, 456, 471
shear band 246, 280, 349, 400, 464, 474
unyielded region 348, 365, 374, 380
Lodge's theory 132
lubrication 200, 214
boundary xii, 129
elastohydrodynamic 313
hydrodynamic (*see* also journal bearing) 218, 222, 225
Reynolds theory 305, 348, 374, 380, 396
lyotropic liquid crystal
see liquid crystal

M (McKinley) group
see dimensionless groups
Maxwell model
see constitutive models
Maxwell-Boltzmann distribution 266
mechanics
see continuum
microgel
see network
microscopy 339
microstructure xiii, 89, 315
migration
see particle
molecular dynamics simulation
see computational techniques
Monte Carlo simulation
see computational techniques

network
  crosslink density 135, 144
  gel 312, 348, 355
  microgel 134, 151
  model 133
  percolating 312, 355
  transient 134, 149, 151, 343
  swelling 135, 151, 193
neutron
  diffraction 316, 331
  scattering 320
Newtonian fluid
  see constitutive equations
no-slip boundary condition
  see wall boundary condition
normal stress
  see stress
normal stress difference
  see stress
nuclear magnetic resonance
  see spectroscopy

Oldroyd number
  see dimensionless groups
Oldroyd B model
  see constitutive models
order parameter 319, 330
orientation distribution function 287, 333
oscillatory flow
  see flow

particle (*see* also plates, polymer latex, rod)
  carbon black 356
  clay 331, 359
  inelastic 273
  hard spheres 249, 266, 304, 310, 317
  migration 246, 280, 361
paste 362, 380, 399
Peclet number
  see dimensionless groups
percolating network
  see network
Phan-Thien Tanner (PTT)
  see constitutive models
phase transition 130, 231, 241, 254, 260, 307
plastic (*see* also yield, viscoplastic)
  elasto-plastic 349, 406, 409, 422, 426, 444, 451, 464, 469
  potential 426, 452
  rigid-plastic xiii, 379
pipe flow
  see flow
plug flow
  see flow
polymer ix
  block copolymer 129, 193, 200
  brush 129, 194, 200, 210, 305, 310
  extrusion 30
  latex
    polymethylmethacrylate 320
  liquid 176
  melt 21, 31, 61, 69, 74, 91, 359
  solution 61, 73, 108, 359
    dilute 8, 62, 93, 128, 156
    concentrated 69, 91, 127, 136, 196
    semi-concentrated 305
potential
  Lennard-Jones 219
  Weeks-Chandler-Anderson 219
probability density 63

radius of gyration 220
random forces 91
relaxation time (*see* also stress relaxation) 18, 48, 132, 137, 180, 196, 239
  spectrum 9, 173
Reynolds number
  see dimensionless groups

Reynolds theory
see lubrication
rheological characterisation
rheology 6, 351
computational 77, 88
extensional 11
simple shear 9, 286, 302, 315
rheometer
capillary 127, 397
cone-and-plate 7, 128, 361
Couette 41, 78, 128, 321, 361
disc cell 321
extensional
filament 7, 20
filament stretching 7
falling ball 145, 151
parallel plate 339, 361
triaxial compression 350, 431, 449, 464
rigid-plastic
see plastic
rigid-viscoplastic
see viscoplastic
rod xii, 246, 288
Rouse model
see constitutive models

scaling 16, 40, 64, 84, 197, 245, 288, 348, 353, 375, 380
sedimentation xii, 245, 248, 325, 361, 450, 466
self-assemble 193
Sellars-Tegart constitutive equation
see constitutive equations
SFEM simulation
see computational techniques
shear
flow (*see* flow)
rate 10, 24, 31, 39, 50, 143, 151, 159, 171, 176, 189, 220, 224, 339, 341
stress (*see* stress)
thickening/hardening xiii, 171, 246, 304, 310, 339, 354, 382
thinning 10, 142, 171, 225, 310, 315, 335, 341, 382
shear bands
see localisation
similarity 64
solution 41
simple shear
see rheology
SIMPLER simulation
see computational techniques
SIMPLEST simulation
see computational techniques
simulation
see computational techniques
singularity
corner 3, 38, 42
edge 3
stress 50, 90
slip
see wall boundary condition
soft solids xiii, 355, 383, 394, 425
spectroscopy
nuclear magnetic resonance 188
sum frequency spectroscopy 129
squeeze flow
see flow
stability 112
numerical 66
start-up flow
see flow
sticker group 200
stochastic 4, 111, 122
differential equation 61, 75, 91, 109
stress x
birefringence (*see* birefringence)
boundary layer 4, 38, 54
deviatoric 382, 449, 471
effective 350, 431, 447, 465

growth 14, 132
normal 45
normal stress difference 3, 8, 50, 97, 128, 156, 172, 185, 296, 464
overshoot (*see* viscoplastic, viscoelastic)
relaxation (*see* also relaxation time) 14, 177, 184
shear 31, 50, 140, 164, 182, 204
tensile 21, 30
tensor 4, 48, 62, 373, 408
structure factor 317
surfactant xii, 128, 130, 190, 214, 222, 230, 239, 316
suspensions (*see* also dispersions) xii, 129, 249, 263, 277, 281, 287, 301, 347, 361, 380
swelling
see network

Taylor-Couette instability
see instability
tensile stress
see stress
thermotropic liquid crystal
see liquid crystal
thixotropy 315, 341, 348, 357
transient flow
see flow
transient network model
see model
Tresca wall boundary condition
see wall boundary condition
triaxial compression
see rheometer
Trouton ratio
see viscosity
turbulent flow
see flow

unyielded region (*see* also localisation)

viscoelasticity x, 3, 10, 47, 58, 73, 88, 101, 106, 247, 347, 422
creep 355
linear 7, 168, 339
stress overshoot 140
viscometric flow
see flow
viscoplastic x
creep 426
flow potential 381, 408
materials 394, 425, 422
elasto-viscoplastic 347, 399, 406, 428
rigid-viscoplastic 347, 354
stress overshoot 140, 350, 381
viscosity x, 3
apparent 450
elongational 12, 23
plastic 373
shear 48, 128, 151, 164, 171, 341, 375, 422
Trouton ratio 3, 12

wall boundary condition xiii
Coulomb 216, 384, 407
frictionless 348, 383
Herschel-Bulkley 383, 394
Navier 383, 395
no-slip 11, 39, 50, 115, 205, 348, 374, 383, 395, 400
stick-slip 31, 50, 58
stochastic Maxwell 278
Tresca 382, 394
slip 31, 127, 134, 145, 349, 355, 361, 381
Weissenberg number
see dimensionless groups

x-ray diffraction 316

yield (*see* also plastic, viscoplastic)
criterion/function 381
Mohr-Coulomb 403, 426

Tresca 349
von Mises 373, 382
surface 348, 354, 364, 373, 381, 452, 471
stress xiii, 356, 372, 380, 394
yield stress fluid
see fluid

Zimm model
see constitutive models